Experimental and Applied Physiology

LABORATORY MANUAL

Eighth Edition

Richard G. Pflanzer, Ph.D.

Purdue University School of Science
Indiana University School of Medicine
Indiana University–Purdue University at Indianapolis

Boston Burr Ridge, IL Dubuque, IA Madison, WI New York San Francisco St. Louis
Bangkok Bogotá Caracas Kuala Lumpur Lisbon London Madrid Mexico City
Milan Montreal New Delhi Santiago Seoul Singapore Sydney Taipei Toronto

EXPERIMENTAL AND APPLIED PHYSIOLOGY LABORATORY MANUAL, EIGHTH EDITION

Published by McGraw-Hill, a business unit of The McGraw-Hill Companies, Inc., 1221 Avenue of the Americas, New York, NY 10020.

This book is printed on recycled, acid-free paper containing 10% postconsumer waste.

1 2 3 4 5 6 7 8 9 0 QPD/QPD 0 9 8 7 6 5

ISBN-13 978-0-07-250077-8
ISBN-10 0-07-250077-8

Editorial Director: *Kent A. Peterson*
Senior Sponsoring Editor: *Michelle Watnick*
Developmental Editor: *Fran Schreiber*
Senior Marketing Manager: *Debra B. Hash*
Project Manager: *Joyce Watters*
Lead Production Supervisor: *Sandy Ludovissy*
Cover Designer: *Rick D. Noel*
Cover Illustration: *Peg Gerrity*
Senior Photo Research Coordinator: *John C. Leland*
Compositor: *Carlisle Communications, Ltd.*
Typeface: *10/12 Times Roman*
Printer: *Quebecor World Dubuque, IA*

Some of the laboratory experiments included in this text may be hazardous if materials are handled improperly or if procedures are conducted incorrectly. Safety precautions are necessary when you are working with chemicals, glass test tubes, hot water baths, sharp instruments, and the like, or for any procedures that generally require caution. Your school may have set regulations regarding safety procedures that your instructor will explain to you. Should you have any problems with materials or procedures, please ask your instructor for help.

www.mhhe.com

CONTENTS

PREFACE

The purpose of this text is to introduce the student to the basic principles of human physiology through laboratory experimentation. The major intents of instruction are to (1) reinforce basic principles of human physiology; (2) improve the student's ability to reason scientifically while strengthening the student's habit of thinking in this manner; (3) develop a sense of inquiry, understanding, and appreciation of the functional aspects of the human organism as they pertain to health and general well-being; and (4) allow the student to gain, from actual laboratory experimentation, an appreciation of laboratory science and its role in advancing knowledge of the human body.

This text is intended for a one-semester (or a one- or two-quarter) course in human physiology at the undergraduate level. It has not been designed to accompany any particular textbook of physiology or textbook of anatomy and physiology; rather, it has been developed to allow for laboratory reinforcement of basic physiologic principles that are covered in most of the undergraduate textbooks of recent years.

The nature of physiology is to investigate and understand how living systems function. A few of the experiments in this book require the use of a live animal such as a frog or a turtle. Such use is essential if certain concepts of physiology are to be demonstrated and mastered. However, most of the laboratory experiments in this text make use of students as experimental subjects rather than nonhuman animals. It has been my experience in teaching physiology to undergraduate, graduate, medical, and dental students that experiments that relate concepts of physiology directly to the functions of the student's own body are far more likely than animal experiments to be successful in terms of illustration and reinforcement.

All of the experiments presented have been performed many times by students and are safe, provided that ordinary precautions and normal laboratory safety procedures are observed.

Each experiment has been designed to be performed within a 2-hour laboratory period, which includes time for the presentation of introductory material by the instructor. The introduction to each chapter provides a review of concepts pertinent to the experimentation, and the objectives stated at the beginning of experimentation indicate to the student the reasons for performing the laboratory exercise and the expected gains from the experience. Most of the experiments require a minimum of equipment and are easily performed by inexperienced students. Some experiments (e.g., electrocardiography) require sophisticated equipment and adequate preparation on behalf of the student and instructor. In general, the experiments have been designed as they are not just because they are easy to perform or inexpensive but also because they illustrate and reinforce some fundamental principles of human physiology.

Experiments that require physiologic recording systems have been written for use with the Lafayette Minigraph system. However, the experiments that require standard physiologic recorders are easily adapted for use with other recording systems, such as the Beckman or Grass recorders.

- Each chapter has been carefully revised with expanded discussions to clarify important physiologic concepts.
- New art has been added to support concept presentations and to reinforce fundamental principles of experimental physiology.
- A section on graphic analysis has been included to help the student use graphing as a tool for data analysis and to aid the student's understanding of the relationships between experimental variables.
- Experimental procedures have been simplified, and calculations involving experimental data have been clarified through the use of examples.
- The recording of experimental data in the report section of many of the lessons has been facilitated by the addition of tables and charts into which data may be entered.
- Questions intended to develop critical-thinking skills as well as to test mastery of subject matter have been included in many of the reports.

ACKNOWLEDGMENTS

I gratefully acknowledge the help of Patricia Clark, IUPUI, and Martin Vaughn, IUPUI, for their invaluable assistance in improving the pedagogy of this edition. I also thank the many professors who took the time to critically review the previous edition and offer constructive criticism on how to improve this edition: Ateegh Al-Arabi, Johnson County Community College; Jonathan Day, California State University–Chico; Susan Mounce, Eastern Illinois University; Virginia Pascoe, Mt. San Antonio College; Donald C. Skinner, Wyoming University; Mary Anne Sydlik, Kalamazoo Valley Community College; Jin-Hui Wang, University of Kansas, Lawrence; Heather Wilson-Ashworth, Utah Valley State College; Michael Youther, Southern Illinois Unviersity–Carbondale. Their comments are greatly appreciated and have been implemented into this eighth edition. I offer special thanks to the editorial and production staff of McGraw-Hill.

TO THE STUDENT

■ REQUIRED SUPPLIES

1. One high-quality dissecting kit containing a metric rule, dropper, forceps, blunt and needle-point probes, large and small surgical scissors, and a fixed-blade scalpel.
2. One plastic or rubber chemical laboratory apron or one long white laboratory coat.
3. Eyeglasses with safety glass or laboratory goggles.

■ LABORATORY SAFETY

1. Infants and children are not permitted in the laboratory under any circumstances. Please make proper arrangements for child care before attending class.
2. Become familiar with the location of exits, chemical hood, emergency shower, fire extinguishers, and first-aid supplies associated with your laboratory room.
3. Upon the sounding of the fire alarm, immediately evacuate the building in an orderly and safe manner.
4. Eating, drinking, smoking, and chewing are not permitted in the laboratory except when dictated by experimental protocol.
5. Follow the directions of the laboratory book and your laboratory instructor regarding experimental precautions and procedures. Do not deviate without permission. If uncertain or unclear, ask for clarification.
6. Use disinfectant to clean laboratory work surfaces *before* and *after* procedures.
7. Whenever feasible, wear disposable hypoallergenic examination gloves when working with chemicals, body fluids, and animal tissues.
8. Eye protection is required for all labs that use chemicals, biologicals, or physically hazardous materials. Safety goggles are preferred. Safety glasses are acceptable, but standard glasses containing corrective lenses are not. Students are responsible for purchasing and using acceptable eye protection. Compliance will be monitored by the laboratory instructor.
9. In experiments dealing with body fluid (blood, saliva, urine) handle only your own to avoid contamination and transmission of disease.
10. Disposable mouthpieces, blood lancets, microhematocrit tubes, and other clean disposable supplies are to be used only once and discarded into appropriately marked containers for disposal. If an experimental procedure must be repeated, use new clean or sterile supplies.
11. Nondisposable supplies (test tubes, hemacytometers, etc.) that come into contact with body fluids must be thoroughly cleaned and sterilized after use. Follow the directions of your laboratory instructor.
12. Report any accidents, spills, or damaged equipment to your instructor immediately.
13. Animal remains (e.g., frog skin, muscles, organs) must be discarded into appropriately marked containers for proper disposal. Wastebaskets and sinks are not appropriate containers for animal remains.
14. At the conclusion of a laboratory period, restore your laboratory table to a clean, orderly condition. Wipe the table clean by using napkins and disinfectant solution, clean and return equipment and supplies to their proper place, ensure that the sink is clean and free of debris, and place your chair beneath the laboratory bench.
15. Wash you hands thoroughly *before* leaving the laboratory.
16. Violation of laboratory rules and regulations may result in your being asked to leave the laboratory for your own benefit and safety as well as for the benefit and safety of your classmates. Repeat violators will be dismissed from the class and awarded a failing grade for the semester.

■ HUMAN LABORATORY EXPERIMENTS

Many of the laboratory experiments require humans as subjects. All of the experiments, if performed properly according to the instructions, are safe. If for any reason you feel that you should not perform as the subject of an experiment, simply inform your laboratory instructor and you will be excused from being the subject without penalty. You are responsible for notifying your laboratory instructor of any disease or physiologic disorder you have or have had that would preclude your being used as a subject. If you are in doubt, please ask. For example, if you have, or have had any of the following conditions, you may be excused as noted:

1. *Hemophilia or other abnormalities of blood coagulation (or you are currently taking anticoagulant therapy):* Do not act as a donor for experiments that require blood (e.g., hematocrit, blood typing).
2. *Chronic or acute respiratory infections:* Do not breathe into spirometers unless they can be sterilized or disinfected before another student's use.
3. *Diabetes mellitus (sugar diabetes) or hypoglycemia (low blood sugar):* Do not act as a subject in the oral glucose tolerance test.
4. *Epilepsy:* Do not hyperventilate or rebreathe carbon dioxide in respiratory experiments.
5. *Cardiovascular disease (high blood pressure, angina, etc.):* Do not perform stressful exercise.

LABORATORY ORGANIZATION

Each experimental laboratory period will be organized and conducted as follows:

Review and Evaluation of Previous Experimentation

The first part of the laboratory period will be devoted to a discussion of prior experimentation. On occasion your laboratory instructor may also administer a short unannounced quiz.

Introduction and Directions for Current Experimentation

Your laboratory instructor will explain the techniques and precautions to take in performing each experiment. In some cases, introductory review material may also be supplied by the instructor.

Setting Up Experimental Apparatus

The laboratory equipment provided for your use in the laboratory is some of the finest available for undergraduate laboratory instruction. It is also very expensive. Handle it with care at all times, following instructions as to its proper care and use. Report any damage or malfunction immediately. Do not attempt to repair it. All equipment used in the laboratory will be assigned by your laboratory instructor. Each person will be held responsible for the proper use and care of the equipment.

Experimentation

Never begin an experiment until you read, hear, and clearly understand all instructions. It is wise to read the experimental protocol before attending the laboratory so that you are familiar with the experimental procedure you will be asked to follow. Do not be reluctant to ask questions if you are unclear about any procedure.

Never prepare a living animal, organ, or tissue until all experimental apparatus has been set up. In any procedure involving pain, the nervous system of the animal must be destroyed by pithing or inhibited by an anesthetic. This rule must not be violated. If there is any doubt about an animal's inability to perceive pain, contact your laboratory instructor immediately before continuing any further experimentation.

Always record all observations immediately, either on the experimental recording or in your report for further reference. Experimental recordings should include the name of the experiment, the date it was performed, your name and the names of other students in the group who performed the experiment, the laboratory section, and the instructor's name.

Your laboratory instructor may require you to complete in full in ink the laboratory reports associated with each experiment and to turn them in for evaluation. Questions in the report are to be answered using complete and grammatically correct statements. Be as clear and concise as possible. Neatness is assumed. Most of the questions can be answered directly on the basis of the results that have been obtained from experimentation. Some require outside reading and thought: that is, an analysis of the facts in light of previous experience. Many of the questions are designed to stimulate, improve, and strengthen your ability to think scientifically. Be as independent as possible in this phase of the work in order to derive maximum benefits.

Most of the laboratory experimentation will involve working in harmony with one or more students in a group. Nevertheless, strive to be independent in both observation and thought. Failing to make independent observations and copying a fellow student's data and answers are unethical and dishonest and will not be tolerated.

Cleanup

After the completion of experimentation, the laboratory bench is to be returned to the condition in which it was found. Dirty glassware is to be properly disposed of according to instructions. Paper and other disposable materials are to be placed in the appropriate containers. Animal, tissue, and organ remains are to be placed in a special container for proper disposal. All electrical equipment is to be unplugged from the electrical outlets at the end of the laboratory period. The equipment is to be wiped clean and returned to its original position. Laboratory benches are to be wiped clean with a disinfectant, and the laboratory chair returned to its proper position.

CHAPTER 1

Metrics, Measurements, and Computations

■ INTRODUCTION

Crude measurements of length, capacity, and weight have existed since prehistoric times. Later, units of measurement were based on body size, such as the length of a king's arm or foot, and upon plant seeds. As civilization became more complex, technological and commercial requirements led to an increased standardization of measurements. But for a long time standards varied greatly between one locale and another, because units were usually fixed by edict of local or national rulers. As late as the eighteenth century, for example, one of the earliest units, the foot, possibly had as many as 280 variants in Europe. Today there are two principal systems of measurement—the **American-British system** (commonly called the *English system*) and the **metric system.**

In 1866 the U.S. government permitted the use of the metric system and established a conversion table based upon the yard and the pound. Because the federal government did not require the country to convert to the metric system, it was not widely adopted by commerce and industry in the United States. In scientific work, however, the metric system has been used for well over a century in the United States and for much longer in Europe.

The American-British system of linear, weight, and volume measurements uses units that are not logically related to one another. For example, 12 inches = 1 foot, 3 feet = 1 yard, 5280 feet = 1 mile, 1760 yards = 1 mile, and so on. The metric system uses units that are based on the decimal system and related to one another by some power of 10, like the American monetary system based on the dollar. Consequently, multiplication and division of metric units can be performed by simply moving the decimal point some number of places to the right or left. The term denoting a metric unit of measurement usually contains a prefix indicating the power of 10. The prefix *centi,* for example, means "one-hundredth," or 10^{-2}; the prefix *milli* means "one-thousandth," or 10^{-3}. Table 1.1 indicates prefixes used in the metric system.

■ UNITS OF LENGTH

The English system of linear measurement is based upon the *yard,* which is the equivalent of 3 feet, or 36 inches. The metric system of linear measurement is based on the *meter,* which is equivalent to 39.37 inches. The meter is defined as 1,650,763.73 wavelengths of the spectral orange-red line of krypton-86, or one-millionth part of the distance measured on a meridian from the equator of the earth to a pole, or the distance light travels in a vacuum in one 300-millionth of a second. Table 1.2 indicates metric units of linear measure.

To convert from a smaller unit of length to a larger unit of length, simply move the decimal point the required number of places to the left. For example, 153.0 mm (millimeters) = 15.30 cm (centimeters). Conversely, to convert from a larger unit of length to a smaller unit of length, simply move the decimal point the required number of places to the right. For example, 15.30 cm = 153.0 mm.

■ UNITS OF WEIGHT

In the English system, two sets of weight are employed: *avoirdupois weights,* based on the 16-ounce pound, are used in general commerce, and *troy weights,* based on the 12-ounce pound, are used for precious metals. Troy units form the basis of apothecary weights. By contrast, metric units of weight are based on the *gram,* defined as the weight of pure water at 4° C and 760 mm Hg pressure contained in a cube whose edge is one-hundredth of a meter (a cubic centimeter). Table 1.3 indicates metric units of weight.

To convert from a smaller unit of weight to a larger unit of weight, simply move the decimal point the required number of places to the left. For example, 246 mg (milligrams) = 0.246 g (gram). Conversely, to convert from a larger unit of length to a smaller unit of length, simply move the decimal point the required number of places to the right. For example, 0.246 g = 246 mg.

TABLE **1.1** Metric prefixes

Prefix	*Abbreviation*	*Meaning*	*Factor*	*Decimal*
tera	T	one trillion	10^{12}	1,000,000,000,000
giga	G	one billion	10^{9}	1,000,000,000
mega	M	one million	10^{6}	1,000,000
myria	my	ten thousand	10^{4}	10,000
kilo	k	one thousand	10^{3}	1000
hecto	h	one hundred	10^{2}	100
deka	da	ten	10^{1}	10
uni	—	one	10^{0}	1.0
deci	d	one-tenth	10^{-1}	0.1
centi	c	one-hundredth	10^{-2}	0.01
milli	m	one-thousandth	10^{-3}	0.001
micro	μ	one-millionth	10^{-6}	0.000001
nano	n	one-billionth	10^{-9}	0.000000001
pico	p	one-trillionth	10^{-12}	0.000000000001
femto	f	one-quadrillionth	10^{-15}	0.000000000000001

TABLE **1.2** Metric units of linear measure

Metric unit	*Definition*	*Metric equivalent*	*English equivalent*
megameter (Mm)	10^{6} meters	1,000,000 meters	621.37 miles
myriameter (mym)	10^{4} meters	10,000 meters	6.2137 miles
kilometer (km)	10^{3} meters	1000 meters	0.62137 mile
hectometer (hm)	10^{2} meters	100 meters	328.0833 feet
dekameter (dam)	10 meters	10 meters	32.80833 feet
meter (m)	basic unit of reference		39.37 inches
decimeter (dm)	10^{-1} meter	0.1 meter	3.937 inches
centimeter (cm)	10^{-2} meter	0.01 meter	0.3937 inch
millimeter (mm)	10^{-3} meter	0.001 meter	0.03937 inch
micrometer (μ)	10^{-6} meter	0.000001 meter	0.00003937 inch
nanometer (nm)	10^{-9} meter	0.000000001 meter	0.00000003937 inch
angstrom (Å)	10^{-10} meter	0.0000000001 meter	0.000000003937 inch
picometer (pm)	10^{-12} meter	0.000000000001 meter	0.00000000003937 inch
femtometer (fm)	10^{-15} meter	0.000000000000001 meter	0.00000000000003937 inch

■ UNITS OF VOLUME

The American (U.S.) system of liquid measure is based on the 16-ounce pint, and the British (Imperial) system is based on the 20-ounce pint. Metric units of liquid measure are based on the *liter,* which is defined as the volume of a kilogram of pure water at 4° C, equal to exactly 0.001 cubic meters. Table 1.4 indicates metric units of volume.

To convert from a smaller unit of volume to a larger unit of volume, simply move the decimal point the required number of places to the left. For example, 437 mL (milliliters) = 0.437 L (liter). Conversely, to convert from a larger unit of volume to a smaller unit of volume, simply move the decimal point the required number of places to the right. For example, 0.437 L = 437 mL.

Although arithmetical conversion between the English and metric systems is simple, many students have difficulty visualizing a distance of 10 meters or sensing the weight of a 2-kilogram object. Figure 1.1 offers a comparison of some common English and metric units of measurement.

TABLE 1.3 Metric units of weight

Metric unit	Definition	Metric equivalents	English equivalent (avoirdupois)
metric ton	10^6 grams	1,000,000 grams	2204.62 pounds
kilogram (kg)	10^3 grams	1000 grams	2.20462 pounds
hectogram (hg)	10^2 grams	100 grams	0.220462 pound
dekagram (dag)	10 grams	10 grams	0.35274 ounce
gram (g)	basic unit of reference		0.035274 ounce
decigram (dg)	10^{-1} gram	0.1 gram	0.0035274 ounce
centigram (cg)	10^{-2} gram	0.01 gram	0.00035274 ounce
milligram (mg)	10^{-3} gram	0.001 gram	0.000035274 ounce
microgram (µg)	10^{-6} gram	0.000001 gram	0.000000035274 ounce
nanogram (ng)	10^{-9} gram	0.000000001 gram	0.000000000035274 ounce
picogram (pg)	10^{-12} gram	0.000000000001 gram	0.000000000000035274 ounce

TABLE 1.4 Metric units of volume

Metric unit	Definition	Metric equivalent	English equivalent (U.S.)
myrialiter (myL)	10^4 liters	10,000 liters	2641.7 gallons
kiloliter (kL)	10^3 liters	1000 liters	264.17 gallons
hectoliter (hL)	10^2 liters	100 liters	26.417 gallons
dekaliter (daL)	10 liters	10 liters	10.567 quarts
liter (L)	basic unit of reference		1.0567 quarts
deciliter (dL)	10^{-1} liter	0.1 liter	0.10567 quart
centiliter (cL)	10^{-2} liter	0.01 liter	0.010567 quart
milliliter (mL)	10^{-3} liter	0.001 liter	0.0010567 quart
microliter (µL)	10^{-6} liter	0.000001 liter	0.0000010567 quart
nanoliter (nL)	10^{-9} liter	0.000000001 liter	0.0000000010567 quart
picoliter (pL)	10^{-12} liter	0.000000000001 liter	0.0000000000010567 quart
femtoliter (fL)	10^{-15} liter	0.000000000000001 liter	0.0000000000000010567 quart

TEMPERATURES: FAHRENHEIT, CELSIUS (CENTIGRADE), KELVIN (ABSOLUTE)

In the average U.S. household, as well as in general U.S. commerce and industry, changes in thermal energy are measured on the **Fahrenheit scale,** rather than the **Celsius (centigrade) scale** used nearly everywhere else. In scientific work, the Celsius scale, or the **Kelvin (absolute) scale,** or both, are always used. Comparisons and conversions of one scale to another conveniently use the freezing and boiling points of pure water.

On the Fahrenheit scale, the freezing point of water is 32° and the boiling point is 212°. On the Celsius scale, water freezes at 0° and boils at 100°. Therefore, one degree on the Fahrenheit scale indicates a smaller change in thermal energy than does one degree on the Celsius scale. On the Kelvin, or absolute, scale, water freezes at 273° and boils at 373°. Thus, one degree on the Kelvin scale measures the same thermal change as one degree on the Celsius scale. The Kelvin scale is primarily used in chemistry and physics and does not appear again in this manual. To convert between Celsius and Fahrenheit scales, use the following formulas:

$$°C = 0.56\ (°F - 32°)$$

$$°F = (1.8 \times °C) + 32°$$

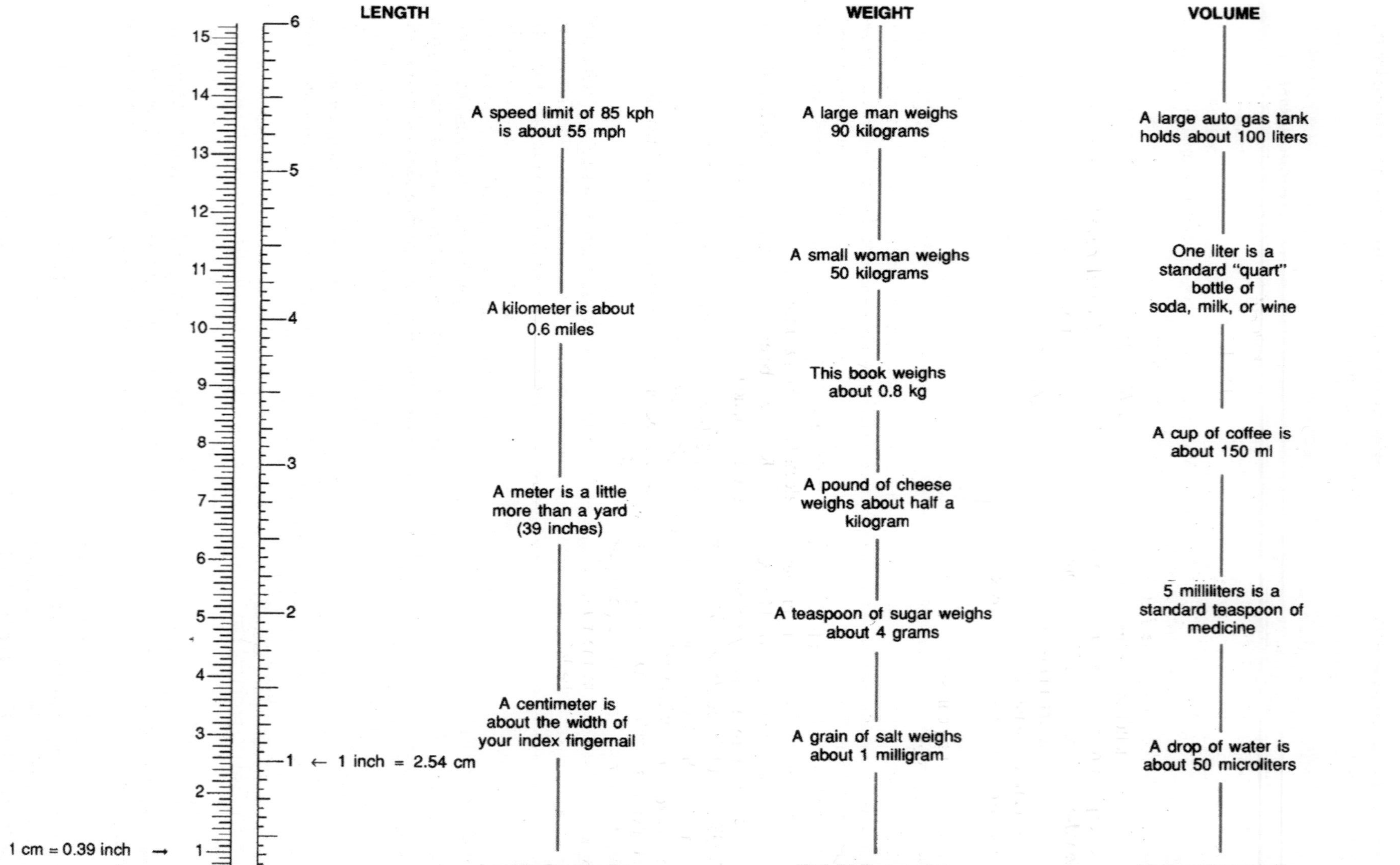

LENGTH

1 cm = 0.39 inch →

← 1 inch = 2.54 cm

A speed limit of 85 kph is about 55 mph

A kilometer is about 0.6 miles

A meter is a little more than a yard (39 inches)

A centimeter is about the width of your index fingernail

Length Conversion

1 in	= 2.54 cm		1 mm	= 0.04 in	
1 ft	= 30.5 cm		1 cm	= 0.39 in	
1 yd	= 0.91 m		1 m	= 39.4 in	
1 mi	= 1.61 km		1 m	= 1.1 yd	
			1 km	= 0.62 mi	

WEIGHT

A large man weighs 90 kilograms

A small woman weighs 50 kilograms

This book weighs about 0.8 kg

A pound of cheese weighs about half a kilogram

A teaspoon of sugar weighs about 4 grams

A grain of salt weighs about 1 milligram

Weight Conversion

1 oz	= 28 g	1 g	= 0.035 oz
1 lb	= 0.45 kg	1 kg	= 2.2 lb

VOLUME

A large auto gas tank holds about 100 liters

One liter is a standard "quart" bottle of soda, milk, or wine

A cup of coffee is about 150 ml

5 milliliters is a standard teaspoon of medicine

A drop of water is about 50 microliters

Volume Conversion

1 tsp	= 5 ml	1 ml	= 0.03 fl oz
1 tbsp	= 15 ml	1 l	= 2.1 pt
1 fl oz	= 30 ml	1 l	= 1.06 qt
1 cup	= 0.24 l	1 l	= 0.26 gal
1 pt	= 0.47 l		
1 qt	= 0.95 l		
1 gal	= 3.8 l		

FIGURE 1.1 Comparison of metric and English units of measurement

■ MEASUREMENT AND COMPUTATION

Scientific Notation

The biological and physical sciences frequently deal with very large numbers and very small numbers when an observation is to be described quantitatively. For example, the hydrogen ion concentration (H^+) of human urine may be close to 0.000001 g/L. The normal adult male concentration of red blood cells is 5,400,000 cells per microliter of whole blood. Very small numbers such as the former or very large numbers such as the latter are cumbersome to manipulate if written or expressed verbally in the form of a decimal number. **Scientific notation** simplifies the expression and manipulation of such numbers and is widely used within and without the scientific community. For example, the red cell count is often expressed clinically as 5.4×10^6 cells per microliter.

Scientific notation is a floating-point system of numerical expression in which numbers are expressed as products consisting of a number between 1 and 10 multiplied by an appropriate power of 10. Consider the number 186,740,000. In scientific notation, this number would be expressed as 1.8674×10^8. Essentially, scientific notation removes the need to write out all of the zeros in this number by using a power of 10 to represent the zeros. Numbers without zeros can also be represented by scientific notation—e.g., $247{,}632 = 2.47632 \times 10^5$ and $78{,}323 = 7.8323 \times 10^4$. When any number greater than 10 is expressed by scientific notation, the decimal point is moved to the left until the number has a value between 1 and 10, and that number is multiplied by some positive power of 10. The positive power of 10 is equal to the number of places the decimal point was moved to the left; for example:

10,263

1.0263 (decimal moved 4 places to left)

$1.0263 \times 10^4 = 10{,}263$

Very small numbers may also be expressed using scientific notation. For example, $0.00008 = 8 \times 10^{-5}$. When any number less than 1 is expressed by scientific notation, the decimal point is moved to the right until the number has a value between 1 and 10, and that number is multiplied by some negative power of 10. The negative power of 10 is equal to the number of places the decimal point was moved to the right; for example:

0.01467

01.467 (decimal moved 2 places to right)

$1.467 \times 10^{-2} = 0.01467$

To convert a number expressed by scientific notation into a single decimal number, simply move the decimal point the appropriate number of spaces to the right or to the left and omit the power of 10 multiplier. The power of 10 indicates the number of spaces to move the decimal point, and the sign of the power indicates the direction of movement.

With positive power, the decimal point is moved to the right; for example:

1.3×10^3

1300 (decimal moved 3 places to right, power of 10 omitted)

$1300 = 1.3 \times 10^3$

With negative powers, the decimal point is moved to the left; for example:

1.76×10^{-4}

0.000176 (decimal moved 4 places to left, power of 10 omitted)

$0.000176 = 1.76 \times 10^{-4}$

Ratios and Proportions

A **ratio** is an expression that compares two numbers or quantities by division. For example, if 400 students had been enrolled in a class at the beginning of a semester and 40 students had withdrawn from class by the end of the semester, the ratio of students withdrawn to students enrolled would be 40/400, or 1/10. That is, 1 student out of 10 elected to withdraw from class.

Ratios can be expressed several ways. Each method of expression, however, means the same thing:

1. 1:250 means 1 part to 250 parts.
2. 1/5 means 1 part out of 5 parts.
3. 2 females to 6 males means a ratio of 1 female to 3 males.

Whenever two quantities are expressed as a ratio, they must have the same units. For example, if the first of two compared animals had a recorded weight of 300 g and the second animal had a recorded weight of 1 kg, it would be necessary to convert one of the units of weight to the other (i.e., grams to kilograms, or kilograms to grams) before expressing the comparison in the form of a ratio. One kilogram equals 1000 grams; therefore, the ratio between the first animal's weight and the second animal's is 300/1000, or 3/10.

A **proportion** is simply a mathematical statement of the equality of two ratios. By arbitrarily using the letters A, B, C, and D to express quantities, we can state a proportion in the following manner:

$$\frac{A}{B} = \frac{C}{D}$$

The statement says A is to B as C is to D. Mathematically, it is also valid to say A times D equals B times C.

For example, if A = 15, B = 25, C = 3, and D = 5, then

$$\frac{A}{B} = \frac{C}{D}, \frac{15}{25} = \frac{3}{5}$$

and

$$A \times D = B \times C,\ 15 \times 5 = 25 \times 3$$

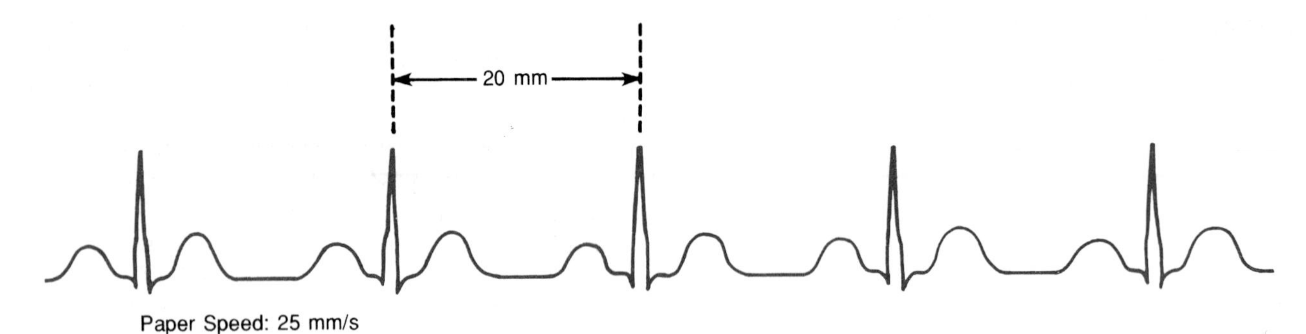

FIGURE **1.2** Electrocardiogram

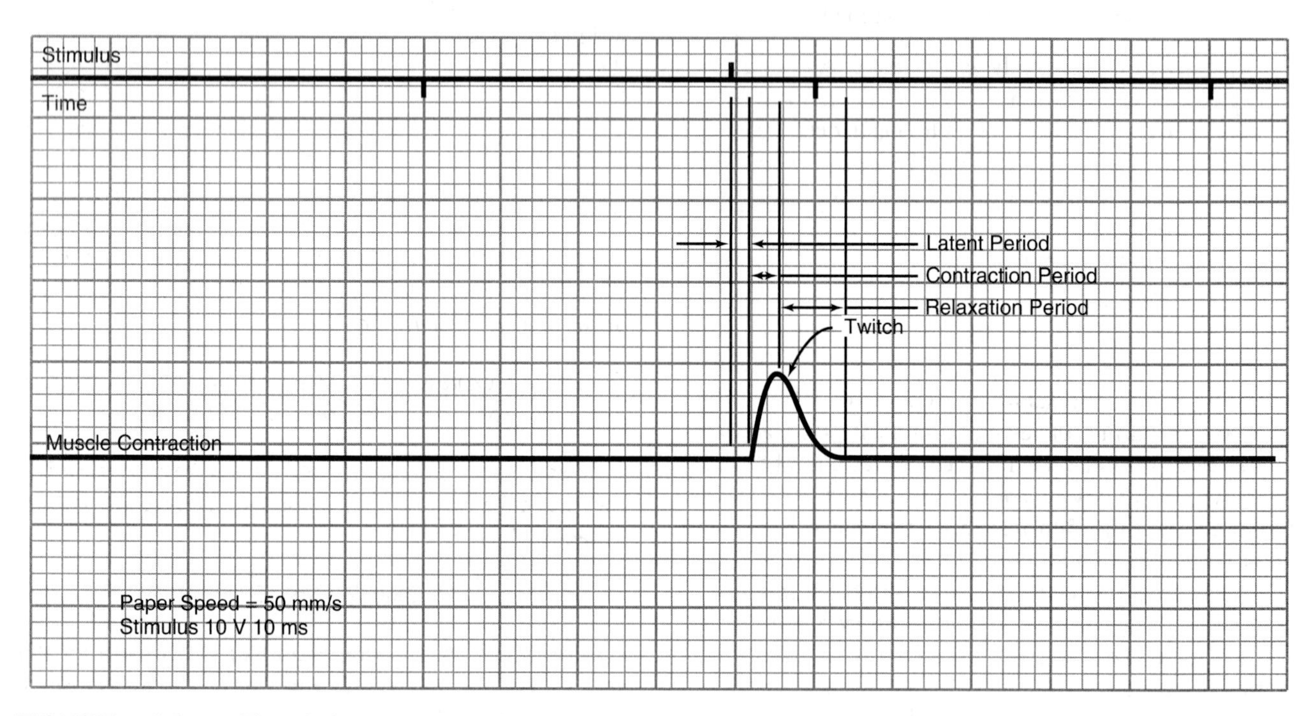

FIGURE **1.3** The skeletal muscle twitch

Therefore, it follows that if three of the quantities are known, the value of the fourth can be determined.

For example, assume that an electrocardiogram is being recorded on a moving strip of paper (figure 1.2). The speed of the moving paper is 25 mm/s. If each repeating cycle of the electrocardiogram represents one heartbeat, how many heartbeats are occurring each minute?

The problem can be solved as follows:

1. Distance between cycles = 20 mm (as measured from record).
2. Time interval between cycles = X seconds.

$$\frac{25 \text{ mm}}{1 \text{ second}} = \frac{20 \text{ mm}}{X \text{ seconds}}$$

$$25X = 20$$

$$X = 20/25 = 0.8 \text{ second}$$

3. If the interval between cycles is 0.8 second, then the number of cycles occurring each minute is

$$60 \text{ seconds} \div 0.8 \text{ second} = 75 \text{ beats/min}$$

Another example of how ratios and proportions can be used to solve problems is the determination of time in the events of muscle contraction. When a single stimulus of sufficient strength and duration is applied to an isolated skeletal muscle, the resultant contraction is known as a *twitch.* The mechanical record of a skeletal muscle twitch (figure 1.3) indicates a period between the time the stimulus is applied and the beginning of the contractile response. This interval of time is known as the *latent period.* Following the latent period, the muscle responds to the stimulus by shortening (the contraction phase) and then returning to its original length (the relaxation phase). Given a paper recording speed of 50 mm/s, determine the duration of the latent period, the contraction period, and the relaxation period as shown in figure 1.3.

The problem can be solved as follows:

1. Determine the time value of 1 mm of distance on the record:

$$\frac{50 \text{ mm}}{1 \text{ second}} = \frac{1 \text{ mm}}{X \text{ seconds}}$$

Cross-multiplying:

$$50X = 1$$

Solving for X:

$$X = 1/50 \text{ seconds} = 0.02 \text{ second}$$

2. Measure the length of the latent period, contraction period, and relaxation period in millimeters and multiply each value by 0.02 s to obtain durations.
   a. Latent period = 3 mm × 0.02 s/mm = 0.06 second.
   b. Contraction period = 5 mm × 0.02 s/mm = 0.10 second.
   c. Relaxation period = 11 mm × 0.02 s/mm = 0.22 second.

Mixtures

Many of the experiments in this book involve fluids, such as blood and urine, or testing chemicals, such as Benedict's reagent, that are mixtures of various chemicals. Thus, an elementary discourse about the nature of mixtures and methods of expressing their concentrations will prove beneficial.

A **mixture** may be defined as a portion of matter consisting of two or more components in varying proportions that retain their own properties. A **solution** is a mixture of a **solute** dissolved in a **solvent.** When sugar is mixed with water, the sugar (solute) dissolves in the water (solvent). Solutions made with water are called **aqueous solutions**.

A solution may contain several kinds of solutes, such as sugar, salt, and so forth, dissolved in the same solvent. Biologically speaking, the universal solvent is water. Water is the dominant chemical compound making up the human body.

A solution is homogeneous. **Homogeneous** means "the same throughout," as opposed to **heterogeneous,** which means "not the same throughout." Homogeneity results from solute molecules becoming uniformly distributed in the solvent so that solute concentration everywhere in the solution is equal.

The number of solute molecules that can dissolve in a fixed volume of solvent is limited. When the maximum number of solute molecules has been dissolved, the solution is described as being **saturated.** Further addition of solute results in the added solutes failing to dissolve and, instead, settling to the bottom of the solution as a **precipitate.**

The concentration of a solution may be expressed in terms of percent or parts per hundred. The percent number is the number of grams of solute present in 100 mL of solution. For example, physiologic saline, a 0.9% chloride solution, contains 0.9 gram of NaCl dissolved in 100 mL of water.

Another way to express the concentration of a solution is to use molarity. **Molarity** is the number of moles of solute dissolved in 1 liter of solution. One **mole** of a compound is the number of grams of that compound equal to its molecular weight. The **molecular weight** of a compound is the sum, expressed in grams, of the atomic weights of all atoms that make up the compound. For example, the molecular weight of glucose ($C_6H_{12}O_6$) is equal to the sum of the atomic weights of 6 carbon atoms plus 12 hydrogen atoms plus 6 oxygen atoms. The atomic weights are: carbon (C) = 12; hydrogen (H) = 1; oxygen (O) = 16. Therefore, 1 mole of glucose = $(6 \times 12) + (12 \times 1) + (6 \times 16) = 180$ grams. A 1-molar (1M) glucose solution contains 180 grams of glucose dissolved in enough water to make 1 liter of solution.

Osmolarity is a method of expressing the solution concentration based upon the *number of particles in solution.* This method of expressing solution concentration is most useful when dealing with compounds that ionize (form charged particles) in water.

Osmolarity (expressed in osmols) = (molarity) × (number of particles per molecule of solute)

For example, one molecule of NaCl, when dissolved in water, completely ionizes to form two particles: a sodium ion and a chloride ion:

$$NaCl \rightarrow Na^+ \; Cl^-$$

Therefore, 1M NaCl solution will have a concentration of 2 osmols, a 0.5 M NaCl solution will have a concentration of 1 osmol, and so forth. For solutes such as glucose that do not ionize in water, the values for osmolarity and molarity are equal because the number of particles per molecule of solute is one.

Human blood and other human body fluids have an osmolarity of approximately 300,000 osmols. It is more practical to express this in terms of **milliosmoles.** One osmol equals 1000 milliosmoles. Thus, the osmolarity of blood is approximately 300 milliosmoles.

A **suspension** is a mixture of insoluble particles in a suspending medium. When sand is placed in water and shaken, a sand–water suspension is produced. When the shaking stops, the sand particles settle. Blood is a suspension of blood cells in blood plasma. If blood is placed in a vertical test tube and prevented from clotting, the blood cells will gradually settle toward the bottom of the tube.

A **colloid** is a suspension of very tiny particles, usually in water. Colloids differ from other suspensions in that colloids do not settle. Proteins in the water of blood plasma exist as colloids.

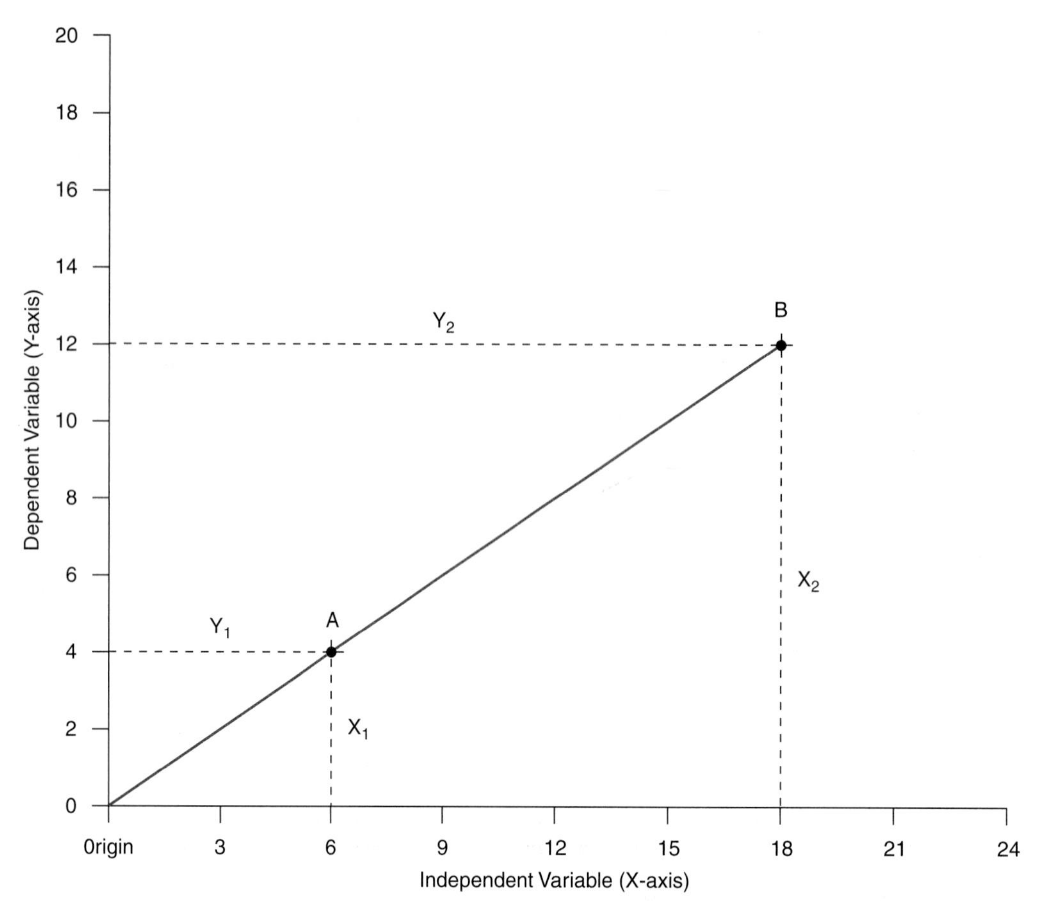

FIGURE 1.4 Independent and dependent variables

Graphic Analysis

A relationship between two or more quantities can often be diagrammatically expressed as a graph. A graph puts into visual form abstract ideas or experimental data so that their relationship becomes apparent. For example, the relation between two kinds of experimental data may not be easy to see when the data are displayed in the form of a table, but when the data are graphed, a relationship such as cause and effect is easier to see. Graphic presentation of data may not explain the reason for the relationship, but the shape of the graph can provide clues.

The related quantities displayed on a graph are called **variables.** A simple graph (figure 1.4) uses a system of coordinates, or axes, one horizontal and one vertical, to represent the values of the variables. Usually, the relative size of the variable is represented by its position along the axis, and the numbers along the axis allow the reader to estimate the values. In general, the value of the variable increases from left to right along the horizontal axis and from bottom to top along the vertical axis.

If the relationship being plotted is one of cause and effect, the variable that expresses the cause is called the **independent variable.** Usually this is represented by the horizontal axis (also sometimes called the *x-axis,* or **abscissa**). The variable that changes as a result of changes in the independent variable is called the **dependent variable.** It usually is expressed on the vertical axis (also called the *y-axis,* or **ordinate**). The two axes are arranged at right angles to each other and cross at a point called the **origin.**

The relationship between x and y variables is shown by the vertical extension of the value on the x-axis (X_1) and the horizontal extension of the value on the y-axis (Y_1). The point, A, at which these lines cross is determined by their relationship (figure 1.4). If another pair of data points (X_2 and Y_2) is chosen, their point of intersec-

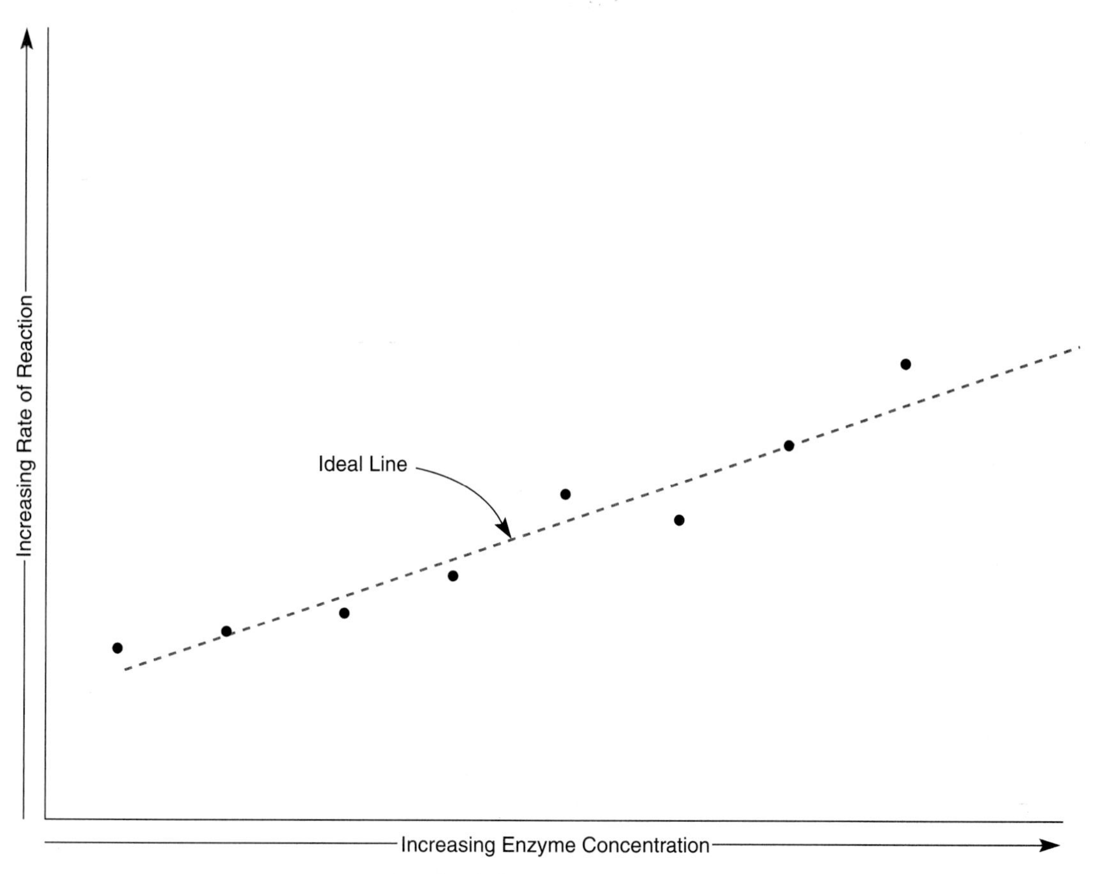

FIGURE 1.5 A direct linear relationship

tion can also be plotted; this is point B. A line drawn between points A and B can then give information about how all other X and Y values on this graph should be related to each other.

One of the simplest kinds of relationship that a graph can represent is called a **direct relationship:** the y values get larger as the x values get larger. An example of this type of graph is shown in figure 1.5. The data plotted here could have come from an experiment in which various concentrations of an enzyme were used to study how fast a particular chemical reaction happened at each concentration.

Data points frequently do not fall exactly on a straight line and appear scattered. In many cases, a mathematical procedure may be carried out to determine the "best-fitting" line to describe the relationship. This is called the **ideal line** in figure 1.5. A relationship that can be described by a straight line is called a **linear relationship.** Inverse relationships are also common. In such relationships, the y values get smaller as the x values get larger. Such relationships may still be linear if straight lines describe them. Other inverse relationships may be **curvilinear,** as in figure 1.6. The graph shown here summarizes the experimental finding that a skeletal muscle maximally contracts with less force as it is stretched beyond its optimum initial length.

If an experimenter had enough confidence in the reliability of the data, two kinds of predictions could be made from such a graph. Predicting data values that fall "between the points" is called **interpolation.** This process is useful if the curve is to be used as a guide for interpreting or testing the reliability of newly obtained data. A riskier procedure is extrapolation, which involves extending the ideal line into ranges where experimental data are not present. If there is good reason to believe that the same relationship should hold outside this range, then this prediction could be valid and might permit useful information to be gained.

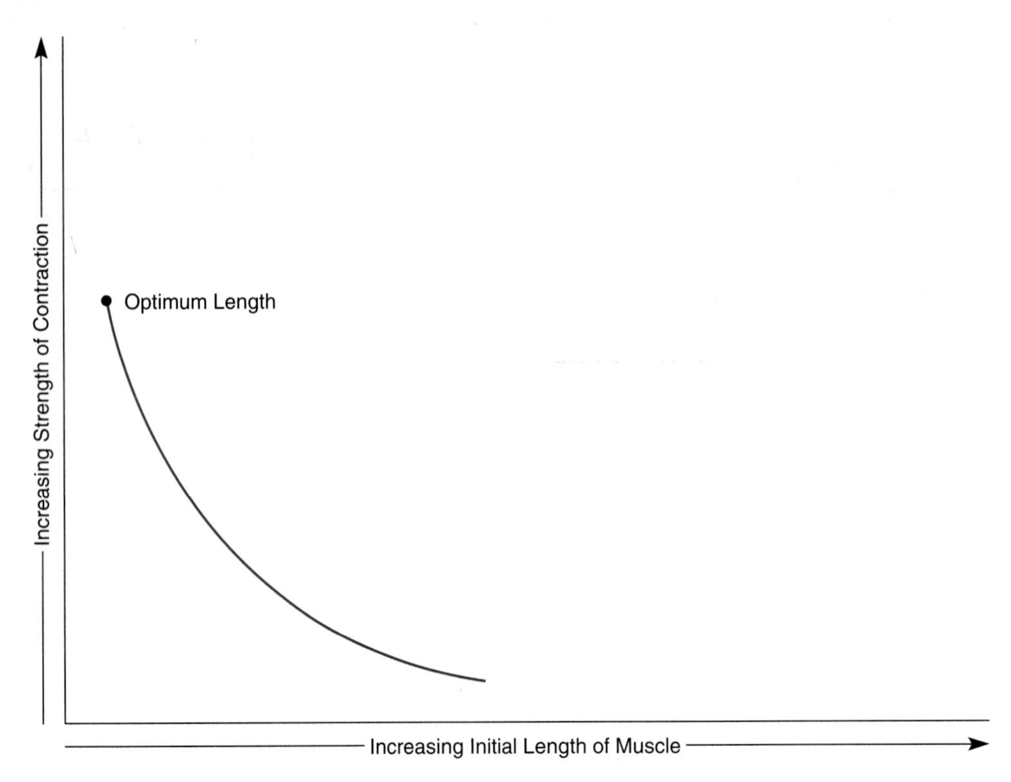

FIGURE 1.6 A curvilinear inverse relationship

Computation of Arithmetic Mean

It is often useful in comparing groups of numerical data to calculate the arithmetic mean, or average. It may be calculated using the following formula:

$$\bar{X} = \frac{\Sigma X}{N}$$

where $\bar{X}$ = the mean of X, ΣX = the sum of all values of X in each group, and N = the number of individual values for X in each group.

For example, assume that as part of a study, resting systolic blood pressure was recorded from 10 male subjects each aged 21 years, and you wish to calculate the mean systolic blood pressure and compare it with that of another group. Table 1.5 indicates the data as recorded and the calculation of the arithmetic mean.

TABLE 1.5 Calculation of arithmetic mean

Subject	***Resting systolic pressure (mm Hg)***
1. RGP	120
2. DKK	135
3. ALB	126
4. RWK	133
5. WGB	127
6. RSS	140
7. FAJ	110
8. NDL	117
9. WSC	125
10. RCS	129
$N = 10$	$\Sigma X = 1262$
	$\bar{X}$ = 1262/10 = 126 mm Hg (rounded to the nearest whole number)

Metrics, Measurements, and Computations

REPORT 1

Name: ______________________________ Date: ______________

Lab Section: ______________________________

1. Use a small metric rule to measure the diameter of the 1-inch and 1½-inch circles below. Record the measurements in millimeters and centimeters.

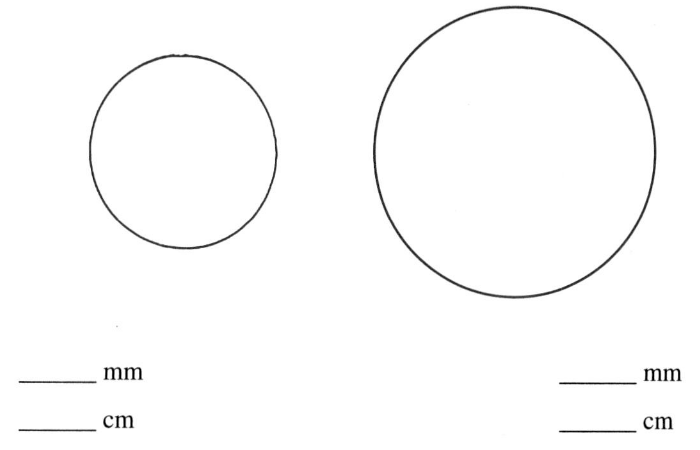

2. Use the laboratory scale to measure your weight and height. Record weight in kilograms and pounds. Record height in centimeters and inches.

weight: ______ kg ______ pounds

height: _______ cm ______ inches

3. Compute the following conversions:

a. 242 mg = ________ g

b. 6 g = ________ cg

c. 4 lb = ________ kg

d. 0.83 cm = ________ mm

e. 3450 mL = ________ L

f. 243 mm = ________ cm

g. 10° C = ________ ° F

h. 72° F = ________ ° C

4. Solve the following proportions for *X:*

a. 6/36 = *X*/48, *X* = __________

b. 9:72 as *X*:64, *X* = __________

c. 24/144 = 18/*X*, *X* = __________

d. *X*/27 = 17/81, *X* = __________

5. At rest, the left ventricle of the heart pumps 5.0 L of blood per minute. Blood flow to the kidneys is approximately 1200 mL/min at rest. Assuming a proportionate increase in renal blood flow, what will be the blood flow to the kidneys if the heart pumps 7.0 L/min?

____

____

____

6. An electrocardiogram is recorded on millimeter grid paper moving at a speed of 25 mm/s. If the distance between cycles as recorded is 15 mm, what is the subject's heart rate in beats per minute?

____

____

____

7. Express the following numbers using scientific notation:

a. 1563 = ______________________________

b. 0.364 = ______________________________

c. 1.000 = ______________________________

d. 5.463 = ______________________________

8. Compute the arithmetic mean (average) of the following body weights:

a. 76 kg   **d.** 59 kg

b. 63 kg   **e.** 68 kg

c. 81 kg   **f.** 74 kg

Mean body weight = ____________________ kg.

9. Mary has a circulating red blood cell count of 4.5×10^6 cells per microliter and a circulating blood volume of 5.5 liters. How many red blood cells are circulating?

10. In order to control a patient's blood pressure, a physician has ordered a drug dosage of 0.20 mg/kg of body weight, to be taken orally twice each day. The patient weighs 220 lb. How large is each dose?

11. A 10% glucose solution will contain __10__ grams of glucose dissolved in __100__ mL of water.

12. Ten moles of water weigh ______ grams.

13. A 2M NaCl solution will have a concentration of ______ osmoles.

14. A 0.1M NaCl solution will have a concentration of ______ milliosmoles.

CHAPTER 2

Bioinstrumentation

■ INTRODUCTION

Laboratory experimentation in physiology often involves detecting, observing, recording, and measuring one or more biological activities. Biological activity is often manifest by change in tension, length, pressure, voltage, current, resistance, or other variables that occur as cells, tissues, or organs function. For example, each beat of the heart begins with a very small bioelectric signal that is generated within the heart and conducted to heart muscles, stimulating them to contract. The electrical activity associated with each heartbeat can be studied by using bioinstrumentation to detect, amplify, and record voltage changes versus time. Mechanical and electrical activity associated with skeletal muscle contraction and electrical activity of the brain are other classic examples of biological phenomena that can be studied using bioinstrumentation. Because many of the following experiments require instrumentation to study physiologic processes, an introduction to bioinstrumentation principles and use is warranted.

Bioinstrumentation systems consist of components that are mechanical, such as springs, levers, and bellows; components that are electronic, such as meters and oscilloscopes; and components that are both mechanical and electronic, such as force transducers. In general, most electronic instrumentation systems contain one or more components that (1) detect the biological signal (e.g., electrodes), (2) amplify or otherwise condition the signal so that it may be recorded, and (3) record the signal in a recognizable form for interpretation and study. Computer-based instrumentation systems have additional advantages of being able to rapidly acquire and store large amounts of experimental data and assist in data analysis. Regardless of the nature of the components, the primary function of all bioinstrumentation is the same: to sense, amplify, and record biological activities. Hence, almost any kind of recording system designed for biological experiments can be used in performing the experiments in this book.

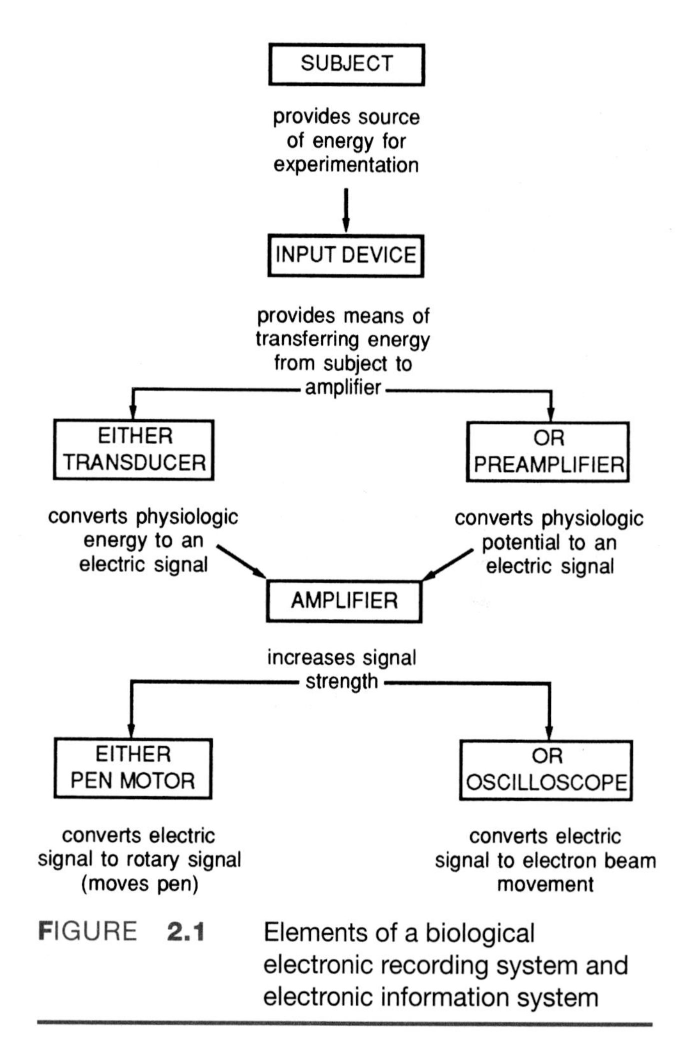

FIGURE **2.1** Elements of a biological electronic recording system and electronic information system

First, let us examine the elements of an electronic recording system. The human body or animal tissue such as the frog gastrocnemius muscle is often used as the source of a biological signal. Elements of an electronic instrumentation system are diagrammed in figure 2.1.

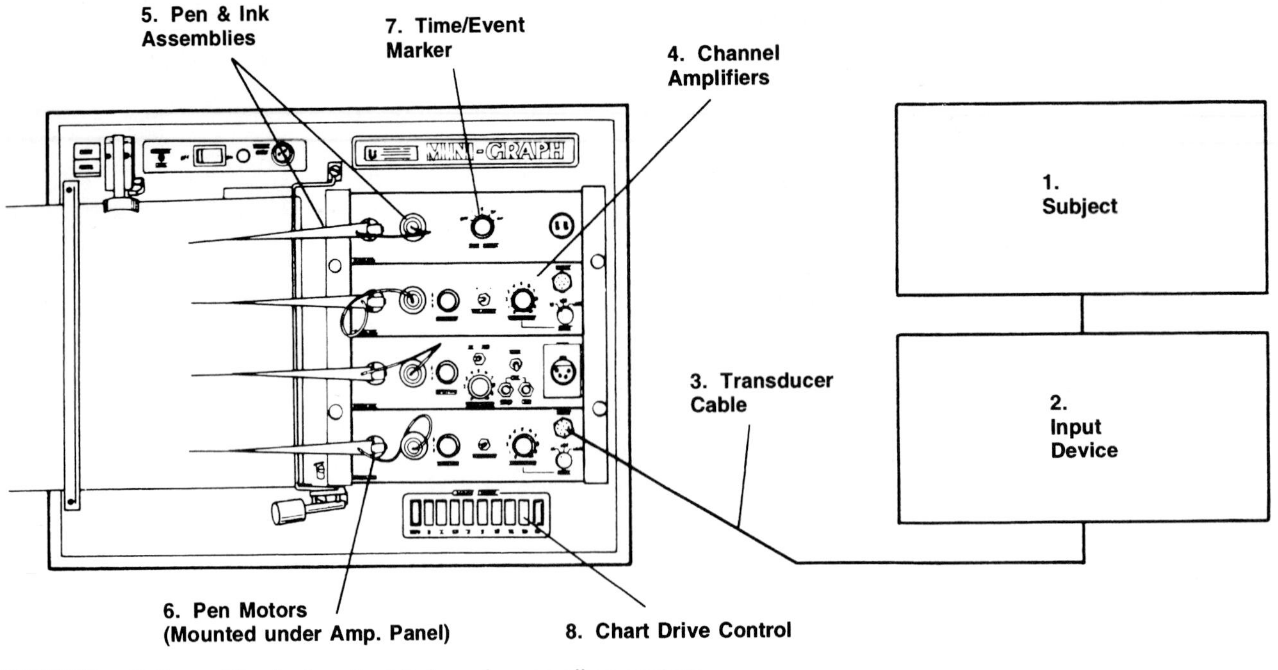

FIGURE 2.2 The Lafayette Minigraph recording system

Signal detectors sense biological signals. In general, there are two forms of detector: electrodes and input transducers. If the signals are electrical in origin (e.g., nerve impulses), electrodes are used. If the signals are mechanical, thermal, sonic, or optical in nature, input transducers are used.

Transducers are devices that convert one form of energy to another form of energy. In a physiologic recording system, input transducers convert various forms of energy (e.g., mechanical) to electrical energy, which then may be processed by the amplifier. In general, transducers are designed to absorb (and thereby lose) as little energy as possible during the conversion.

Amplifiers increase the strength of the electric signals generated by input transducers or detected by electrodes. Signal amplification is necessary because the output signal from the latter devices is small and generally not of sufficient strength to drive the pen motor. Sometimes a preamplifier is required in addition to an amplifier to match the input requirements of the amplifier. Input transducers and preamplifiers are often combined into a single unit. Most amplifiers allow the user to adjust or control the degree of signal amplification much like the control of volume on a radio.

After the signals have been amplified, they must be displayed visually or in some other examinable manner; this is the function of the output transducer. Output transducers may display signals on cathode-ray oscilloscopes, loudspeakers, electrical meters, or other devices. Frequently, an output transducer records electric signals as deflections of a pen on a strip of moving paper or acetate film.

Pen motors are forms of output transducers. The pen motor converts electrical energy coming from the amplifier to rotary motion of the recording stylus. The excursions of the stylus across the moving chart paper or acetate film are proportional to the physiologic activity, thereby providing a permanent record of an experiment.

Each of the elements discussed so far can be found in a standard strip-chart recording system such as the Lafayette Minigraph recording system (figure 2.2). Elements of a computer-based data acquisition/analysis system are diagrammed in figure 2.3.

Before electric signals can be processed by a computer, they must be converted from *analog* form to *digital* form. A clock with a numbered face and hands that point to the correct hour and minute presents time in an analog form, whereas a clock with a window showing only the number of the hour, a colon, and the number of the minute (e.g., 10:32) presents time in a digital form.

The computer's brain, called the **central processing unit (CPU),** works only with information in digital form. **Analog-to-digital converters (ADCs)** are devices that receive an input voltage (e.g., a 2-millivolt signal) and convert it to digital form (e.g., 0010, a binary number). ADCs may be inside the computer, or they may be external to the computer, as shown in figure 2.3. An external ADC is usually part of a device called a **signal conditioner,** which may amplify, filter, and otherwise condition the signal before sending it to the CPU.

The CPU, or simply **processor,** is part of the computer's **hardware.** Other parts of the hardware are the **computer memory, disk storage,** and **input/output (I/O) devices.**

The computer's memory represents the place used by the processor to do its work. Memory is equivalent to a

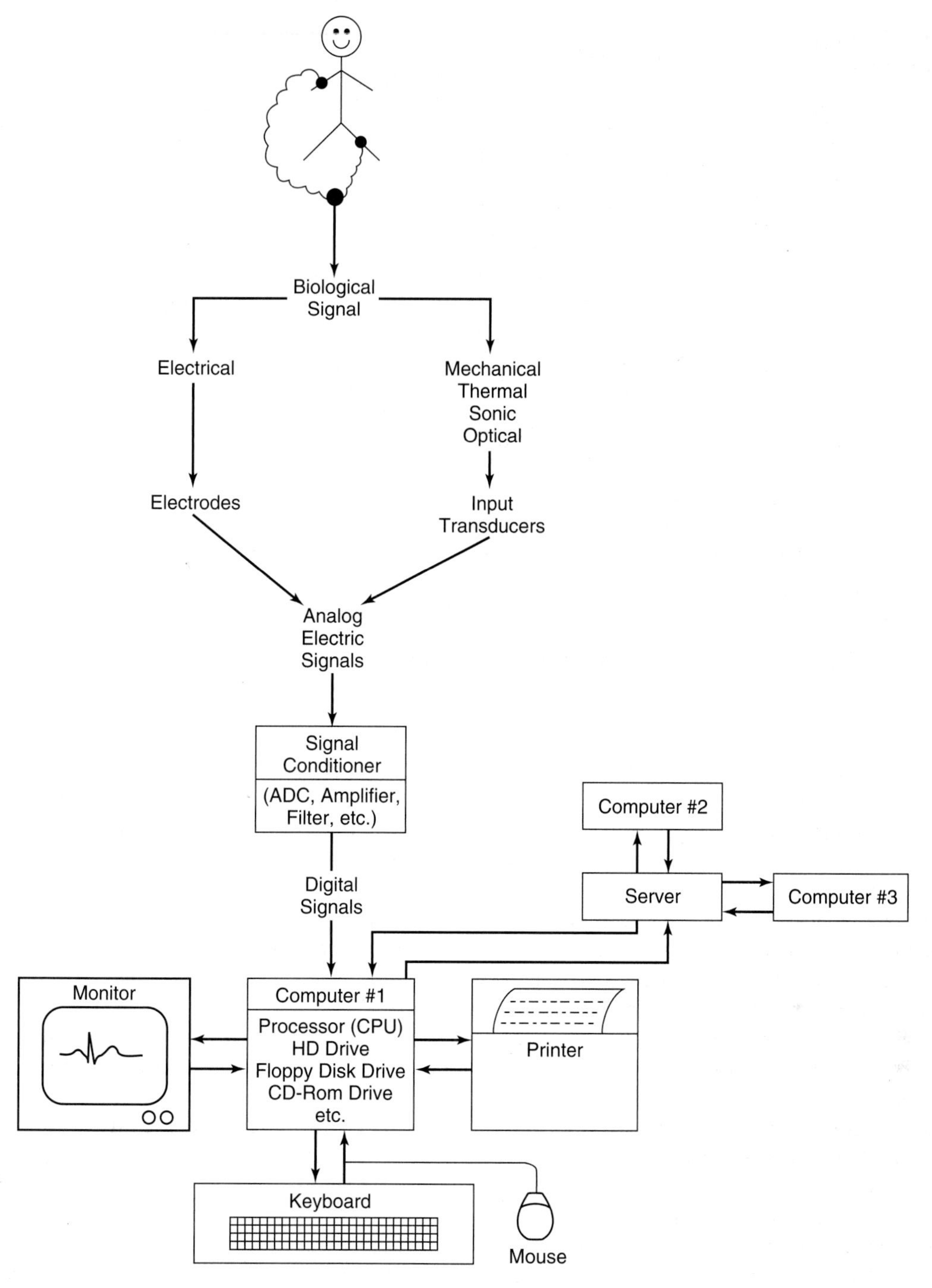

FIGURE 2.3 Elements of a computer-based data acquisition/analysis system

work space, such as a laboratory bench. Memory is where the computer places information it is working on at a given time. The more complicated the work, the greater the amount of memory (work space) required. Computer memory is contained within electronic chips inside the computer. There are two types of memory chip: ROM and RAM. **ROM** means "read-only memory." ROM chips contain important data that can be read but not changed. ROM chips always retain their data, even when the computer is turned off. **RAM** means "random-access memory." RAM chips lose their data when the power is turned off, but you can read and write to these chips and therefore change the information they contain. RAM is more properly called read/write memory.

Disk storage is where information is placed when the computer is not actually working on it. There are two principal forms of disk storage: removable **diskettes** (floppy and rigid) and **hard disks** (HD), which may or may not be removable. Disk storage is analogous to a filing cabinet. Within disk storage, information is organized into **files** contained within **folders.** A file may contain data (e.g., written text, tables, etc.) or sets of instructions (programs) for the computer. When the computer needs information from a file, it opens the file, reads it or writes to it, and then closes the file. The computer performs these tasks by means of a **disk drive.**

I/O devices send information to the CPU, receive information from the CPU, or do both. A signal conditioner, a keyboard, a monitor (CRT display screen), a mouse, and a modem with a communications adapter are examples of I/O devices. Because these devices are attached to the computer by cables, they are also called **peripheral devices.** Another example of an I/O device is the CPU of another computer.

A **server** (figure 2.3) is a computer that is connected by cable to several other computers and provides services to them. Servers provide additional memory and can store data and programs, thereby increasing the capabilities of each machine connected to the server.

Software refers to the **computer programs** a computer uses; software is the set of precise detailed instructions that tell the computer exactly what and when to do something. The most important program is the master program, called an **operating system (OS).** The OS tells the computer how to get itself ready to perform tasks and how to perform ordinary tasks, such as operating the disk storage, that all programs need to carry out. The OS tells the computer how to run other programs such as word-processing or data-analysis programs or a program for acquiring, storing, and analyzing physiologic data.

The BIOPAC Student Lab (figure 2.4) is a computer-based data acquisition and analysis system that consists of hardware and software. Hardware components include electrodes, input transducers, connecting cables, and the MP30 or MP35 data acquisition unit. Bioelectric signals received by the MP30/MP35 are converted from analog to digital form, amplified, and further processed before being sent to the computer's CPU. Software in this system consists of the BIOPAC Student Lab programs installed on the computer's hard drive. BIOPAC programs are sets of instructions the CPU can refer to (as the computer sets itself up to function in each lesson or physiology experiment) regarding data acquisition, data analysis, and data storage.

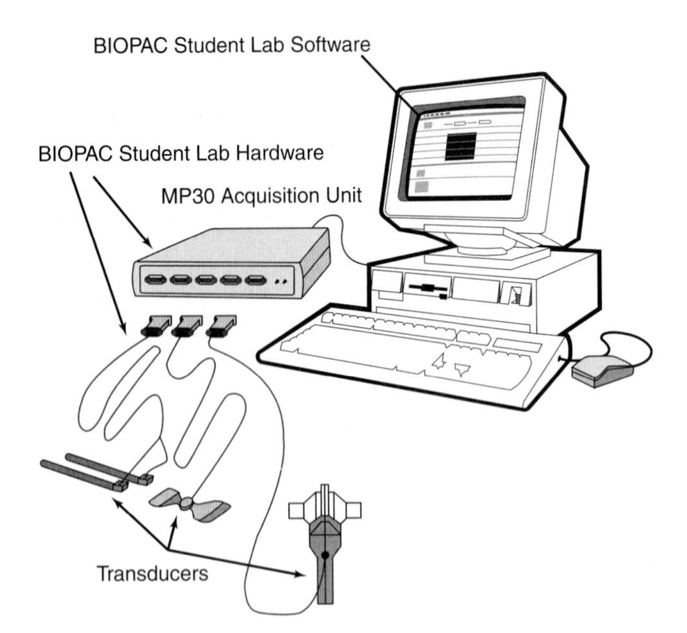

FIGURE **2.4** Software and hardware components of the BIOPAC Student Lab

Courtesy of BIOPAC Systems, Inc.

In the sections that follow, general principles regarding the use of electrodes and stimulators will be outlined. These peripheral hardware devices are common to many types of recording systems, computerized as well as standard pen writing systems. A brief presentation of the Lafayette Minigraph, a noncomputerized recording system remaining in use will also be included.

ELECTRODES

Electrodes are used in physiology experiments to conduct an electric current to or away from body cells or tissue. Electrodes may take many forms, depending on the magnitude of the electric current conducted and the location of the electrodes in or on cells or tissue. All electrodes must be good electrical conductors. They should also be generally nontoxic to body tissues. Many electrodes are composed of gold, nickel-silver, platinum, or paladium, biologically inert metals that are good electrical conductors.

When using electrodes, keep several factors in mind:

1. Good, low-resistance contact must be made between the electrode surface and the biological tissue. Achieving this contact requires that the electrodes be clean and free of oil, dirt, oxidation, and so on. If electrodes are applied to the surface of the skin, the skin must be free of dirt, oil, cosmetics, and other substances that tend to increase contact resistance.
2. Electrodes must be in firm contact with the tissue but should not mechanically interfere with tissue function and should not produce recording artifacts.
3. Electrode cables may pick up interference from fluorescent lights, power lines, and other electrical devices. Properly shielded cables and grounding of electrical devices to a water pipe should eliminate such interference.

CAUTION

Always use electrodes designed for human use when stimulating human skeletal muscle or nerve. These electrodes have current-limiting features, enhanced isolation, and a user-operated "dead man's switch" for optimum safety. Even with proper electrodes, **never create an electrical path across the heart** (e.g., by touching an active tip in each hand), and never use stimulating electrodes on subjects with a pacemaker.

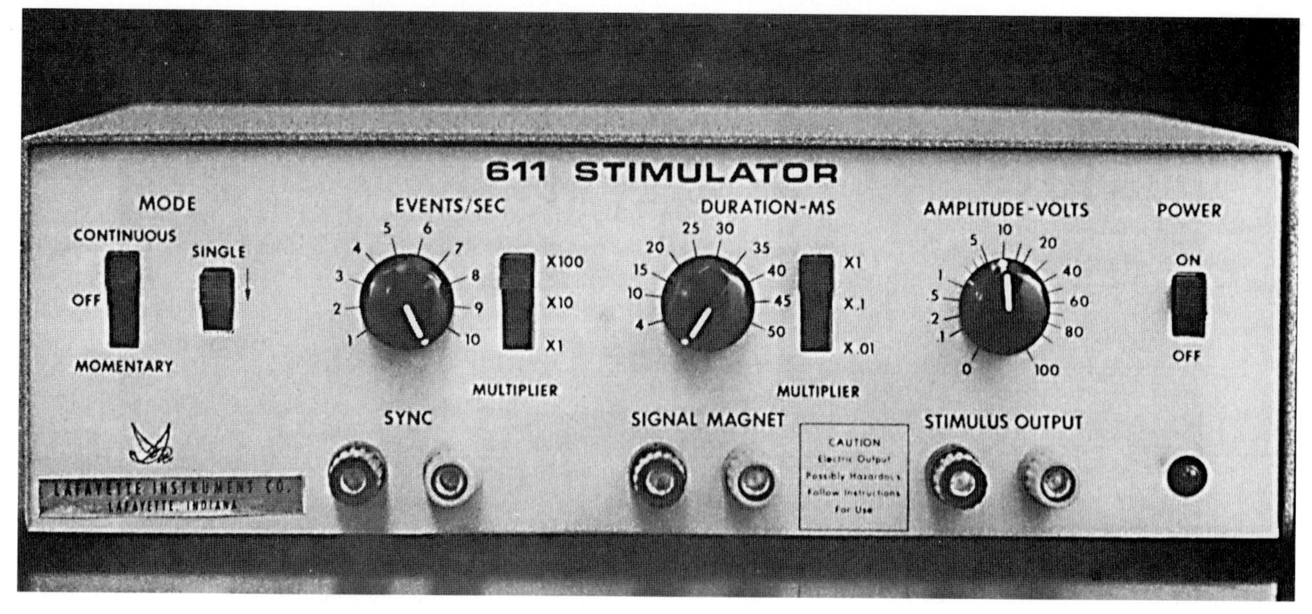

FIGURE **2.5** Photo of model 82415 Lafayette 611 square-wave stimulator

■ STIMULATORS

Use of a device that electrically stimulates tissue may be necessary to elicit a biological event for recording. Figure 2.5 illustrates the model 82415 Lafayette 611 square-wave stimulator. Other stimulators such as the Grass or Harvard student models function much like the Lafayette model.

The **stimulator** not only allows a stimulus to be generated to command the tissue but also allows the operator to control the stimulus frequency, duration, and strength. *Frequency* is the number of stimuli per unit of time (pulses per second). *Duration* is the length of time during which the electric pulse exists (milliseconds). *Strength* is the amplitude, or intensity, of the stimulus pulse (volts).

Additional controls on the Lafayette 611 stimulator include the following:

1. *Power switch:* This switch turns power on or off. An LED (light-emitting diode) indicates when power is on.
2. *Mode switches:* Single stimuli may be produced by placing the left mode switch in the "off" position and depressing the single switch once for each stimulus desired. Continuous stimuli (series of stimuli in a continuous train) may be produced by placing the left mode switch in the "continuous" position. In the "momentary" position, the stimuli will continue only while the mode switch is held down.
3. *Stimulus output terminals:* These terminals are used to connect the stimulator to the biological specimen via electrode cables. The variable output from these terminals is determined by adjusting the frequency, duration, and amplitude controls.
4. *Signal magnet terminals:* These terminals are used to connect the signal marker circuit of the stimulator to a signal magnet or time/event marker in a recording system. This connection allows the recording of stimulus delivery. A connecting cable is required. Although the characteristics of the output stimulus may be varied, the output from the signal magnet terminals does not vary; thus, the event record marks are of uniform amplitude and follow the stimulus frequency up to 20 Hz.
5. *Voltage reducer toggle switch:* Located on the rear panel, this switch reduces the value of the stimulus output voltage to one-tenth the value indicated on the front panel. For example, when the amplitude = volts knob is set to 0.2 V and the reducer switch is set to ×0.1, the strength of the actual stimulus output is 0.02 V.

Bipolar electrodes used for stimulating skeletal and cardiac muscles are connected to the output and ground terminals. When using the stimulator, always check to ensure that the stimulator is properly connected to the event/time module, that the power switch is on, and that the mode switch is in the "off" position (except when delivering a stimulus).

■ LAFAYETTE INSTRUMENT COMPANY MINIGRAPH

The Minigraph consists of a mainframe that houses channel amplifiers, an electronic timer and event marker, a chart drive and speed control, and recording pen assemblies (see figure 2.2).

The Mainframe

Two Minigraph mainframes are available. Model 76107 is equipped with a solid-state push-button–controlled chart drive, and Model 76107VS is equipped with a continuously variable chart drive. Both mainframes are housed in a fully portable cabinet and have a removable cover. To set up the Minigraph:

1. Open and remove the cover, unlatching it at the hinges.
2. Ensure that the power switch and the chart drive switch are off.
3. Attach the power cord to the power input receptacle located on the top left center of the mainframe and to a suitable AC outlet.
4. Turn on the power switch. Model 76107 has a pilot light indicating when the power is on. The power-on indicator for model 76107VS is integrated into the switch.
5. If there is no response when the power switch is turned on, check connections at both ends of the power cord, the AC outlet, and the mainframe fuse located on the upper left mainframe panel.
6. Check the ground-loss indicator light. If it is on, turn off the power supply and consult a qualified electrician before attempting to use the recorder. Loss of ground may result from a miswired or improperly grounded AC outlet.

Chart Paper

The Minigraph uses model 76702 chart paper available in an 8-in. × 115-ft roll marked with a millimeter grid. The following procedure for installing chart paper is suggested:

1. Release the friction drive wheel, located on the upper left mainframe panel, by pushing the vertical release lever toward the rear of the cabinet. The lever will rotate to the horizontal position as the drive wheel elevates.
2. Gently push the counterweight on the pen lift bar and rotate it to the right, lifting the pens. The lift bar will lock in place, holding the pens in an elevated position.
3. Remove the rectangular panel directly in front of the pen housing.
4. Insert a new roll of paper so that the number 76702 printed on the paper is toward the rear of the cabinet. Unwind approximately 1 foot of paper.
5. Replace the rectangular panel with the rounded edge pointed toward the paper slot.
6. Guide the paper under the paper cutoff bar located at the left edge of the cabinet. Align the paper and reengage the friction drive wheel.
7. Gently push the counterweight on the pen lift bar and rotate it to the left, returning the pens to recording position.

Recording Pens

The Minigraph utilizes individual captive inking systems, which permit the use of different-colored inks and individual flow control. The following procedure for inking the pen is recommended:

1. Remove the bottle from the well in the recording channel and completely remove the cap and capillary tube.
2. Fill the bottle with the desired color of ink to approximately one-quarter to one-half full.
3. Replace the cap and capillary tube by *rotating the bottle,* not the cap. Snug the cap.
4. Squeeze the bottle until the pressure forces ink to the tip of the pen. Maintain the squeeze while loosening the cap by slight rotation of the bottle, then release the pressure on the bottle.
5. Reinsert the bottle in the well in the recording channel. Ink will now flow by capillary action from the bottle to the pen tip as the pen moves across the chart paper.
6. Adjust the flow of ink by raising or lowering the bottle within the well. The higher the bottle position, the heavier the flow of ink.

For maintaining proper pen operation, it is recommended that the rubber pen pad be placed under the pens whenever the unit is not being used. If the instrument is to remain idle for more than a day or two, the ink should be removed and the pens cleaned in the following manner:

1. Remove the pen together with the capillary tube and ink bottle. Do not disconnect the total inking system; doing so will eventually stretch the capillary tube seal or the seal at the ink bottle.
2. Unscrew the bottle from the cap and empty the ink bottle.
3. Fill and rinse the bottle several times with warm tap water. Repeat until clean.
4. Fill the bottle with distilled water and screw the bottle to the cap. Squeeze the bottle and flush the ink from the capillary tube and pen. Repeat until clear water flows from the pen tip.
5. Unscrew and empty the bottle. Reattach the bottle and squeeze to force residual water out of the system.
6. Return the inking system to the Minigraph mainframe.

Chart Drive Control

Model 76107 offers solid-state push-button control of chart speed with the following selections: 0.5, 1, 2.5, 3, 5, 10, 25, 30, and 50 mm/s. Depressing any chart-speed push button will select that speed and simultaneously supply power to the time marker (if installed). Depression of the "off" push button stops the chart drive and cuts off power to the time marker.

Model 76107VS offers a continuously variable DC drive from 2.5 to 25 mm/s. Depression of the chart-drive push button (top left center of the mainframe) activates the chart drive and the time marker (if installed). The chart speed is controlled by the chart-speed selector knob (lower right corner of the mainframe). Turn the knob clockwise to increase speed or counterclockwise to decrease speed. Because the control is continuously variable, the exact chart speed desired or in use must be computed using the time marker for calibration. This procedure will be explained later.

Modules: Installation, Removal, Rearrangement

All systems come fully assembled with modules in place; however, sometimes it is desirable to add, delete, or reposition the recording channels in the mainframe. The following procedure is suggested:

1. Disconnect the power cord from the AC outlet.
2. Remove the retainer screws (4) on the front and rear module retainer bars and slide each bar away from the modules.
3. Tighten the cap on the ink bottle to minimize spillage, and remove the bottle from the well in the recording channel. Insert index finger into the well, grasp the module, and carefully lift it out of the mainframe, exposing the wire harness. Reinsert the ink bottle into the well.
4. If the module is to be removed entirely, disconnect the 4-pin plug beneath the inkwell (3-pin plug near transformer on the time/marker module). Replace the removed channel with a blank panel.
5. If the position of the modules is to be altered but the modules are not to be removed, it may not be necessary to disconnect the wiring harnesses.
6. After rearrangement of modules or their removal and replacement with blanks, replace the module retainer bars and tighten the four screws.

Modules: Type and Operation

Three basic recording modules have been designed for the Minigraph: the time/event marker (76322MG), the biopotential amplifier (76402MG), and the basic amplifier (76406MG) (figure 2.6).

The *time/event marker* is used to mark intervals of time and may be used simultaneously to mark events. The module may be set to mark off intervals of 1, 5, 30, or 60 seconds. Time is marked by a downward deflection of the pen. Although it has its own separate "off" position, once turned on, it will also be automatically deactivated whenever the chart drive is stopped. This feature prevents any pulling of ink between operations. An event is marked by an upward deflection of the pen whenever contact closure is provided via the auxiliary 2-pin Cinch/Jones connector. This feature may be utilized with a handheld remote push button connected to the module, or it may be utilized by connecting the signal marker output of the square-wave stimulator directly to the module.

NOTE: Special cable may be required.

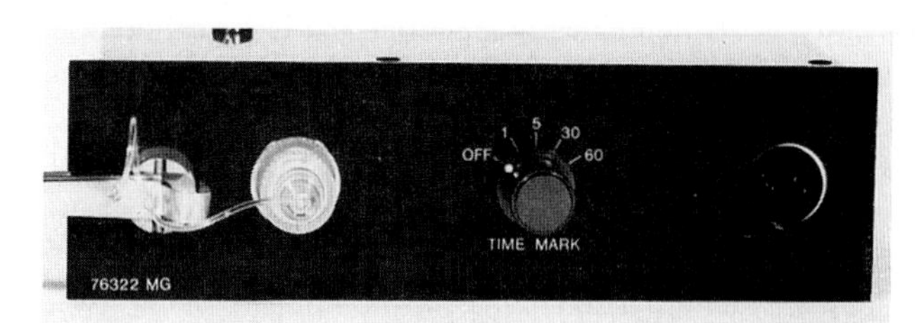

Time/Event Marker **Model 76322MG**

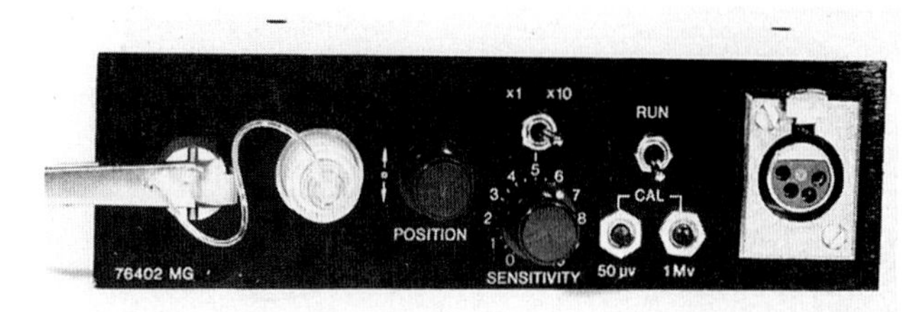

Biopotential Amplifier **Model 76402MG**

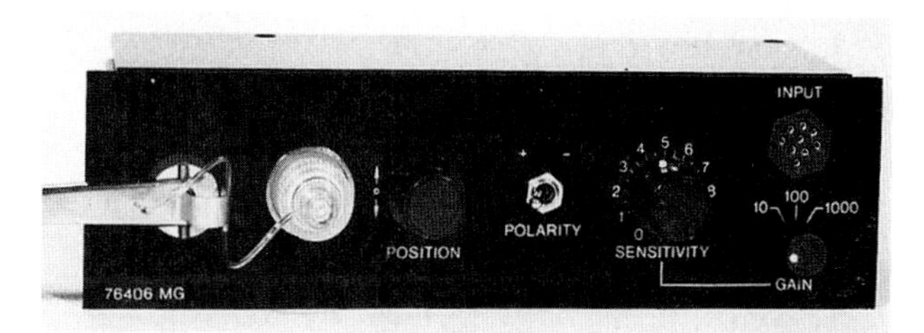

Basic Amplifier **Model 76406MG**

FIGURE **2.6** Lafayette Minigraph recording modules

The **biopotential amplifier** has a high-gain differential input used for recording small biopotential signals such as in electroencephalography (EEG), electrocardiography (EKG), and electromyography (EMG). The 4-pin amphenol socket allows for connection of electrodes, leads, or lead selector devices used in recording the EEG, EKG, and EMG. The run/cal switch connects the incoming signal to the amplifier and pen motor ("run" position) or allows a calibration signal ("cal" position) of 50 microvolts (µV) or 1 millivolt (mV) to be recorded when the corresponding button is depressed. The amplification switch (×1 or ×10) and the vernier sensitivity control (0–10) are used to adjust the amplifier gain (the degree of pen excursion per unit of incoming signal strength). The maximum amplifier gain of this channel is approximately 1.5 mm/µV. The position control simply allows the operator to adjust up or down the baseline recording position of the pen. For most experiments in this manual, the position control will be used to center the pen on the chart paper.

The *basic amplifier* has been designed to record a wider range of input signals coming from electrodes, transducers, and/or preamplifiers. It is more versatile but not as sensitive as the biopotential amplifier and lacks calibration circuitry. The 9-pin amphenol socket provides for the input signal. Amplifier gain of the input signal is controlled via the gain selector switch (10, 100, 1000) and a vernier sensitivity control (0–10). Maximum amplifier gain of this channel is approximately 2.5 cm/mV. The polarity switch determines whether a positive input signal will cause an upward deflection of the pen (+ position) or a downward deflection (– position). Although normally left in the + position, this control simplifies correcting an inverted waveform when electrodes or transducers have been mistakenly misapplied. The position control is used to center the pen, adjust the baseline, and ensure proper pen travel on the chart paper. Examples of the use of the basic amplifier will be given later, along with a discussion of the transducers and preamplifiers where appropriate.

Biointrumentation

REPORT 2

Name: ______________________ Date: ______________

Lab Section: ______________________

1. What is the principal function of each of the following components of a recording system?

   a. Transducer ______________________

   b. Preamplifier ______________________

   c. Amplifier ______________________

2. List two factors that should be kept in mind concerning the proper use of electrodes.

   a. ______________________

   b. ______________________

3. Diagram the basic components of an electronic system, and indicate the relationship of one component to another.

4. The recording paper on a physiologic recorder is moving at a constant speed of 2.0 cm/s. The distance on the record between the beginning of an event and the end of an event is 5.0 mm. Calculate the duration of the event. (Refer to chapter 1 if necessary.)

   ____

5. Describe the difference between an analog signal and a digital signal. ______________________

6. Define the following components of a computerized data acquisition/analysis system:

   a. Central processing unit ______________________________

   ____

   ____

   b. Signal conditioner ______________________________

   ____

   ____

   c. Computer memory ______________________________

   ____

   ____

   d. RAM ______________________________

   ____

   ____

   e. ROM ______________________________

   ____

   ____

   f. Hard disk ______________________________

   ____

   ____

   g. Server ______________________________

   ____

   ____

   h. Operating system ______________________________

   ____

   ____

CHAPTER 3

Compound Light Microscopy

INTRODUCTION

The **microscope** is an optical instrument used to increase the visual (apparent) size of an object. An ordinary magnifying glass, which is a double-convex lens with a short focal length, is a simple microscope. The hand lens used by the fictional Sherlock Holmes is an example of an instrument of this type. This is the principle involved: When an object is placed within the focal length of such a lens, an image of the object (called a virtual image)—erect and larger than the original object—is produced. The magnification is commonly expressed in diameters. For example, if a lens magnifies an object 10 times, the power of magnification is 10 diameters, or "10×."

The compound microscope differs from the simple microscope in that it has two or more such lenses fixed in the extremities of a hollow metal cylinder. The lens nearest to the object—the *objective*—magnifies the specimen a definite amount. Then the lens nearest to the eye—the ocular, or eyepiece—further magnifies the image formed by the objective. The total magnification of the compound microscope is computed by multiplying the magnifying power of the ocular times the magnifying power of the objective.

The compound microscope has been and continues to be extremely important in the development of biology and of medicine. Its invention has been credited to Galileo (1564–1642), the great Italian astronomer, mathematician, and physicist.

EXPERIMENTAL OBJECTIVES

1. To become familiar with the structure of the compound light microscope.
2. To become acquainted with the function of the major parts of the microscope.
3. To learn to use the compound light microscope in a competent and responsible manner.
4. To acquire an appreciation for the value of the microscope as an important tool of the scientist, educator, and student.

Materials

compound light microscope
lens paper
clean microscope slides
clean coverslips
toothpick
dilute methylene blue
prepared slide of microfilm
prepared slide of human blood
immersion oil
xylene

EXPERIMENTAL METHODS

Care of the Microscope

Microscopes are costly and sensitive instruments that must be handled with care at all times. When carrying a microscope, place one hand beneath the base and grip the upright, curved arm with the other. Always carry the microscope in an upright position, close to the body to avoid bumping it against other objects or dropping it. When not in use, the microscope should be placed in its case or properly covered to protect it from dust.

Exposed glass surfaces of the microscope, such as those found in the ocular, objectives, and condenser, should be kept meticulously clean. Optical glass is generally softer than window glass and may be easily scratched by cleaning it with paper towels, tissue paper, or handkerchiefs. Use only the special lens paper designed for cleaning optical glass. Always clean the ocular, objectives, slides, and coverslips before using the microscope. Dust on the ocular is seen as specks that rotate when you turn the ocular while looking through it. Dirt on the objective prevents clear vision; the object may look foggy. Always remove the oil from the tip of the oil-immersion objective with lens paper moistened with xylene immediately after use; otherwise, the immersion oil may become gummed on the lens.

Structure of the Microscope

With the assistance of figure 3.1, locate the following components on your microscope:

1. *Ocular (eyepiece):* This uppermost lens of the microscope is closest to the eye when the microscope is in use. It has a magnification power of 10× signifying that if it were used alone as a hand lens, it would magnify an object 10 times its actual size. Some microscopes have a single ocular (monocular); other microscopes have two oculars (binocular).
2. *Body tube:* This inclined cylinder supports the lens system of the ocular. On some microscopes, the body tube is vertical, supporting the ocular at one end and the revolving turret bearing the objectives at the other. On microscopes with a vertical body tube, the tube is attached to the arm of the microscope by means of a rack-and-pinion gear. The gear allows the body tube to be raised or lowered, thus bringing the specimen into focus. On microscopes with an inclined body tube, the specimen is brought into focus by raising or lowering the stage.
3. *Revolving turret (nosepiece):* To this revolving part of the microscope the objectives are attached. Any objective may be used by revolving the turret until the objective locks into line with the body tube. Always lower the stage (or raise the body tube) before revolving the turret to prevent the objectives from striking the stage.
4. *Objectives:* This system of one, two, three, or four mounted lenses is attached to the turret. The numbers on the objectives indicate the magnifying power of the objective. The number on the objective multiplied by the power of the ocular is the total magnification of what is seen in the circular field of view. Here are some examples.
   a. Scanning objective = 4×, ocular = 10×; total magnification = (4×)(10×) = 40×.
   b. Low-power objective = 10×, ocular = 10×; total magnification = (10×)(10×) = 100×.
   c. High-power objective = 45×, ocular = 10×; total magnification = (45×)(10×) = 450×.
   d. Oil-immersion objective = 100×, ocular = 10×; total magnification = (100×)(10×) = 1000×.
5. *Arm:* This curved portion of the microscope holds the body tube and stage.
6. *Stage:* This platform supports the microscope slide while it is being viewed. A slide may be held in place for viewing by either a pair of stage clips or a mechanical stage clamp.
7. *Diaphragm and condenser:* Located below the stage on the substage, the diaphragm is used to adjust the amount of light passing through the slide into the lens system of the microscope, thus controlling the contrast. The iris diaphragm lever is located just below the stage on the front of the substage assembly. The condenser focuses light from the in-base illuminator onto the specimen being viewed. The condenser may be adjusted (raised or lowered) by turning the adjustment knob located on the substage assembly. For most viewing purposes, the condenser should be adjusted to the fully raised position. Normally, the greater the total magnification, the wider the diaphragm should be opened. A greater amount of light is needed to properly illuminate a smaller field of view and to provide adequate contrast for differentiating structures. Some microscopes have a rheostat instead of an iris diaphragm to control light intensity.
8. *Coarse adjustment:* This knob is used to bring the specimen into view before final focusing.
9. *Fine adjustment:* This knob is used for final focusing after the specimen has been brought into view with the coarse adjustment.
10. *In-base illuminator:* This is the source of light necessary for viewing.
11. *Base:* This basic support for the entire microscope may contain a built-in light source.

Use of the Microscope

1. Place the microscope on a stable surface such as a desk or laboratory table. If your microscope is equipped with an in-base illuminator, plug the microscope into an electrical outlet.
2. Clean the ocular, objectives, and condenser lenses with lens paper.
3. Using the coarse adjustment knob, move the stage downward as far as it will go. (If your microscope has a movable body tube, move the body tube upward.)
4. Revolve the turret until the low-power objective (or scanning objective, if your microscope has one) snaps into correct alignment with the body tube.
5. Adjust the condenser to maximum height.
6. Turn on the in-base illuminator (or use the rheostat), and open the iris diaphragm to admit the maximum amount of light.
7. Looking through the ocular, make certain there is a bright uniform circle of light in the field of view.
8. Take the prepared slide of microfilm and clip it onto the stage, making sure the coverslip is facing up. If your microscope is equipped with a mechanical stage, place the slide between the jaws of the stage clamp. Viewing the stage laterally, adjust the position of the slide so that the specimen is located over the hole in the stage.
9. Check that the low-power objective is in correct alignment with the body tube. While viewing the

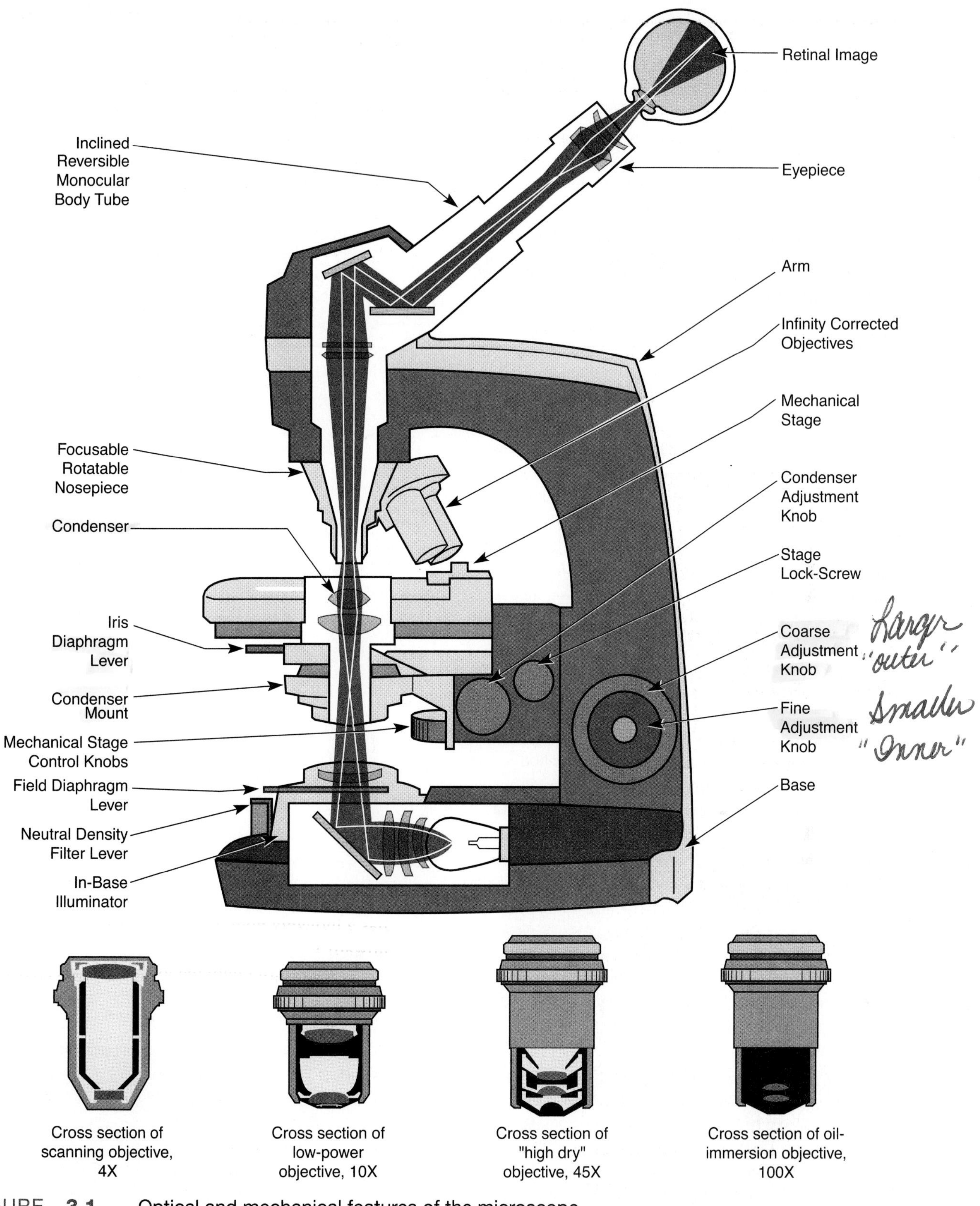

FIGURE 3.1 Optical and mechanical features of the microscope

microscope slide and objective from the side, use the coarse adjustment and move the stage up (or the body tube down) as far as it will go. *Note:* If, by mistake, the oil-immersion lens is in alignment, this position will result in a broken slide and a damaged objective.

10. While looking through the ocular with either eye, lower the stage (or raise the body tube) gradually by using the coarse adjustment until the specimen begins to come into view. Always obtain initial focus by moving the objective away from the stage. Never obtain initial focus by moving the objective toward the stage.
11. Complete the focusing by use of the fine adjustment. When using a monocular microscope, keep your nonviewing eye open to avoid tiring its muscles by trying to hold it tightly shut. At first, you may be bothered by seeing with the eye not looking through the microscope, but a little practice will permit you to keep both eyes open and see with one. Eyeglasses used to correct for visual defects may be worn when using the microscope.
12. Readjust the light so that you can see the object without strain and yet avoid using unnecessary light that can cause fatigue from glare.
13. Objects in the field of vision should be sharp and clear if the focus, intensity of light, and contrast have been properly adjusted.
14. With the specimen in focus beneath the low-power objective, view the stage laterally and revolve the turret until the high-power objective snaps into correct alignment. Looking through the ocular, adjust the focus, if necessary, only using the fine adjustment knob. To avoid a broken slide or a damaged objective, be careful not to raise the stage (or lower the body tube) farther than necessary.
15. Modern compound light microscopes are parfocal (a set of lenses having their focal point in the same plane); that is, after the initial focus is obtained, generally with the low-power objective, the focus is maintained without further adjustment when a switch to one of the other objectives is made.
16. Notice that the field of view decreases and the apparent size of the object increases as objectives having higher magnification power are switched into place (figure 3.2).
17. Revolve the turret until the low-power objective snaps into alignment with the body tube. Remove the slide containing the microfilm, and place the prepared slide of stained blood cells on the stage. Focus on the slide, using the low-power objective at first, and then using the high-power objective. Repeat steps 8–12, if necessary.

When viewing a prepared slide of Wright's stained blood with the low-power objective, first reduce the

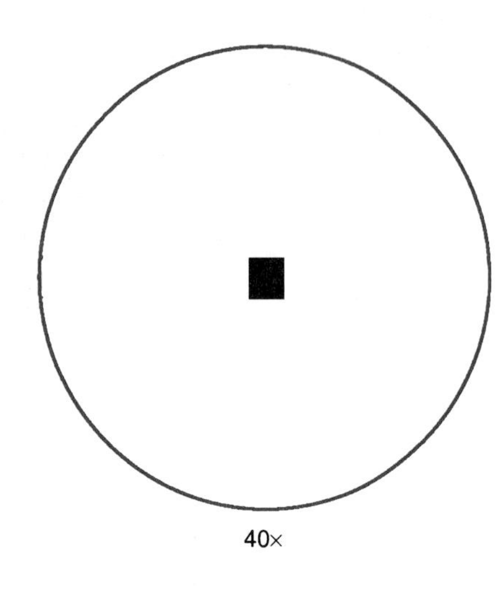

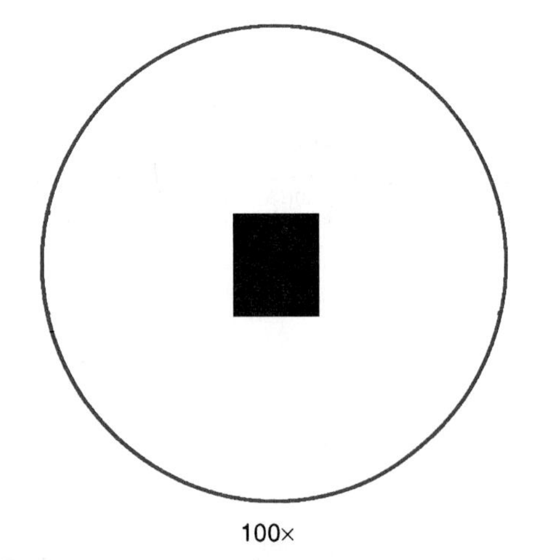

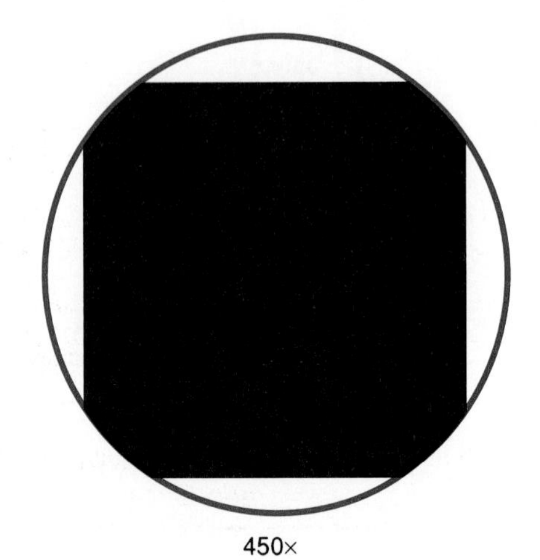

FIGURE 3.2 The apparent size of a 0.1-mm square as seen with a microscope using the scanning objective (4×), the low-power objective (10×), and the high-power objective (45×). As magnification of the square increases, the actual field of view of the area surrounding the square decreases.

Actual size of viewed square	Total magnification	Image size of viewed square	Actual field of view
0.10mm	40x	4mm	55mm
0.10mm	100x	10mm	22mm
0.10mm	450x	45mm	4.9mm

light intensity to increase the contrast of the lightly stained, small, almost translucent red blood cells. The white blood cells will appear as slightly larger, dark spots interspersed among the mass of red cells.

18. Revolve the turret until the oil-immersion objective and the high-power objective straddle the coverslip. Place a drop of special immersion oil on the coverslip, then rotate the turret until the oil-immersion objective snaps into correct alignment with the body tube and comes in contact with the immersion oil. Using only the fine adjustment knob, carefully and slowly bring the specimen into clearer focus. Usually, when using the oil-immersion objective, you should adjust the light source to maximally illuminate the specimen.
19. Clean both the slide and the oil-immersion objective with xylene and lens paper after use.

CAUTION

In following parts of the exercise, you will remove, stain, and examine cells from the inside surface of your cheek. Handle only your own saliva and cells. As soon as practical, discard all disposable items that have come into contact with your saliva (e.g., slides, coverslips, toothpicks) in an appropriately marked hazardous waste container. Disinfect your laboratory table or bench top by wiping it clean with a bleach solution immediately after you have completed the exercise.

20. Taking the flat end of a wooden toothpick, scrape the inside of your cheek. Place the scraping in the center of a clean, dry microscope slide, and spread it into a thin, even smear. Allow the smear to air-dry. Add one drop of dilute methylene blue to the air-dried smear, and place a clean coverslip over the prepared specimen. Gently blot away excess moisture with blotting paper. Place the slide on the stage and view under low power and high power. Repeat steps 8–12, if necessary. You should be able to see flat, irregularly shaped, scalelike cells (simple squamous epithelial cells) containing small, oval, darker-stained nuclei.
21. As you focus the microscope up and down on a specimen, notice that only a very thin layer of the specimen is in focus at a time. The greater the magnification, the thinner the layer in focus at one time. If you were to look at a small, pyramidal object between the microscope slide and the coverslip, the base of the object would appear as a large square. You would see smaller and smaller squares as you focused the microscope upward, until at the very top you would see only a dot. Figure 3.3 illustrates the depth of a pyramidal object seen as successive layers.

    Pull out a hair from your scalp. Using scissors, cut a piece 2 cm long and place it on a clean microscope slide. Cover the specimen with a clean coverslip, and view it using the low- and high-power objectives. Slowly focus up and down on the hair shaft, noting the layered appearance. To understand the nature of a specimen, always focus up and down until you achieve a perception of the depth within the specimen.
22. You have now completed an exercise designed to introduce you to the compound light microscope. Remove the slide from the microscope stage and place it, along with others you may have used, in their appropriate containers. Carefully cleanse the

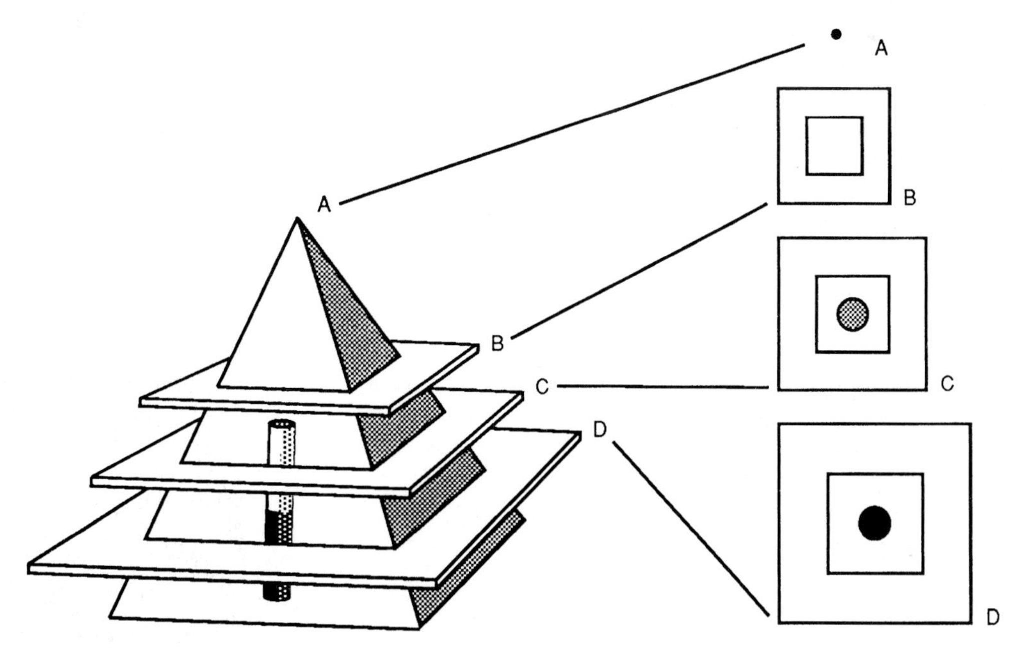

FIGURE 3.3 Depth of a pyramidal object seen as successive layers represented by planes A, B, C, and D

lens of the ocular and objectives with lens paper. Use xylene to remove oil from the slides and the oil-immersion lens. Rotate the turret until the low-power or scanning objective locks into alignment with the body tube, and adjust the body tube to its lowest position. Turn off the stage light, unplug and wrap the cord around the base, and cover the microscope. Follow the directions of your laboratory instructor regarding storage of the microscope. Repeat this step each time you have finished using the microscope for the laboratory period. You should find the microscope clean and ready for use at the beginning of the laboratory period, and you should leave it that way at the end of the period.

Physical Processes of Biological Importance

CHAPTER 4

■ INTRODUCTION

An average 70-kilogram adult human consists of approximately 100 trillion cells organized into tissues, organs, and systems. Each cell is separated from its extracellular environment by a plasma membrane, the function of which is to regulate the continual exchanges of chemical substances between the fluid inside the cell *(intracellular fluid)* and the fluid outside the cell *(extracellular fluid).* For cellular metabolism to proceed normally, the cell must acquire chemicals from the extracellular fluid. The processes of cellular metabolism, in turn, produce waste products, which must be eliminated by the cell into the extracellular environment. The nature of the chemical exchanges between intracellular and extracellular fluid compartments involves the physical processes of filtration, diffusion, and osmosis in addition to the biochemical processes of carrier-mediated transport.

A substance that enters into solution or becomes dissolved in a liquid is called a *solute.* Liquids that allow solutes to become dissolved are called *solvents.* The process whereby a solute enters into solution with a solvent is called *dissolution.*

Filtration is the removal of particles suspended in a solution by allowing the solution to pass through a selectively permeable membrane. Membranes that are selectively permeable allow the passage of certain substances and inhibit the passage of others. All plasma membranes are selectively permeable. The driving force in the process of filtration is a pressure differential across the membrane.

An example of the process is the filtration of blood as it passes through specialized capillaries (thin-walled blood vessels) of the kidney. Large molecules, such as most proteins and formed elements such as erythrocytes (red blood cells), remain in the capillaries, while smaller molecules, such as simple sugars (e.g., glucose) or particles such as ions (e.g., Na^+, K^+, etc.), leave the capillaries and enter the urine-forming structures (nephrons) of the kidney.

Simple diffusion is the continual random movement of particles among one another in a liquid or a gas. *Net diffusion* is the measurable movement of particles from an area of greater concentration to an area of lesser concentration until the concentration of the diffusing substance is uniform. When the concentration of the diffusing substance becomes uniform, diffusion equilibrium exists and net diffusion ceases, although simple diffusion continues.

The rate of net diffusion is directly proportional to the *concentration gradient,* which is the difference in concentration divided by the distance separating that difference, the cross-sectional or surface area of the diffusion pathway, and the temperature of the diffusing substance. The rate of net diffusion is inversely proportional to the molecular weight of the diffusion substance. Factors that influence the rate of net diffusion of a substance may be summarized as follows (Fick's law):

$$\text{Rate of diffusion of Y from point A to point B} \propto \frac{[\Delta C][\text{x-area}][\text{temp.}]}{[d][mw]}$$

where:

ΔC = Concentration of Y at point A minus the concentration of Y at point B. (i.e., the concentration difference)

x-area = the cross-sectional area of the diffusion pathway between point A and point B

temp. = the temperature of Y molecules

d = the distance of the diffusion pathway from point A to point B

mw = the molecular weight of substance Y

Additional factors that influence the rates of diffusion across biological membranes are atomic or molecular size and configuration, the ability of the diffusing solute to dissolve in lipids, and the presence or absence of an electrical charge on the diffusing solute.

The *plasma membrane* presents a barrier to the movement of materials into and out of the cell. Materials that diffuse through the membrane must either dissolve in the membrane and then diffuse from one side to the other or pass through pores in the membrane. The plasma membrane is a complex

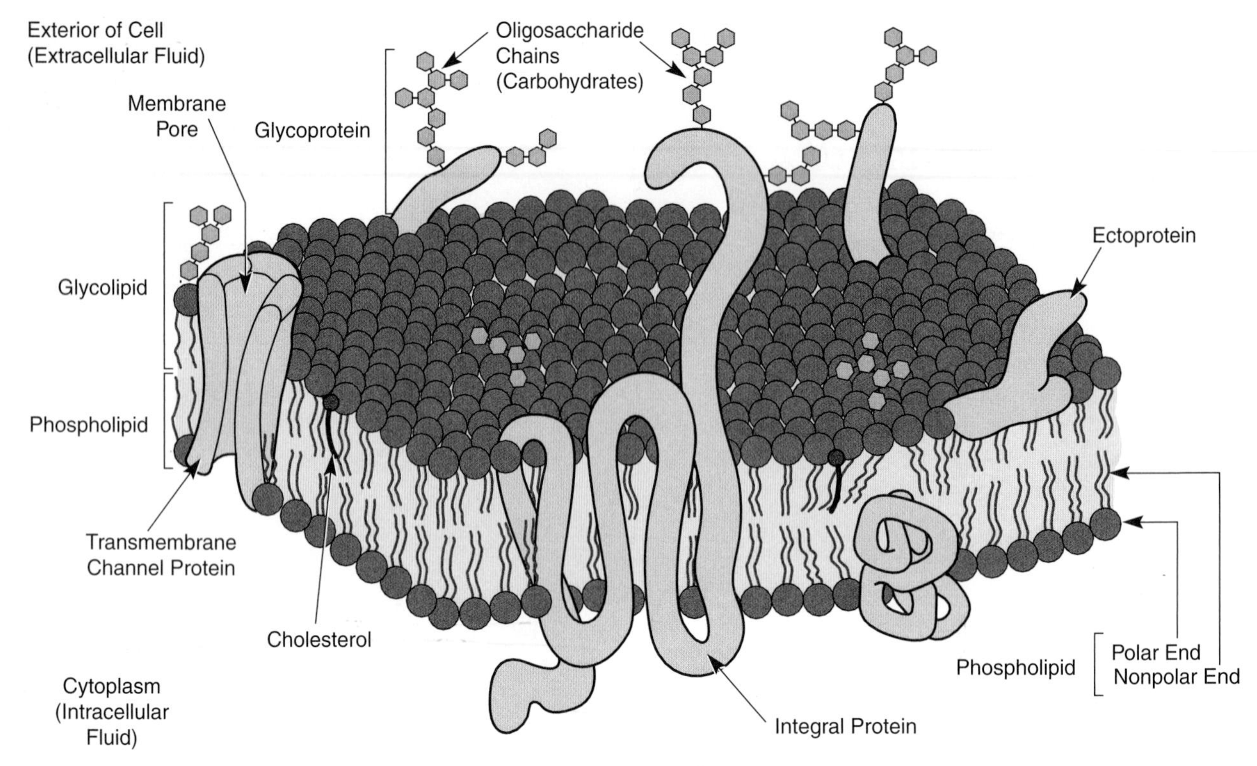

FIGURE 4.1 Fluid mosaic model of the plasma membrane

of lipid, protein, glycoprotein, glycolipid, phospholipid, and other types of molecules. The core of the membrane structure (figure 4.1) is a bilayer of fluid lipids. Proteins are found on each surface of the bilayer as well as within it. The membrane is not continuous but contains tiny channels, commonly called pores, for the diffusion of substances into or out of the cell. In general, substances with a diameter greater than 8 angstroms pass through the pores with difficulty. Uncharged particles pass through the membrane more readily than charged particles (ions), and positively charged particles (cations) pass through more readily than do negatively charged particles (anions). Molecules with a higher degree of lipid solubility, such as lipid-soluble drugs and vitamins A, D, E, and K, tend to diffuse through plasma membranes more readily than do molecules that are less soluble in lipids.

The movement of oxygen molecules into cells or carbon dioxide molecules out of cells is an example of the process of net diffusion. Oxygen molecules are being continually consumed by metabolic processes within the cell, so the concentration gradient favors movement of this gas into the cell. Carbon dioxide molecules are being continually produced during cellular metabolism, so the concentration gradient favors diffusion of this gas out of the cell.

Osmosis is the net diffusion of a solvent through a selectively permeable membrane. Within body fluids, the solvent is water, and therefore biologists consider osmosis to be the net diffusion of water through a selectively permeable membrane.

When the concentration of nondiffusible solutes is greater on one side of the membrane than on the other, net diffusion of water will occur through the membrane toward the area of greater solute concentration until the solute/solvent ratio is equal on both sides of the membrane or until a force of equal magnitude opposing the force created by the movement of water is applied. That is, water molecules will diffuse from an area of greater water concentration through the selectively permeable membrane to an area of lesser water concentration.

Consider a 500-mL beaker containing 280 mL of pure water divided into two equal compartments by a selectively permeable membrane (figure 4.2). For purposes of discussion, consider the system to be unaffected by atmospheric pressure. Compartment A contains 140 mL of water, as does compartment B. If 8 g of a nondiffusible solute (a solute to which the membrane was not permeable) were dissolved in the water in compartment A and 6 g of nondiffusible solute were dissolved in the water in compartment B, net diffusion of water would occur from compartment B to compartment A until the solute/solvent ratios of the two compartments became equal. At equilibrium, the solute/solvent ratio of compartment A would be 8 g/160 mL and that of compartment B would be 6 g/120 mL. A net diffusion of 20 mL of water from compartment B to compartment A would have occurred.

In addition to the concentration gradient of water, the rate of osmosis in living systems depends on temperature, electri-

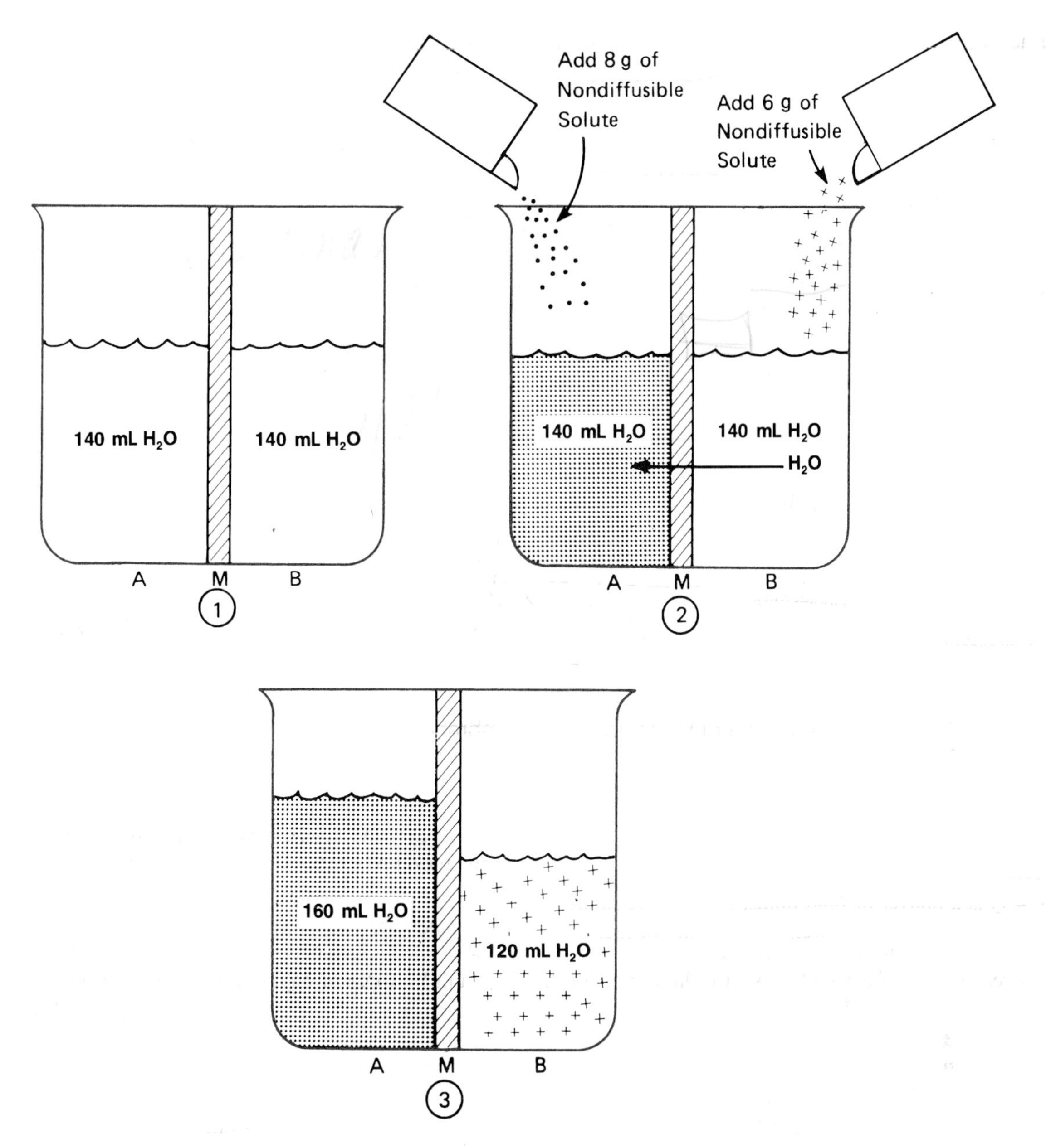

FIGURE 4.2 Osmosis

cal potential across the membrane, membrane porosity, the surface area of the membrane, and the volume of the cell. In the body, osmosis is important in the maintenance of plasma volume, interstitial and intracellular fluid volumes, and the volumes of other body fluid compartments.

Osmotically active particles are solutes that either do not diffuse or diffuse with difficulty through membranes. Fluids with osmotically active particles in the same concentration as found in the plasma of blood are *isotonic*. If a human red blood cell is placed in an isotonic solution, it neither swells nor shrinks, because the net diffusion of water into or out of the cell is zero. An example of an isotonic solution is 0.9% sodium chloride in water.

Although 0.9% sodium chloride in water is isotonic, the concentration of sodium chloride inside the red blood cell is not 0.9%. Inside the red blood cell, there are other kinds of osmotically active solutes (e.g., hemoglobin) that contribute to the intracellular fluid osmotic pressure; thus, the concentration of sodium chloride must be less than 0.9%.

Fluids containing a higher concentration of osmotically active particles than blood plasma are *hypertonic*. Red blood cells in a hypertonic solution will shrink and shrivel (crenation), because net diffusion of water out of the cells occurs. *Hypotonic* solutions contain a lower concentration of osmotically active particles than does the plasma; therefore, red blood cells will swell when placed in hypotonic solutions, because the net diffusion of water will be into the cell. As the red blood cell expands in a hypotonic solution, the plasma membrane becomes leaky, allowing hemoglobin to escape into the surrounding fluid. This process is known as hemolysis. The red cell that has lost hemoglobin becomes paler and is known as a ghost.

Dialysis is the net diffusion of a solute through a selectively permeable membrane. The permeability of an artificial membrane is related to the size of its pores. Molecules larger than pore size fail to pass through the membrane even though a concentration gradient for the molecule exists across the membrane. Molecules that are smaller than pore size pass through the membrane, diffusing down their concentration gradient. By selecting a membrane with desired permeability characteristics, dialysis can be used to separate molecules in solution.

■ EXPERIMENTAL OBJECTIVES

1. To become familiar, through experimentation and observation, with the physical processes of dissolution, filtration, diffusion, osmosis, and dialysis.
2. To gain an understanding and appreciation of the relationship between physical processes and the functions of biological systems.

Materials

Equipment

compound light microscope
coverslips
microscope slides
sterile blood lancet
droppers
sterile gauze (or alcohol prep)
test-tube rack
test tubes (15 mL)
plastic funnel
filter paper disk
25 mm × 100 cm cellulose dialysis tubing
nylon thread
meterstick
burette clamp
ring stand
Erlenmeyer flask (100 mL)
beakers (50 mL, 500 mL, 1000 mL)
glass stirring rod
agar-filled petri dishes
wax pencil
metric rule
agar-filled plastic pipettes (10 mL, 1 mL)
egg osmometer
graduated cylinder (50 mL)
thistle tube

Chemicals

distilled water
saline (0.9%, 5.0%)
crystalline quartz
potassium permanganate crystals
sodium chloride crystals
methylene blue solution
congo red solution
potassium hydroxide (5.0 molar)
70% alcohol
30% sucrose solution
phenol red
agar
10% acetic acid
human blood substitutes

■ EXPERIMENTAL METHODS

Dissolution

Add small amounts (enough to just cover the bottom of the test tube) of crystalline quartz, potassium permanganate, and sodium chloride to three dry 15-mL test tubes (quartz to one, potassium permanganate to another, sodium chloride to the third). Add 10 mL of distilled water to each tube, shake it gently, place it in a test-tube rack, and observe it after 3 minutes. Which substance(s) dissolved? Which substance(s) did not? How can you tell? Record the observations in the report.

Filtration

Fit a plastic funnel with a disk of filter paper (a selectively permeable artificial membrane), and insert the funnel spout into a clean, dry 100-mL Erlenmeyer flask. Pour the contents of the crystalline quartz test tube into the prepared funnel and collect the filtrate in the flask. Note the appearance of both the filtrate and the residue. Did the quartz pass through the selectively permeable disk? Replace the filter paper and repeat the filtration procedure separately for the sodium chloride test tube and the potassium permanganate test tube. Which substances passed through the selectively permeable disk? Which substances did not? Why do some substances pass through the filter paper while other substances do not? Record these observations in the report.

Diffusion: Influence of Concentration, Temperature, Solubility, Molecular Weight

Obtain six agar-filled petri dishes from the laboratory instructor. Each dish has a 1-cm well (a circle of gel removed from the center). All dishes are of equal volume and size and contain identical volumes of agar. Refer to the agar plate diffusion data table in the report section prior to setting up the plates. Mark each dish with a wax pencil so you can identify the dishes as yours. Using clean droppers, fill two wells with congo red (molecular weight = 696.68) solution and the other two with methylene blue (molecular weight = 319.86) solution. Be careful not to exceed the capacity of the well. Note the time. Place one of the dishes containing congo red and one of the dishes containing methylene blue in the refrigerator. Record the temperature in the refrigerator.

Dilute the congo red and the methylene blue solutions to 50% strength and fill the wells in the remaining two dishes. Place the four dishes ($2 \times 100\% + 2 \times 50\%$) on the laboratory bench and record the room temperature. Allow the dishes to remain undisturbed for at least 24 hours.

During the next laboratory period, or at a time designated by the laboratory instructor, use a metric rule to measure the distance from the edge of the well to the diffusion edge in each of the four dishes. Note the time, and record all measurements in the report. Calculate the rate of linear diffusion (millimeters per hour) for each of the four dishes. Are the rates of diffusion the same for methylene blue (lighter) and congo red (heavier)? Why or why not? (*Hint:* Congo red is more soluble in agar.) How does temperature influence the rate of diffusion? Does the more concentrated solution diffuse at a faster rate? Record all observations and calculations in the report.

Diffusion: Influence of Cross-Sectional Area

Your laboratory instructor has prepared a 10-mL and a 1-mL agar-filled plastic pipette and a 1-L beaker containing 300 mL of 5.0 molar potassium hydroxide (see appendix A). The agar in each pipette has been acidified with a drop of acetic acid and contains phenol red, a color indicator that imparts a clear yellow color to the agar at an acid pH. When an alkaline solution diffuses through the agar, the color of the agar will turn to purple. The rate of diffusion of the alkaline solution (potassium hydroxide) through the acidified agar can be measured by calculating the rate of color change along the length of the pipette.

Measure the internal diameter of each pipette before your laboratory instructor places them in the potassium hydroxide. Calculate the cross-sectional surface area of the bore of each pipette (area of a circle = πr^2, where π = 3.1416 and r = radius) and record the data in the report.

After student measurements have been made, the laboratory instructor will place both pipettes into the beaker containing potassium hydroxide and note the time of day. Toward the end of the laboratory period, measure the linear diffusion distance in millimeters along the length of each pipette and record the time of day. Subtract, starting from the ending time to obtain the duration of diffusion in minutes. Calculate the rate of diffusion of potassium hydroxide in each pipette using the following formula:

$$(\pi r^2)(d) \div (t) = \text{vol/time} = \text{mm}^3\text{/min}$$

where: $\pi = 3.1416$

r = radius of the pipette bore

d = distance of diffusion

t = duration of diffusion

Are the rates of diffusion the same in the two pipettes? How does the cross-sectional surface area of the diffusion pathway influence the rate of diffusion? Compare your observation with that predicted by Fick's law in the introduction to this chapter.

■ OSMOSIS AND DIALYSIS

Artificial Membrane: Osmosis

Obtain the following items from the laboratory instructor:

1. thistle tube
2. flat-base stand
3. burette clamp
4. 4-inch wetted dialysis tubing
5. 3 feet of nylon thread
6. 500-mL beaker of distilled water
7. 50 mL of 30% sucrose solution

Grasp one end of the dialysis tubing and fold it back upon itself a short distance. As you hold it, instruct your laboratory partner to tie the folded end tightly shut using a short piece of nylon thread for the ligature. Tie tightly so as to prevent leakage. Open the other end of the dialysis tubing, insert the stem of the thistle tube about 1 inch into the tubing, and tightly secure the dialysis tubing to the stem of the thistle tube using nylon thread. To prevent leakage, wrap the thread around the stem and tubing several times before securing with a knot. Holding the dialysis sac and thistle tube upright, instruct your laboratory partner to use a plastic squeeze bottle filled with sucrose solution to fill the dialysis sac via the bowl end of the thistle tube. The use of a plastic squeeze bottle will minimize air trapping in the sac and thistle tube. If air becomes trapped in the sac while filling, instruct your partner to gently squeeze and release the sac several times to force the bubbles to the surface. Check for sucrose leaks. No leakage must occur; otherwise, the experiment will fail. If

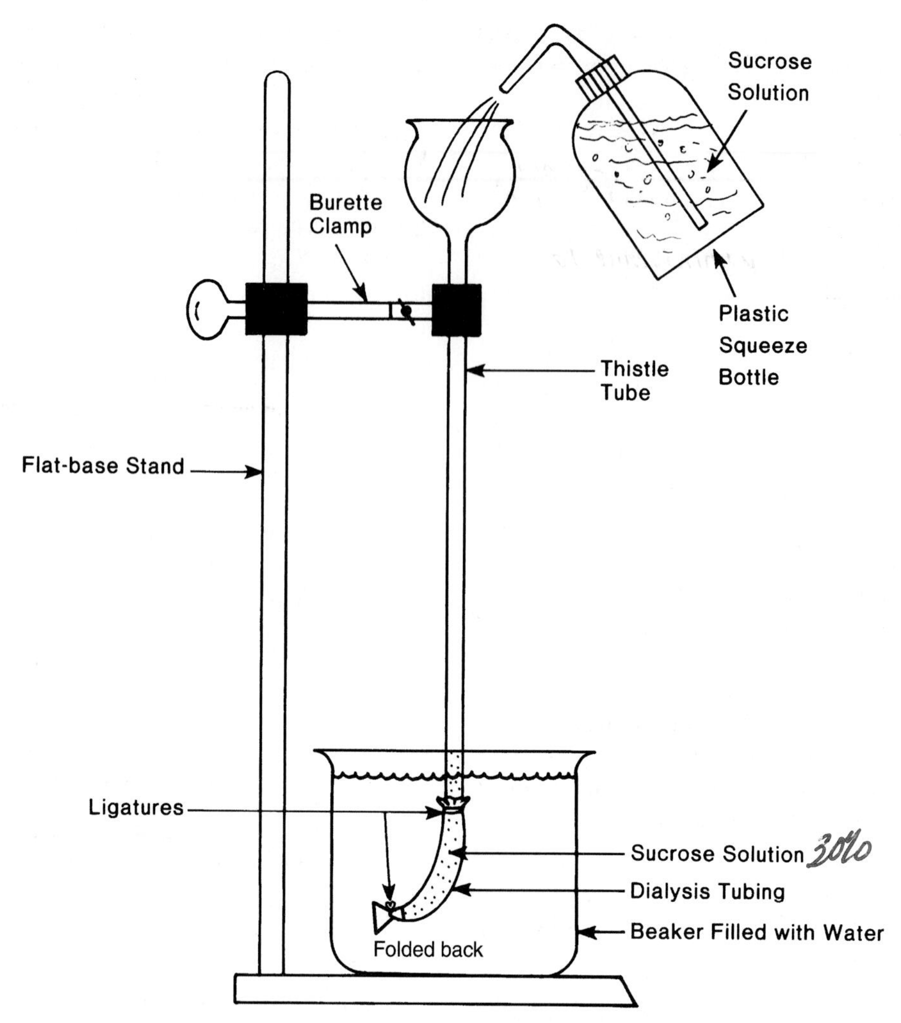

FIGURE **4.3** Artificial osmometer

leakage is detected, tighten the nylon ligatures. After filling the sac, immerse the sac in a beaker of distilled water and support the thistle tube in an upright position using the flat-base stand and burette clamp (figure 4.3).

Add sucrose solution to the sac via the thistle-tube bowl until the fluid in the stem is about level with the edge of the beaker. Tie a piece of nylon thread (or use colored tape) to mark the meniscus (top of the column of sucrose solution), and record the time. Record the changes in the height of the meniscus every 15 minutes thereafter for one hour. Plot the data on the graph in the report.

Artificial Membrane: Dialysis

Obtain the following items from the laboratory instructor:

1. thistle tube
2. flat-base stand
3. burette clamp
4. 4-inch wetted dialysis tubing
5. 3 feet of nylon thread
6. 500 mL beaker of distilled water
7. 50 mL of an aqueous mixture of 10% NaC1 + 1% starch
8. disposable test tubes

Grasp one end of the dialysis tubing and fold it back upon itself a short distance. As you hold it, instruct your laboratory partner to tie the folded end tightly shut using a short piece of nylon thread for the ligature. Tie tightly so as to prevent leakage.

Open the other end of the dialysis tubing, carefully fill the dialysis tubing with the salt/starch solution until it is 1/2 full, insert the stem of the thistle tube about 1 inch into the tubing, and tightly secure the dialysis tubing to the stem of the thistle tube using nylon thread. To prevent leakage, wrap the thread around the stem and tubing several times before securing with a knot.

Holding the dialysis sac with one hand and the thistle tube upright with the other hand, instruct your laboratory partner to carefully pour the salt/starch mixture down the inner side of the thistle tube bowl and into the sac until the sac is full and the solution rises in the tube about an inch above the sac. If air becomes trapped in the sac or tube while filling, gently squeeze and release the sac several times to force the bubbles to the surface and draw fluid into the sac.

After filling, immerse the sac in a beaker of distilled water and support the thistle tube in an upright position using the flat-base stand and the burette clamp (figure 4.3). Use thread or tape to mark the top of the column of the salt/starch mixture and record the time. Record changes in the height of the fluid column every 15 minutes thereafter for 1 hour. Plot the data on the same graph used to plot the osmosis data from the previous section.

After 1 hour, carefully remove the dialysis sac from the water, save the beaker of water, empty the dialysis sac into the laboratory sink, and remove and discard the tubing. Perform the following control tests for NaCl and starch prior to testing the water in the saved beaker.

1. *Control test for chloride:* Add 10 mL of 10% NaCl to a test tube. Add three drops of 3% silver nitrate. A milky-white precipitate forms when silver nitrate reacts with chloride (a positive test for chloride). Discard the tube and its contents.
2. *Control test for starch:* Add 5 mL of 1% starch solution to a test tube. Add three drops of Lugol's iodine. Starch turns blue when mixed with iodine (a positive test for starch). Discard the tube and its contents.

Add 10 mL of the saved beaker water to a clean, dry test tube. Test for the presence of chloride. After the test, discard the tube and its contents. Record the result in the report.

Add 5 mL of the saved beaker water to a clean, dry test tube. Test for the presence of starch. After the test, discard the tube and its contents. Record the result in the report.

Biological Membrane: Osmosis

The laboratory instructor has prepared an egg osmometer by removing the shell from one end of the egg, exposing the inner membrane (see appendix B). A glass tube was inserted into the other end (through both the shell and inner membrane) and sealed in place. The egg was then immersed in a beaker of water (figure 4.4).

The space inside the egg between the yolk and the inner membrane contains a large amount of albumin, a protein that cannot diffuse through the inner membrane. The inner membrane is permeable to water; therefore, water will diffuse from the beaker through the inner membrane into the albumin space of the egg. As pressure in the albumin space increases, water and albumin will rise in the glass tube. The rate of the rise is directly proportional to the rate of osmosis at the exposed inner membrane. Record the rise of fluid in the glass tube at 30-minute intervals for 90 minutes, and plot the data on the graph in the report. Allow the egg osmometer to remain intact until your next laboratory period, and again measure the height of the fluid column. Is the rate of osmosis rectilinear? Why or why not? If the egg osmometer is allowed to remain intact for 2–3 days, yolk material will be observed rising in the glass tube. Since the yolk is separated from the albumin space by an unbroken membrane, how do you account for the observation? Eventually, the rise of fluid in the glass tube will cease, although the solute/solvent ratios on the two sides of the exposed inner membrane are not equal. Why?

Hemolysis (Optional)

The following experiments require a small sample of mammalian blood or an appropriate blood substitute. Ward's artificial blood (36 W 0019), available from Ward's National Science Establishment, or fresh nonhuman mammalian blood (from a mouse, sheep, etc.) treated with anticoagulant may be substituted for human fingertip blood. Follow your instructor's directions regarding the use of human blood substitutes.

CAUTION

In this and sequential experiments on blood, you may be required to obtain a sample of blood from your fingertip. If you have any history of a blood-clotting disorder or you are being treated with anticoagulant medicine, inform the laboratory instructor and you will be excused from obtaining samples of your blood. Do not use samples from another student. Avoid contact with the blood of another student. Handle only your own blood. Properly dispose of all lancets, slides, gauze pads, and other disposable supplies that have come into contact with your blood by discarding them into appropriately marked containers. Disinfect your laboratory table by wiping it clean with a bleach solution before obtaining a fingertip blood sample and also at the end of this experiment.

The following technique can be used in obtaining fingertip blood: Clean the palmar surface of a finger (third or fourth) with a sterile gauze pad soaked with 70% alcohol or a sterile disposable alcohol prep. Allow the skin to dry; do not blow on it to make it dry faster. Remove the sterile lancet from its container and use it to make a quick stab wound through the cleansed surface of the fingertip. The wound should be located slightly to the side of the central part of the fingerprint with the long axis of the cut running diagonally. Discard the lancet. (Always use a new sterile lancet when required.) After obtaining the blood sample, compress the gauze pad over the cut until bleeding ceases.

Your instructor may require the use of an automatic lancet or other device for obtaining a blood sample. If so, follow directions carefully regarding procedures.

Use three clean microscope slides and three coverslips. Label the slides 1, 2, and 3, and draw a large wax circle in the center of each slide. Obtain a sample of fingertip blood,

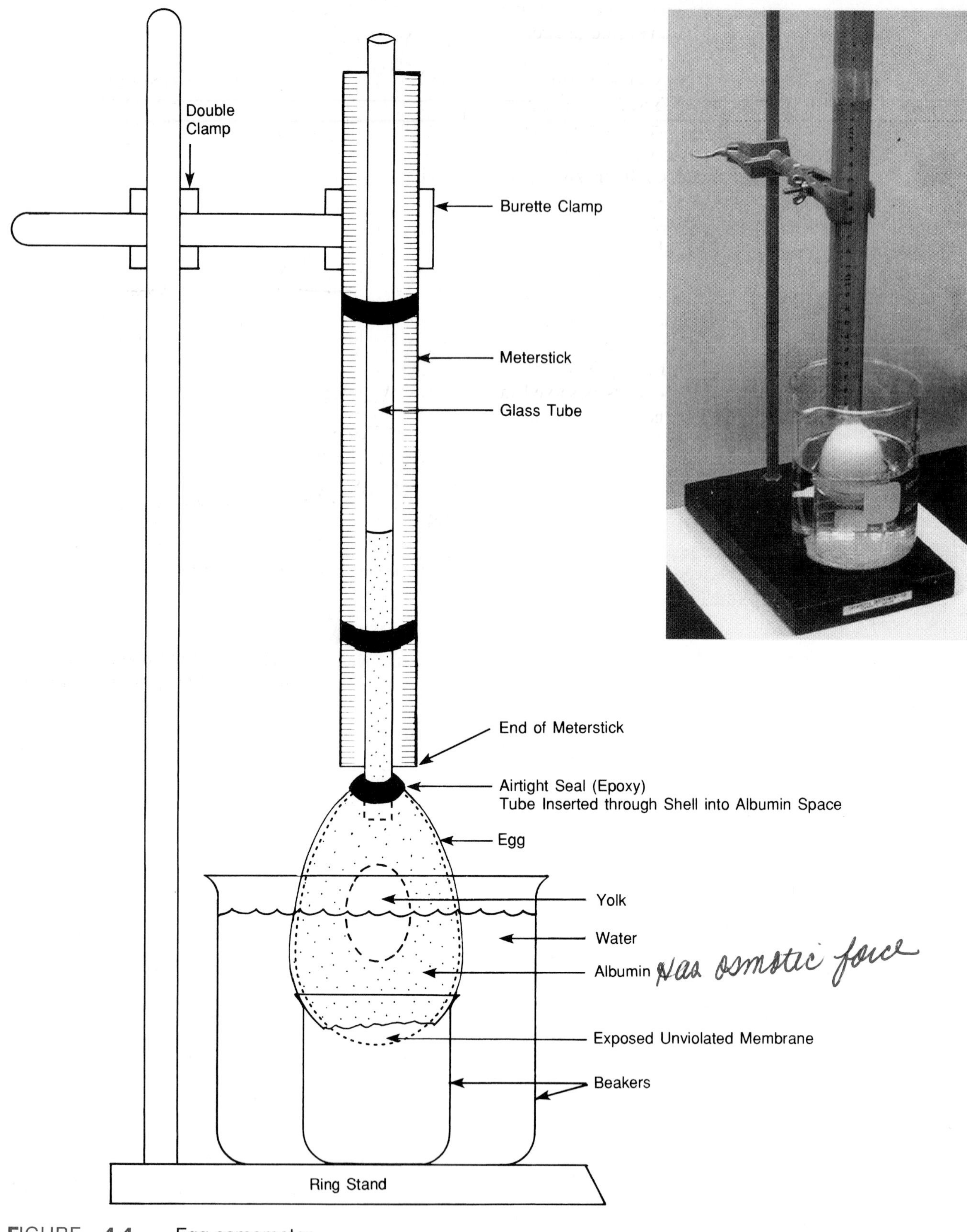

FIGURE **4.4** Egg osmometer

and use the finger to place a small amount in each of the three wax circles. Immediately dilute the blood in circle 1 with a drop of 5% saline, and apply a coverslip. Dilute the blood in circle 2 with one drop of physiologic saline and apply a coverslip. Dilute the blood in circle 3 with one drop of distilled water, and apply a coverslip. It is important to perform the dilutions immediately after obtaining the blood samples; otherwise, the blood samples will quickly dry on the slide and the results will be erroneous.

Examine each mixture under the compound light microscope. Reduce the brightness of the field of view by using the iris diaphragm or rheostat control and focus near the periphery of the blood smear. Observe the size, shape, and appearance of the cells in each circle. Compare them with the cells shown in figure 4.5. In which circle has crenation occurred? In which circle has hemolysis occurred? Explain the changes in the red blood cells observed in each of the three circles, and record these observations in the report.

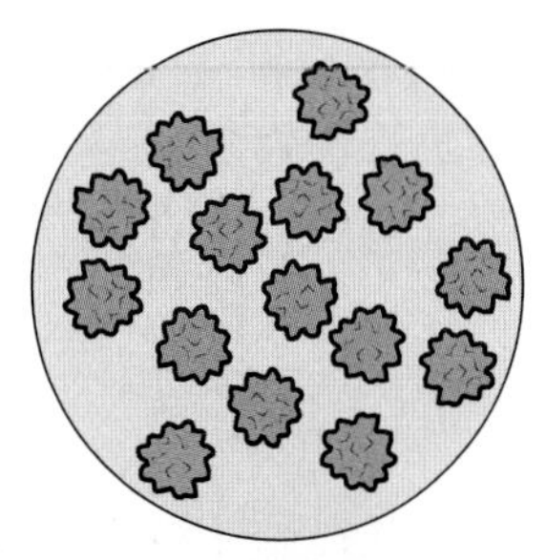

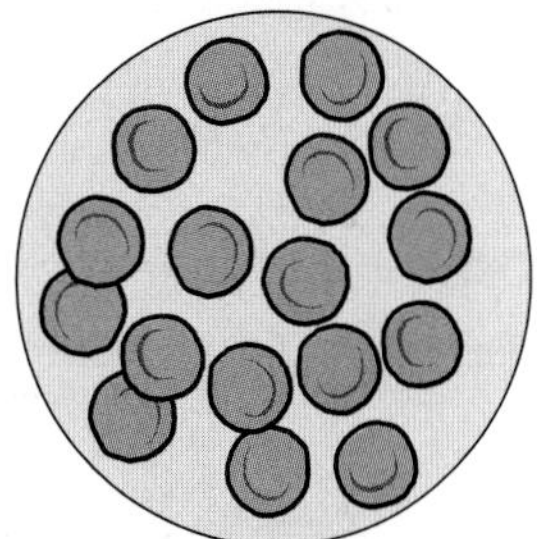

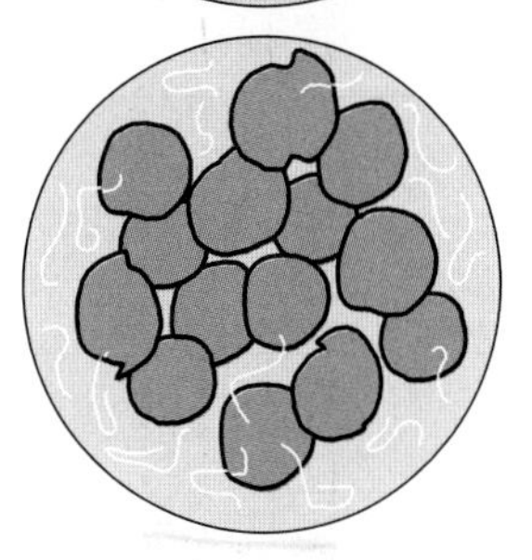

FIGURE 4.5 Erythrocytes in hypertonic, isotonic, and hypotonic solutions

Neural Control of Skeletal Muscle

CHAPTER 5

■ INTRODUCTION

Vertebrate skeletal muscle consists of hundreds of individual cylindrical cells, called **fibers,** bound together like straws in a broom by connective tissue. In the body *(in vivo)* skeletal muscles are stimulated to contract by somatic motor nerves, which carry signals, in the form of nerve impulses, from the brain or spinal cord to the skeletal muscles (figure 5.1). Nerve cells that innervate skeletal muscle are called **somatic motor neurons.** They are located in the gray matter of the spinal cord and brain. Axons, or nerve fibers, which are long cylindrical extensions of the neurons, leave the spinal cord or brain via spinal or cranial nerves and are distributed to appropriate skeletal muscles in the form of a peripheral nerve, a cable-like collection of individual nerve fibers. Upon reaching the muscle, each nerve fiber branches and innervates several individual muscle fibers. The connection between a nerve fiber and a skeletal muscle fiber is called the **neuromuscular junction.**

Although a single motor neuron can innervate several muscle fibers, each muscle fiber is innervated by only one motor neuron. The combination of a single motor neuron and all the muscle fibers it controls is called a **motor unit** (figure 5.1). When a somatic motor neuron is activated, all

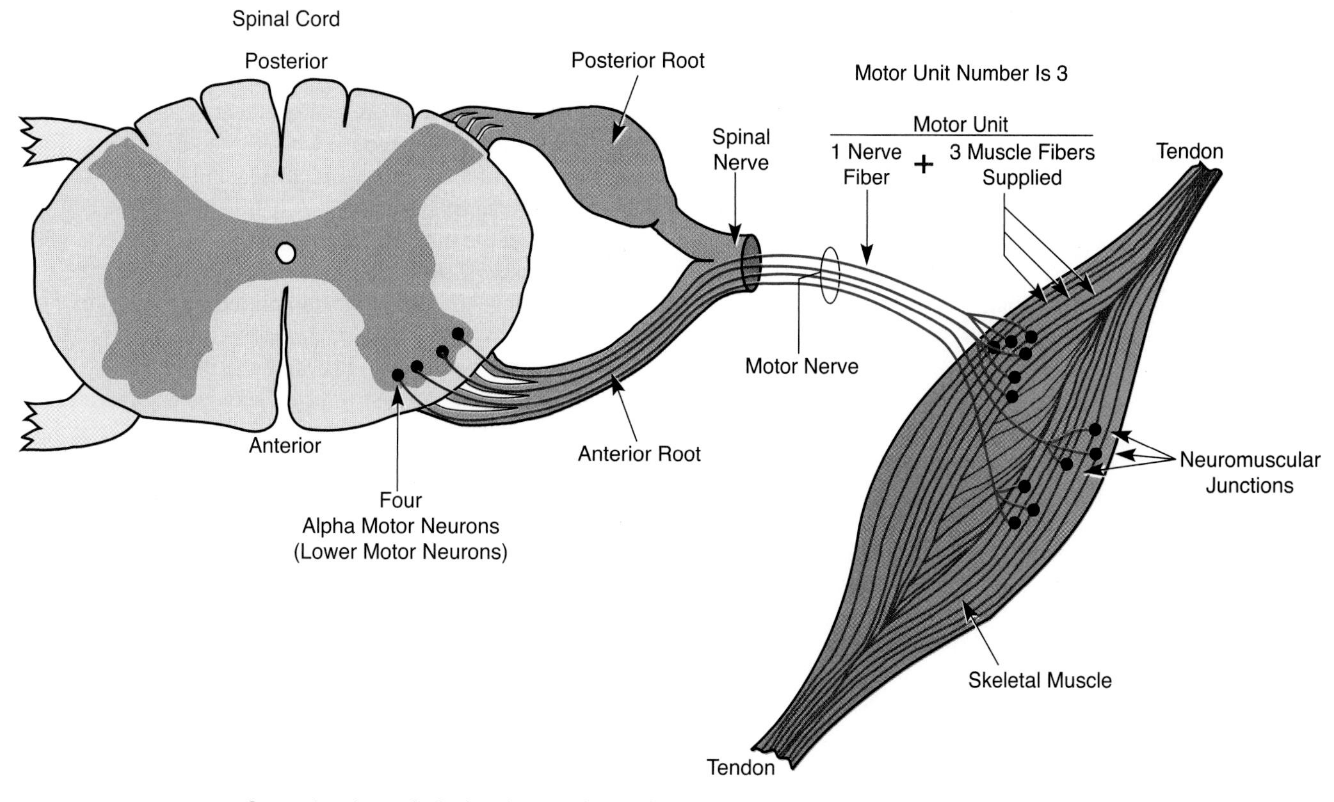

FIGURE 5.1 Organization of skeletal muscle and nerve

the muscle fibers it innervates respond to the neuron's impulses by generating their own electrical signals that lead to contraction of the activated muscle fibers.

Physiologically, the degree of skeletal muscle contraction is controlled by:

1. Activating a desired number of motor units within the muscle.
2. Controlling the frequency of motor neuron impulses in each motor unit.

When an increase in the strength of a muscle's contraction is necessary to perform a task, the brain increases the number of simultaneously active motor units within the muscle. This process is known as **motor unit recruitment.**

In the laboratory, we can simulate motor unit recruitment by either stimulating the motor nerve with increasingly stronger electrical stimuli, thereby activating more motor nerve fibers, or by directly stimulating the muscle with increasingly stronger electrical stimuli, thereby activating more muscle fibers and obtaining a stronger contraction. In this exercise you will learn how to prepare the frog gastrocnemius muscle and its motor nerve, the sciatic nerve, for investigating properties of nerve and muscle, and compare nerve and muscle threshold stimuli.

The minimum applied voltage necessary to elicit a response is called the **threshold voltage.** The value of the threshold voltage depends upon the duration of the stimulus and the characteristics of the nerve or muscle. In general, the threshold for a motor nerve is less or lower than for the skeletal muscle because nerve fibers are more sensitive to electrical stimuli and are surrounded by less connective tissue through which the electric current must penetrate.

■ EXPERIMENTAL OBJECTIVES

1. To become familiar with the procedure for preparing the frog gastrocnemius muscle–sciatic nerve for neuromuscular experimentation.
2. To determine and compare the values of threshold voltage for a skeletal muscle and its motor nerve.

■ EXPERIMENTAL METHODS

Lafayette Minigraph

Materials

frog
glass plate
glass probe
dissecting kit
dropper
thread
50-mL beaker
frog Ringer's solution
flat-base stand
double clamps
Minigraph model 76107 or 76107VS
model 76322 time/marker channel + remote marker button
model 76406MG basic amplifier channel
model 76613 semi-isometric force transducer
model 82415 square-wave stimulator and cable
model 76613-T tension adjuster
model 76802 transducer stand
model 76804 muscle clamp or equivalent femur clamp
fine nylon thread (24 inches)
model 76805 scalepan and weights
model 76632 pin electrodes

Preparation of the Recorder

Prepare the Minigraph for two-channel recording. This experiment will require use of the 76322MG time/marker channel and the 76406MG basic amplifier channel. Check to make sure there is a sufficient amount of recording paper and the inking system is working properly. Refer to chapter 2, if necessary, and to figure 5.2.

1. Attach the model 76613 force transducer to the model 76613-T tension adjuster, and then secure the apparatus to a ring stand or flat-base stand. This semi-isometric force transducer is a silicon-strain gauge, which converts the mechanical movement of a leaf spring into an electric signal. Muscle movement is transmitted by a loop of thread attached to a hole in the leaf spring. The strength of the signal is proportional to the force exerted on the spring by the contracting muscle.
2. Connect the transducer to the balance control box via the standard 1/4-inch phone jack, and connect the balance control box to the amplifier via the 9-pin amphenol connector. Swing the top four leaves of the transducer to the side, exposing the thinnest leaf spring. If later in the exercise the muscle preparation produces excessive deflection of the transducer, it may become necessary to add one additional leaf. When doing so, always add the leaf closest to the sensing leaf (the thin leaf).
3. Turn on the mainframe power, make sure that the polarity switch is at + and the sensitivity control is fully counterclockwise, and center the basic amplifier pen using the pen-position control.
4. Connect the square-wave stimulator to a suitable AC outlet, and connect the signal report terminals to the time/event marker. Set the timer to mark off 1-second intervals.

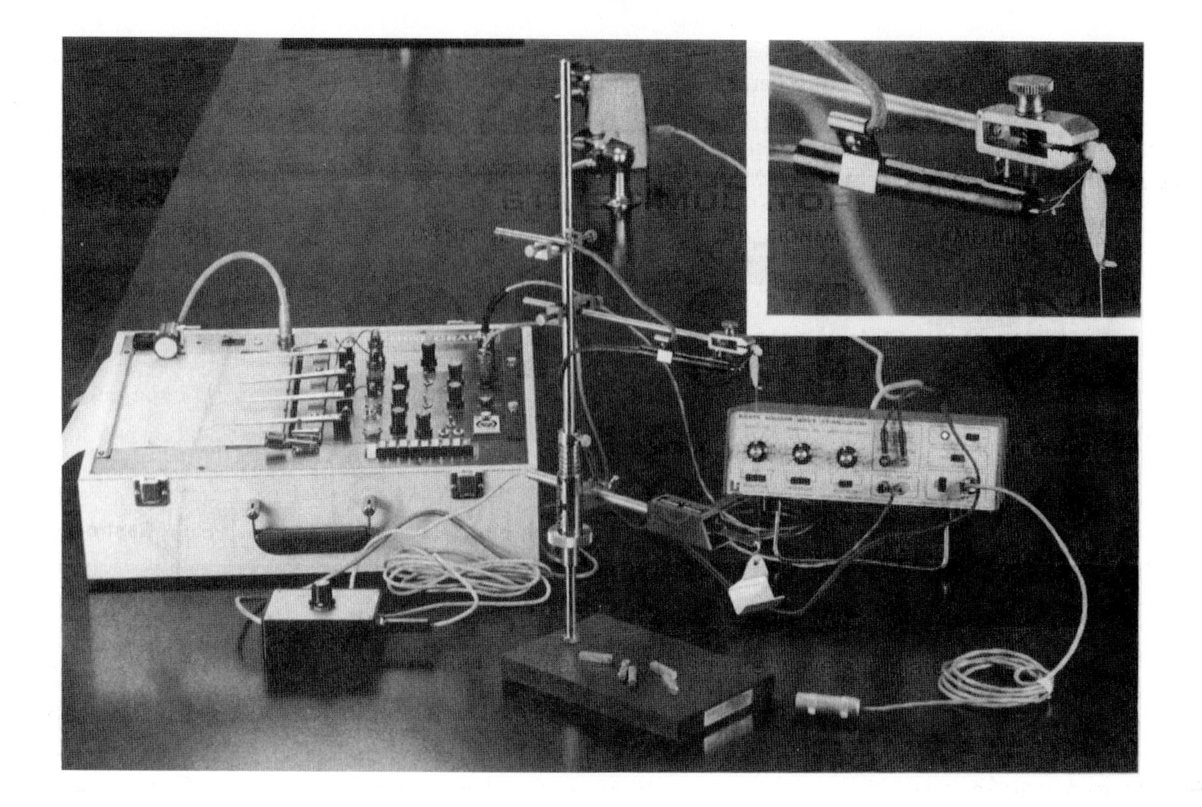

FIGURE 5.2 Lafayette Minigraph setup for isotonic skeletal muscle experiments

5. Set the basic amplifier gain switch to 100 and the sensitivity to 2. Recenter the pen using the balance control box. Whenever the gain and sensitivity controls are adjusted, the balance control box must be used to recenter the pen.
6. Turn on the chart drive power. Check for proper operation of the timer and event marker. Check for proper operation of the transducer-amplifier channel by gently deflecting the leaf spring upward. The amplifier pen should move upward and return to center when the leaf spring returns to its original position. Turn off the chart drive and power.

Preparation of the Animal

Obtain a frog from the laboratory instructor and destroy the brain and spinal cord by pithing. The correct procedure for pithing is given in appendix C. Following the pithing procedure, place the frog on a clean glass plate. Proceed to isolate the gastrocnemius muscle–sciatic nerve and prepare it for recording experimental data.

1. Remove the skin from one of the hind legs by making an incision in the skin around the thigh where it joins the body and peeling the skin down and off the toes with forceps.
2. Using heavy surgical scissors, amputate the thigh by dividing the femur and the muscles of the thigh near the pelvis.
3. Using a glass probe, gently free the sciatic nerve from surrounding muscle and connective tissue in the thigh. The sciatic nerve appears as a white thread deep between muscle groups on the anterior of the thigh. Free the nerve down to the knee and gently reflect it onto the surface of the gastrocnemius.
4. Bare the femur by removing most of the muscular and connective tissue from the thigh. Do not cut structures of the knee joint or damage the origin of the gastrocnemius or the sciatic nerve.
5. Free the Achilles tendon and the gastrocnemius from surrounding tissue by using a blunt probe. Cut the Achilles tendon as close as possible to its attachment at the heel.
6. Divide the tibia at the knee joint and cut away the remaining part of the lower leg. The preparation should now consist of the gastrocnemius muscle with severed Achilles tendon, the sciatic nerve attached to the muscle, and the knee joint with parts of the femur and tibia. Refer to figure 5.3.
7. Place the frog on the glass plate and cover it with a napkin soaked in frog Ringer's solution. The remaining leg may be used by other students in the laboratory.

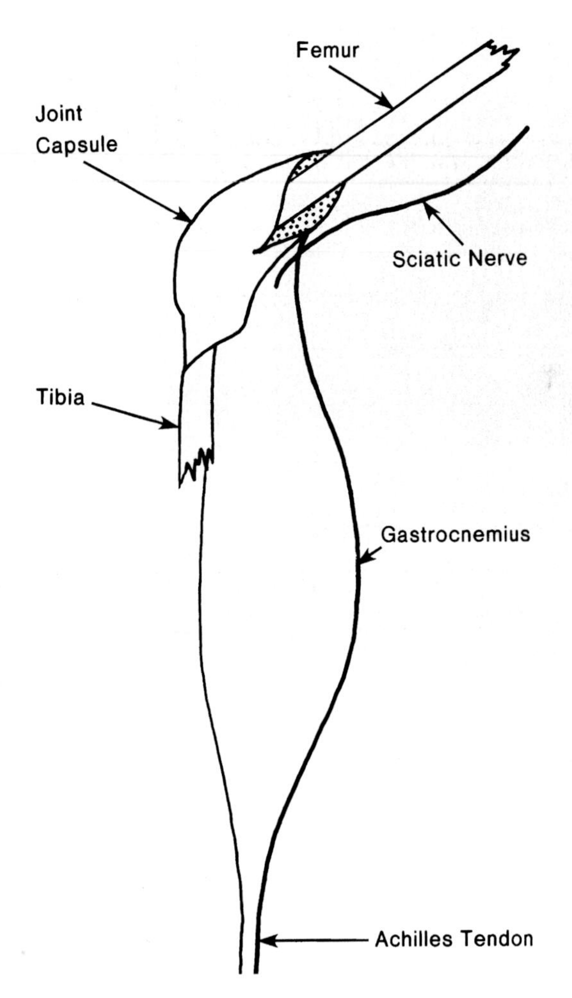

FIGURE **5.3** Gastrocnemius muscle—sciatic nerve preparation

Experimental Procedure

Refer to figure 5.2 and secure the frog gastrocnemius preparation to the recording apparatus as follows:

1. Place the femur in the femur clamp located above the transducer on the ring stand or flat-base stand. Make certain that the jaws of the femur clamp are parallel with the laboratory table and that the muscle is suspended vertically above the point where it will be attached to the leaf spring of the force transducer.
2. Tie a piece of thread tightly around the Achilles tendon and attach the other end to the tip of the thinnest leaf spring of the force transducer.
3. Adjust the position of the tension adjuster on the flat-base stand so that the thread is nearly taut; then turn the knurled knob on the tension adjuster so that the thread is taut and the leaf spring is barely beginning to bend.
4. Connect stimulator electrodes to the output terminals of the stimulator. Gently place the nerve over the electrodes (see figure 5.2 insert). The electrodes must not touch the muscle, nor should they be in contact with anything but the freed sciatic nerve.
5. Keep the muscle and the nerve moist by occasionally dripping frog Ringer's solution on them.
6. Turn on mainframe power. Use the balance control box to recenter the pen if necessary.
7. Gently turn the tension adjuster knob until the amplifier pen begins to move upward. Recenter the pen.
8. Turn on stimulator power and set the stimulator to deliver single stimuli of 20 milliseconds duration. Adjust the voltage to 0.01 V.
9. Turn on chart drive power. A paper speed of 1.0 mm/s will be used when recording data. If a continuously variable speed recorder is to be used, a paper speed of 1.0 mm/s may be approximated by adjusting the speed control until two 1-second time marks fall within a single small square (2 mm) on the recording paper.
10. Stimulate the nerve with single stimuli of increasing voltage. Deliver the stimuli about 2–4 seconds apart, and each time gradually increase voltage (i.e., 0.01, 0.015, 0.02, 0.025, etc.) until the muscle responds by contracting. Note the threshold voltage and compare with figure 5.4.
11. Continue to gradually increase the voltage with each successive stimulus above threshold. Note that the strength of contraction does not change appreciably.
12. Turn off the chart drive power. Remove the sciatic nerve by cutting it close to the knee with fine scissors. Note the muscle contraction when the nerve is severed.
13. Turn off mainframe power. Place the electrodes on the surface of the muscle near the knee so that they will not interfere with muscle contraction.
14. Turn on mainframe power and recenter the pen (if necessary).
15. Turn on chart drive power, adjust the stimulator voltage to 0.01 V, and repeat the determination of threshold voltage. Compare the value with the previously determined threshold for the sciatic nerve.

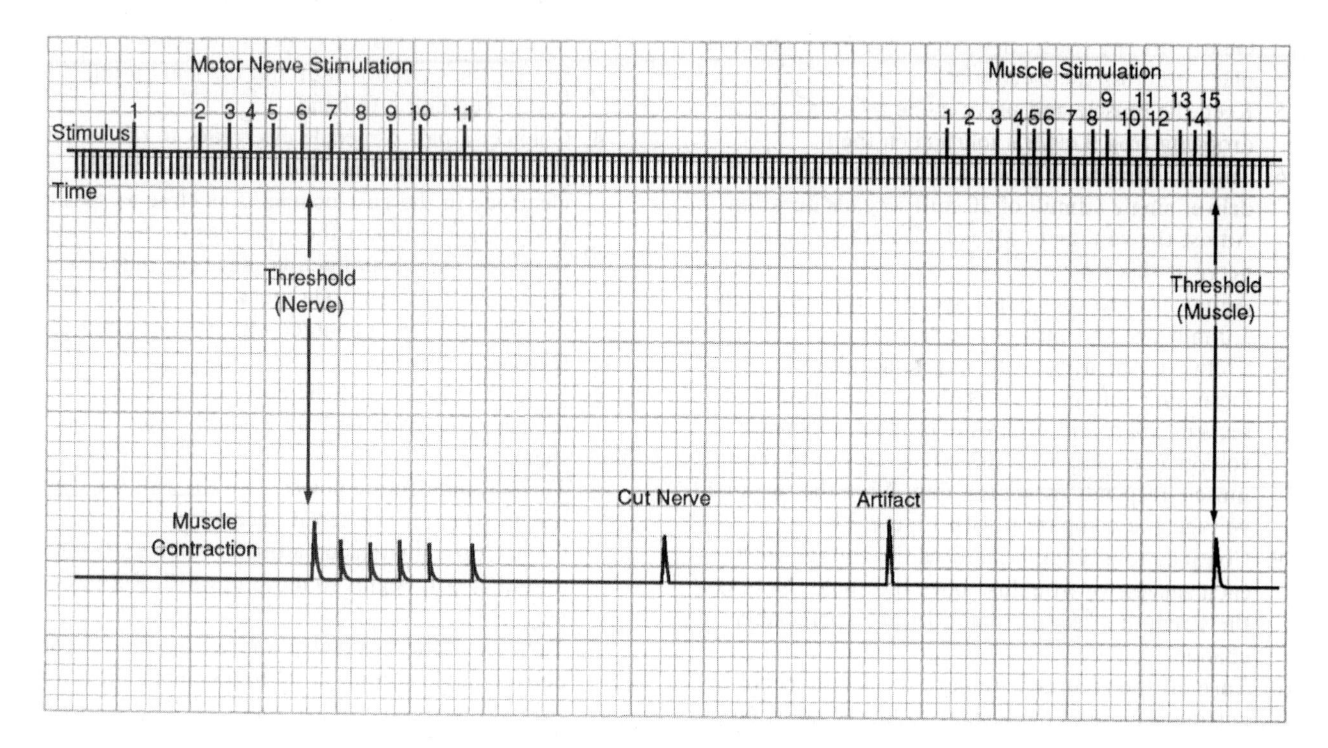

FIGURE 5.4 Threshold stimulation of nerve and muscle

Neural Control of Skeletal Muscle

REPORT 5

Name: ______________________ Date: ______________

Lab Section: ______________________

1. Data:
   a. Threshold voltage for nerve ______________________
   b. Threshold voltage for muscle ______________________

   Append copy of experimental record to this report.
2. Why is the threshold voltage for nerve lower than the threshold voltage for muscle? ______________________
3. Why did the muscle contract when the motor nerve was severed? ______________________
4. Stimulation of the nerve with stimuli of increasing strength above threshold voltage produced no significant increase in the strength of muscle contraction. Why? ______________________

Contractility of Skeletal Muscle I: The Twitch and Motor Unit Recruitment

CHAPTER 6

■ INTRODUCTION

The human body contains three kinds of muscle tissue, each performing specific tasks to maintain homeostasis. **Cardiac muscle** is found only in the heart. When it contracts, blood circulates, delivering nutrients to cells and removing their wastes. **Smooth muscle** is located in the walls of hollow organs such as the intestines, blood vessels, and lungs. Contraction of smooth muscle changes the internal diameter of hollow organs and is thereby used to regulate the passage of material through the alimentary canal, control blood pressure and flow, and regulate airflow during the respiratory cycle. **Skeletal muscle** derives its name from the fact that it is usually attached to the skeleton. Contraction of skeletal muscle moves one part of the body with respect to another part, as in flexing the forearm. Contraction of several skeletal muscles in a coordinated manner moves the entire body in its environment, as in walking or swimming. Regardless of kind, the primary function of muscle is to *convert chemical energy to mechanical work,* and in doing so, the muscle shortens as it contracts.

Each skeletal muscle of the body consists of bundles, or **fasciculi,** of individual skeletal muscle cells (fibers). A fasciculus consists of several muscle fibers separated from one another and bound by connective tissue called **endomysium** (figure 6.1). Blood vessels invade the endomysium to supply skeletal muscle fibers. The connective tissue that covers and binds the fasciculi together is called the **perimysium;** the heavy outermost fibrous connective tissue sheath enclosing the fasciculi that form the muscle is called the **epimysium.**

The skeletal muscle fiber is an elongated, cylindrical cell having many nuclei near its periphery. The plasma membrane of the cell is called the *plasmalemma* or *sarcolemma,* and the intracellular fluid that surrounds the nuclei and other organelles is called *sarcoplasm.* Embedded in the cell sarcoplasm are hundreds of smaller cylindrical elements called **myofibrils,** which begin at one end of the cell and terminate at the other. Myofibrils are the contractile elements of the skeletal muscle cell. Each myofibril is composed of a linear series of repeating segments called *sarcomeres,* the smallest units of the muscle cell capable of contracting. The sarcomeres, bound at each end by the Z disks, are composed of complex protein filaments called *myofilaments.* The thick myofilaments are made of large molecules of the fibrous protein called **myosin.** The thin myofilaments consist primarily of the protein **actin,** plus the smaller proteins **troponin** and **tropomyosin.**

When a skeletal muscle fiber is stimulated with an adequate stimulus (either by way of the motor nerve or by applying the stimulus directly to the muscle), calcium ions (Ca^{++}) are released from the sarcoplasmic reticulum (internal storage sites within the fiber) and contraction of the sarcomeres occurs. The free Ca^{++} binds to troponin, shifting the position of tropomyosin, thereby exposing active sites on the actin filament. The myosin heads engage the overlapping actin filaments and pull them toward the center of the sarcomere, causing the Z disks to move closer to one another and the sarcomere to shorten (figure 6.2). When the Ca^{++} ions are taken back up by the sarcoplasmic reticulum, actin and myosin can no longer interact and the sarcomeres relax.

When a single stimulus of sufficient strength and duration is applied to a skeletal muscle, the resultant contraction is known as a **twitch.** The record of a skeletal muscle twitch (figure 6.3) indicates a period between the time the stimulus is applied and the beginning of the contractile response. This interval of time is known as the *latent period.* The latent period represents the time required for the electric current to spread through connective tissue and muscle and stimulate the release of Ca^{++} within the muscle fibers and the time required for the muscle to overcome inertia (internally and externally) and begin to contract. Following the latent period, the muscle responds to the stimulus by shortening (contraction phase) and then returning to its original length (relaxation phase).

Skeletal muscle is made up of three types of fibers (table 6.1). Fiber type is determined by myoglobin content, or color, and speed of contraction. Myoglobin is a red-colored

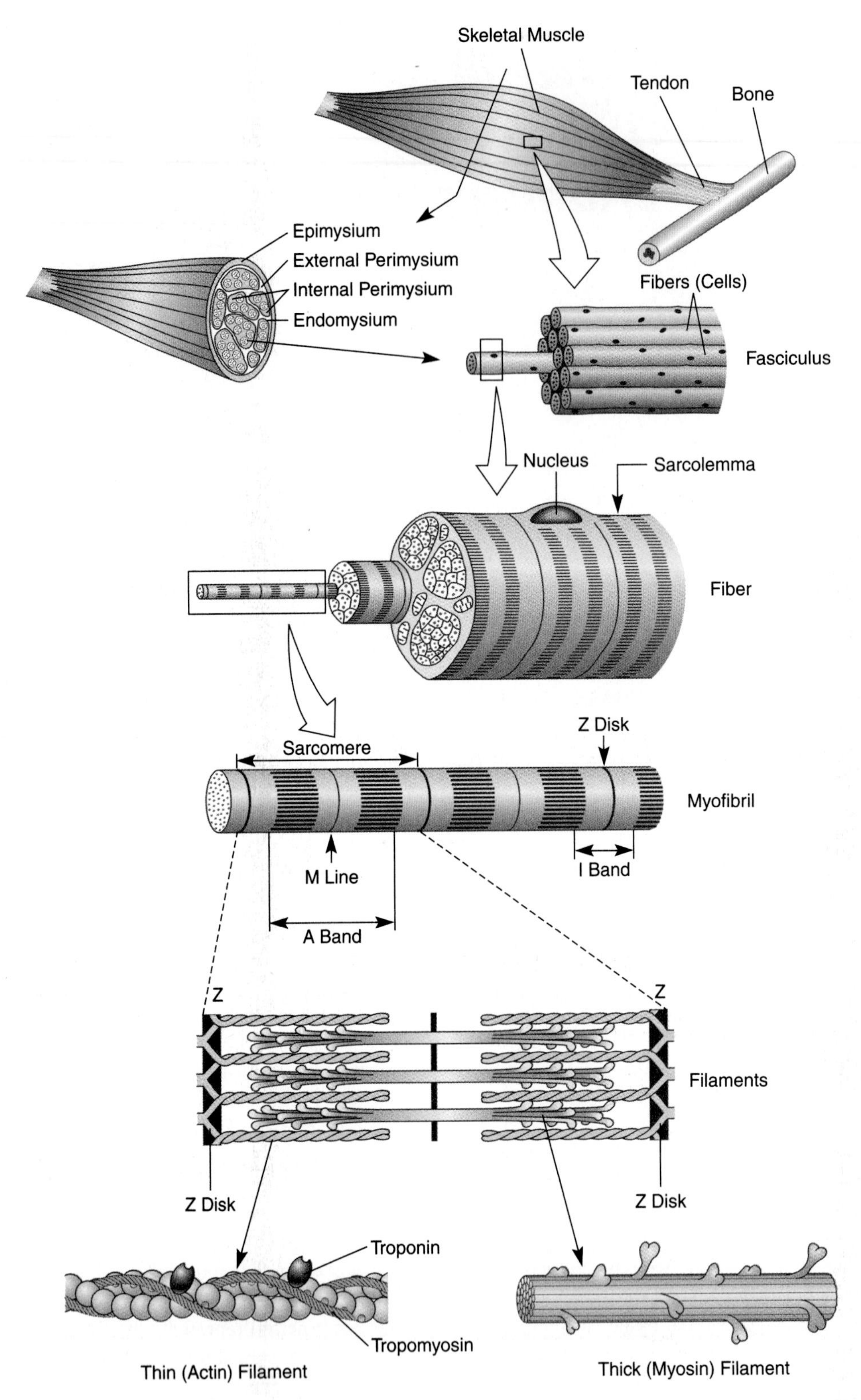

FIGURE 6.1 Skeletal muscle organization

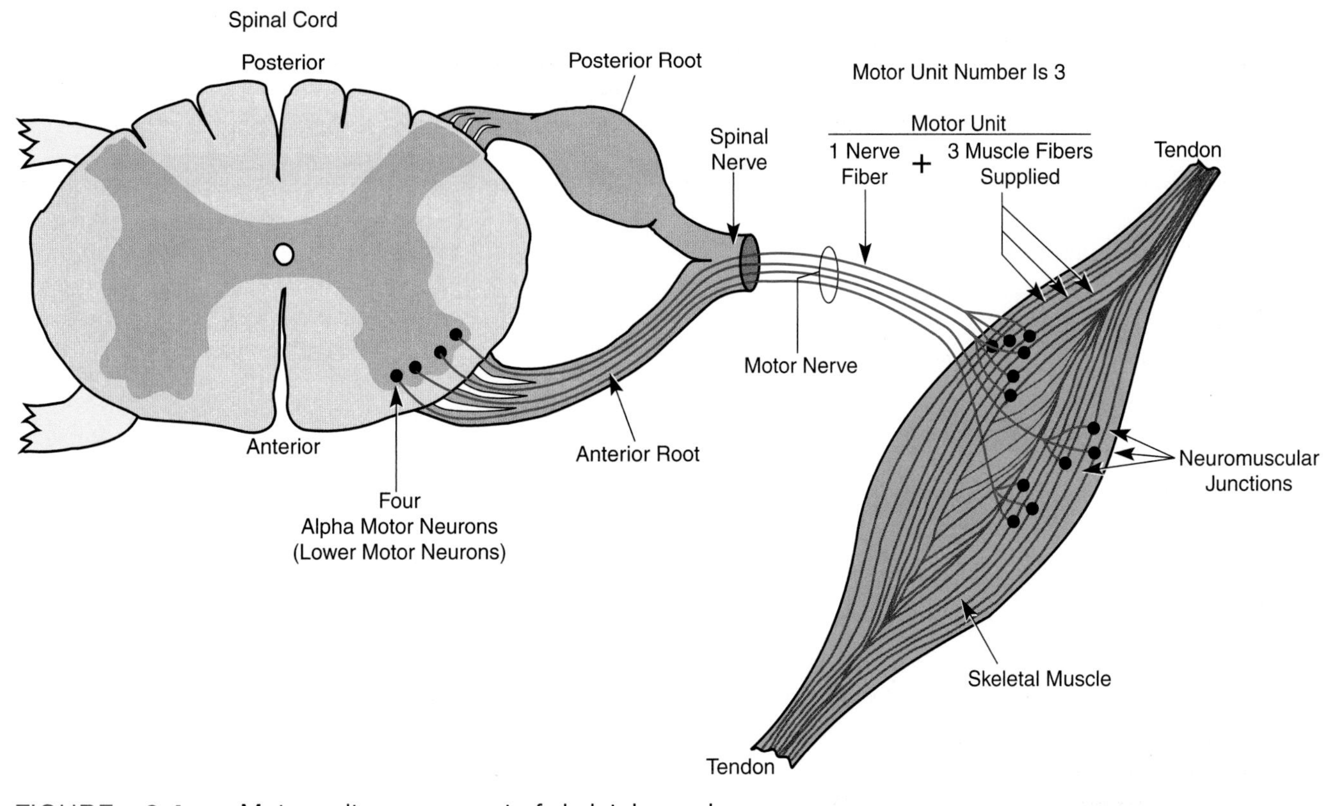

FIGURE 6.4 Motor unit arrangement of skeletal muscle

tain posture of the spine have very large motor units because precise control over the extent of shortening is not necessary.

Smooth, controlled movements of the body, such as in walking, swimming, or jogging, are produced by graded contractions of skeletal muscle. **Grading** means changing the strength of muscle contraction or the extent of shortening in proportion to the load placed on the muscle. Skeletal muscles are thus able to react to different loads accordingly. For example, the effort of muscles used in walking on level ground is less than the effort those same muscles expend in climbing stairs. Physiologically, the degree of skeletal muscle contraction is controlled by activating a desired number of motor units within the muscle and by controlling the frequency of motor neuron impulses in each motor unit.

The strength of skeletal muscle contraction is directly proportional to the number of motor units that are simultaneously active. When an increase in the strength of a muscle's contraction is necessary to perform a task, the brain increases the number of simultaneously active motor units within the muscle. This process is known as **motor unit recruitment.**

In motor unit recruitment, the force of contraction of newly recruited motor units is added to the strength generated by the muscle's previous contraction.

In the laboratory, motor unit recruitment, and hence graded contractions of skeletal muscle, can be simulated by stimulating the muscle directly with stimuli of increasing strength until all of the component muscle fibers are responding (figure 6.5).

EXPERIMENTAL OBJECTIVES

1. To record a skeletal muscle twitch and determine times for the latent, contraction, and relaxation periods.
2. To record the effect of motor unit recruitment on the degree of skeletal muscle contraction.

EXPERIMENTAL METHODS

Lafayette Minigraph

Materials

frog
glass plate
glass probe
dissecting kit
dropper
thread
50-mL beaker
frog Ringer's solution
flat-base stand
double clamps
Minigraph Model 76107 or 76107VS
model 76322 time/marker channel + remote marker button
model 76406MG basic amplifier channel

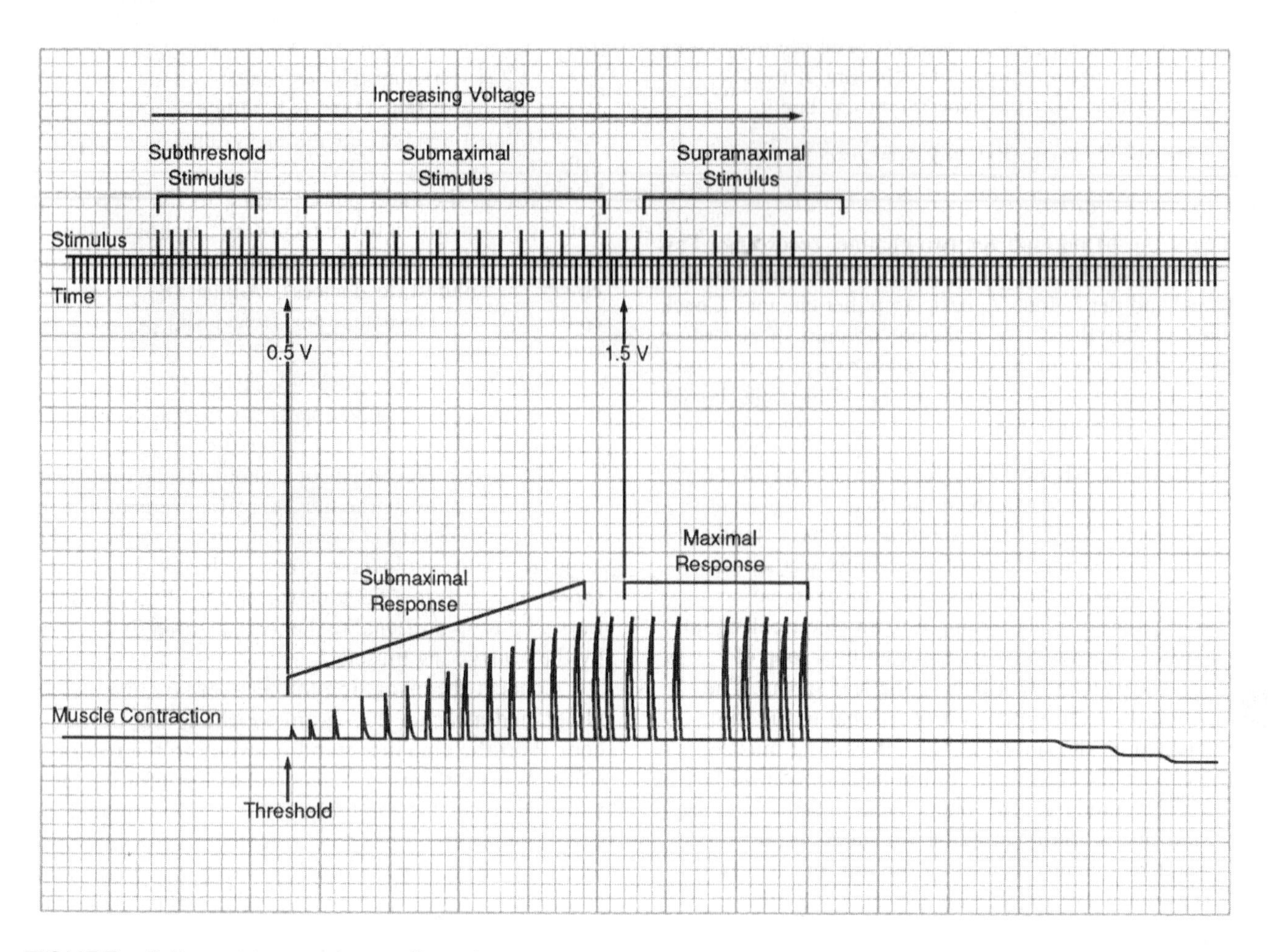

FIGURE 6.5 Motor unit recruitment

model 76613 semi-isometric force transducer

model 82415 square-wave stimulator and cable

model 76613-T tension adjuster

model 76802 transducer stand

model 76804 muscle clamp or equivalent femur clamp

model 76805 scalepan and weights

model 76632 pin electrodes

bipolar electrodes

Preparation of the Recorder

Prepare the Minigraph for two-channel recording. This experiment will require use of the 76322 time/marker channel and the 76406MG basic amplifier channel. Check to make sure there is sufficient recording paper and the inking system is working properly. Refer to the previous section on use of the Minigraph, if necessary, and to figure 6.6.

1. Attach the model 76613 force transducer to the model 76613-T tension adjuster, and then secure the apparatus to a ring stand or flat-base stand. This semi-isometric force transducer is a silicon-strain gauge, which converts the mechanical movement of a leaf spring into an electric signal. Muscle movement is transmitted by a loop of thread attached to a hole in the leaf spring. The strength of the signal is proportional to the force exerted on the spring by the contracting muscle.
2. Connect the transducer to the balance control box via the standard 1/4-inch phone jack, and connect the balance control box to the amplifier via the 9-pin amphenol connector. Swing the top four leaves of the transducer to the side, exposing the thinnest leaf spring. If later in the exercise the muscle preparation produces excessive deflection of the transducer, it may become necessary to add one additional leaf. When doing so, always add the leaf closest to the sensing leaf (the thin leaf).
3. Turn on the mainframe power, make sure that the polarity switch is + and the sensitivity control is fully counterclockwise, and center the basic amplifier pen using the pen-position control.
4. Connect the square-wave stimulator to a suitable AC outlet, and connect the signal report terminals to the time/event marker. Set the timer to mark off 1-second intervals.

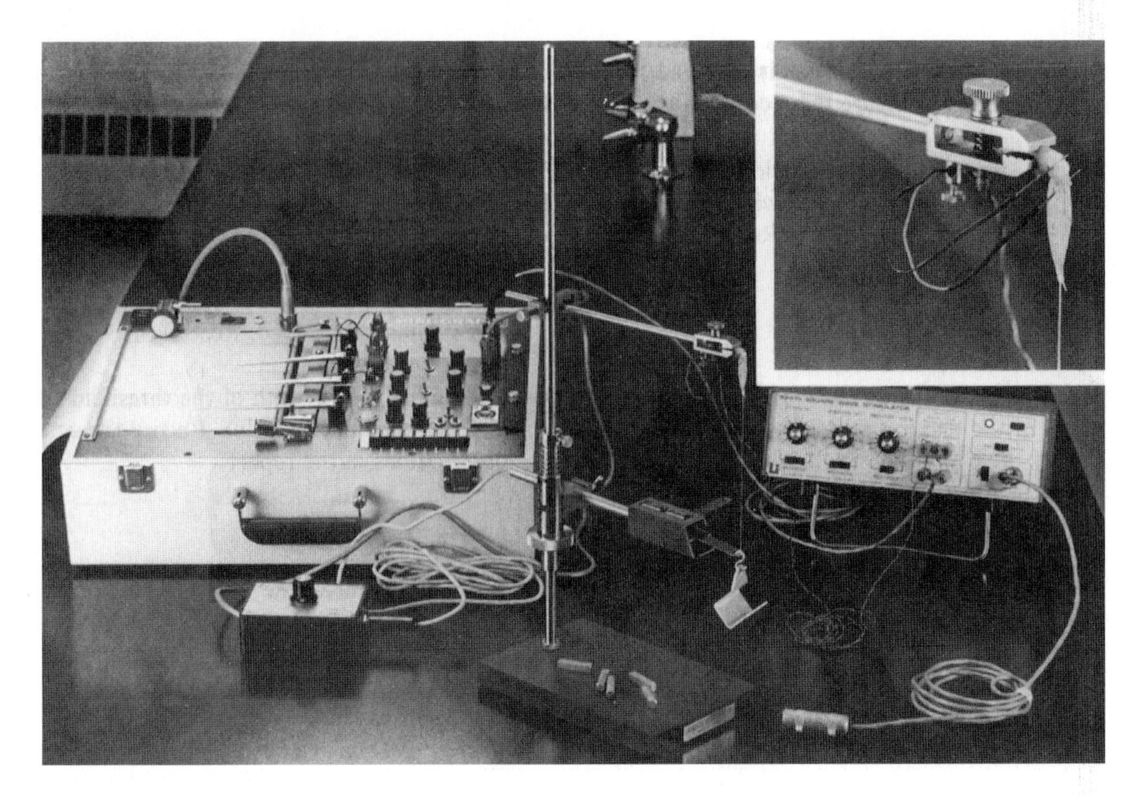

FIGURE 6.6 The Lafayette Minigraph setup for isotonic contraction

5. Set the basic amplifier gain switch to 100, and the sensitivity to 2. Recenter the pen using the balance control box. Whenever the gain and sensitivity controls are adjusted, the balance control box must be used to recenter the pen.
6. Turn on the chart drive power. Check for proper operation of the timer and event marker. Check for proper operation of the transducer-amplifier channel by gently deflecting the leaf spring upward. The amplifier pen should move upward and return to the center when the leaf spring returns to its original position. Turn off the chart drive and power.

Preparation of the Animal

Obtain a frog from the laboratory instructor and destroy the brain and spinal cord by pithing. The correct procedure for pithing is given in appendix C. Following the pithing procedure, place the frog on a clean glass plate. Proceed to isolate the gastrocnemius muscle and prepare it for recording experimental data.

1. Remove the skin from one of the hind legs by making an incision in the skin around the thigh where it joins the body and peeling the skin down and off the toes with forceps.
2. Using heavy surgical scissors, amputate the thigh by dividing the femur and the muscles of the thigh near the pelvis.
3. Bare the femur by removing all muscular and connective tissue from the thigh. Do not cut the structures of the knee joint or damage the origin of the gastrocnemius.
4. Free the Achilles tendon and the gastrocnemius from the surrounding tissue by using a blunt probe.
5. Cut the Achilles tendon as close as possible to its attachment at the heel.
6. Divide the tibia at the knee joint, and cut away the remaining part of the lower leg. The preparation should now consist of the gastrocnemius muscle with severed Achilles tendon and the knee joint with parts of the femur and tibia. Refer to figure 6.7.
7. Keep the muscle moist by occasionally dripping frog Ringer's solution (at room temperature) on it.
8. Place the frog on the glass plate, and cover it with a napkin soaked in frog Ringer's solution. The remaining leg will be used later.

Experimental Procedure

1. Place the femur in the femur clamp located above the transducer on the ring stand or flat-base stand. Make certain that the jaws of the femur clamp are parallel with the laboratory table and that the muscle is suspended vertically above the point where it will be attached to the leaf spring of the force transducer.
2. Tie a piece of thread tightly around the Achilles tendon and attach the other end to the tip of the thinnest leaf spring of the force transducer.
3. Adjust the position of the tension adjuster on the flat-base stand so that the thread is nearly taut; then turn

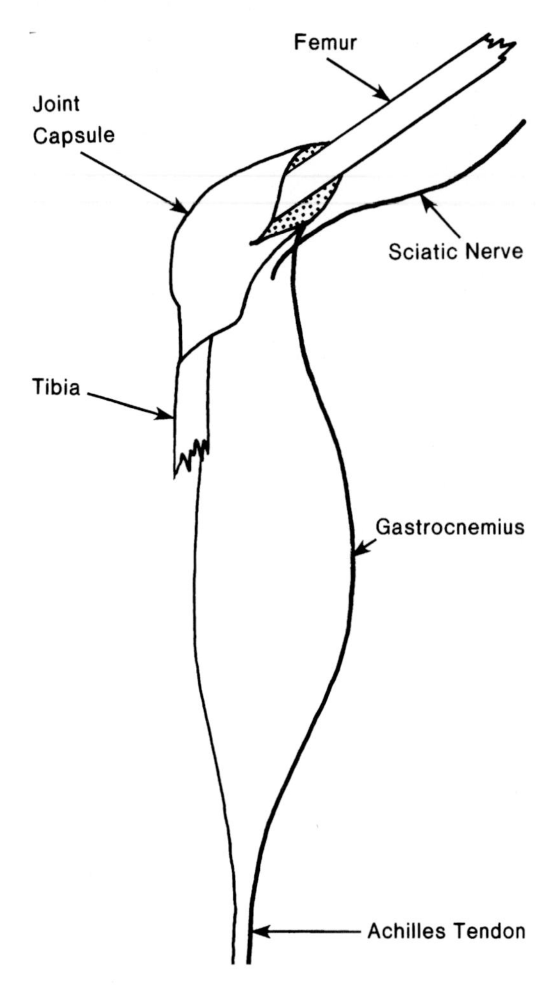

FIGURE 6.7 Gastrocnemius muscle—sciatic nerve preparation

the knurled knob on the tension adjuster so that the thread is taut and the leaf spring is barely beginning to bend.

4. Connect a pair of needle electrodes to the output terminals of the stimulator. Insert one needle electrode in the gastrocnemius as close as possible to the knee. Support the other needle electrode near the sciatic nerve and gently place the nerve over the electrode. This electrode must not touch the muscle, nor should it be in contact with anything but the freed sciatic nerve.
5. Keep the muscle and the nerve moist by occasionally dripping frog Ringer's solution on them.
6. Turn on mainframe power and advance the sensitivity control to 2. Use the balance control box to recenter the pen.
7. Gently turn the tension adjuster knob until the amplifier pen begins to move upward. Recenter the pen.
8. Turn on stimulator power and set the stimulator to deliver single stimuli of 20 milliseconds duration. Adjust the voltage to 10 V.
9. A paper speed of 50 mm/s will be used when recording the twitch. If a continuously variable speed recorder is to be used, turn the speed control to its maximum setting (fully clockwise) of approximately 30 mm/s.
10. Turn on the chart drive power. Immediately stimulate the muscle with a single stimulus of 10 V, 20 milliseconds. Turn off the chart drive power and compare the recorded twitch with figure 6.3. Depending upon the size of the muscle, it may be necessary to increase or decrease the sensitivity setting in order to approximate the record in figure 6.3. Measure and compute the time of the latent period, contraction period, and relaxation period.
11. Set the stimulator to deliver single stimuli, 0.01 V, 20 milliseconds.
12. A paper speed of 1.0 mm/s will be used to record motor unit summation. If a continuously variable speed recorder is used, approximate a paper speed of 1.0 mm/s by adjusting the speed control until two 1-second time marks fall within a single small square.
13. Turn on the chart drive power. Stimulate the muscle with single stimuli of increasing voltage. Deliver the stimuli about 2–4 seconds apart and each time gradually increase voltage (i.e., 0.01, 0.015, 0.02, 0.025, etc.) until the contractile response is maximum. Determine the voltage of the threshold stimulus and the maximal stimulus. Compare the record with figure 6.5.

Contractility of Skeletal Muscle I: The Twitch and Motor Unit Recruitment

REPORT 6

Name: ______________________ Date: ______________________

Lab Section: ______________________

TWITCH

1. Data:

a. Latent period __________ seconds

b. Contraction period __________ seconds

c. Relaxation period __________ seconds

MOTOR UNIT SUMMATION

a. Threshold stimulus __________ volts

b. Maximal stimulus __________ volts

Append a copy of the experimental record to this report.

2. Would you expect an athlete proficient in running the 100-yard dash to have a greater proportion of "slow-twitch" red fibers or "fast-twitch" white fibers in

muscles of the leg? Why? ______________________

____

____

3. Define the following:

a. Maximal stimulus ______________________

____

b. Threshold stimulus ______________________

____

c. Submaximal stimulus ______________________

____

4. A skeletal muscle can be stimulated, under laboratory conditions, with a supramaximal stimulus. Is it possible for the muscle to respond with a supramaximal contraction? Explain. ______________________

____

____

5. Define a motor unit. Of what physiological advantage is a small motor unit arrangement in skeletal muscle?

6. Explain why a latent period is observed between the application of a stimulus and contraction of the stimulated muscle.

7. Graded contractions of skeletal muscle may be produced in the laboratory by stimulating the muscle directly with stimuli of increasing strengths between threshold and maximum. How are graded contractions produced physiologically in skeletal muscles in their normal location in the body *(in situ)?*

Contractility of Skeletal Muscle II: Mechanical Summation, Contracture, Tetanus, and Fatigue

CHAPTER 7

INTRODUCTION

The contractile response of skeletal muscle to a single stimulus of sufficient strength to excite the component muscle fibers is known as a *twitch.* The duration of the twitch (less than 1 second) is too short to be useful for most work the muscle normally must perform. However, if the muscle is restimulated before it has completely relaxed, the second twitch will add its mechanical effect (shortening) to the first, producing a stronger contraction than that produced by a single stimulus. The process is called **mechanical summation,** or **frequency summation** (figure 7.1). If the interval between successive stimuli is short enough, repeated stimulation will produce a fusion of individual twitches, resulting in a state of continued contraction without relaxation known as **tetanus** (figure 7.2). Complete tetanus of skeletal muscle leads rapidly to complete fatigue. The usual physiological pattern of skeletal muscle activation is that of incomplete tetanus of an appropriate duration because the mechanical summation generated during tetanic contraction produces more force than does a twitch.

Tetanus should not be confused with treppe (staircase phenomenon). When a muscle is maximally stimulated to isotonically contract at a low frequency (e.g., 2 twitches per second) after a long period of rest, the first several twitches may be submaximal, gradually increasing in amplitude (like a staircase) until reaching a plateau. Historically, treppe was thought to be a "warm-up" phenomenon in which several initial contractions were required for the muscle to overcome viscoelastic forces of inertia. A more probable cause is the gradual increase in free Ca^{++} concentration within the sarcomeres during repetitive stimulation until a balance of Ca^{++} uptake and release by the sarcoplasmic reticulum occurs with each contraction-relaxation cycle.

A continued, rapid sequence of stimuli will also result in a gradual decrease in maximum contraction and progressively less relaxation at the same time, resulting in a condition in which the muscle fails to react maximally before the next stimulus becomes effective. The condition is known as **contracture** and is an early sign of skeletal muscle fatigue.

Fatigue (figure 7.3) of skeletal muscle in situ is caused by a reversible depletion of the muscle's fuel supply. The immediate source of energy is adenosine triphosphate (ATP), generated in the muscle fiber's mitochondria using nutrient molecules brought by the blood. Excitation-contraction coupling and cross-bridge cycling are processes of skeletal muscle contraction that are dependent on Ca^{++} release and ATP. Relaxation of skeletal muscle also is dependent on ATP and the energy-dependent uptake of Ca^{++} by the sarcoplasmic reticulum. If muscle fibers use ATP faster than it can be generated by cellular metabolism, fatigue occurs.

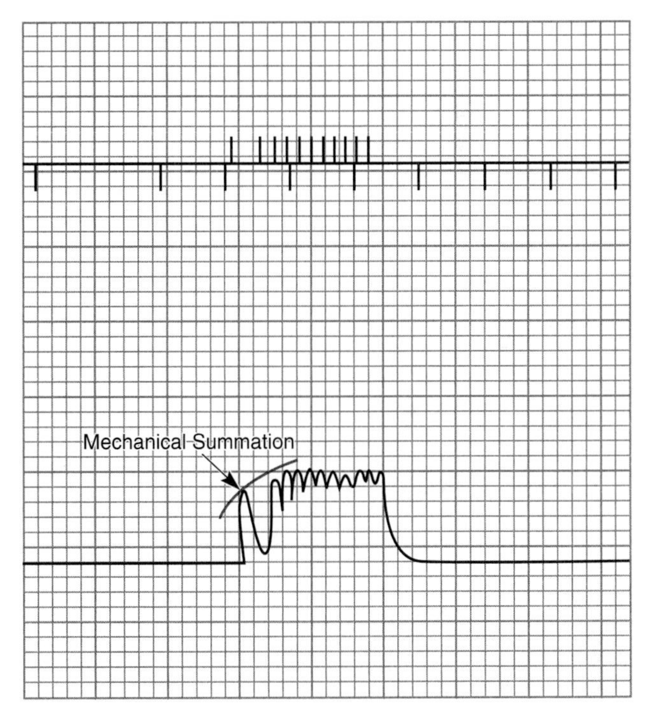

FIGURE 7.1 Mechanical summation in the frog gastrocnemius

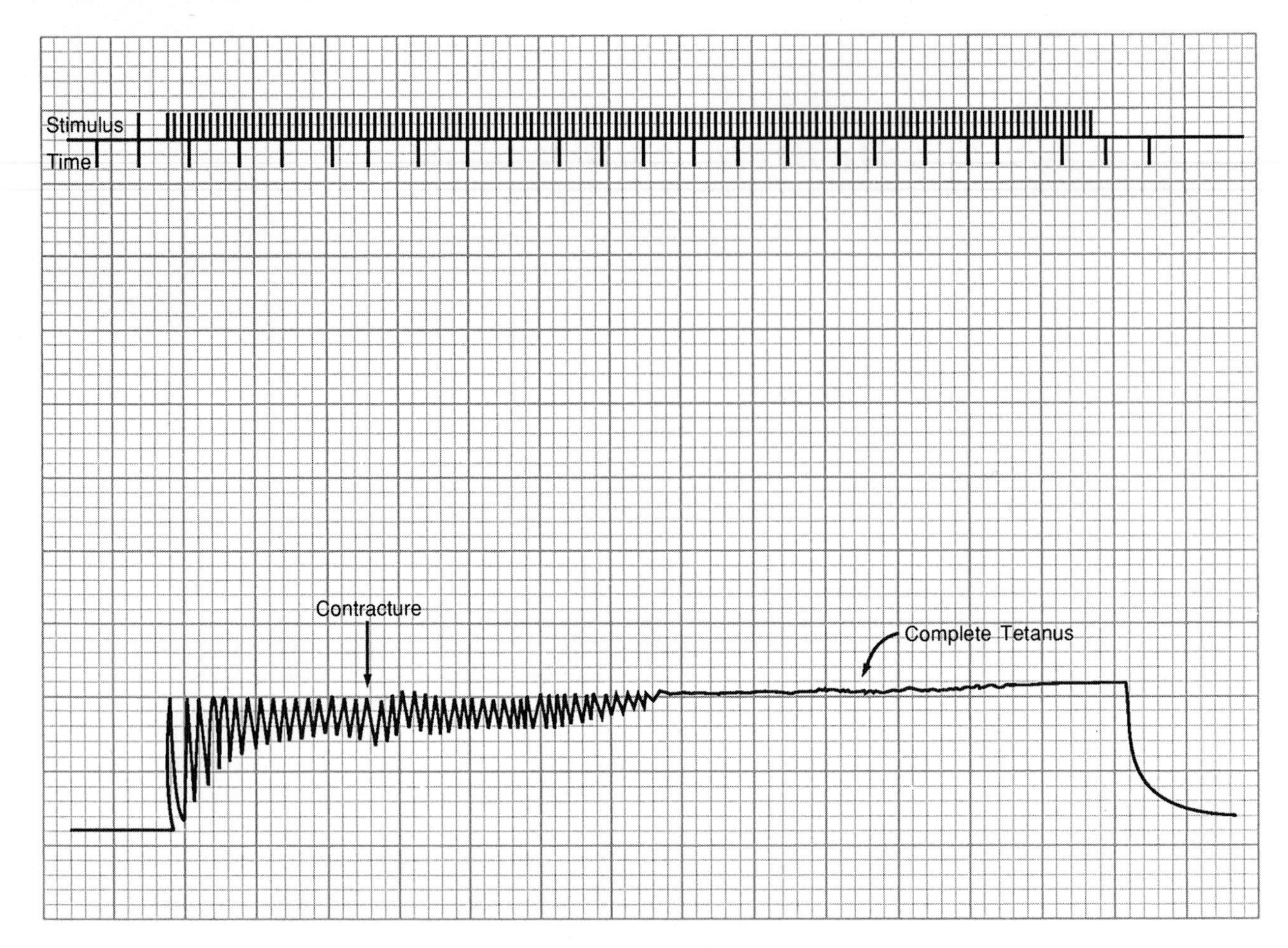

FIGURE 7.2 Contracture and complete tetanus in the frog gastrocnemius

During contraction, muscle cells convert chemical energy into thermal and mechanical energy and, in the process, produce chemical waste products. Normally, the waste products are removed from the muscle by the circulatory system as the blood brings nutrients to the muscle for energy transformation. If certain waste products (metabolites) are not removed at an adequate rate they will accumulate and chemically interfere with the muscle's contraction and relaxation, thereby hastening the onset of fatigue. Some accumulated waste products also stimulate pain receptors and induce cramping of skeletal muscle.

■ EXPERIMENTAL OBJECTIVES

1. To observe and record the phenomenon of mechanical summation in skeletal muscle.
2. To observe and record contracture and complete tetanus of skeletal muscle.
3. To observe and record the onset and completion of skeletal muscle fatigue.
4. To gain, through experimentation, an understanding of skeletal muscle contraction and the factors that influence its rate, strength, and ability to do work.

■ EXPERIMENTAL METHODS

Lafayette Minigraph

Materials

frog
glass plate
glass probe
dissecting kit
dropper
thread
50-mL beaker
frog Ringer's solution
flat-base stand
double clamps
Minigraph model 76107 or 76107VS
model 76322 time/marker channel + remote marker button
model 76406MG basic amplifier channel
model 76613 semi-isometric force transducer
model 82415 square-wave stimulator and cable

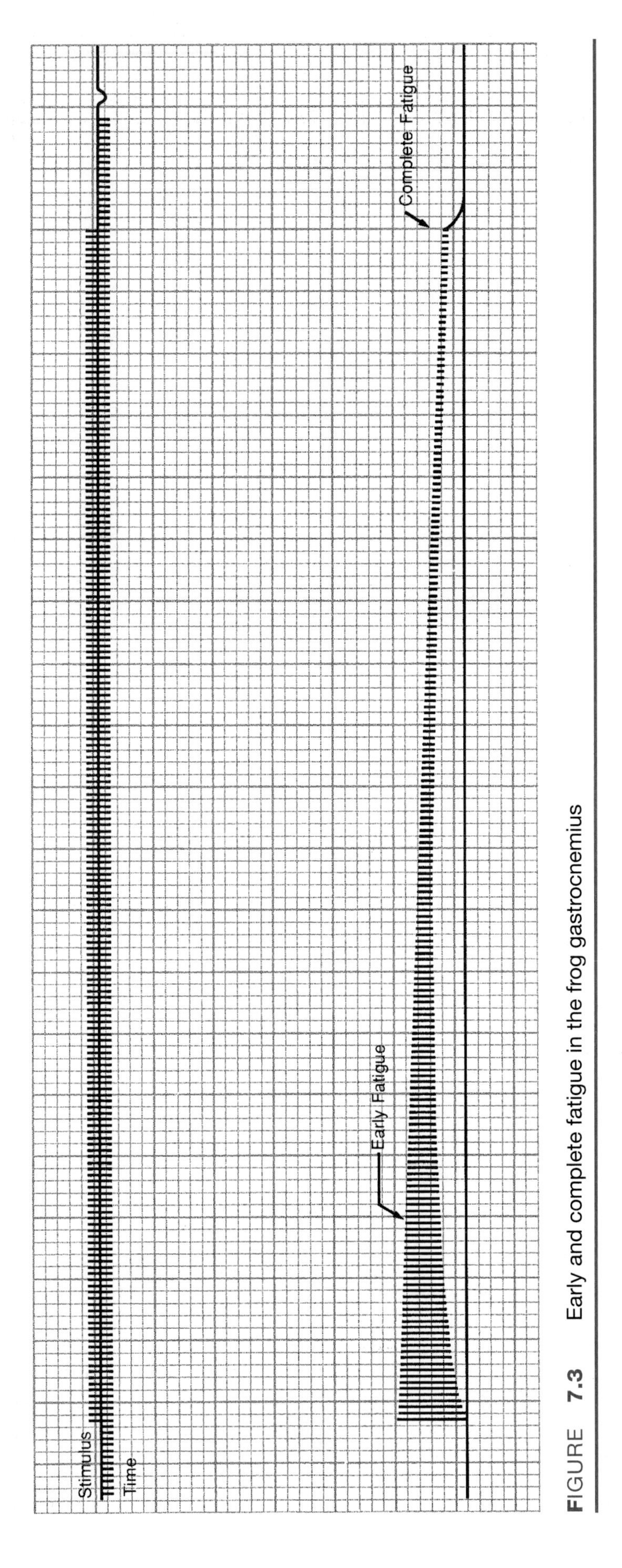

FIGURE 7.3 Early and complete fatigue in the frog gastrocnemius

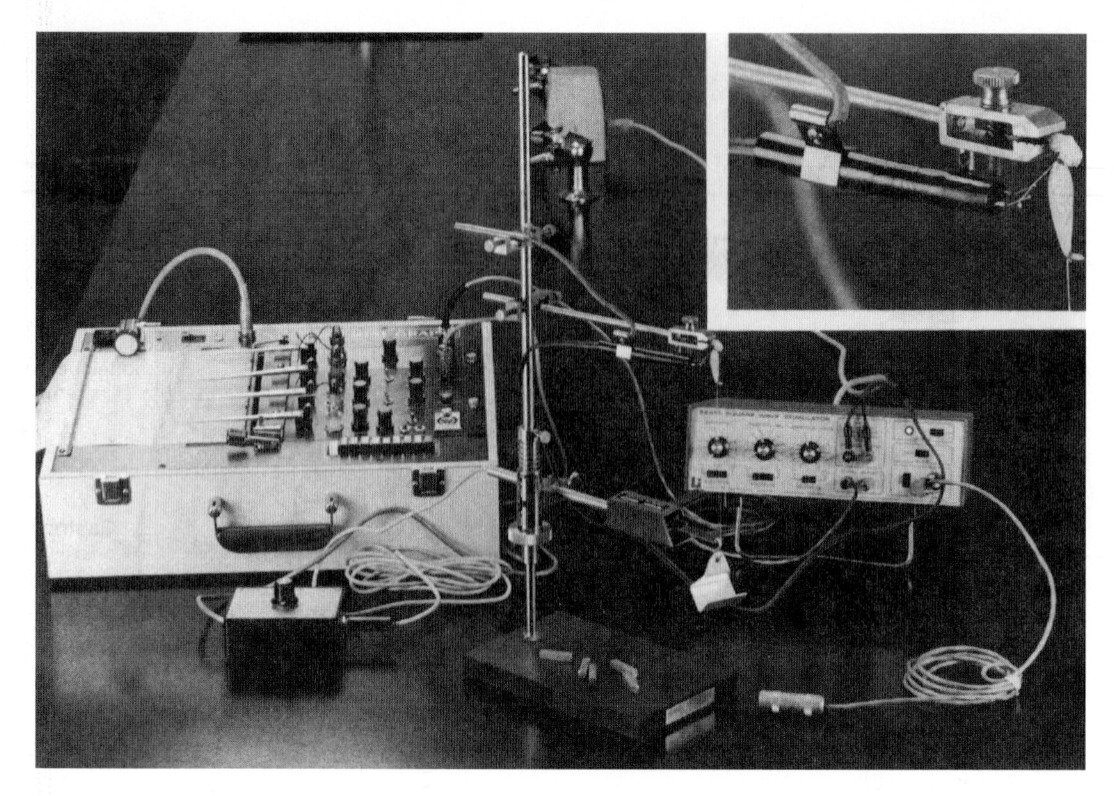

FIGURE 7.4 Lafayette Minigraph setup for isotonic skeletal muscle experiments

model 76613-T tension adjuster

model 76802 transducer stand

model 76804 muscle clamp or equivalent femur clamp

fine nylon thread (24 inches)

model 76805 scalepan and weights

model 76632 pin electrodes

Preparation of the Recorder

Prepare the Minigraph for two-channel recording. This experiment will require use of the 76322 time/marker channel and the 76406MG basic amplifier channel. Check to make sure there is sufficient recording paper and the inking system is working properly. Refer to chapter 2, if necessary, and to figure 7.4.

1. Attach the model 76613 force transducer to the model 76613-T tension adjuster and then secure the apparatus to a ring stand or flat-base stand. This semi-isometric force transducer is a silicon-strain gauge, which converts the mechanical movement of a leaf spring into an electric signal. Muscle movement is transmitted by a loop of thread attached to a hole in the leaf spring. The strength of the signal is proportional to the force exerted on the spring by the contracting muscle.
2. Connect the transducer to the balance control box via the standard 1/4-inch phone jack, and connect the balance control box to the amplifier via the 9-pin amphenol connector. Swing the top four leaves of the transducer to the side, exposing the thinnest leaf spring. If later in the exercise the muscle preparation produces excessive deflection of the transducer, it may become necessary to add one additional leaf. When doing so, always add the leaf closest to the sensing leaf (the thin leaf).
3. Turn on the mainframe power, make sure that the polarity switch is + and the sensitivity control is fully counterclockwise, and center the basic amplifier pen using the pen position control.
4. Connect the square-wave stimulator to a suitable AC outlet and connect the signal report terminals to the time/event marker. Set the timer to mark off 1-second intervals.
5. Set the basic amplifier gain switch to 100, and the sensitivity to 2. Recenter the pen using the balance control box. Whenever the gain and sensitivity controls are adjusted, the balance control box must be used to recenter the pen.
6. Turn on the chart drive power. Check for proper operation of the timer and event marker. Check for proper operation of the transducer-amplifier channel by gently deflecting the leaf spring upward. The amplifier pen should move upward and return to the center when the leaf spring returns to its original position. Turn off the chart drive and power.

Preparation of the Animal

Obtain a frog from the laboratory instructor and destroy the brain and spinal cord by pithing. The correct procedure for pithing is given in appendix C. Following the pithing procedure, place the frog on a clean glass plate. Proceed to isolate the gastrocnemius muscle and prepare it for recording experimental data.

1. Remove the skin from one of the hind legs by making an incision in the skin around the thigh where it joins the body and peeling the skin down and off the toes with forceps.
2. Using heavy surgical scissors, amputate the thigh by dividing the femur and the muscles of the thigh near the pelvis.
3. Bare the femur by removing all muscular and connective tissue from the thigh. Do not cut the structures of the knee joint or damage the origin of the gastrocnemius.
4. Free the Achilles tendon and the gastrocnemius from the surrounding tissue by using a blunt probe.
5. Cut the Achilles tendon as close as possible to its attachment at the heel.
6. Divide the tibia at the knee joint, and cut away the remaining part of the lower leg. The preparation should now consist of the gastrocnemius muscle with severed Achilles tendon and the knee joint with parts of the femur and tibia. Refer to figure 7.5.
7. Keep the muscle moist by occasionally dripping frog Ringer's solution (at room temperature) on it.
8. Place the frog on the glass plate, and cover it with a napkin soaked in frog Ringer's solution. The remaining leg will be used later.

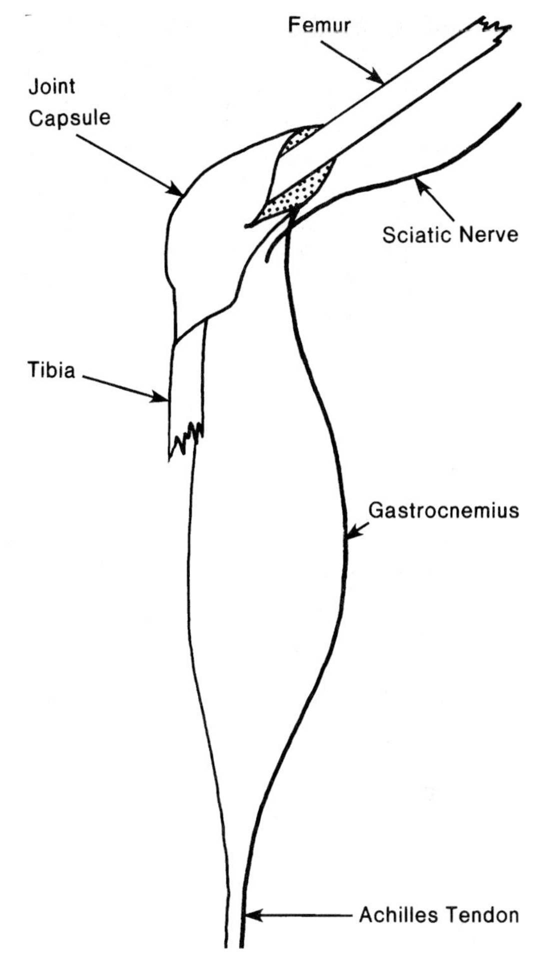

FIGURE 7.5 Gastrocnemius muscle–sciatic nerve preparation

Experimental Procedure

1. Place the femur in the femur clamp located above the transducer on the ring stand or flat-base stand. Make certain that the jaws of the femur clamp are parallel with the laboratory table and that the muscle is suspended vertically above the point where it will be attached to the leaf spring of the force transducer.
2. Tie a piece of thread tightly around the Achilles tendon and attach the other end to the tip of the thinnest leaf spring of the force transducer.
3. Adjust the position of the tension adjuster on the flat-base stand so that the thread is nearly taut; then turn the knurled knob on the tension adjuster so that the thread is taut and the leaf spring is barely beginning to bend.
4. Connect a pair of needle electrodes to the output terminals of the stimulator. Insert one needle electrode in the gastrocnemius as close as possible to the knee. Support the other needle electrode near the sciatic nerve and gently place the nerve over the electrode. This electrode must not touch the muscle, nor should it be in contact with anything but the freed sciatic nerve.
5. Keep the muscle and the nerve moist by occasionally dripping frog Ringer's solution on them.
6. Turn on mainframe power and advance the sensitivity control to 2. Use the balance control box to recenter the pen.
7. Gently turn the tension adjuster knob until the amplifier pen begins to move upward. Recenter the pen.
8. Turn on stimulator power and set the stimulator to deliver a train of stimuli, 5.0 V, 20 milliseconds each, at a frequency of 1 per second.
9. A paper speed of 1.0 mm/s will be used to record mechanical summation. If a continuously variable speed recorder is used, adjust the speed control so that two 1-second time marks fall within one small square on the recording paper.
10. Turn on the chart drive power. Depress the stimulator switch and deliver repetitive single stimuli to the muscle. The stimulator initiate switch must be held down to deliver a train of stimuli. Individual contractions of the same magnitude should be recorded with each stimulus delivery.

11. Continue to stimulate the muscle with repetitive stimuli, and, using the frequency control, gradually increase the frequency of stimulus delivery until mechanical summation is observed. The record should resemble figure 7.1.
12. Discontinue stimulation and turn off the chart drive power. Moisten the muscle with frog Ringer's solution.
13. Return the frequency control on the stimulator to 1 second.
14. A paper speed of 10 mm/s will be used to record contracture and complete tetanus. If a continuously variable speed recorder is used, adjust the speed control so that the one-second time marks fall 1 cm apart (one large square).
15. Turn on the chart drive power. Depress the stimulator switch and stimulate the muscle with repetitive stimuli. As a train of stimuli are being delivered, gradually increase the frequency of stimulus delivery until contracture and complete tetanus are observed.
16. Discontinue stimulation and turn off the paper-speed control. The record should resemble figure 7.2.
17. Remove the previously used muscle and replace it with the gastrocnemius from the remaining leg. Be sure to keep the muscle moist with frog Ringer's solution.
18. A paper speed of 1 mm/s will be used to record fatigue of skeletal muscle.
19. Turn on the chart drive control. Set the stimulator frequency at 1 per second. Depress the stimulator switch and continue to deliver repetitive stimuli to the muscle until complete fatigue is observed. The record should resemble figure 7.3.

Contractility of Skeletal Muscle II: Mechanical Summation, Contracture, Tetanus, and Fatigue

REPORT 7

Name: ______________________ Date: ______________________

Lab Section: ______________________

1. Data:
   a. Append a copy of the experimental record. Label mechanical summation, contracture, tetanus, and fatigue.
   b. Frequency producing complete tetanus ______________________
   c. Time required for complete fatigue ______________________
   d. Amount of tension ______________________ g
2. Define mechanical summation. Of what physiologic benefit is mechanical summation? ______________________
3. Define contracture of skeletal muscle. What does contracture signify? ______________________
4. Explain, giving two causative factors, why skeletal muscle displays fatigue. ______________________

5. How does mechanical summation differ from motor unit summation? ____________________

____

____

6. Does complete or incomplete tetanus occur physiologically? Explain. ____________________

____

____

____

7. The gastrocnemius muscle, as prepared for laboratory experimentation, generally fatigues faster than it would in situ (in its regular body location). Why? ____________________

____

____

____

Contractility of Skeletal Muscle III: Isotonic Contraction and Initial Length versus Work

CHAPTER 8

■ INTRODUCTION

In common usage, **contraction** means "to shorten." In the terminology of muscle physiology, it means any form of muscle activity in response to stimulation, whether it is accompanied by shortening or not. When a muscle contracts it develops **force** or **tension.** The relation between the generated force or tension and the muscle's length is used to define several kinds of skeletal muscle contraction.

If the tension increases as the muscle contracts but the muscle's external length (distance between origin and insertion) remains constant, the contraction is called **isometric** (*iso* = equal, *metric* = measure). In this kind of contraction, the maximum force developed by the contracting muscle is equal to the force opposing the shortening of the muscle: for example, the simultaneous contraction of flexors and extensors at a common joint so as to immobilize the joint (figure 8.1).

If the muscle's tension remains constant but its length changes as it contracts, the contraction is called **isotonic** (*iso* = equal, *tonic* = tension). When a weight is lifted by a contracting muscle, the force developed by the muscle need only exceed the gravitational force opposing the muscle's

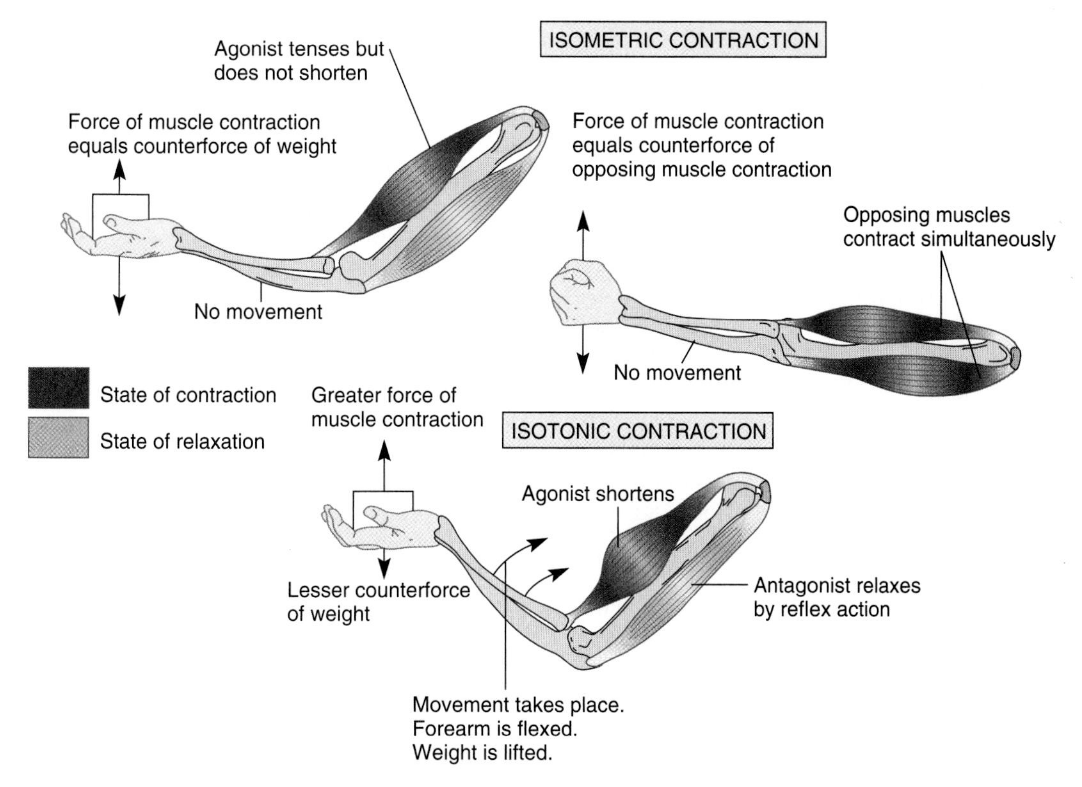

FIGURE 8.1 Isometric versus isotonic contraction of skeletal muscle

shortening. Once the required level of tension has been attained, the muscle tension then becomes constant as the weight is lifted. Contraction of skeletal muscle in which the developed tension exceeds a constant force opposing shortening is called *concentric.*

In an **auxotonic** contraction the developed force continually increases as the muscle shortens. This type of contraction occurs when a muscle is contracting against an opposing force that is continually increasing, as when a person is pulling back a bowstring. In a **meiotonic** contraction, the developed force lessens as the muscle contracts, as when gradually releasing the bowstring tension to prevent shooting the arrow. Contraction of skeletal muscle in which the developed tension is less than an opposing force, thereby allowing the muscle's length to passively increase, is called *eccentric.* This kind of contraction occurs as you sit down from a standing position.

All of the aforementioned kinds of skeletal muscle contraction occur in the course of daily living. The usual skeletal muscle contraction is a mix of isometric and isotonic contraction, as, for example, in walking up a flight of stairs. In the laboratory study of muscle physiology, it is useful to examine contractions that are, for example, only isometric or primarily isotonic. In the following exercises, we will examine the ability of the isotonically contracting skeletal muscle to do mechanical work.

Mechanical work, in the physical sense, refers to the application of a force that results in the movement of an object in the direction of the applied force. Mechanical work is the product of the applied force and the distance of movement (Work = Force × Distance). A skeletal muscle performs mechanical work when it isotonically contracts and moves an attached object, such as a bone or a weight. When a skeletal muscle isometrically contracts, as in attempting to move an immovable object, it does not perform mechanical work, even though it develops tension or force, because the distance factor is zero.

In the laboratory, a skeletal muscle performs work by lifting a load. The mechanical work done is equal to the load lifted times the distance the load was lifted:

$$\text{Work (g·mm)} = \text{Weight lifted (g)} \times \text{Distance weight was lifted (mm)}$$

For example, a skeletal muscle attached to a recording lever (figure 8.2) lifts a 10.0-g weight, and the lever records the height of contraction as 24 mm. The actual distance the weight was lifted is determined, using geometry, to be 8.0 mm. Therefore, the amount of work performed is 10.0 g × 8.0 mm = 80.0 g·mm.

If a number of isometric contractions are made with a muscle adjusted to a different length (called the initial length) before each contraction, it will be found that there is a length at which maximum force can be generated during contraction. This length is known as the **optimum length.** In the laboratory, the initial length of the muscle is controlled by adding weight to stretch the muscle. The amount of weight added to stretch the muscle to its optimum length is known as the **optimum load.**

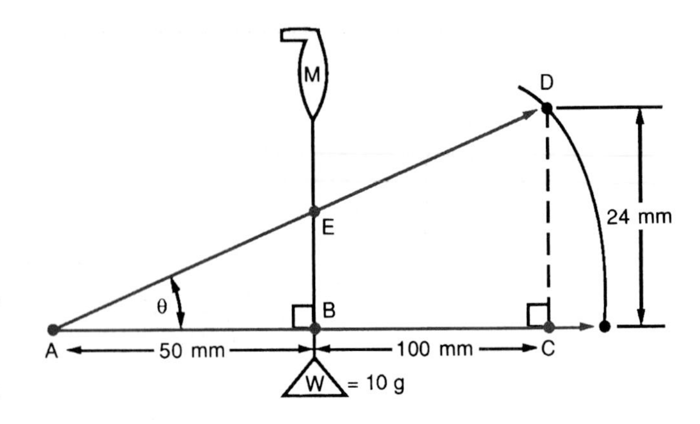

Triangles ABE and ACD are right triangles having θ (theta) as a common angle. Thus

$$\frac{EB}{DC} = \frac{AB}{AC} \text{ and therefore}$$

EB, the distance weight was lifted,

$$= (DC)\left(\frac{AB}{AC}\right) = 24\left(\frac{50}{150}\right) = 8 \text{ mm.}$$

Work performed = (10 g)(8 mm) = 80 g•mm.

FIGURE 8.2 Skeletal muscle work geometry involving similar right triangles

When a muscle is longer or shorter than its optimum length, it will develop less than its maximum force when stimulated to contract. The initial length of a muscle and the force generated during contraction are functionally related to the change in the amount of myofilament overlap at different muscle lengths (figure 8.3). Up to the optimum length of the muscle, the greater the stretch or initial length, the greater the force developed when the muscle contracts. In the following experiments, we will observe the effect of initial length on the ability of a skeletal muscle to perform work and calculate the work performed by an isotonically contracting muscle.

■ EXPERIMENTAL OBJECTIVES

1. To record and compute the mechanical work performed by an isotonically contracting skeletal muscle.
2. To observe and record the effects of initial length and load on the ability of a skeletal muscle to do work.
3. To gain, through experimentation, an understanding of skeletal muscle contraction and the factors that influence its strength and ability to do work.

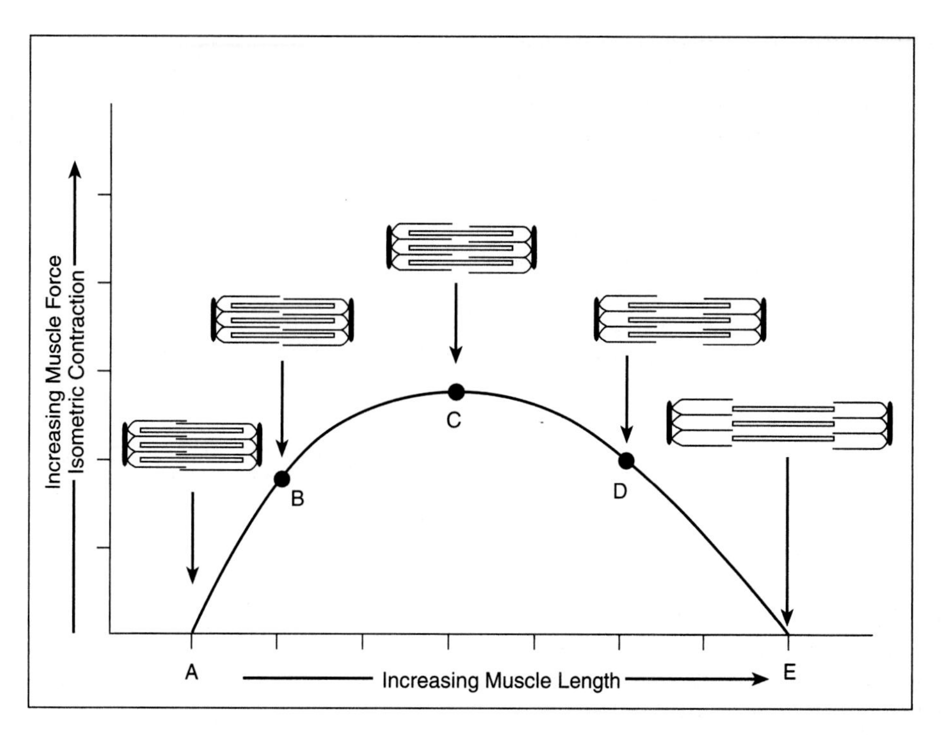

FIGURE **8.3** The structural basis for the length-force relationship in skeletal muscle. When the muscle is so highly stretched that the thin myofilaments are pulled out from the thick filaments (E), no active tension is developed. With partial overlap (D), partial force results. Optimal overlap (C) produces the most tension. As overlap increases (B), thin filaments from opposite sides of the sarcomere interfere with each other and tension decreases. Force can no longer be developed when overlap is so great (A) that the Z lines are pushed against the thick filaments. For these reasons, force-generating capability passes through a maximum.

EXPERIMENTAL METHODS

Lafayette Minigraph

Materials

frog
glass plate
glass probe
dissecting kit
dropper
thread
50-mL beaker
frog Ringer's solution
flat-base stand
double clamps
Minigraph model 76107 or 76107VS
model 76322 time/marker channel + remote marker button
model 76406MG basic amplifier channel
model 76614 semi-isotonic force transducer
model 76614-A muscle lever
model 82415 square-wave stimulator and cable
model 76613-T tension adjuster
model 76802 transducer stand (2)
model 76804 muscle clamp or equivalent femur clamp (2)
fine nylon thread (24 inches)
model 76805 scalepan and weights
model 76632 pin electrodes

Preparation of the Recorder

Prepare the Minigraph for two-channel recording. This experiment will require use of the 76322 time/marker channel and the 76406MG basic amplifier channel. Check to make sure there is sufficient recording paper and the inking system is working properly. Refer to chapter 2, if necessary, and to figure 8.4.

1. Attach the model 76614 force transducer (with 76614-A muscle lever) to the model 76613-T tension adjuster,

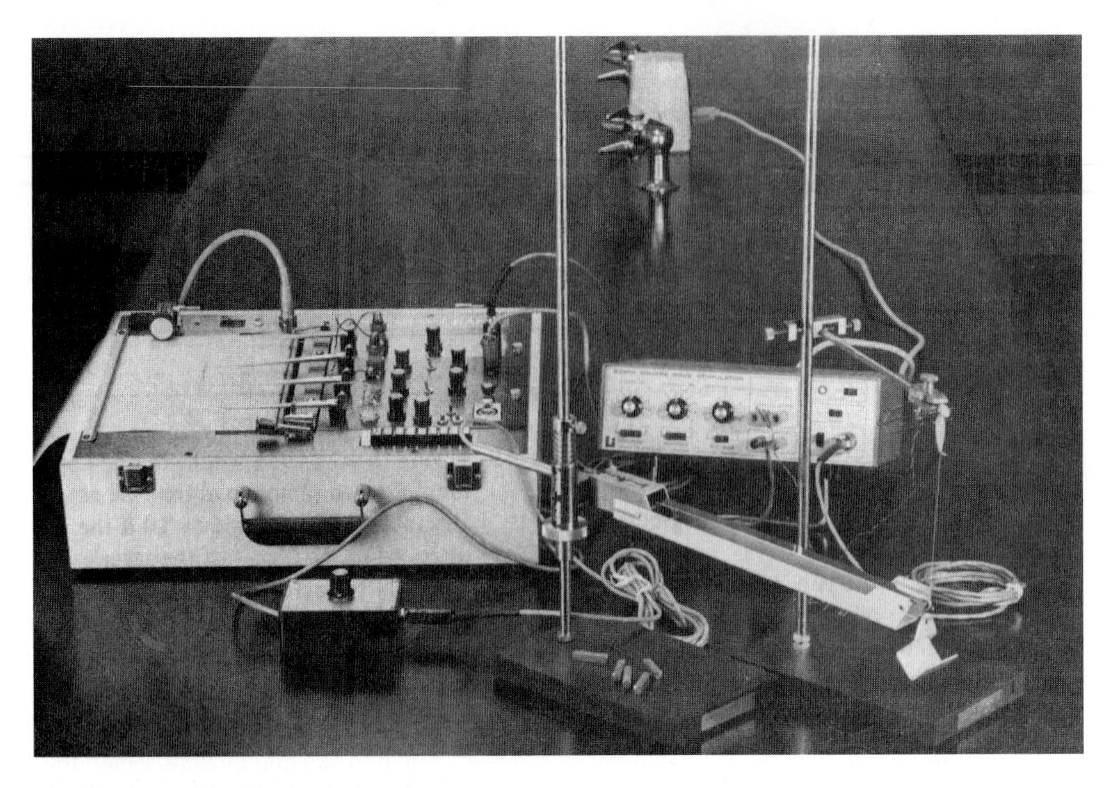

FIGURE 8.4 Lafayette Minigraph setup for skeletal muscle work experiments

and then secure the apparatus to a ring stand or flat-base stand.

2. Connect the transducer to the balance control box via the standard 1/4-inch phone jack, and connect the balance control box to the amplifier via the 9-pin amphenol connector.
3. Turn on the mainframe power, make sure that the polarity switch is + and the sensitivity control is fully counterclockwise, and center the basic amplifier pen using the pen-position control.
4. Connect the square-wave stimulator to a suitable AC outlet, and connect the signal report terminals to the time/event marker. Set the timer to mark off 1-second intervals.
5. Set the basic amplifier gain switch to 100 and the sensitivity to 2. Recenter the pen using the balance control box. Whenever the gain and sensitivity controls are adjusted, the balance control box must be used to recenter the pen.
6. Turn on the chart drive power. Check for proper operation of the timer and event marker. Check for proper operation of the transducer-amplifier channel by gently deflecting the leaf spring upward. The amplifier pen should move upward and return to center when the leaf spring returns to its original position. Turn off the chart drive and power.

Preparation of the Animal

Obtain a frog from the laboratory instructor and destroy the brain and spinal cord by pithing. The correct procedure for pithing is given in appendix C. Following the pithing procedure, place the frog on a clean glass plate. Proceed to isolate the gastrocnemius muscle and prepare it for recording experimental data.

1. Remove the skin from one of the hind legs by making an incision in the skin around the thigh where it joins the body and peeling the skin down and off the toes with forceps.
2. Using heavy surgical scissors, amputate the thigh by dividing the femur and the muscles of the thigh near the pelvis.
3. Bare the femur by removing all muscular and connective tissue from the thigh. Do not cut the structures of the knee joint or damage the origin of the gastrocnemius.
4. Free the Achilles tendon and the gastrocnemius from the surrounding tissue by using a blunt probe.
5. Cut the Achilles tendon as close as possible to its attachment at the heel.
6. Divide the tibia at the knee joint, and cut away the remaining part of the lower leg. The preparation should now consist of the gastrocnemius muscle with severed

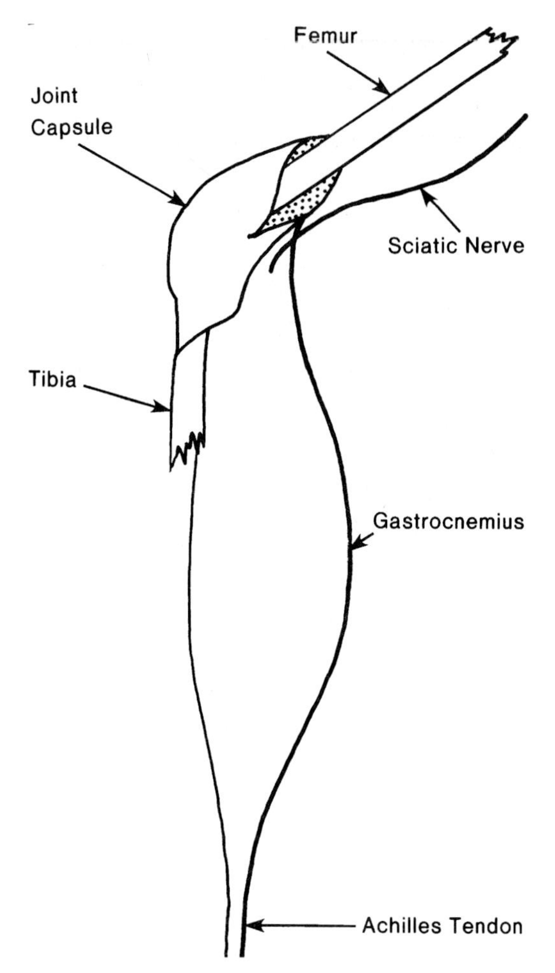

FIGURE 8.5 Gastrocnemius muscle–sciatic nerve preparation

Achilles tendon and the knee joint with parts of the femur and tibia. Refer to figure 8.5.

7. Keep the muscle moist by occasionally dripping frog Ringer's solution (at room temperature) on it.
8. Place the frog on the glass plate, and cover it with a napkin soaked in frog Ringer's solution. The remaining leg will be used later.

Experimental Procedure

Isotonic Contraction and Work

1. Place the femur in the femur clamp located above the transducer on the ring stand or flat-base stand. Make certain that the jaws of the femur clamp are parallel with the laboratory table and that the muscle is suspended vertically above the point where it will be attached to the leaf spring of the force transducer. Because of the large size of the transducer, the muscle must be supported by a femur clamp attached to a separate flat-base stand (see figure 8.4). Support the muscle directly above the hole in the transducer leaf spring, and connect the muscle to the leaf spring as in previous experiments. Arrange to stimulate the muscle directly by inserting one needle electrode into the muscle near the knee and the other needle electrode into the muscle near the Achilles tendon. Model 76614 semi-isotonic (displacement) force transducer is a silicon-strain gauge, which converts the mechanical movement of a long leaf spring into an electric signal and sends it to the channel amplifier. Muscle movement is transmitted by a loop of thread attached to a hole in the leaf spring. The strength of the signal is proportional to the distance the tip of the spring is moved by a contracting muscle.

   Model 76614-A muscle lever is attached to the body of the transducer and provides for adjustment and calibration of leaf spring displacement. Rotation of the knurled knob under the base of the muscle lever displaces the tip of the leaf spring a distance of 1 mm for each mark on the knurled knob.
2. Tie a piece of thread tightly around the Achilles tendon and attach the other end to the tip of the thinnest leaf spring of the force transducer.
3. Adjust the position of the tension adjuster on the flat-base stand so that the thread is nearly taut; then turn the knurled knob on the tension adjuster so that the thread is taut and the leaf spring is barely beginning to bend.
4. Keep the muscle and the nerve moist by occasionally dripping frog Ringer's solution on them.
5. Attach a scalepan to the tip of the leaf spring.
6. Adjust the muscle lever, via the knurled knob under the base, such that the leaf spring rests on top of the end of the muscle lever but the spring is not greatly displaced. The leaf spring and muscle lever should be parallel to the tabletop.
7. Adjust the amplifier gain to 1000 and the sensitivity control to 4. The precise setting of sensitivity is determined by the size and strength of the muscle. The larger or stronger the muscle, the lower the sensitivity setting.
8. Turn on mainframe power. Using the balance control box, position the pen about 2 cm below center.
9. Turn on stimulator power. Set the stimulator to deliver single stimuli of 5.0 V, 20 milliseconds.
10. Add 10 g of weight to the scalepan. Stimulate the muscle, record the contraction, and adjust sensitivity until a 3–4 cm pen deflection is obtained.
11. A paper speed of 1.0 mm/s will be used when recording data. If a continuously variable speed recorder is to be used, adjust the speed control until two 1-second time marks fall within a single small square.
12. Turn on the chart drive power. Rotate the knurled muscle lever control knob so that two marks pass the index point and the leaf spring tip is elevated 2 mm. Note the pen deflection on the recording channel and mark this deflection as the calibration line: 2 mm displacement. Rotate the knob in the reverse direction back to its original setting.

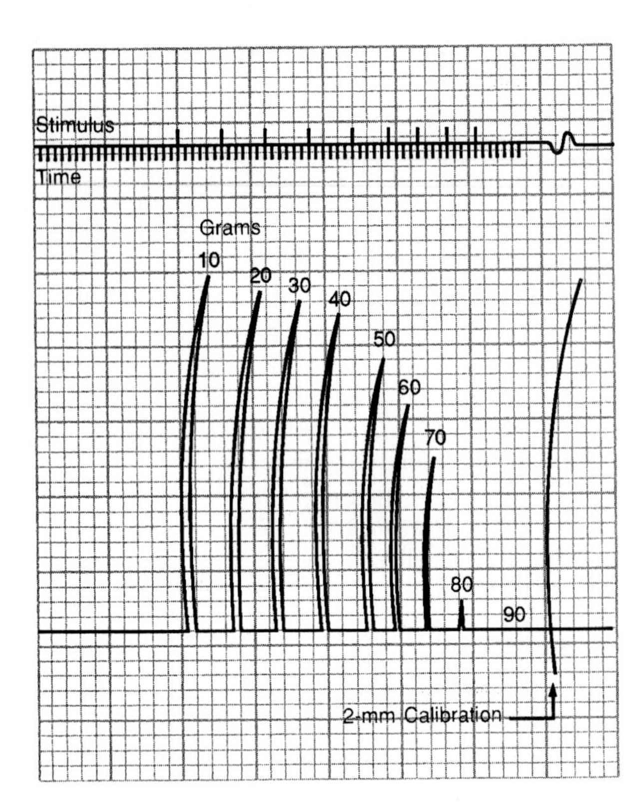

FIGURE 8.6 Minigraph recording of work performed by an unstretched skeletal muscle

13. Stimulate the muscle and record the height of contraction as the muscle lifts the scalepan and 10-g weight. Add another 10-g weight to the scalepan and stimulate the muscle. Continue to add 10-g weights to the scalepan. After each additional weight, stimulate the muscle and record the height of contraction. Repeat until the muscle is no longer able to lift the accumulated weight. Mark each contraction with the total amount of weight lifted, excluding the weight of the scalepan. The record should resemble figure 8.6.
14. Turn off the chart drive power. Remove all but one of the 10-g weights from the scalepan. Moisten the muscle with frog Ringer's solution.

Initial Length versus Work

1. Adjust the muscle lever, via the knurled knob, so that the leaf spring is free to be displaced (i.e., the muscle is stretched) when weight is added to the scalepan.
2. If necessary, reposition the amplifier pen about 1.5–2.0 cm below center.
3. If amplifier gain and sensitivity have not been changed since leaf spring displacement was calibrated, continue to the next item. If changed, repeat calibration procedure.
4. Turn on the chart drive power and repeat step 13, this time allowing the muscle to be stretched each time weight is added before contraction. The record should resemble figure 8.7.

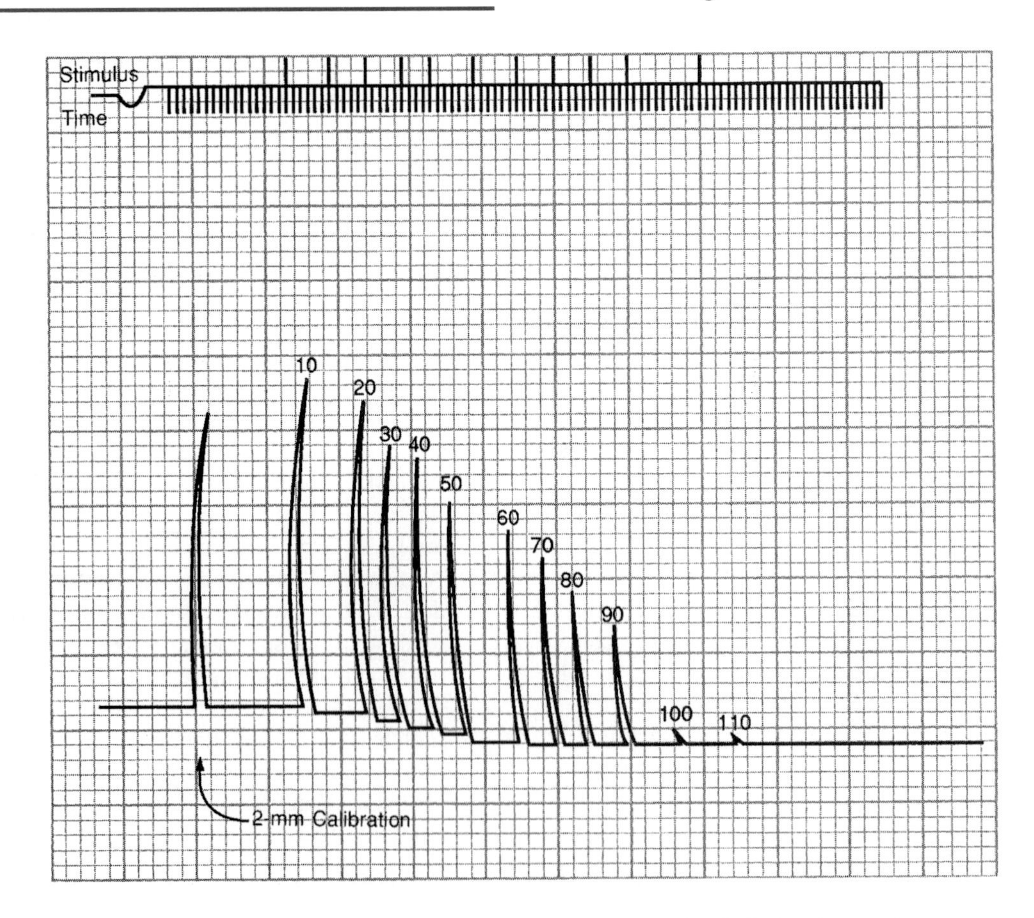

FIGURE 8.7 Minigraph recording of work performed by a stretched skeletal muscle

Contractility of Skeletal Muscle III: Isotonic Contraction and Initial Length versus Work

REPORT
8

Name: ______________________ Date: ______________________

Lab Section: ______________________

1. Append a copy of the experimental record to this report.
2. Calculate the amount of work performed during each contraction by the unstretched and stretched muscle. The amount of work performed during each contraction is computed by multiplying the weight lifted times the distance of the lift. To determine distance, compare the height of contraction on the record with the calibration line. Use a simple ratio calculation to compute the actual distance of lift based on the recorded height of contraction. Enter the data below:

Weight (g)	*Unstretched Distance*	*Work*	*Weight (g)*	*Stretched Distance*	*Work*
5			5		
10			10		
15			15		
20			20		
25			25		
30			30		
35			35		
40			40		
45			45		
50			50		
55			55		
60			60		
65			65		
70			70		
75			75		
80			80		
85			85		
90			90		
95			95		
100			100		

3. Plot the data shown above on the millimeter grid graph on the next page. Indicate maximum load for the unstretched muscle and optimum load for the stretched muscle.
4. Explain the effect of stretch (initial fiber length) on the ability of a skeletal muscle to do work.

____

____

____

5. Define the following:

a. Isometric contraction ____

____

b. Isotonic contraction ____

____

c. Optimum length (load) ____

____

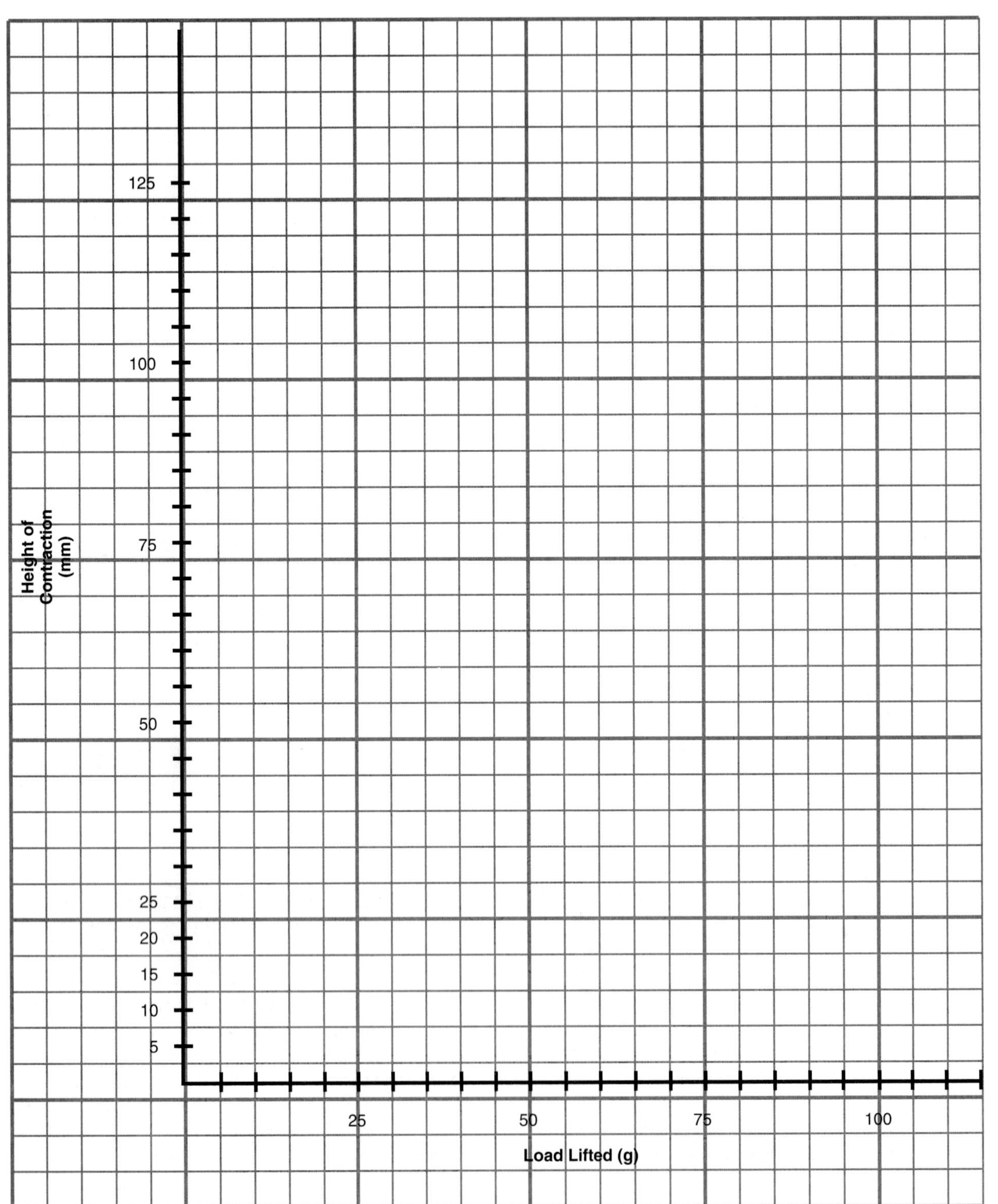

Electromyography and Dynamometry

CHAPTER 9

■ INTRODUCTION

As observed in chapter 5, skeletal muscles are physiologically activated by motor nerve impulses. Effective transmission at the neuromuscular junction results in the muscle fiber's generating and conducting its own electrical impulse, which ultimately results in contraction of the fiber. Although the electrical impulse generated and conducted by each fiber is very weak (less than 100 μV), many fibers conducting simultaneously induce voltage differences in the overlying skin that are large enough to be detected by a pair of surface electrodes. The detection, amplification, and recording of changes in skin voltage produced by underlying skeletal muscle contraction is called **electromyography.** The recording thus obtained is called an *electromyogram (EMG).*

Dynamometry means the measurement of force or power (*dyno* = power, *meter* = measure). In previous experiments, the force generated by an isotonically contracting skeletal muscle was indicated by the degree of shortening and the amount of work performed by lifting weight. In this experiment, the force of skeletal muscle contraction will be determined by a **hand dynamometer,** a device used to measure grip strength. As observed in previous experiments, the strength of skeletal muscle contraction is proportional to the number of motor units that are simultaneously active. Combining dynamometry and electromyography makes it possible to observe and record the recruitment of motor units as the muscle performs increased work or contracts with greater power.

■ EXPERIMENTAL OBJECTIVES

1. To determine maximum grip strength for right and left hands and compare differences between male and female.
2. To observe, record, and correlate motor unit recruitment with increased power of skeletal muscle contraction.
3. To record the dynagram and electromyogram of grip muscles as they perform repetitive work and to determine the endurance and time to fatigue of these muscles.

■ EXPERIMENTAL METHODS

Lafayette Minigraph

Materials

Minigraph model 76322 time/marker channel + remote marker button

two model 76406MG basic amplifier channels

model 76422 biopotential preamplifier

model 76623 EMG electrodes

model 76618 or 76619 hand dynamometer

model 76716 adhesive collars

model 76621 biogel

NOTE: A model 76402MG biopotential amplifier may be substituted for one of the basic amplifiers and the biopotential preamplifier.

Preparation of the Recorder

1. Prepare the Minigraph for three-channel recording. Check to make sure there is sufficient recording paper and the inking system is working properly. Refer to chapter 2, if necessary, and to figure 9.1.
2. Connect the dynamometer to the channel amplifier via the 9-pin amphenol connector. Model 76618 (0–100 kg) and model 76619 (0–50 kg) hand dynamometers contain an electronic transducer that converts mechanical movement of the dynamometer grip into an electric signal that is sent to the channel amplifier. The magnitude of the electric signal, and hence the degree of amplifier pen movement, is

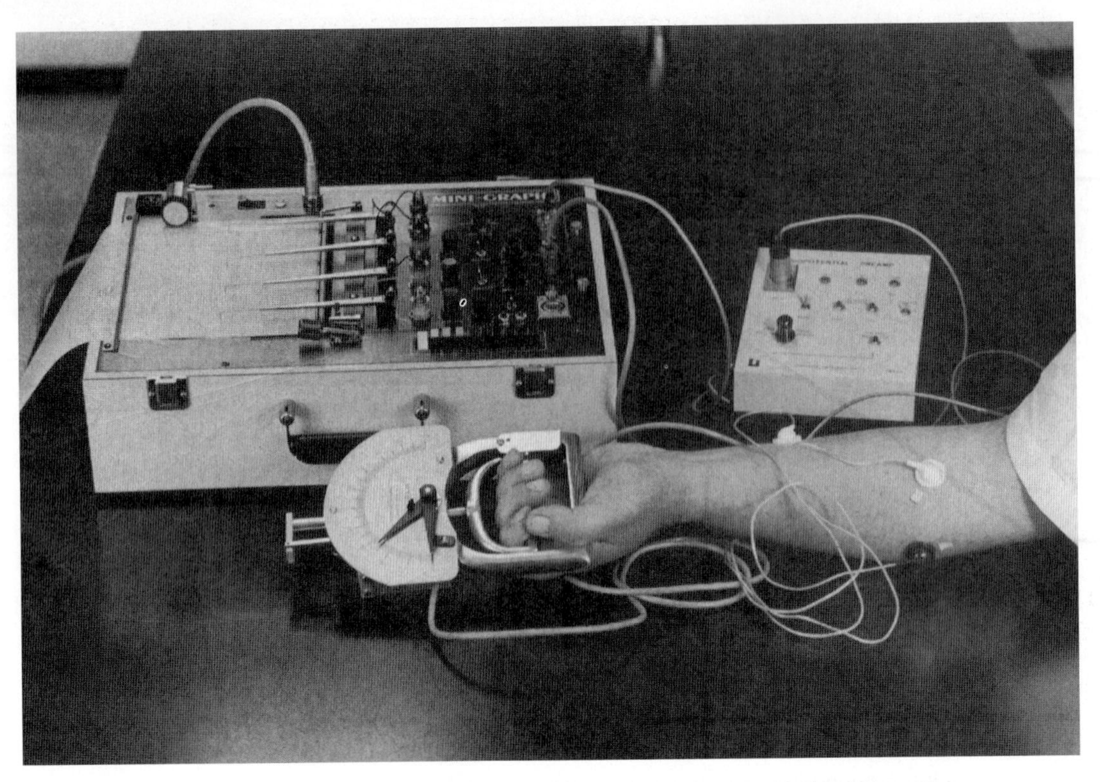

FIGURE **9.1** Lafayette Minigraph setup for electromyography and dynamometry

proportional to the displacement of the dynamometer grip.

3. Attach the EMG electrode cable to the biopotential preamplifier via the 4-pin connector. Connect the preamplifier to the channel amplifier via the 9-pin amphenol connector. *Optional:* Connect the EMG electrodes directly to the 76402MG biopotential amplifier.
4. Set the timer to mark off 1-second intervals.
5. Set the following controls on the preamplifier: mode—EMG; standby/cal—on; filter—in.
6. Set the amplifier polarity switch to +, the gain switch to 100, and the sensitivity to 5. *Optional:* Set the 76402MG sensitivity to ×l and 3.
7. Set the dynamometer channel gain switch to 100 and the sensitivity to 5.
8. Turn on mainframe power, center the amplifier pens using the pen-position controls, and then turn off mainframe power.

Experimental Procedure

1. Set the adjustable pointer on the hand dynamometer to zero. Determine maximal grip strength for the right hand. Record the value (kg), reset the adjustable pointer to zero, and repeat for the left hand. The determinations should be made with mainframe power off.
2. Using cotton or a paper towel soaked in alcohol, cleanse the skin on the medial aspect of the anterior forearm (see figure 9.2) where EMG electrodes will be attached. Use the extremity that demonstrated the strongest grip strength for placement of EMG electrodes.
3. Apply the self-sticking paper washers to the raised plastic area around the electrode plates. Squeeze electrolyte gel onto the metal electrode plates. Use a paper towel to smooth the gel so that it completely fills the well between the metal plates and the surrounding plastic.
4. Remove the protective coverings over the adhesive area of the washers and apply the electrodes to the skin of the medial forearm (see figure 9.2). The (+) and (–) electrodes (same color) are placed on the medial forearm in a straight line, 5–6 inches apart, over underlying flexors of the hand and digits. The ground electrode (odd color) is placed on the lateral forearm, about midway between the elbow and wrist.
5. A paper speed of 1.0 mm/s will be used when recording data. If a continuously variable speed recorder is to be used, a paper speed of 1.0 mm/s may be approximated by adjusting the speed control until two 1-second time marks fall within a single small square on the recording paper.
6. Turn on mainframe and chart drive power. Squeeze the dynamometer grip as strongly as possible and adjust the amplifier sensitivity to obtain a 2.5–3.0 cm pen

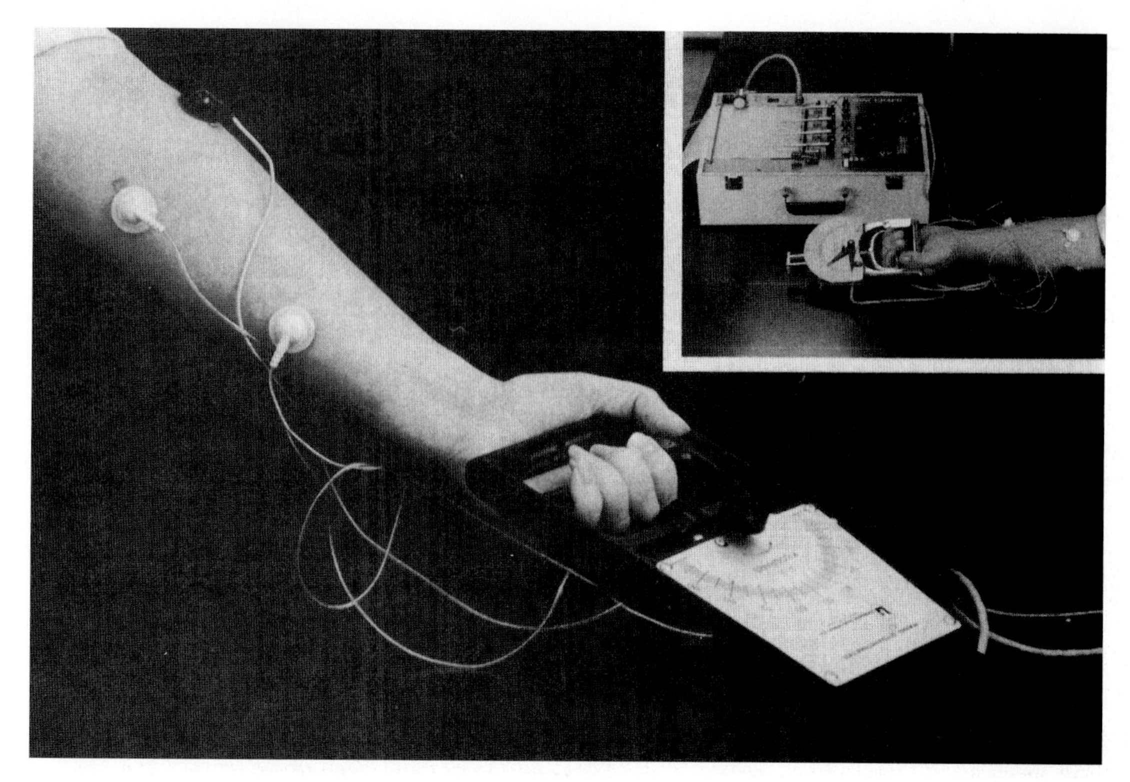

FIGURE 9.2 Electrode placement for electromyography

deflection with each maximum squeeze. Adjust the EMG channel amplifier sensitivity so that a pen excursion of 1.5–2.0 cm on either side of the center line occurs when grip muscles contract maximally. When recording EMG, flip the standby-cal/run switch on the preamplifier (or amplifier) to the "run" position.

7. Squeeze the dynamometer repetitively at a frequency of one squeeze per second, with a grip force of approximately 30 kg, for about 12 seconds. Turn off mainframe power. The recording should resemble figure 9.3.
8. Turn on mainframe power and the paper-speed control. Squeeze the dynamometer grip repetitively at a frequency of one squeeze per second with a grip force of 10 kg for 5 seconds, 20 kg for 5 seconds, and 30 kg for 5 seconds. Turn off mainframe power and speed control. Note that with increased work, a greater number of motor units are activated as evidenced by increased EMG activity.
9. Turn on mainframe power and the paper-speed control. Squeeze the dynamometer grip repetitively at a frequency of one squeeze per second with a grip force of 30 kg until the grip muscles fatigue. Note any changes in the EMG as fatigue develops. Repeat the procedure for the opposite hand and compare the records. Record time to fatigue in the report.

 9.4

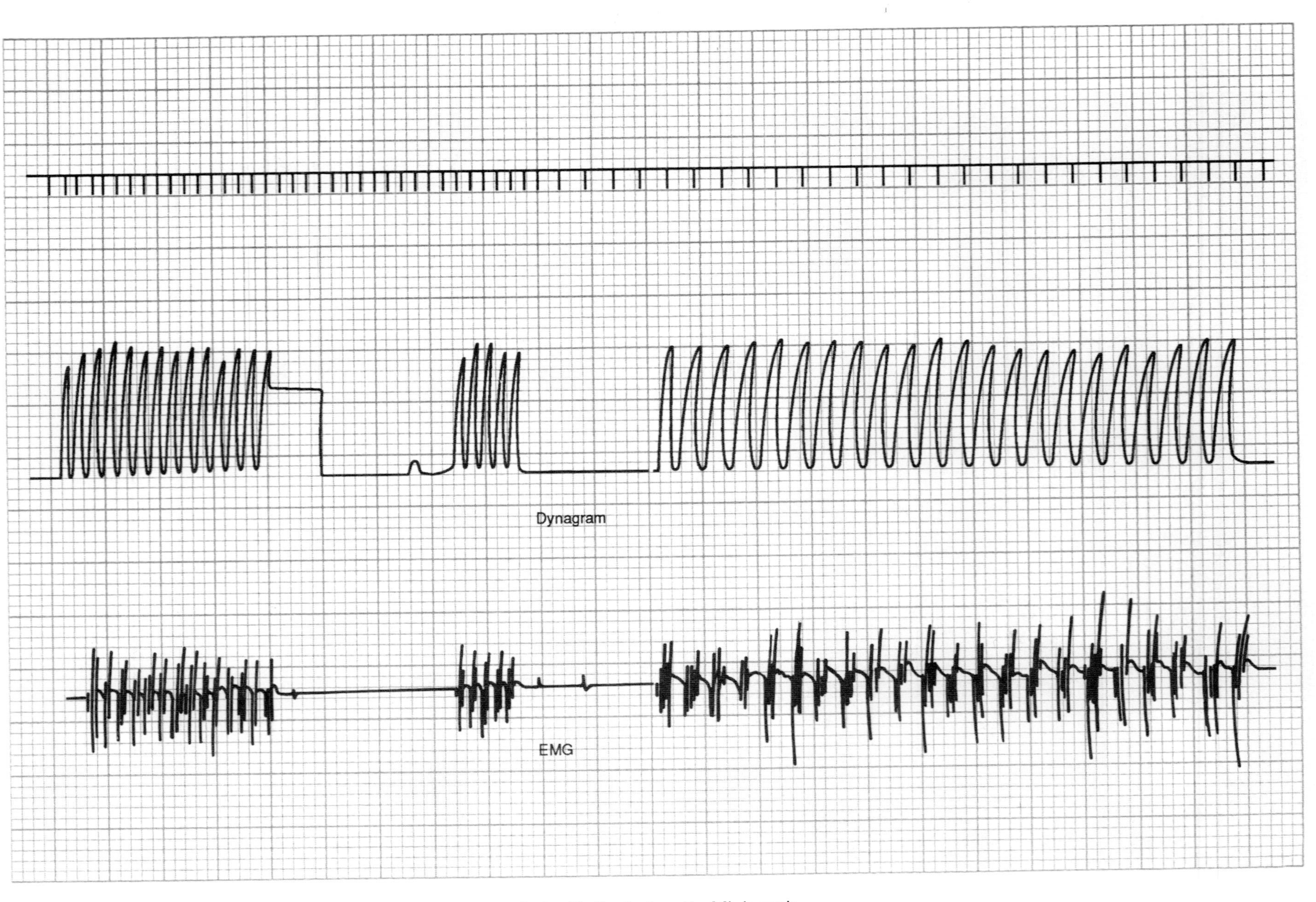

FIGURE 9.3 An electromyogram and dynagram recorded with the Lafayette Minigraph

Electromyography and Dynamometry

REPORT 9

Name: ______________________ Date: ______________________

Lab Section: ______________________

1. Data: Append a copy of the experimental record to this report.
   - **a.** Maximum grip strength:

     Right hand ______________________

     Left hand ______________________
   - **b.** Time to fatigue:

     Right hand ______________________

     Left hand ______________________
2. Define the following:
   - **a.** Electromyography ______________________
   - **b.** Dynamometry ______________________
3. Explain the source of electric signals detected by the EMG electrodes.______________________
4. What does the term *motor unit recruitment* mean?______________________
5. Are you right-handed, left-handed, or ambidextrous? Which hand took the longest time to fatigue? Explain.

Neuromuscular Reflexes of the Spinal Cord and Brain Stem

CHAPTER 10

INTRODUCTION

The nervous system and the endocrine system are primarily responsible for the integration and control of body functions. Of all organ systems within the human body, the central nervous system (the brain and spinal cord) has perhaps the most complex physiology. Functions of the central nervous system range from single activation and reflex control of skeletal muscle, smooth muscle, and glandular secretory cells to extremely complex functions of memory, abstract thought, association, and language.

The **reflex arc** is a basic structural and functional unit of the nervous system. The reflex arc allows the body to react automatically and involuntarily to a variety of internal and external stimuli so as to maintain homeostasis. Nearly every function of the body involves reflexes. Some reflexes are structurally and functionally simple, such as the withdrawal reflex illustrated in figure 10.1. Other reflexes, such as those involved with controlling heart function, respiration, and body fluid and electrolyte balance, are more complex. Nevertheless, all reflex arcs have common features.

Each reflex arc consists of the following components:

Receptor: a specialized structure at the beginning of a sensory neuron that receives the original stimulus

Afferent neuron: the sensory neuron that relays sensory information from the receptor into the brain or spinal cord

CNS (central nervous system) center: a center in the spinal cord or brain where information is relayed across one or more synapses from the afferent neuron to the efferent neuron

Efferent neuron: the motor or secretomotor neuron that transmits information out of the central nervous system to an effector

Effector: smooth, cardiac, skeletal muscle cells or secretory cells (in glands) that respond to the application of a stimulus to the receptor

Most reflex arcs involve several synapses (**multisynaptic**), but a few involve only one synapse (**monosynaptic**), such as that between the afferent and efferent neurons. Some reflexes involve afferent and efferent neurons on the same side of the brain and spinal cord (**ipsilateral reflexes**), whereas others involve afferent neurons on one side of the brain and spinal cord and efferent neurons on the other side (**contralateral reflexes**). The simple spinal reflex activity shown in figure 10.1 is an example of multisynaptic ipsilateral and contralateral reflexes and serves to illustrate the functional aspects of the spinal reflex arc.

The receptor is stimulated (in this case, pain receptors in the finger), and a neural impulse is generated. The afferent neuron conducts the impulse into the posterior horn of the spinal gray matter by way of the posterior root of the spinal nerve. The afferent neuron synapses with **association neurons** (internuncial neurons, interneurons), which relay the information to efferent neurons on the same side of the spinal cord (ipsilateral) and on the opposite side (contralateral). The efferent neurons, in this case, are large alpha motor neurons (motoneurons) located in the anterior horn of the spinal gray matter, which supply skeletal muscles. The efferent neuron transmits impulses out of the spinal cord, by way of the anterior root of the spinal nerve, to skeletal muscles (in this case, flexor muscles of the forearm). Both ipsilateral flexion and contralateral flexion of the forearms occur, thereby removing the stimulus (hammer) as well as the injured finger from the area. The reflex activity illustrated is an example of an ipsilateral-contralateral flexion reflex.

Reflex arcs that use skeletal muscles as effectors are usually not composed exclusively of excitatory neurons. Some association or interneurons are inhibitory. When flexion is reflexively initiated in an extremity, as at the knee, contraction of the opposing extensor muscles is simultaneously inhibited (figure 10.2). Conversely, contraction of extensors also involves reflex inhibition of flexors at the same joint. This pattern of skeletal muscle control at a movable

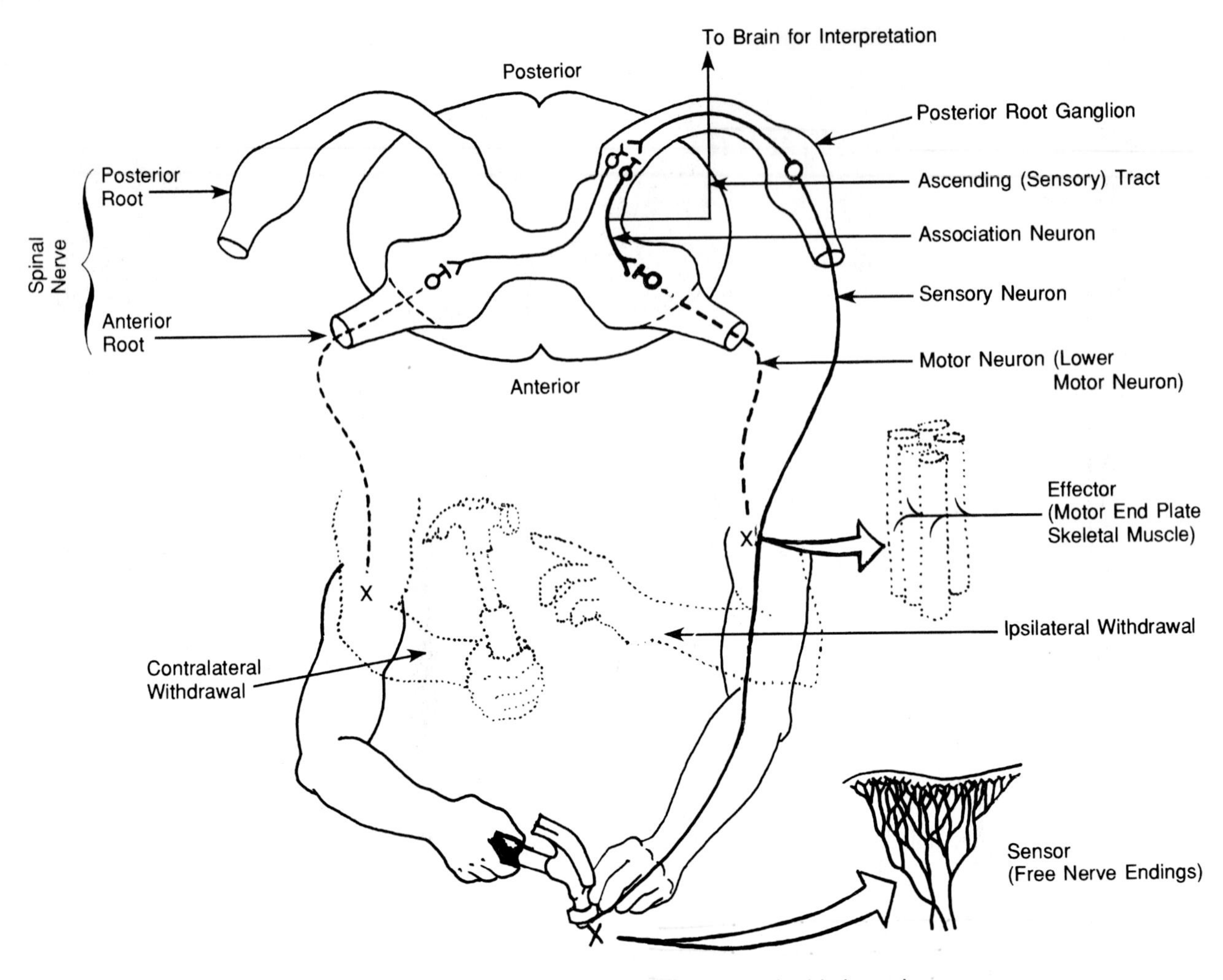

FIGURE 10.1 A simple spinal reflex: ipsilateral and contralateral withdrawal

joint is called **reciprocal inhibition.** When a person steps on a tack with the right foot, the flexors of the right knee reflexively contract as the extensors of the right knee are simultaneously inhibited, causing the injured foot to be withdrawn. However, for the person to remain standing, the contralateral knee must be extended, so the contralateral pattern of reciprocal inhibition is reversed; that is, extensors contract as flexors are inhibited.

It is important to remember that skeletal muscles are not directly inhibited, because neuromuscular junctions are excitatory. Inhibition of skeletal muscle contraction occurs as a result of inhibiting the alpha motor neurons that innervate the muscle fibers and not inhibiting the muscle fibers at the neuromuscular junctions.

Impulses generated in the reflex arc are also relayed up and down the spinal cord and to the brain. This relaying of impulses permits a coordination of reflex activities and informs the brain so that interpretation of the stimulus (in this case, localization of pain) takes place.

Reflexes may be classified by the activity that results from the passage of impulses over the reflex arc. *Postural reflexes* maintain posture and skeletal muscle tonus. *Extensor reflexes* involve the contraction of the extensor muscles in response to the muscle stretch; the *patellar reflex* (knee jerk) is an example. *Flexor reflexes* (e.g., withdrawal reflex) involve the contraction of the flexor muscles, usually in response to cutaneous stimuli. Other reflexes involve visceral rather than somatic components; the *pupillary reflex, accommodation reflex, vasopressor reflexes,* and *carotid sinus reflex* are examples.

■ EXPERIMENTAL OBJECTIVES

1. To observe the occurrence of spinal reflexes in the absence of the brain.
2. To demonstrate the reflex origin of skeletal muscle tonus.
3. To observe the effects of spinal shock on reflex activity.
4. To become familiar with human neuromuscular reflexes and their usefulness in physical diagnosis.

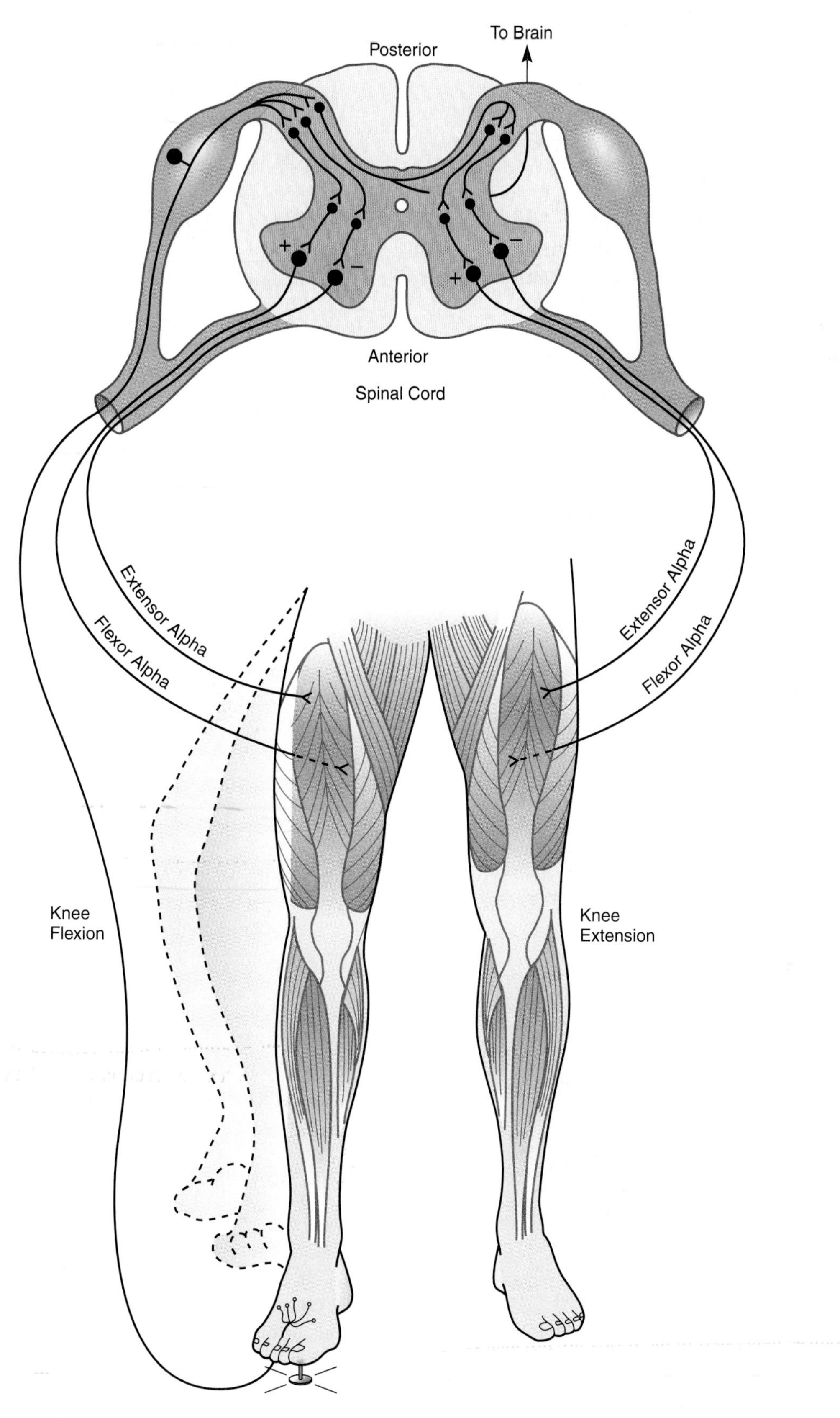

FIGURE 10.2 Coupled ipsilateral flexion–contralateral extension reflexes

■ EXPERIMENTAL METHODS

Spinal Shock

Obtain an intact frog from the laboratory instructor. Pith or decapitate the frog following the correct procedure outlined in appendix C.

Because the spinal cord is left intact, the preparation is referred to as a spinal frog. The spinal frog will display skeletal muscle activity of a reflex nature such as sitting and jumping, but it should not be interpreted that the frog is aware and can still perceive pain. Without the brain, awareness or perception is impossible.

After brain-pithing the frog or removing the top half of the head, note the time, immediately place the animal on its belly, and stimulate various areas of the skin by pinching with forceps. No reflexes should result at first. Continue to stimulate at 15-second intervals, and record the time at which reflex contractions are first observed.

Spinal shock in the frog usually lasts about 3 minutes. Spinal shock in the human may last up to 6 months. Record your observations in the report.

Spinal Cord Reflexes

1. Suspend the frog by the lower jaw from a brass hook attached to a ring stand with a double clamp (figure 10.3). Saturate a 1-cm^2 piece of filter paper with 5.0% acetic acid, and place it on the skin of the upper lateral thigh. Record the time that elapses before the foot is raised in an attempt to remove it (scratch reflex). Record an average of three trials. Immediately after each trial, remove the paper and immerse the suspended frog in a beaker of water several times to wash away the acid. Record the observations.
2. Pour 5 mL of 0.2% sulfuric acid into a 50-mL beaker. Immerse the toes of one foot of the spinal frog in the acid, and record the time that elapses before the foot is withdrawn (withdrawal reflex). Record an average of three trials. Wash the toes in water after each trial. Observe the other leg during the withdrawal reflex. It should extend (contralateral-extension reflex). Record all observations.
3. Place one foot of the spinal frog in sulfuric acid, and at the same time pinch the opposite hind foot. Record the time of the withdrawal from the sulfuric acid. Again, rinse the toes thoroughly with water immediately after observing the reflex. Is the withdrawal reflex time prolonged when the opposite foot is pinched?

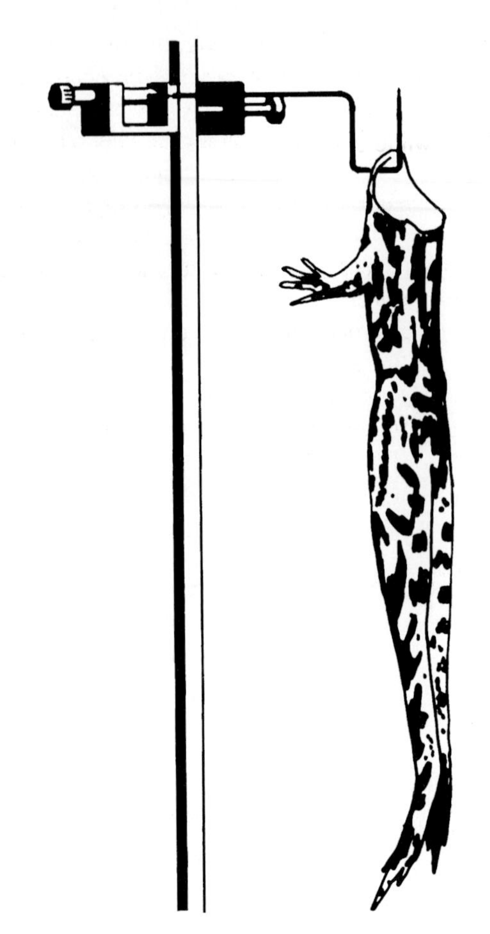

FIGURE **10.3** Spinal reflex frog preparation

Reflex Origin of Skeletal Muscle Tonus

Skeletal muscle in vivo exhibits a phenomenon known as tonus, a slight state of constant tension that serves to maintain the muscle in a state of readiness. Tonus is due to the continual low-frequency firing of alpha motor neurons that supply the muscle fibers. The low-frequency firing of the alpha motor neurons is due in part to sensory impulses generated in the stretch receptors of the muscle and in part to impulses generated in the motor centers of the brain.

If the brain and spinal cord are intact and the sensory nerve from the muscle is cut, tonus will be significantly reduced. If the motor nerve to the muscle is cut, tonus will be abolished and the muscle will become flaccid.

Paralysis of skeletal muscle may occur because of damage to the upper motor neurons (located in the somatic motor cortex) or because of damage to the lower motor neurons (alpha motor neurons). Paralysis due to an upper motor neuron lesion (spastic paralysis) is characterized by tonus in the affected muscle and by the muscle's ability to respond reflexly. Paralysis due to a lower motor neuron lesion (flaccid paralysis) is characterized by the absence of both tonus and reflex activity.

In the following experiment, the reflex origin of skeletal muscle tonus will be demonstrated using the gastrocnemius of the frog and the sciatic nerve, a mixed (motor and sensory) nerve that supplies the muscle.

Preparation of the Animal

Place the brain-pithed frog or frog with its brain removed on a clean glass plate, and remove the skin from one of the lower extremities.

Using blunt dissection, separate the muscles of the thigh and locate the sciatic nerve. The sciatic nerve is found between the dorsal muscles of the thigh and appears as a silvery white cord. Do not permit steel instruments to touch it. Free the nerve from surrounding tissue with a glass probe, being careful not to pinch or pull the nerve. Lift the nerve with the glass probe, and loop a piece of thread around the nerve. Allow the nerve to settle back among the muscles.

Free the gastrocnemius muscle from surrounding tissue, and sever the Achilles tendon. Tie an 18-inch piece of thread securely around the tendon.

Proceed to the appropriate following section for recording skeletal muscle tonus.

Lafayette Minigraph

Materials

frog
glass plate
glass probe
scissors
forceps
scalpel
dropper
thread
dissecting pan
three 50-mL beakers
flat-base stand
double clamps
brass hook
0.2% sulfuric acid
5.0% acetic acid
filter paper
sodium chloride crystals
patellar hammer
straight pins
Minigraph model 76107 or 76107VS
model 76322 time/marker channel + remote marker button
model 76406MG basic amplifier channel
model 76613 semi-isometric force transducer
model 76613-T tension adjuster
model 76802 transducer stand
model 76805 scalepan and weights

Preparation of the Recorder

Prepare the Minigraph for two-channel recording. Check to make sure there is sufficient recording paper and the inking system is working properly. Refer to chapter 2, if necessary, and to figure 10.4.

1. Attach the model 76613 force transducer to the model 76613-T tension adjuster and then secure the apparatus to a ring stand or flat-base stand.
2. Connect the transducer to the balance control box via the standard 1/4-inch phone jack, and connect the balance control box to the amplifier via the 9-pin amphenol connector. Swing the top four leaves of the transducer to the side, exposing the thinnest leaf spring.
3. Turn on the mainframe power, make sure that the polarity switch is + and the sensitivity control is fully counterclockwise, and center the basic amplifier pen using the pen-position control.

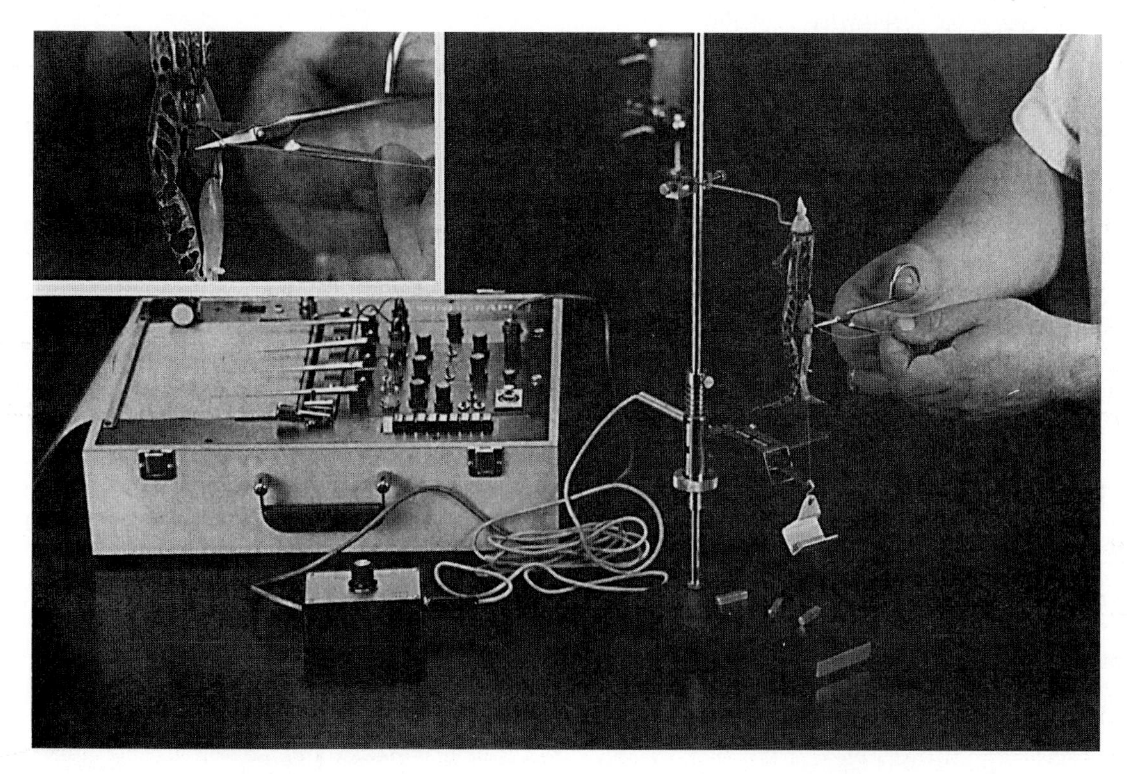

FIGURE **10.4** Lafayette Minigraph setup for demonstrating skeletal muscle tonus

4. Set the basic amplifier gain switch to 100 and the sensitivity to 2. Recenter the pen using the balance control box.
5. Turn on chart drive power and check for proper operation of the transducer-amplifier channel by gently deflecting the leaf spring upward. The amplifier pen should move upward and return to center when the leaf spring returns to its original position. Turn off mainframe power.
6. Suspend the frog, via the brass hook, on the flat-base stand above the transducer (figure 10.4). Attach the other end of the thread from the Achilles tendon to the tip of the thinnest leaf spring of the transducer.
7. Adjust the position of the tension adjuster so that the thread is taut and the leaf spring is barely beginning to bend.
8. Turn on mainframe power. Use the balance control box to recenter the pen.
9. Gently turn the tension adjuster knob until the amplifier pen begins to move upward. Recenter the pen.

Experimental Procedure

1. Turn on the paper-speed control, and record a baseline approximately 10 cm long. Turn off the paper-speed control.
2. Carefully lift the sciatic nerve by its thread away from the thigh, without moving the thigh or recording setup.
3. Sever the sciatic nerve with scissors. Record a new baseline approximately 10 cm long. Note that when the nerve was cut, the muscle was stimulated to contract. Also observe that, after the nerve was cut, the baseline dropped, indicating abolishment of tonus and the establishment of a flaccid state of paralysis.

Clinical Neuromuscular Reflexes in the Human

The functional integrity of reflex arcs is tested to obtain objective evidence regarding the function of muscles, peripheral nerves (motor and sensory), and the central nervous system. Conclusions drawn from testing reflexes are not the sole considerations in diagnosis but rather are considered in conjunction with other symptoms and signs of pathophysiology. The following are several reflexes commonly tested in physical diagnosis. Methods of eliciting some of these reflexes are shown in figure 10.5. Additional reflexes are listed in table 10.1.

1. *Plantar reflex:* Have your laboratory partner assume a supine position on the laboratory table. Remove the sock and shoe from one foot. Press on the heel to firmly brace his or her foot. Using a blunt dissecting needle, stroke lightly the outer border of the sole of the foot from the heel to the origin of the toes, then across the ball of the foot medially. (Never stroke downward from the toes to the heel.) If the reflex is normal, the toes will adduct and plantar flex. If the reflex response is abnormal, the toes will abduct instead of adduct and dorsiflex instead of plantar flex. This abnormal reflex response is called a complete Babinski reflex. An incomplete Babinski reflex is characterized by dorsiflexion of the big toe and plantar flexion of the other four toes. The Babinski reflex is normally obtained in infants up to the age of 6 months and sometimes in children up to age 4 (physiologic Babinski). The Babinski reflex in an adult may occur temporarily during sleep and in epileptics immediately after a seizure, but usually it signifies a lesion in the corticospinal tracts or peripheral nerve damage.
2. *Achilles tendon reflex* (ankle jerk): Instruct your laboratory partner to sit on the edge of the laboratory table with feet relaxed. Strike the Achilles tendon sharply with a patellar hammer or the side of the hand. This strike will cause the tendon to stretch the gastrocnemius muscle, initiating a reflex contraction of the muscle and a resultant plantar flexion of the foot.
3. *Patellar reflex* (knee jerk): As your laboratory partner sits on the edge of the laboratory table with legs relaxed, strike his or her knee sharply with a patellar hammer (or the side of the hand) just below the patella (kneecap). This strike will cause the tendon of the quadriceps femoris muscle to stretch, initiating a reflex contraction of the muscle and resulting in extension of the leg.
4. *Corneal reflex* (blink reflex): Touch the cornea of your laboratory partner's eye by approaching it from the side with a wisp of cotton. Bilateral blinking of the lids should occur. In some lesions of the brain stem and lesions of the trigeminal nerve in which the cornea becomes anesthetic, the reflex is absent.
5. *Ciliospinal reflex:* Have your laboratory partner sit and look straight ahead. Using a dissecting needle or pin, lightly scratch the skin of your partner's neck while observing his or her pupils. Ipsilateral pupillary dilation should occur in response to the scratch. Repeat the test for the reflex, using the opposite side of the neck.
6. *Biceps tendon reflex* (biceps jerk): Ask your laboratory partner to sit at the edge of the laboratory table with an arm relaxed on top of the table. Gently press the biceps tendon in the antecubital fossa (anterior elbow) with your forefinger to stretch the biceps a bit. Maintaining a gentle pressure, strike this finger with the patellar hammer. If the reflex is normal, the forearm will flex. This reflex tests for the functional integrity of the musculocutaneous nerve and the lower cervical segments of the spinal cord.

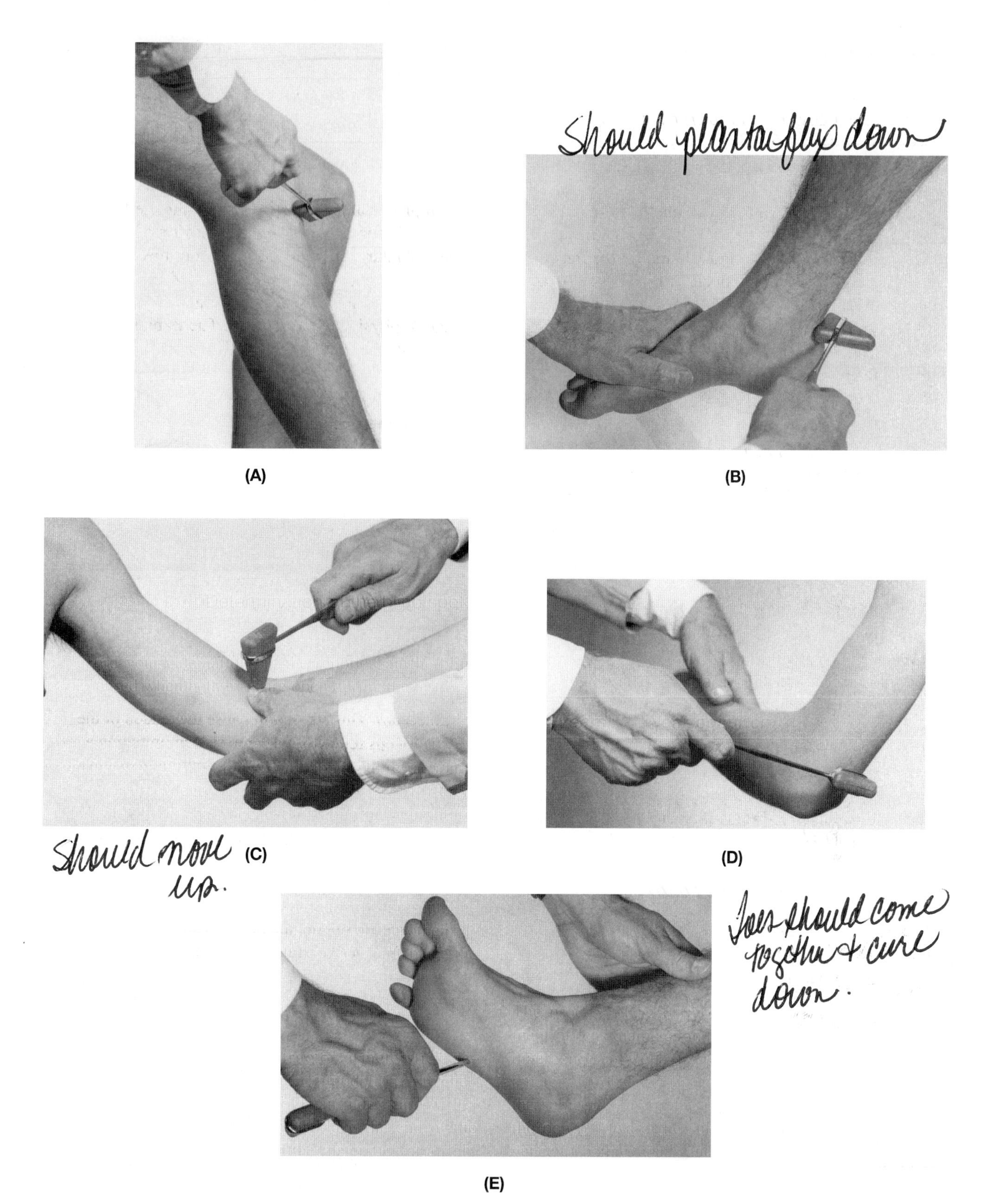

FIGURE **10.5** Some reflexes of clinical importance: (A) knee (patellar) reflex; (B) ankle (achilles) reflex; (C) biceps reflex; (D) triceps reflex; (E) plantar reflex

TABLE 10.1 Spinal cord and brain stem reflexes

Name of reflex	Method of eliciting	Afferent nerve	Center	Efferent nerve	Effect
Corneal	Touching cornea	Cranial V	Pons	Cranial VII	Closure of eyelids
Palatal	Touching soft palate	Cranial IX	Medulla	Cranial X	Elevation of palate
Upper abdominal	Stroking abdomen	T7 to T10	T7 to T10	T7 to T10	Drawing in of abdominal wall
Lower abdominal	Stroking abdomen	T10 to T12	T10 to T12	T10 to T12	Drawing in of abdominal wall
Plantar	Stroking sole of foot	Tibial	S1 to S2	Tibial	Plantar flexion of toes
Jaw jerk	Striking middle of chin	Cranial V	Pons	Cranial V	Closure of jaw
Biceps	Striking biceps tendon	Musculocutaneous	C5 to C6	Musculocutaneous	Contraction of biceps
Triceps	Striking triceps tendon	Radial	C7	Radial	Extension of forearm
Knee jerk	Striking patellar tendon	Femoral	L2 to L4	Femoral	Extension of lower leg
Achilles	Striking Achilles tendon	Tibial	S1 to S2	Tibial	Plantar flexion of foot
Light	Shining light on retina	Cranial II	Midbrain	Cranial III	Constriction of pupil
Ciliospinal	Causing pain	Sensory nerve	T1 to T2	Cervical sympathetics	Dilation of pupil
Oculocardiac	Pressure on eyeball	Cranial V	Medulla	Cranial X	Slowing of heart rate
Carotid sinus	Pressure on carotid sinus area	Cranial IX	Medulla	Cranial X	Slowing of heart rate and lowering of blood pressure

Somatic Sensation

■ INTRODUCTION

For the body to react in a purposeful manner to changes in the external and internal environment, the central nervous system (brain and spinal cord) needs information concerning the nature of the environmental change. Such information is generated by specialized structures at the beginning of sensory nerve fibers. These structures receive stimuli (changes in the environment) and are therefore called receptors.

Receptors are sensory transducers. They convert information about the environmental change (temperature, pressure, stretch, etc.) into nerve signals that are transmitted to the brain for interpretation. Sensory perception is a function of the brain and involves the recognition of stimulus type and intensity as well as localization of stimulus input. In part, sensory perception is dependent upon the type, number, and distribution of sensory receptors in the body. Several types of cutaneous receptor are listed in table 11.1.

Receptors may be classified into two major categories:

1. **Exteroceptors:** These receptors are sensitive to changes in the body's external environment. They include cutaneous receptors sensing pain, touch, temperature, pressure and hair movement (figure 11.1), taste receptors, olfactory receptors, visual receptors, and auditory receptors.
2. **Interoceptors:** These receptors are sensitive to changes in the body's internal environment and are divided into two categories: visceroreceptors and proprioceptors.

Visceroreceptors are found within viscera (body organs) and provide information regarding visceral pain and pressure, blood pressure (baroreceptors), and osmotic pressure (osmoreceptors). **Proprioceptors** are found in skeletal muscles, tendons, ligaments, joint capsules, and the inner ear (labyrinthinoreceptors). They provide information regarding position and movement of the body.

Although the body's various receptors provide different kinds of information about the internal and external environment, many common physiologic properties are shared by receptors. In the experiments that follow, some of these properties will be explored.

TABLE **11.1** Types of cutaneous receptor

Receptor	***Type***	***Modality***
Hair follicle plexus	Mechanoreceptor	Hair movement
Pacinian corpuscle	Mechanoreceptor	Deep pressure, vibration, acceleraion of stimulus
Merkel's disks	Mechanoreceptors	Light touch (location, intensity)
Meissner's corpuscle	Mechanoreceptor	Light touch (velocity of stimulus)
Krause's bulb	Mechanoreceptor	Touch (location, intensity)
Ruffini's organ	Mechanoreceptor	Pressure (location, intensity)
Free nerve endings		
Unmyelinated fibers	Nociceptor	Pain
Unmyelinated fibers	Warm	Increasing temperature
Myelinated fibers	Cold	Decreasing temperature

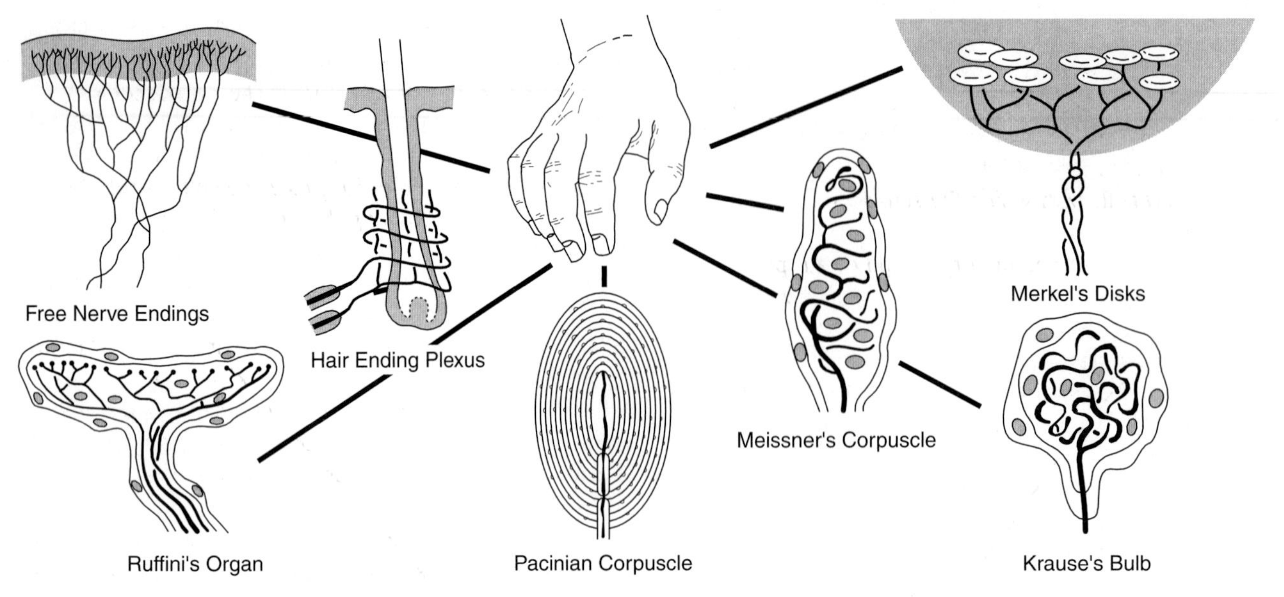

FIGURE 11.1 Cutaneous receptors

■ EXPERIMENTAL OBJECTIVES

1. To gain an understanding of the role that general somatic and special sensory receptors play in the modification of body activity.
2. To recognize common physiologic properties of all receptors, such as the laws of adequate stimulus and specific nerve energies, minimal stimulus strength, adaptation, and afterimage.
3. To explore the distribution of cutaneous receptors and experience sensory phenomena such as two-point discrimination, projection, localization, and shifting of "physiologic zero."

Materials

four 50-mL beakers
disposable paper cups
sucrose solutions:
- 1:1000
- 1:800
- 1:500
- 1:300
- 1:200
- 1:100

salt (NaCl) solutions:
- 1:1000
- 1:800
- 1:500
- 1:300
- 1:200
- 1:100

droppers
dress pins
red, black, green, and blue felt marking pens
500-mL beaker of ice
three 500-mL beakers of water at 20° C, 30° C, and 40° C
alcohol preps
rubber band
brass rod (1/16-inch diameter)
porcelain plate
dividers
10-g weight
water bath
cotton
tuning forks
metric rule
surgical gloves

■ EXPERIMENTAL METHODS

Law of Specific Nerve Energies and Adequate Stimulus

Specific types of receptors are more sensitive than others to certain types of stimuli (e.g., the eye to light, taste buds to chemical stimuli). The type of stimulus to which the receptor is most sensitive is called the *adequate stimulus*. Nearly all receptors will respond to a stimulus other than the adequate stimulus if the stimulus strength is great enough. Cones of the retina may respond to mechanical stimuli even though their adequate stimulus is light. Regardless of the type and strength of a stimulus applied to a given receptor, the sensation perceived when the receptor responds is always the same. This phenomenon is known as the *law of specific nerve energies.*

So that you may experience the law of specific nerve energies, the laboratory room will be darkened at the beginning of the laboratory period. In the darkened room, with your eyes closed, turn your eyes as far as possible to the left. Using your right index finger, gently press the outer part of the right eyeball. A dark circle surrounded by a bright white ring near the ridge of the nose will soon be "seen." This visual sensation of light is caused by stimulation of the rods and cones of the retina—receptors for vision—by mechanical pressure against the eyeball.

For a receptor to be aroused, the stimulus strength must exceed a certain minimal value. The minimally required strength of a stimulus is the lowest for the adequate stimulus and higher for other types of stimuli. The following procedure will be used to determine and compare the minimally required stimulus strengths for tasting a sugar solution and a salt solution.

To reach threshold may have to ↑ solutions

Place 1 mL of the following dilutions of sugar water into the depression of the porcelain plate: 1:1000, 1:800, 1:500, 1:300, 1:200, 1:100. Thoroughly rinse your mouth with distilled water and taste a small amount of each dilution by placing one drop on the midline of the tip of the tongue. Start with the weakest dilution. (Rinse your mouth thoroughly with distilled water after each sample.) Record the dilution strength with which the sensation of sweetness is first perceived.

Carefully wash, rinse, and dry the porcelain plate, and repeat the taste test using the dilutions of salt instead of sugar. Rinse your mouth thoroughly with distilled water after each sample, and taste by placing one drop at the tip of the tongue to the right or left of the midline near the lateral border. Record the dilution strength with which the sensation of saltiness is first perceived.

Compare the lowest sugar and salt concentrations you are able to perceive. Are they the same? How do smokers compare with nonsmokers regarding taste sensitivity?

Adaptation and Afterimage

When a stimulus of minimally adequate strength or greater is applied to a receptor, the receptor discharges, resulting in a train of impulses conducted along the afferent neuron to the central nervous system. If the stimulus strength remains constant and the stimulus is not removed from the receptor, the receptor will become increasingly less sensitive to the presence of the stimulus. This phenomenon is known as *adaptation.* Some receptor types adapt rapidly to the presence of stimuli (e.g., Meissner's corpuscle to light touch), whereas others adapt very slowly (e.g., taste receptors) or not at all (e.g., some stretch receptors in skeletal muscle, cutaneous pain receptors).

Place a 10-g weight on your forearm, and let it remain for a few minutes. The temperature of the weight must be near skin temperature (about 35° C), and the forearm must be kept motionless. How long does the sensation of pressure persist? Record your observations.

Using the same area of the forearm, repeat the adaptation experiment using an ice cube for determining adaptation time for "cold receptors," a 10-g weight warmed to 50° C for determining adaptation time for "warm receptors," and the point of a divider (gently) for determining adaptation time for pain receptors. Record your observations.

Most cutaneous receptors (excepting pain receptors) adapt rapidly. If a stimulus to which the receptor has adapted is removed from the receptor, the receptor will again discharge, giving rise to the same sensation perceived when the stimulus was first applied, even though the original stimulus is no longer present. The sensation perceived after removal of the stimulus is known as the *afterimage.*

Place a rubber band around your head, allow it to remain for a few minutes, and then remove it. Compare the sensation perceived after removal of the rubber band with that perceived on application. Record your observations.

Receptor Distribution

Receptors found in the skin (integumentary, or cutaneous, receptors) are not distributed uniformly. The type, location, number, and density of receptors in the skin depend on the probability of the receptor's coming into contact with an adequate stimulus. Thus, touch receptors such as Meissner's corpuscle are found in greater density in the skin of the extremities (particularly the digits), the lips, and tip of the tongue.

Select a spot on the inner, relatively hair-free aspect of your laboratory partner's forearm, and draw a 3 cm × 3 cm square using a water-soluble black ink marker pen. Using the points of a divider, touch the skin in various spots within the square while the subject (whose eyes are closed) indicates when contact is felt. Use the same pressure each time. Contact should be felt as pressure and not pain. Mark each spot with green ink.

Explore the 3 cm × 3 cm area in a similar manner for the detection of pain. Use a dress pin to elicit pain sensation without puncturing the skin. Before using the pin, sterilize the tip with 70% alcohol or an alcohol prep. Mark each spot from which pain is elicited with black ink.

Cool a small brass rod (1/16-inch in diameter) by placing the tip in a small beaker of ice water for 5 minutes. Then explore the designated skin area for "cold" receptors by gently touching the cool tip to the skin. The subject may also perceive pressure, but the spot should not be marked unless a cold sensation is perceived. Mark the "cold" receptors with blue ink.

Warm the brass rod by placing the tip in a beaker of hot tap water for 5 minutes. Repeat the experiment above and mark the location of "heat" receptors with red ink.

Count the number of each type of receptor found, and record observations in the report.

Discrimination

Somatic sensory information (pain, temperature, touch, and pressure) originating at the receptor is relayed via the afferent nerve to the spinal cord or brain stem, and from there to the thalamus and then the somatic sensory cortex of the parietal cerebrum, where sensing or perception of the stimulus occurs (figures 11.2 and 11.4). The stimulus applied to the receptor results not only in an evoked sensation (e.g., pain, pressure) but also in localization of the applied stimulus (e.g., hand, head). All areas of the body surface are represented in the primary somatic sensory cortex (figure 11.2). Localization of a stimulus depends on the magnitude of cortical representation of the area on the body surface (greatest for hands and fingers, least for trunk) and overlapping sensory fields in the skin. Several of one type of receptor (e.g., a Pacinian corpuscle) may be connected to one afferent nerve fiber; therefore, adequate stimulation of any of the receptors would result in localization of the stimulus to the sensory field covered by all of the interconnected receptors, not just the sensory field of the stimulated receptor.

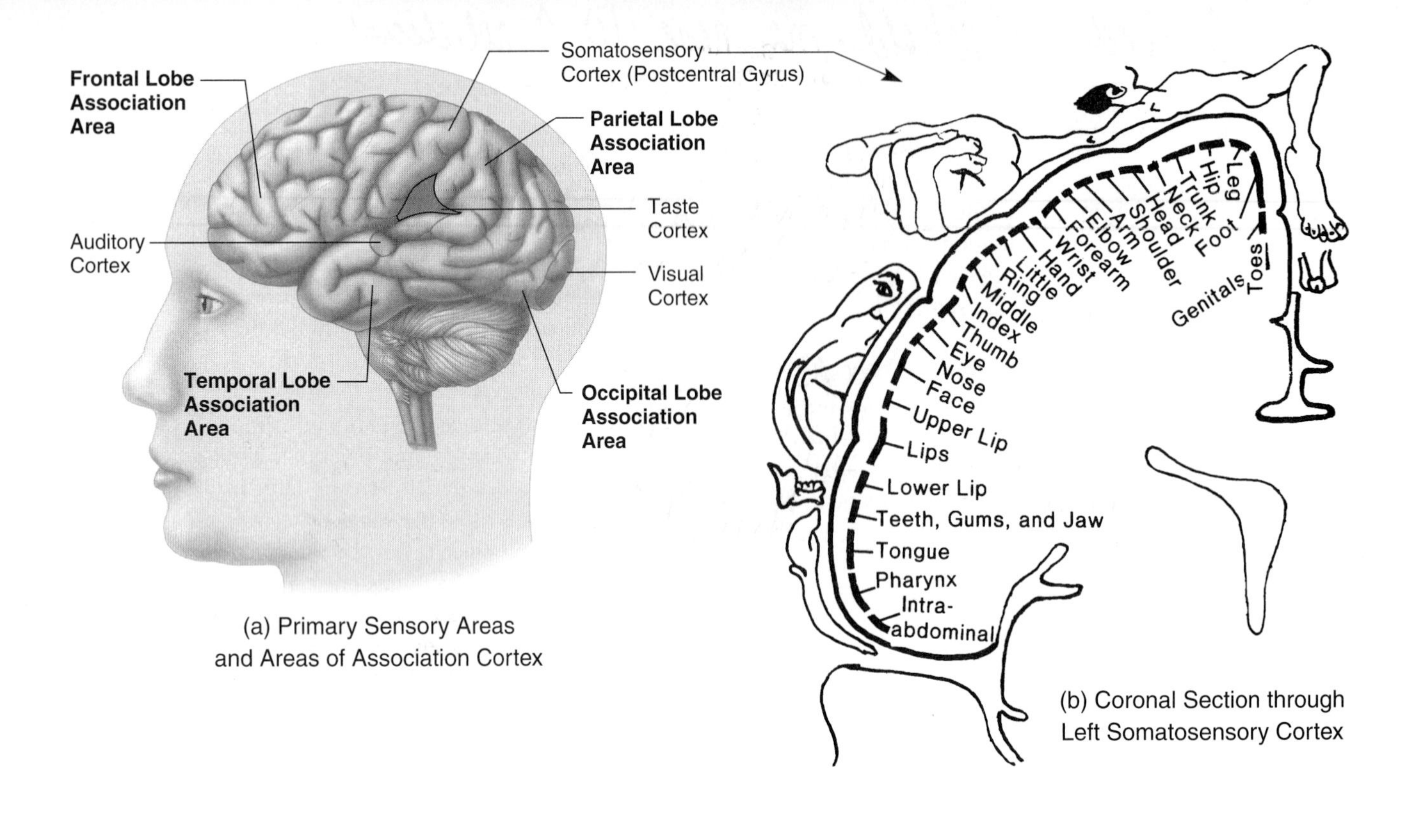

FIGURE 11.2 The somatic sensory cortex of the postcentral gyrus of the parietal lobe

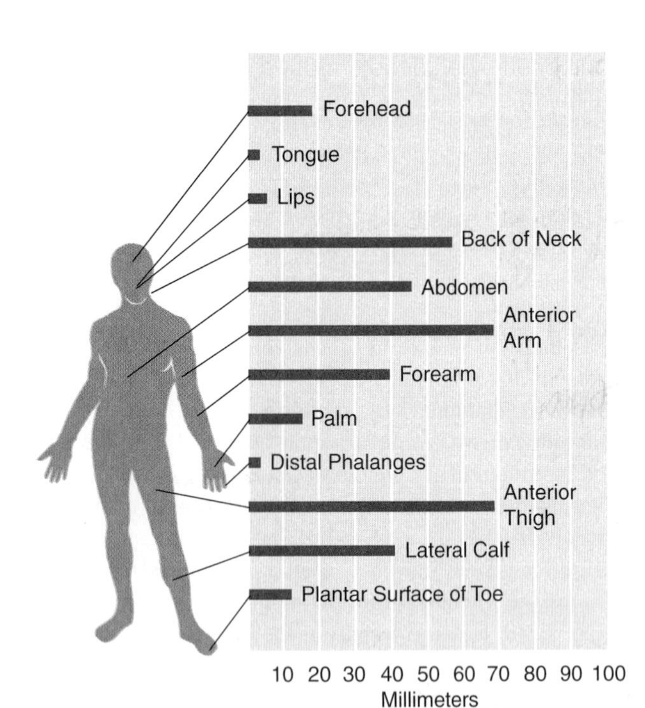

FIGURE 11.3 Two-point discrimination on the body surface: the minimum distance between two points of simultaneously applied light pressure stimuli at which the stimuli are perceived as two, not one

Using the points of a divider, touch various parts of the subject's skin, such as that on the hand, fingertip, arm, face, and neck, and determine the distance apart that the points must be before they are felt as two. Repeat each determination three times, and record the average distance in millimeters for each part stimulated. Where on the body surface is two-point discrimination the greatest (i.e., distance between two points smallest)? Compare your data with the data shown in figure 11.3.

While the subject's eyes are closed, touch the skin on the middle of one arm with a point of the divider. Then have the subject (with eyes still closed) try to place the point of a pencil on the place stimulated. Measure in millimeters by how far the spot was missed. Determine this distance for the fingertip, back of hand, forearm, and cheek. Record all observations.

Shifting of Physiologic Zero

Thermal receptors (certain free nerve endings) respond to changes in temperature rather than a fixed value of temperature. The "cold" receptors increase their rate of firing as their temperatures are lowered from a stable temperature referred to as "physiologic zero." The "warm" receptors increase their rate of firing as their temperatures are raised. Both types of thermal receptors adapt moderately to new levels of temperature.

Prepare three beakers with water at 20° C in one, 30° C in the second, and 40° C in the third. Place the right hand in the 40° C beaker and the left hand in the 20° C beaker for

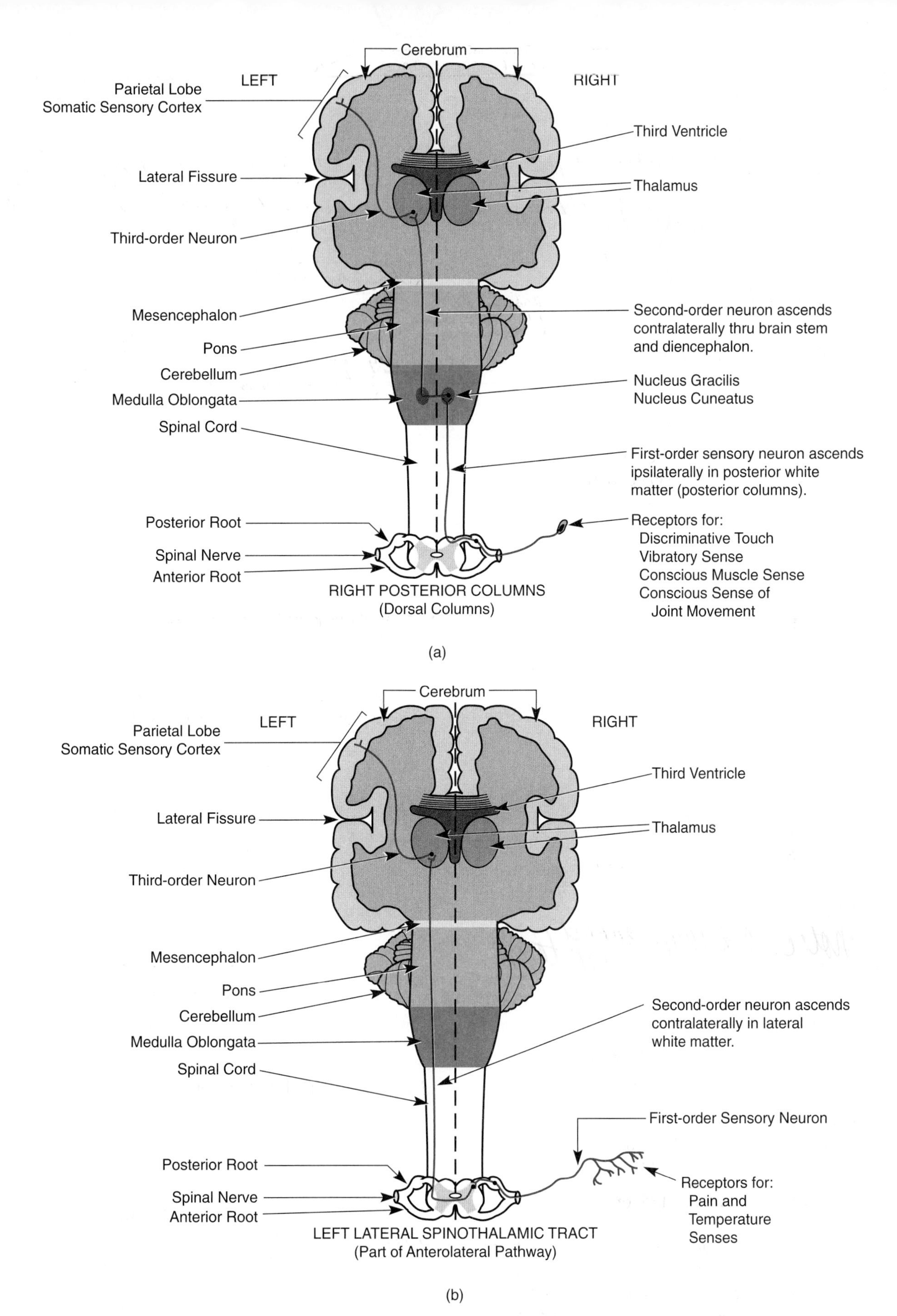

FIGURE 11.4 Posterior view of three ascending (sensory) tracts of the spinal cord and brain: (a) right posterior columns; (b) left lateral spinothalamic tract; (c) left anterior spinothalamic tract

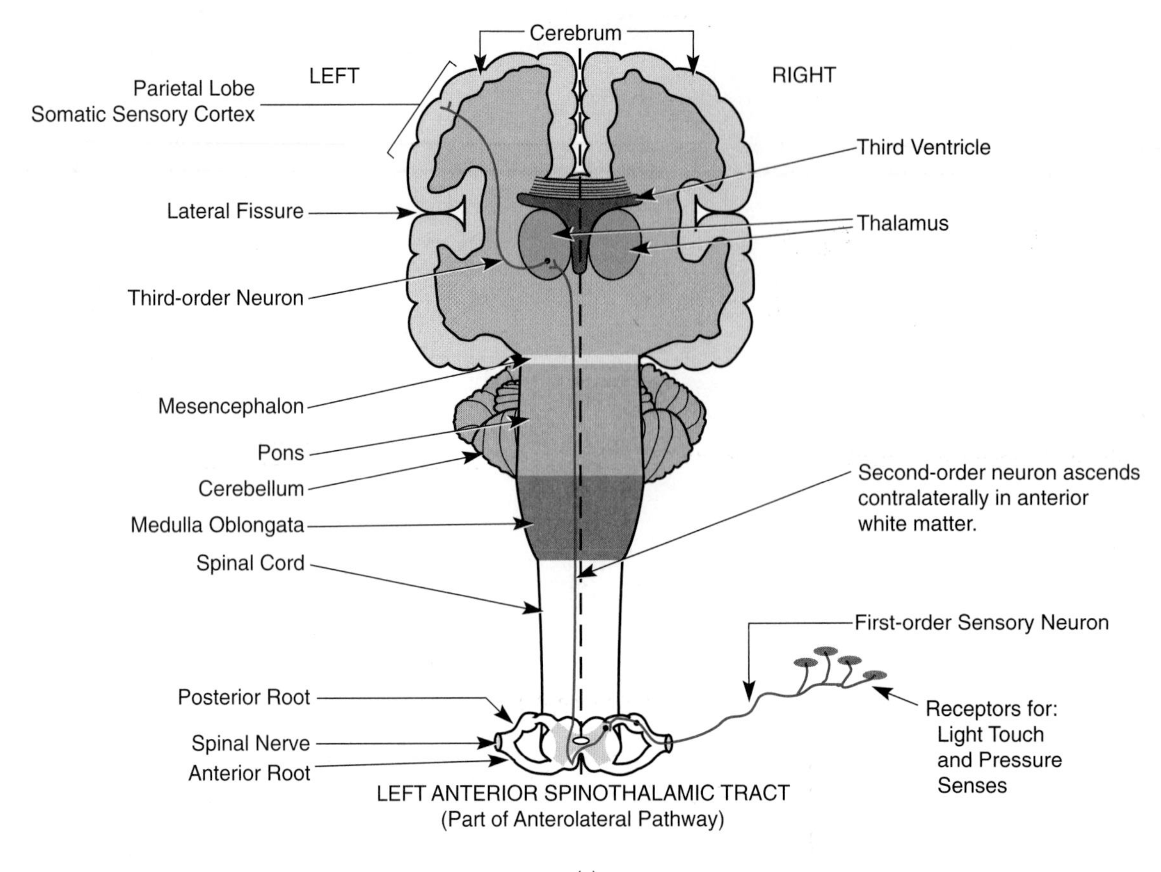

FIGURE 11.4 Continued

2 minutes. 40°C and 20°C become "physiologic zero" for the right hand and the left hand, respectively. Then place both hands into the water at 30°C. What sensations are felt in each hand? Record all observations. If both hands remain in 30°C water, sensations of warm and cold disappear and 30°C becomes the new "physiologic zero."

notice Δ in temp but not exact temp.

Sensory Interpretation

Cutaneous receptors are rarely stimulated singly. Usually the nature of stimuli is such that several different types of receptor are stimulated simultaneously. Cerebral association and interpretation of stimuli produce a broad spectrum of complex sensations from combinations of simple basic sensations.

Tap the end of a tuning fork firmly against the heel of your shoe to set the instrument vibrating, and place its handle to your elbow. Describe the sensation of vibration.

Place your hand in a thin, watertight surgical glove. Dip the hand into a beaker of cold tap water. Although the hand remains dry, the sensation of wetness is felt because its two components (temperature and pressure) are sensed.

CHAPTER 12

Cranial Nerves: Assessment of Functions

INTRODUCTION

The **central nervous system (CNS)** consists of the brain and the spinal cord. The CNS receives sensory information from other parts of the body or the body's external environment and transmits motor information to other parts of the body by way of the **peripheral nervous system (PNS).** The PNS of the human includes 31 pairs of spinal nerves and 12 pairs of cranial nerves. Some nerves contain only motor nerve fibers (efferent fibers); some nerves contain only sensory nerve fibers (afferent fibers); and some nerves contain both sensory and motor nerve fibers (mixed). All spinal nerves are mixed. Cranial nerves I (olfactory), II (optic), and VIII (vestibulocochlear) are entirely sensory. Cranial nerves III (oculomotor), IV (trochlear), VI (abducens), XI (accessory), and XII (hypoglossal) are classified as motor, although they do contain proprioceptive afferent fibers. Cranial nerves V (trigeminal), VII (facial), IX (glossopharyngeal), and X (vagus) are mixed. All cranial nerves except the olfactory nerves are connected to the brain stem (medulla, pons, mesencephalon), and all are distributed in the head and neck except the vagi, which also supply structures in the thorax and abdomen. Figure 12.1 shows cranial nerves and their origins and terminations.

Cranial nerve function is commonly assessed as part of a general physical examination of the head, eyes, ears, nose, throat, and neck by a physician, physician's assistant, nurse, or paramedic. More comprehensive examination of cranial nerve functions is usually done by specialists such as neurologists, ophthamologists, optometrists, and audiologists. Basic cranial nerve tests are included in this book because they provide a convenient framework for reinforcing the learning of cranial nerve numbers, names, and functions. More important, they hone skills of observation and reason.

EXPERIMENTAL OBJECTIVES

1. To learn the name and number of each pair of cranial nerves and to identify the region of the brain to which each pair is attached.
2. To learn and perform simple tests of function for each pair of cranial nerves.

Materials

preserved brain of human (whole) or sheep
model of human brain
anatomical charts: CNS
glass probe
needle probe or straight pin
dissecting pan
tongue depressors
long-stem cotton applicators
cotton balls
penlight
256-Hz tuning fork
droppers
turpentine, oil of cloves, peppermint, wintergreen, and vanilla
10% glucose solution
10% NaCl solution

EXPERIMENTAL METHODS

With the aid of figure 12.1 and anatomical charts available in the laboratory, identify and verify the location of each of the 12 pairs of cranial nerves on a model of the human

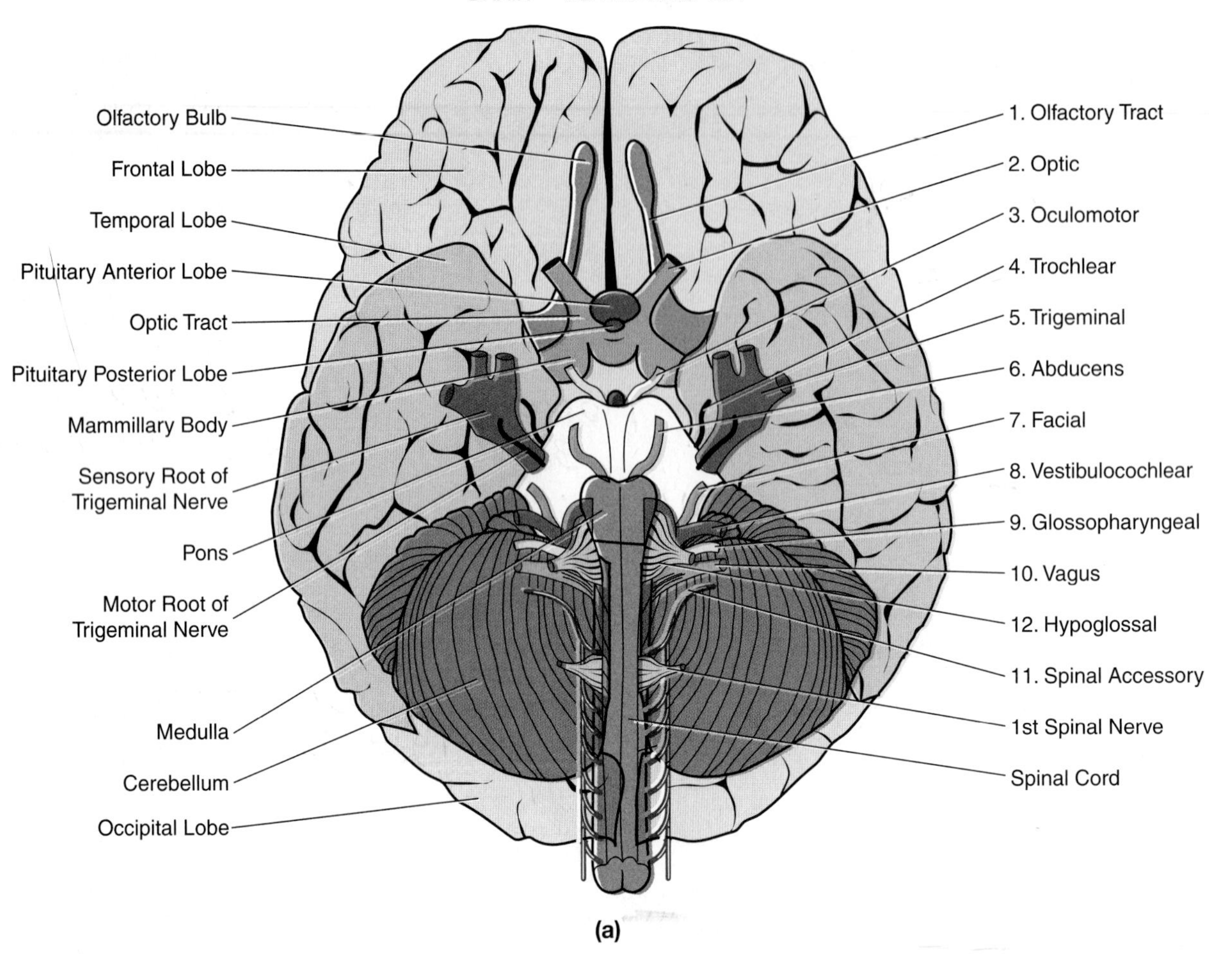

FIGURE 12.1 (a) Inferior aspect of the brain; (b) cranial nerves

brain or a preserved brain or both. The preserved brain should be placed with the inferior surface in view in a dissecting pan for study. Use only the glass probe to study structures on the preserved brains. Do not cut, pick, probe, or tear structures with sharp or pointed instruments. The tissues have been chemically preserved and should be periodically remoistened to prevent desiccation during the period of study. After reviewing the names and locations of the cranial nerves, perform the following assessments of cranial nerve functions.

Cranial Nerve I. Olfactory Nerve (Sensory)

Olfactory nerve fibers pass from olfactory cells in the upper part of the nasal cavities through the cribriform plate of the ethmoid bone and into the olfactory bulb, where they synapse with second-order neurons. Neurons of the olfactory bulb transmit olfactory information along the olfactory tract to the olfactory cortex of the cerebrum. Damage to the olfactory epithelium, olfactory nerves, olfactory bulbs and tracts, or the olfactory cortex produces a loss of the ability to smell *(anosmia)*.

Ask your lab partner to identify, with his or her eyes closed, each of the following common nonirritating odors: turpentine, oil of cloves, peppermint, wintergreen, and vanilla extract. Test each nostril separately. The odor should be familiar to the subject so as to differentiate between an inability to smell a substance and an inability to identify it.

The most common cause of anosmia is damage to the nasal olfactory epithelium as from excessive smoking, cocaine use, or inflammation due to infection. Neurologic causes include tumors of the frontal lobe near the olfactory bulbs and tracts and head injuries. As do other special senses (e.g., vision, hearing), the sense of smell diminishes with increasing age.

Cranial Nerve II. Optic Nerve (Sensory)

The optic nerve arises from cells in the retina of the eye and conveys visual information to the brain. Fibers from the nasal half of each retina cross at the optic chiasma and are distributed to the contralateral brain stem, thalamus, and occipital cerebrum along with fibers from the ipsilateral temporal retina (see figure 12.2). Both eyes are represented

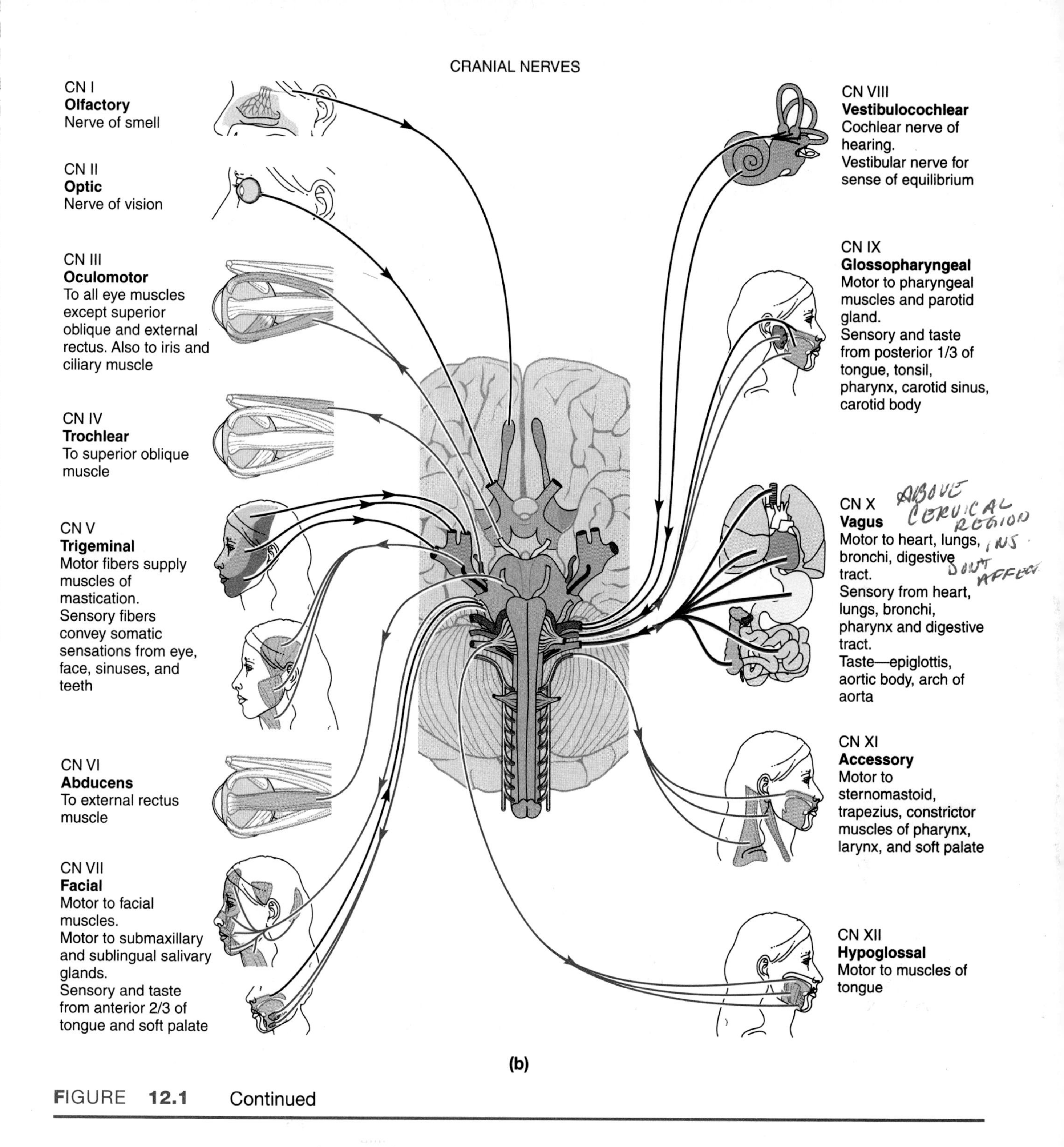

FIGURE 12.1 Continued

in the occipital cortex of each cerebral hemisphere. In addition to providing sensory information regarding vision, the optic nerve provides the sensory component of visual reflexes such as the pupillary light reflex or accommodation of the lens.

Examination of the optic nerve involves determination of visual acuity, peripheral vision, and appearance of the optic fundus.

Ask the subject if there has been any change in his or her vision. Visual acuity or sharpness of vision when looking at near as well as distant objects can best be tested with a standardized chart such as a Snellen test letter chart (see figure 13.5). If a chart is not available, you may use any printed material (e.g., newspaper, magazine, etc.) and compare the subject's visual acuity with your own. If the subject normally wears eyeglasses or contact lenses, these

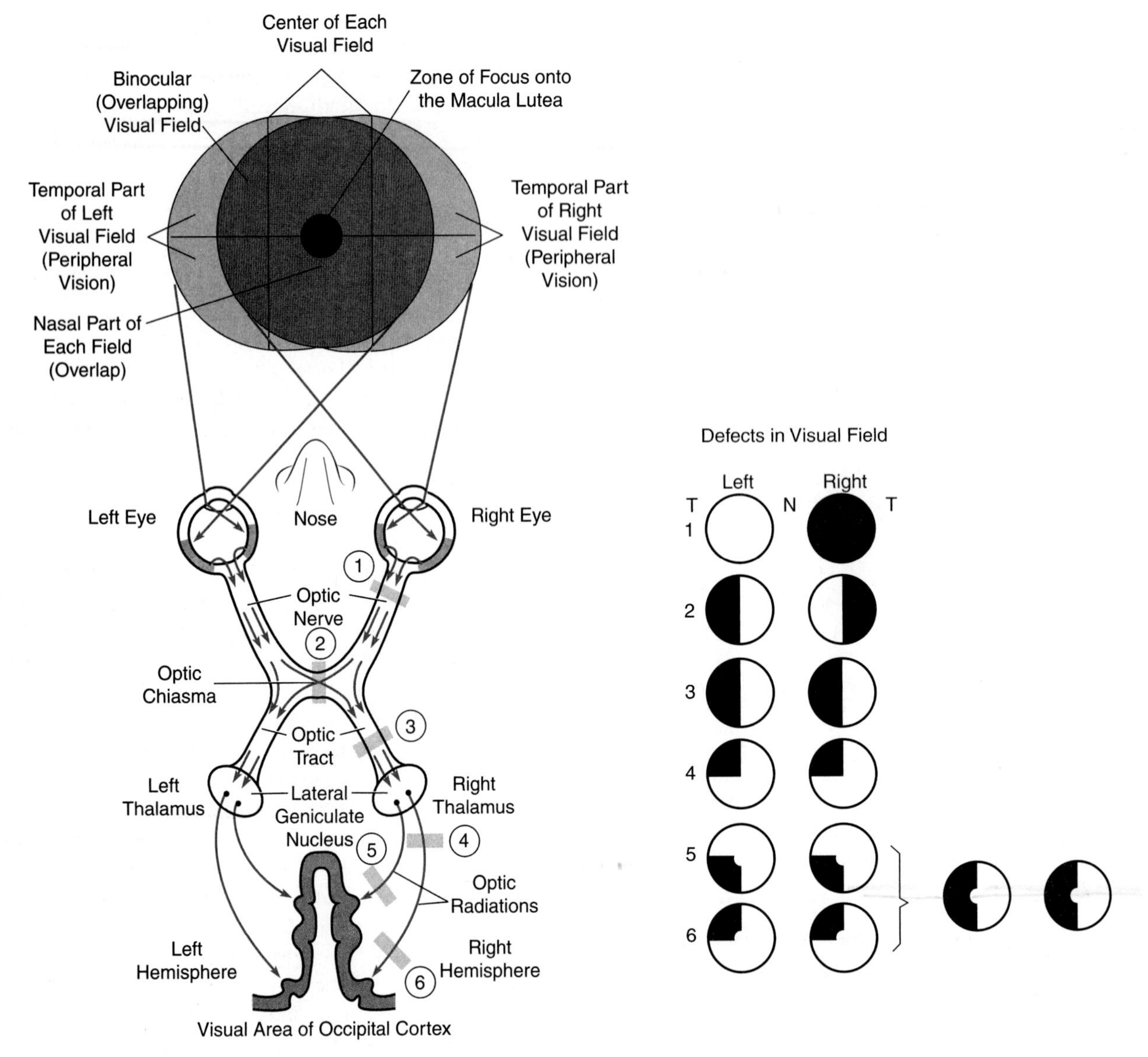

FIGURE **12.2** Visual pathways, lesions, and defects in the visual field

should be worn during the test. If a subject's visual acuity is diminished to the point where even large print cannot be read, ask the subject to count extended fingers at various distances from the eye. Test the subject's ability to detect motion and to distinguish light from dark.

Assess the subject's peripheral vision by asking him or her to sit in front of you about 2 feet away, cover one eye, and with the other eye look straight into your eye. Cover your eye opposite the subject's covered eye. In a plane perpendicular to the gaze and midway between your eye and the subject's, slowly bring a small test object such as a pencil or a cotton applicator into view. As the test object is moved toward the line of gaze, the subject will let you know when he or she first sees it. Compare your visual field (assuming it is normal) with the subject's visual field. All four quadrants should be tested at a 45-degree angle from the vertical and horizontal (figure 12.3). Test each eye separately.

In chapter 13, you will assess peripheral vision in a more exact manner using a perimenter. Normal visual fields for right and left eyes are shown in figure 13.10. In chapter 13, you will also use the ophthalmoscope to assess the appearance of the optic fundi.

Loss of vision in all or part of the visual field is caused by specific lesions in the visual pathway (figure 12.2). Blindness in one half of each visual field is called *hemianopia.* Lesions of the optic tract or optic radiation produce the same hemianopia (homonymous hemianopia) for both eyes. A tumor of the pituitary gland may exert pressure on the optic chiasma (see figure 12.2) causing bilateral loss of the temporal fields of vision.

Cranial Nerve III. Oculomotor Nerve (Motor)

Cranial Nerve IV. Trochlear Nerve (Motor)

Cranial Nerve VI. Abducens Nerve (Motor)

Six extrinsic muscles move the eyeball within the orbit (figure 12.4). The inferior, superior, and medial recti muscles and the inferior oblique muscle are controlled by the oculomotor nerve. The lateral rectus is controlled by the abducens nerve, and the superior oblique muscle is controlled by the trochlear nerve.

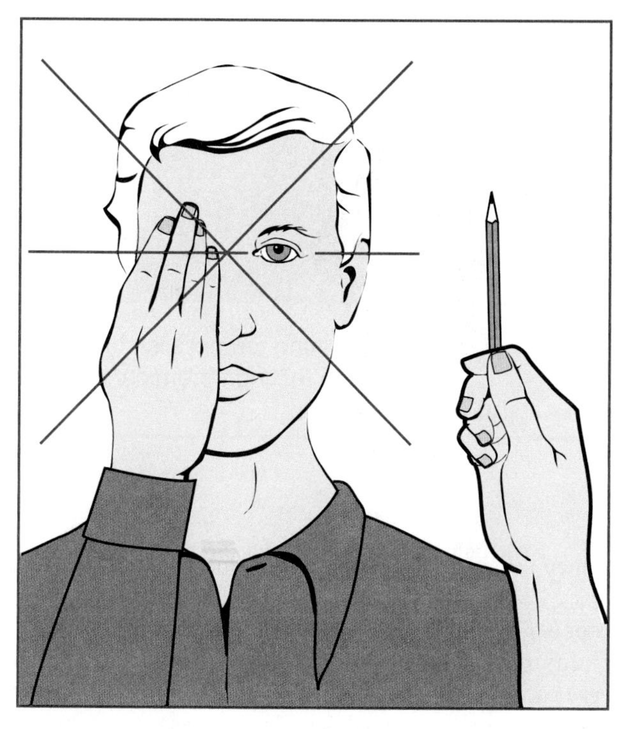

FIGURE **12.3** Quadrants for visual field testing

The oculomotor nerve also controls the elevator muscle of the upper eyelid and the involuntary internal muscles of the eye that control pupil diameter and lens thickness.

Because the oculomotor, trochlear, and abducens nerves control related functions of the eye, they are tested as a unit.

First, observe how much of the subject's iris (colored part around pupil) is covered by the eyelid. Normally about one-third will be covered. Drooping of the eyelid *(ptosis)* occurs in paralysis of the oculomotor nerve, myasthenia gravis, and other diseases.

Second, examine the pupils. Are they both the same size? Are they round or oval? Are they in the center of the eye facing forward, or are they deviated to the side? Normally the pupils should be round, equal in size (about 2–3 mm in diameter), and in the center of the eye. Unequal pupils *(anisocoria)* may be congenital and have no pathologic significance or may occur as a result of a variety of abnormalities, including syphilis, multiple sclerosis, and sympathetic paralysis. If both pupils are markedly smaller or larger than normal, medication may be the cause. A unilateral dilated pupil often occurs with increased intracranial pressure. The pupil becomes fixed and unresponsive to light on the side of the brain where the pressure has increased.

Next, ask the subject to focus on a distant object. Hold a penlight about 20 cm to the side of the eye and shine it directly into the pupil. Check for pupillary constriction in this eye (direct reflex) and simultaneous constriction in the untested eye (consensual reflex). Grade the pupillary response to light as brisk (4+), less than brisk (3+), slow (2+), very slow (1+), or absent (0).

Move a test object (fingertip, pencil, cotton-tipped applicator) from a distance of 1 m in front of the subject to within 3 cm of his or her nose. As the subject gazes at the moving object, note convergence of the eyes (the eyes become crossed) and the pupillary constriction that normally

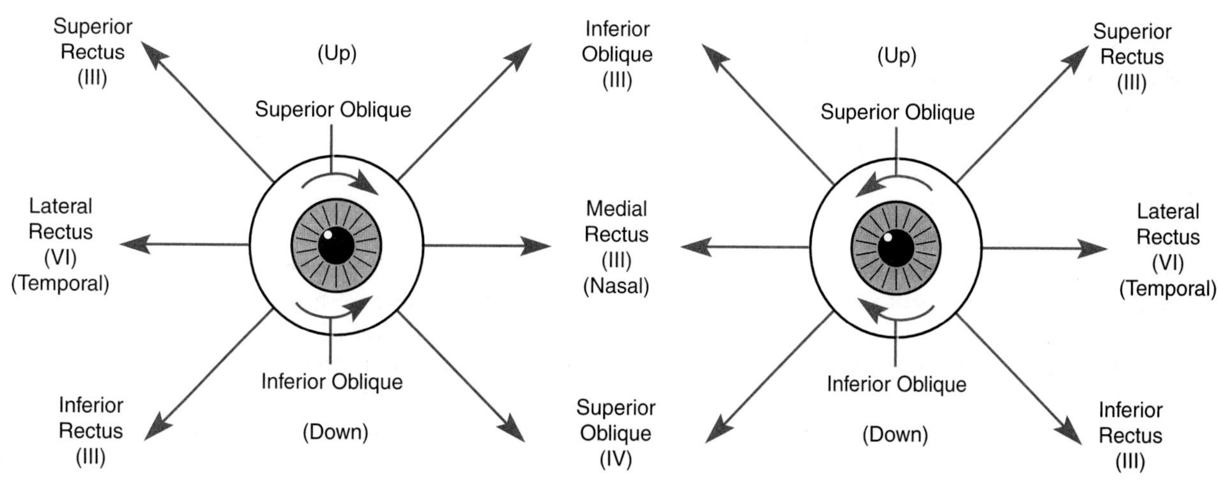

FIGURE **12.4** Extrinsic eye muscles, cranial nerve innervation, and eyeball movement

accompanies the convergence. Convergence should occur smoothly and equally, as should pupillary constriction.

Finally, ask the subject to cover one eye and keep his or her head motionless as you move an object (fingertip, pencil, cotton-tipped applicator) within the visual field about 30 cm from the eye. Ask the subject to follow the object with his or her eye as you move it to each of the following positions: up, down, right, left, up right, down right, up left, down left. Eye movement should occur smoothly and precisely. Check for nystagmus (involuntary, rapid, rhythmic movement of the eyeball) and weakness of eye muscle. Test the other eye using the same procedure. There should be no nystagmus on elevation or depression of the eyes, but a small amount of nystagmus at the extreme of any lateral or medial gaze is normal.

Cranial Nerve V. Trigeminal Nerve (Mixed)

The trigeminal nerve has two roots, one motor and the other sensory, arising from the pons (figure 12.1). The motor fibers supply muscles of mastication (chewing), such as the temporalis and the masseter. The sensory fibers convey pain, temperature, touch, and pressure information from the eye, face, nasal and oral mucosa, gums, teeth, and anterior two-thirds of the tongue.

The motor function may be tested by placing your fingertips on the temporalis muscles at each temple (figure 12.5) of the subject and asking the subject to clench his or her teeth several times. Compare the strength of muscle contraction on each side. They should be the same. Test the strength of masseter muscles using the same technique. The masseter can be palpated just above and to the front of the angle of the lower jaw. Contractions should be symmetrical and equal. Check the strength of jaw closure by asking the subject to grip a tongue depressor with his or her teeth on each side while you try to extract the depressor. Strength of closure should be bilaterally good. Ask the subject to open his or her mouth. Note any deviation of the jaw to the right or left. Normally there is no deviation. Ask the subject to move the lower jaw side to side to assess medial and lateral pterygoid muscle function. Normally the mandible moves smoothly and equally to each side.

Fibers in the sensory root come from the trigeminal ganglion, which receives input from three divisions of sensory nerves: the ophthalmic (I), the maxillary (II), and the mandibular (III). The facial distribution of these divisions is shown in figure 12.6. Each division should be tested separately and bilaterally (on both sides).

First, explain to the subject that you are going to test the corneal reflexes. Then pull a small piece of cotton from a cotton ball, form it into a long thin strand, and ask the subject to look up and away from you. Lightly touch the cornea with the piece of cotton, avoiding the eyelashes. The eyes (both) should blink. Repeat the test on the other eye.

Next, test the areas of the skin on each side of the midline supplied by the three divisions of the trigeminal nerve for their sensitivity to light touch (cotton ball) and pain (pin prick). Before stimulating the skin, ask the subject to close his or her eyes and tell you when the stimulus is applied.

FIGURE 12.5 Palpating temporalis contraction when testing trigeminal nerve function

Injury to the sensory components of the trigeminal nerve causes anesthesia in the area of the affected division. *Tic douloureux* (trigeminal neuralgia) is caused by irritation of the trigeminal nerve and is marked by excruciating pain that follows the distribution of the sensory fibers.

Cranial Nerve VII. Facial Nerve (Mixed)

The facial nerve arises from the pons lateral to the abducens nerve (figure 12.1) and contains motor and sensory fibers. The motor fibers innervate muscles of facial expression, and parasympathetic fibers stimulate salivary glands. Sensory fibers convey taste information from the anterior two-thirds of the tongue (figure 12.7).

Test for facial nerve function by asking the subject to show teeth and smile, lift the eyebrows, frown, and close the eyes tightly. All facial movements should be equal bilaterally. Note any asymmetry of facial movements and features, but keep in mind that some persons habitually smile and talk more out of one side of the mouth than the other and that every face is somewhat asymmetrical.

Disease of the facial nerve results in peripheral facial paralysis (Bell's palsy) on the side of the lesion. Causes of

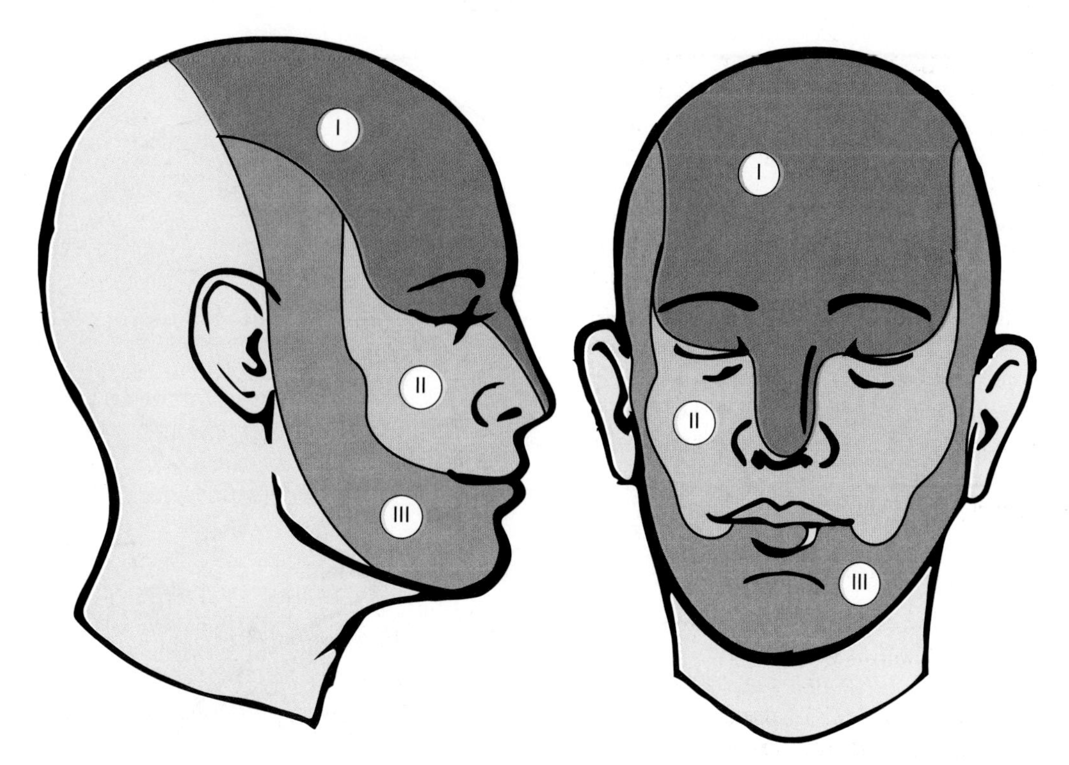

FIGURE 12.6 Cutaneous distribution of ophthalmic (I), maxillary (II), and mandibular (III) divisions of the trigeminal sensory nerves

Bell's palsy include compression of the nerve by a tumor and infections that inflame the nerve and surrounding tissue. The affected individual will be unable to close the eye on that side, wrinkle his or her forehead, or show teeth. Loss of muscle tone on the side of the lesion allows the corner of the mouth to droop.

Test taste using a sugar or salt solution. Place a few drops on half of the anterior two-thirds of the protruded tongue and instruct the subject to keep the tongue out until he or she has tasted the substance. Test each side of the tongue separately.

Cranial Nerve VIII. Vestibulocochlear Nerve (Sensory)

The eighth cranial nerve contains fibers of the vestibular nerve and the cochlear nerve. The *vestibular nerve* is sensory from receptors in the inner ear that provide information concerning movement of the body, balance, and body position in relation to gravitational force. The *cochlear nerve* is sensory from auditory (hearing) receptors in the cochlea of the inner ear.

Ask the subject to stopper one ear canal with his or her finger while you test hearing in the other ear. At a distance of 50 cm directly lateral to the tested ear, whisper a two-digit number (e.g., 29, 35) and ask the subject to identify the number by writing it on a sheet of paper. Repeat the test on the other ear.

The inability to hear some or all of the normally audible sounds could be caused by one or more of several disorders such as blockage of the outer ear canal (with wax, fluid, etc.); damage to the eardrum or ear ossicles; blockage of the internal auditory meatus and inflammation of the middle ear (common in upper respiratory infections); damage to the inner ear, auditory receptors, and auditory (cochlear) nerve; or damage to auditory pathways in the brain. The whisper test does not provide a means of differentiating one type of hearing loss from another.

A more useful test of hearing requires a 256-Hz tuning fork. Strike the prong of the tuning fork gently against the heel of your shoe to start it vibrating. Note the time and immediately place the tip of the tuning fork handle on the mastoid process (the bony bump behind and below the auricle). Be careful not to touch the prongs of the fork against the ear or hand. The subject should hear sound. As soon as the subject says he or she no longer can hear sound, note the time and immediately move the vibrating ends of the fork to within 3 cm of the external ear canal. The subject should again be able to hear the sound. Normally the subject's hearing should be equal in both ears, and air conduction hearing time should be about twice the duration as that for bone conduction. Repeat the test for the other ear.

A shortening of hearing time with air conduction coupled with preservation of hearing time with bone conduction suggests interference with sound transmission in the

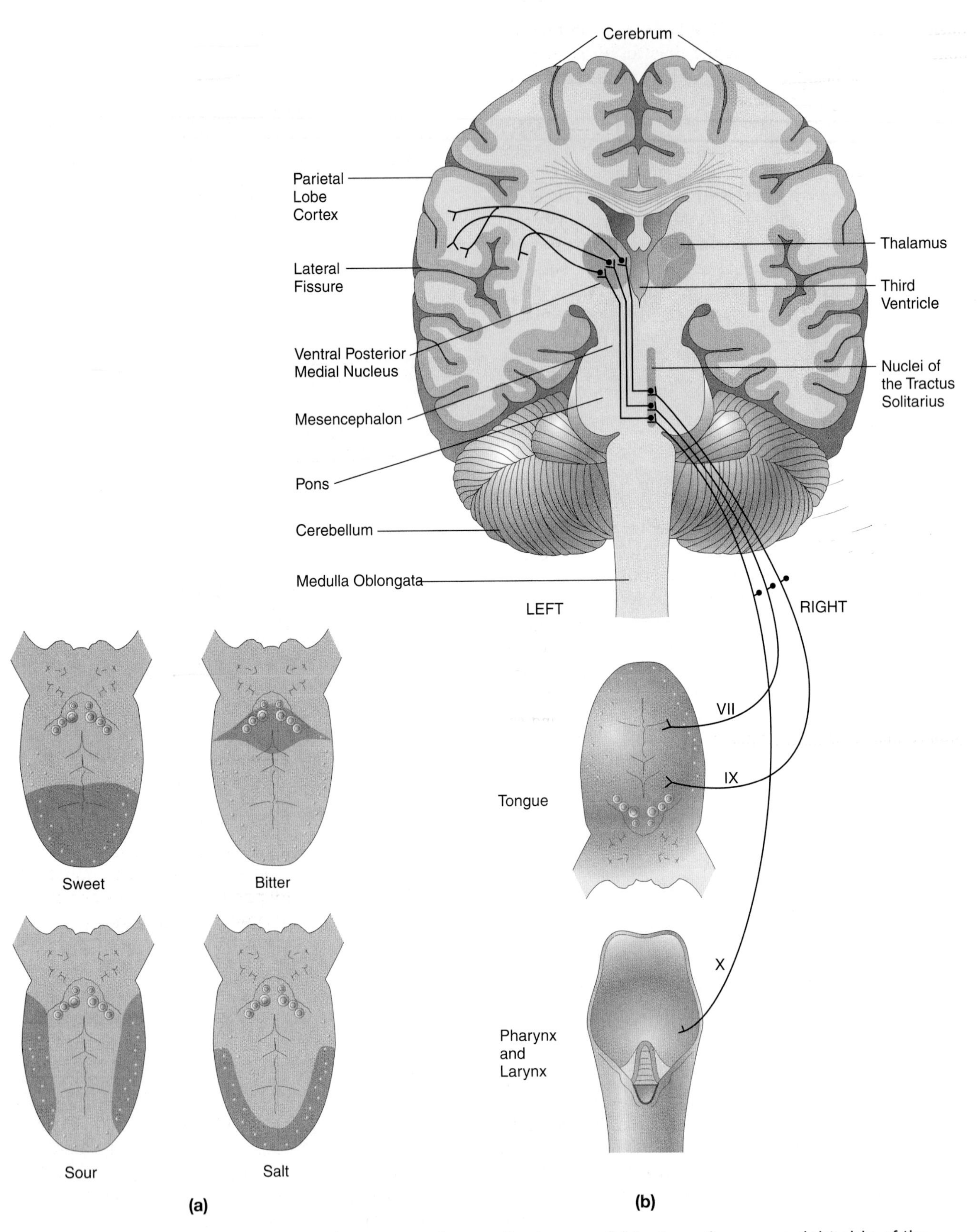

FIGURE 12.7 (a) Distribution of taste receptors on the tongue; (b) taste pathways on right side of the body; pathways on left side (not shown) are similar

outer ear or the middle ear. If both the hearing time for bone conduction and the hearing time for air conduction are reduced, the hearing loss most likely involves the inner ear or the auditory nerve or both.

Strike the prong of the tuning fork gently against the heel of your shoe and place the tip of the handle on the middle of the subject's forehead. If both ears are normal the subject will hear the sound equally in both ears. With unilateral nerve deafness, the subject will hear the tuning fork with the good ear but not the affected ear. If there is blockage or damage to the outer ear or middle ear, or both, the sound will be heard more loudly in the affected ear.

More comprehensive tests of auditory and vestibular nerve function are discussed in chapter 14.

Cranial Nerve IX. Glossopharyngeal Nerve (Mixed)

Cranial Nerve X. Vagus Nerve (Mixed)

The glossopharyngeal nerve supplies motor fibers to the parotid salivary gland and muscles in the pharynx (throat), larynx (voice box), and soft palate. Sensory fibers convey taste information from the posterior one-third of the tongue (figure 12.7) and information pertaining to blood pressure and blood chemistry from the carotid artery.

The vagus nerve supplies motor fibers to constrictor muscles of the pharynx, intrinsic muscles of the larynx, and involuntary muscles of the bronchi, heart, esophagus, stomach, small intestine, and part of the large intestine. Secretory motor fibers of the vagus supply the pancreas and secretory glands of most of the alimentary canal. The vagus is sensory from the laryngeal mucosa, heart, lungs, esophagus, stomach, small intestine, and part of the large intestine. In addition, vagal sensory fibers convey taste from the epiglottis and blood pressure and chemistry information from the aorta. The ninth and tenth cranial nerves are tested together because their functions overlap.

Begin by asking the subject to open his or her mouth and say "ahh." Note the position of the soft palate and uvula at rest and with phonation. Normally, the uvula and palate rise in the midline with phonation (figure 12.8). Paralysis of cranial nerves IX and X on one side of the brain will cause the palate and uvula to deviate to the unparalyzed side during phonation (figure 12.8).

Next, ask if the subject has had any difficulty speaking or swallowing. Note the quality of the voice (hoarse, nasal, etc.) and ask the subject to swallow. Note any difficulty. Tell the subject you are going to test for the gag reflex. Depress the tongue and touch the back of the throat lightly on each side using a long-stem cotton applicator (figure 12.9). Absence of the gag reflex may result from a lesion of either the glossopharyngeal (sensory component) or vagus nerve (motor component) on the same side as the loss. If an abnormality is suspected, test for taste on each side of the posterior one-third of the tongue (figure 12.7).

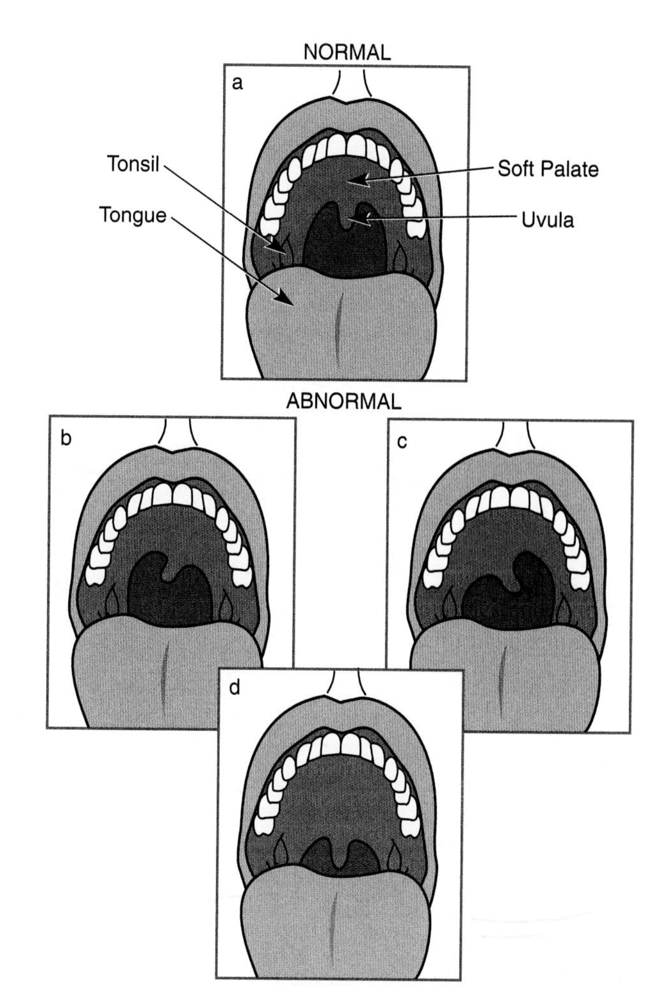

FIGURE 12.8 Tests of uvular deviation (cranial nerves IX and X): (a) normal; (b) deviation to right, paralysis of cranial nerves IX, X on left side; (c) deviation to left, paralysis of cranial nerves IX, X on right side; (d) paralysis on both sides due to brain stem damage

Cranial Nerve XI. Spinal Accessory Nerve (Motor)

The eleventh cranial nerve supplies some motor fibers to the muscles of the larynx and pharynx via the pharyngeal plexus (C.N. IX-X-XI), but its principal distribution is motor to the sternocleidomastoid and trapezius muscles. The *sternocleidomastoid* turns the head to the opposite side, and the *trapezius* muscle elevates the shoulder on the same side. Put one hand on the subject's cheek and tell the subject to turn his or her head against your hand as you resist the movement. Meanwhile, use your other hand to palpate the opposite sternocleidomastoid. Repeat the test for the other side. The two muscles should be equal in strength, and no fasciculations (twitching) should occur.

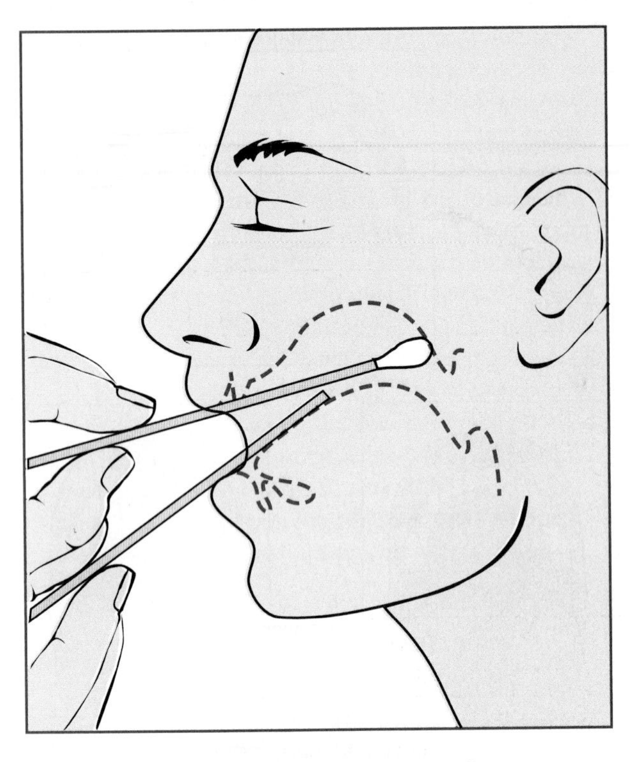

FIGURE 12.9 Testing the gag reflex

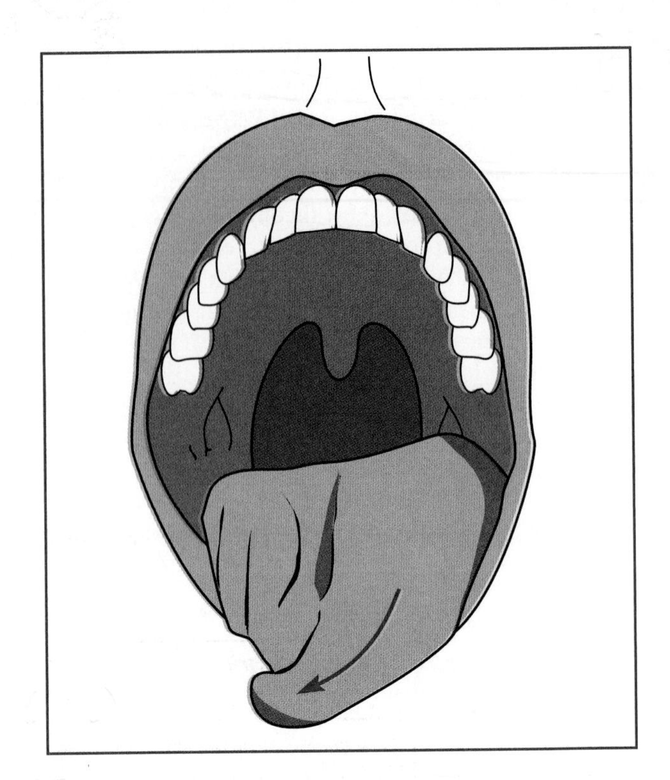

FIGURE 12.10 Deviation of the tongue to the right due to right hypoglossal nerve damage

Place both hands on the subject's shoulders and test the strength of the subject's "shrug" as he or she attempts to elevate the shoulders while you press down with both hands. Normally the strengths of the trapezius muscles are equal and there are no fasciculations.

Cranial Nerve XII. Hypoglossal Nerve (Motor)

The hypoglossal nerve supplies motor fibers to muscles of the tongue. Muscles in the right half of the tongue are supplied by the right hypoglossal nerve, and muscles in the left half are supplied by the left hypoglossal nerve.

First, assess the subject's speech by asking the subject to read aloud. Movement of the tongue is important in forming the vowel and consonant sounds of speech. Note any deficiencies.

Next, ask the subject to protrude his or her tongue in the midline. Check for deviation of the tongue tip to the right or left and for atrophy of muscle in the right or left half of the tongue. Normally the tip of the protruded tongue will not deviate from the midline and the tongue will appear symmetrical. A lesion of the hypoglossal nerve will cause the protruded tongue to deviate toward the affected side, and the appearance of the tongue will be asymmetrical owing to atrophy of muscle on the affected side (figure 12.10). Often, fasciculations of the tongue muscle occur on the affected side.

Visual Acuity, Accommodation, Peripheral Vision, Color Vision, and Ophthalmoscopy

CHAPTER 13

INTRODUCTION

Emmetropia is defined as normal vision. For normal vision, light rays from an object in the external environment are refracted (bent) by the cornea and lens of the eye and focused on the rods and cones of the retinal layer inside the eye. The rods and cones are photoreceptors containing light-sensitive pigments that are altered when exposed to visible wavelengths of light, leading to changes in ion flow, which generate neural signals (receptor potentials). The signals are then transmitted through several neural layers to the ganglion cells where they elicit neural impulses, which leave the eye by way of the optic nerve. The *optic nerves* relay information to the midbrain (involved with light-induced reflexes) and to the lateral geniculate bodies of the thalami. From the thalamus, by way of the geniculocalcarine tracts, the information is relayed to the visual cortex of the occipital cerebrum (figure 13.1). The processing of visual information leading to the perception of visual stimuli is complex, occurring first in the retina, then in the thalamus, and finally in the occipital cortex.

Normal visual acuity requires the light rays from an object in the environment to be focused properly on the retina. Normal focusing is dependent on proper curvature of both the cornea and the lens to match the length of the eyeball.

Irregularity in the curvature of the cornea results in errors of refraction, where some light rays coming from an object are bent more than others. As a result, all parts of the image cannot be in focus at any one time: when one part of the image is focused on the retina, another part will be focused either in front of or behind the retina and, therefore, perceived as a blurred image. Imperfect vision resulting from a lack of equal curvatures of the cornea (or lens) at all directions perpendicular to its axis of symmetry is known as *astigmatism.* Astigmatism may be compensated for by placing a properly ground corrective lens in front of the eye so that all light rays passing through the corrective lens, cornea, and lens of the eye are properly focused on the retina.

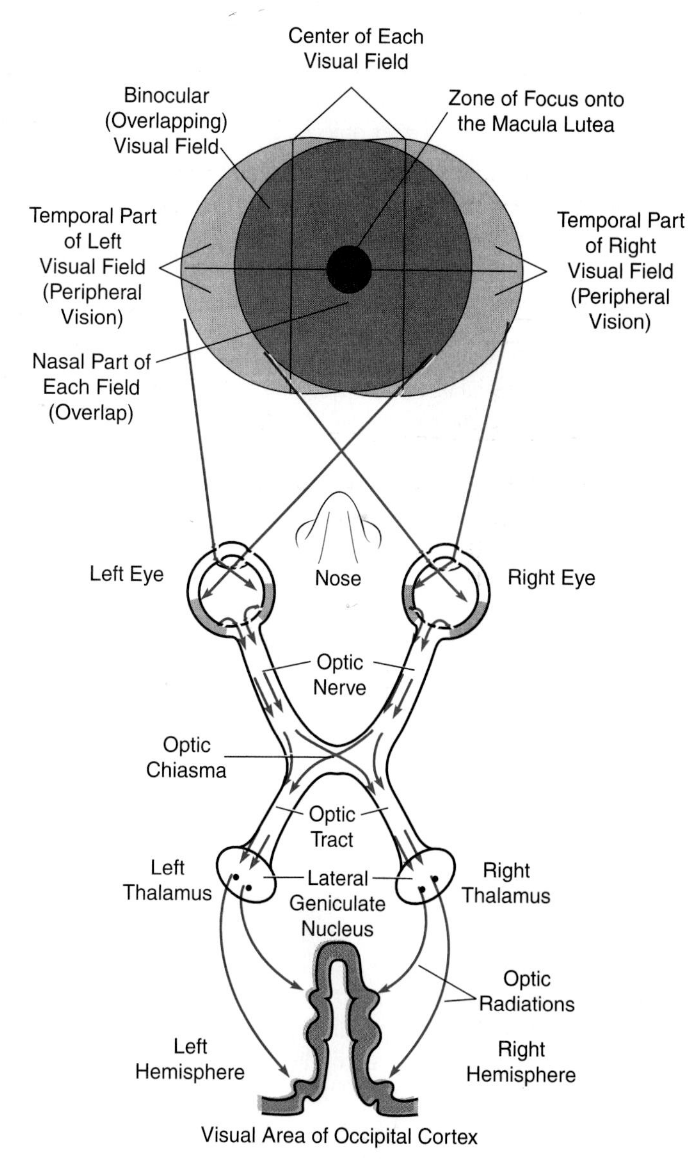

FIGURE 13.1 Visual pathways

Ocular accommodation, or focusing of the lens, is a function of the ciliary muscle, the suspensory ligament, and the lens capsule. Contraction of the ciliary muscle (figure 13.2) reduces tension on the suspensory ligament and lens capsule. The lens, by virtue of its own elasticity, tends to assume a more spherical shape, thus increasing its refractive power to focus on an object closer to the eye. Conversely, relaxation of the ciliary muscle increases tension on the suspensory ligament and lens capsule, flattening the lens and causing the eye to become focused on a more distant object.

As an object is brought closer to the eye, a point is reached where accommodation of the lens becomes maximum. This is known as the **near point of accommodation.** Objects brought closer to the eye than the distance of the near point cannot be properly focused on the retina and, therefore, appear blurred.

With advancing age, the lens of the eye gradually loses its elasticity and capacity for accommodation. As a result, the near point of accommodation recedes (near-point distance becomes larger), and it becomes increasingly difficult to focus properly on near objects, as when reading. The loss of near vision due to the gradual decline of lens elasticity is known as *presbyopia* (old-age vision).

In addition to astigmatism and presbyopia, other common visual defects of the eye that involve focusing include myopia (nearsightedness) and hypermetropia (farsightedness). *Myopia* usually results either from an imperfection in which the eyeball is longer than normal or from imperfections in the lens or cornea whereby the image is focused in front of the retina rather than on it. Myopia can be compensated for by placing in front of the eye a properly ground concave lens, which diverges the light rays just enough for the cornea and lens to focus them accurately on the retina (figure 13.3).

Hypermetropia, like myopia, is also the result of an improperly shaped eyeball, but in this case the eyeball is shorter than normal. This defect results in the image's being focused at a point behind the retina instead of on it. If accommodation of the lens is normal, a person can partially compensate for hypermetropia by continuous partial contraction of the ciliary muscle; however, this solution generally leads to further complications (eyestrain, headaches). Hypermetropia can be corrected by placing in front of the eye an appropriate convex lens, which converges the light rays enough to permit the image to be sharply focused on the retina (figure 13.3).

Normal binocular vision requires that each eye accurately focus an image on its retina and also that the images are focused on corresponding points on the retinas of the two eyes. Failure to do so results in *double vision*—a double image of a viewed object. The reflex adjustment of the position of the eyes to shift the images of an object onto corresponding parts of the retina is called the *convergence reflex.*

The retina contains two types of photoreceptor cells: cones and rods. **Cones** are responsible for acute vision and color vision. **Rods** are insensitive to color but provide for vision under conditions of dim or poor light at the expense of visual acuity. Rods are most numerous in the periphery of the retina, providing for peripheral vision and motion detection. Cones are most concentrated in the central fovea, a thin spot in the retinal layer directly in line with the pupil. The central fovea is the retinal area of maximum visual and color acuity.

Normal human color vision is trichromatic, involving combinations of three primary colors: red, green, and blue. According to the *trichromatic theory,* three classes of cones, each containing a different pigment sensitive to a particular range of light wavelengths, are found in the retina. The red-sensitive cone is sensitive to wavelengths from 475–650 nanometers (nm; maximum sensitivity at 550 nm). The green-sensitive cone is sensitive to a wavelength range of 424–625 nm (maximum sensitivity at 525 nm). The blue-sensitive cone responds to a wavelength range of 400–525 nm (maximum sensitivity at 450 nm). All colors of the visible spectrum of light can be produced by stimulating different combinations of the three classes of cones. Although the ability to discriminate color begins with the cones in the retina, the perception of color is a subjective phenomenon requiring the activity of neuronal circuits in the retina, the lateral geniculate body, and the visual cortex.

In the experiments that follow, color vision, visual acuity, and other aspects of the physiology of vision will be assessed.

EXPERIMENTAL OBJECTIVES

1. To review the functional morphology of the eye.
2. To test for visual acuity, astigmatism, accommodation, and the blind spot.
3. To assess color vision and to characterize common forms of color vision disturbances.
4. To observe the visual reflexes of convergence and divergence.
5. To become familiar with the ophthalmoscope and its use in physical diagnosis.

Materials

Snellen test letter chart
Green's astigmatic chart
accommodation meterstick
perimeter with pointer and colored disks
perimetric charts
Ishihara color-blindness test booklets
ophthalmoscope
reference illustrations of the fundus

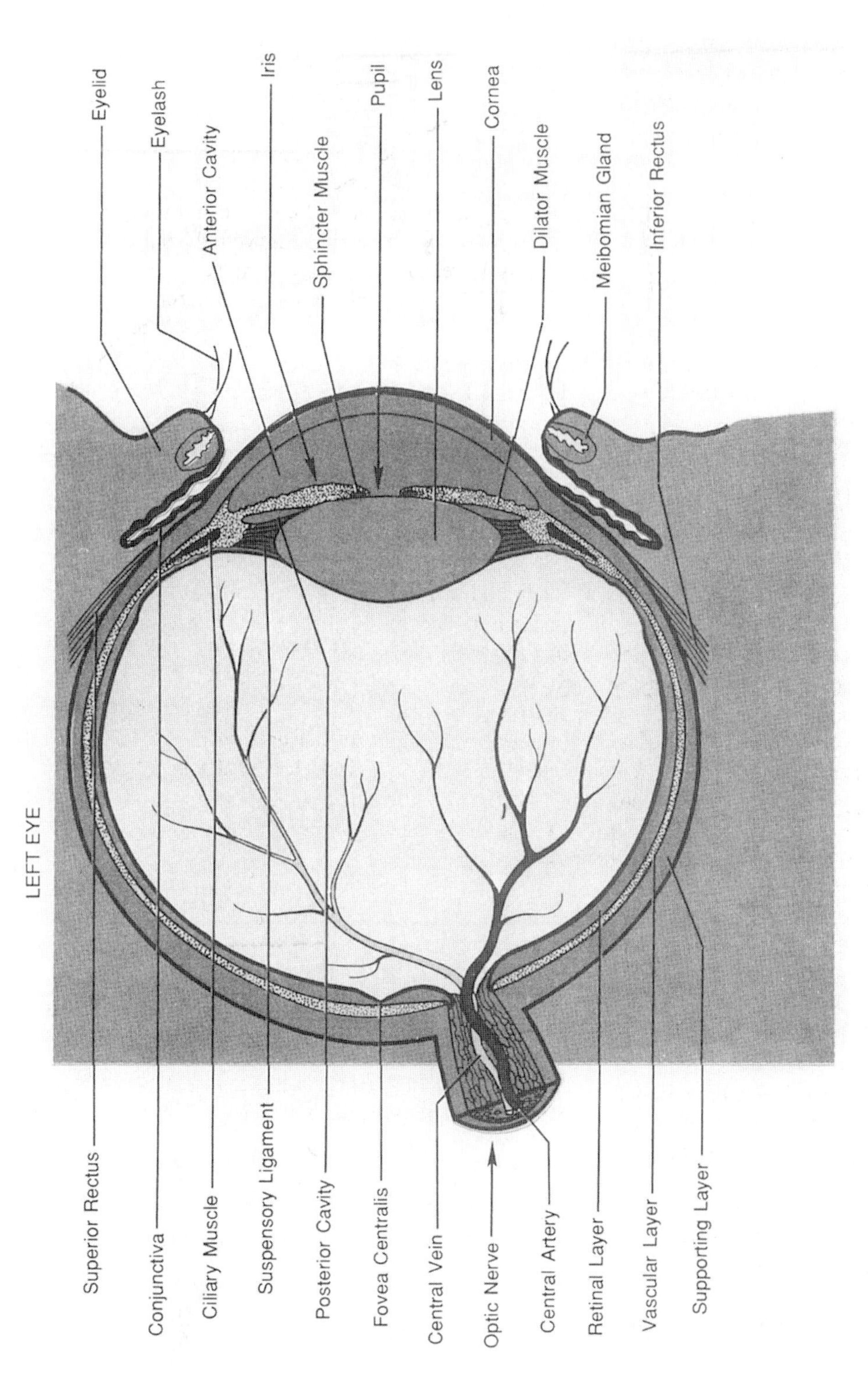

FIGURE 13.2 General structure of the eye

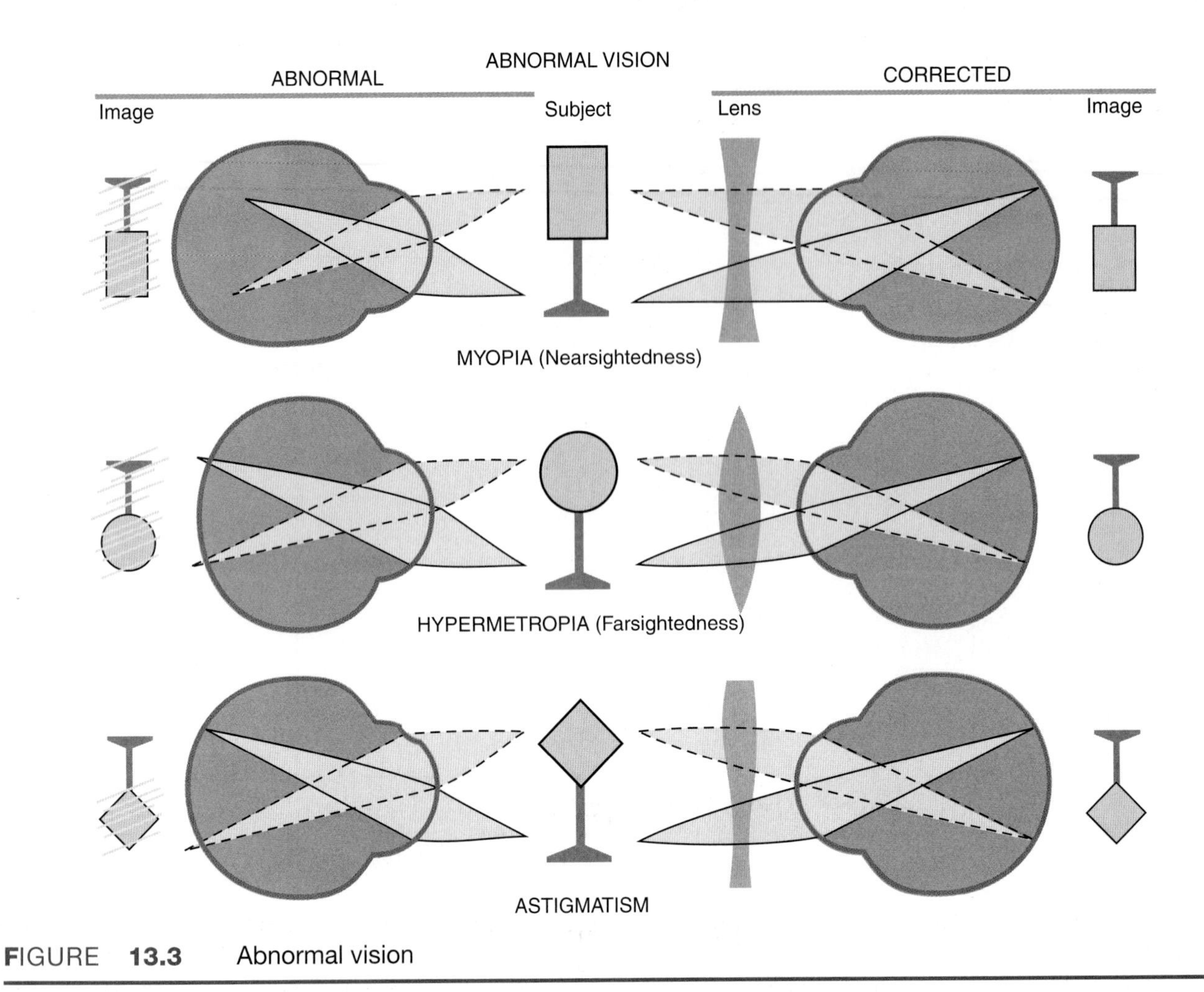

FIGURE 13.3 Abnormal vision

EXPERIMENTAL METHODS

Visual Acuity

To measure visual acuity, one calculates the size of the retinal image by drawing lines from the extremities of the object through the nodal point of the eye to the retina. The angle formed at the nodal point is called the *visual angle* (figure 13.4). As an object becomes smaller or recedes farther from the eye, the visual angle will become smaller, and eventually the object will disappear from view. The smallest visual angle at which the object can still be seen by the normal eye equals about 1 minute of arc. On the retina, this is equivalent to 0.004 mm, the width of one cone. The size of the smallest visual angle is a measurement of visual acuity.

The Snellen test letter chart (figure 13.5) used for the measurement of visual acuity is based on a visual angle of 1 minute. The letters, at the distance given on the chart, subtend at the nodal point a visual angle of 5 minutes. The lines that form the letters subtend an angle of 1 minute.

A ratio of two numbers, expressed as a fraction, is located at the left margin of each line of test letters on the Snellen chart. The numerator (20) is the distance in feet between the chart and the subject's eye. The denominator is the farthest distance in feet from the chart that the line can be read by the normal eye. Thus, if line 6 is the smallest line of letters that can be read by a subject's eye at a distance of 20 feet from the chart, the visual acuity for the tested eye would be expressed as 20/30 = 0.666. A visual acuity of less than 1.00 is below normal. A visual acuity of 20/30 means the smallest letters the subject's eye sees clearly at 20 feet can be seen by the normal eye at 30 feet.

Stand a distance of 6.10 m (20 feet) from the chart of test letters (the full-sized chart on the wall, not the reduced chart in the book), and cover one eye with an index card. With the other eye, endeavor to read line 8, marked "20 feet." If you succeed, your visual acuity for that eye is said to be normal and is expressed as:

$$V = d/D = 20/20$$

where:

V = visual acuity

d = distance from chart

D = distance at which letter should be read

At this distance, $V = 20/20 = 1$. If at a distance of 20 feet, you can read the letters on line 9 marked 15 feet,

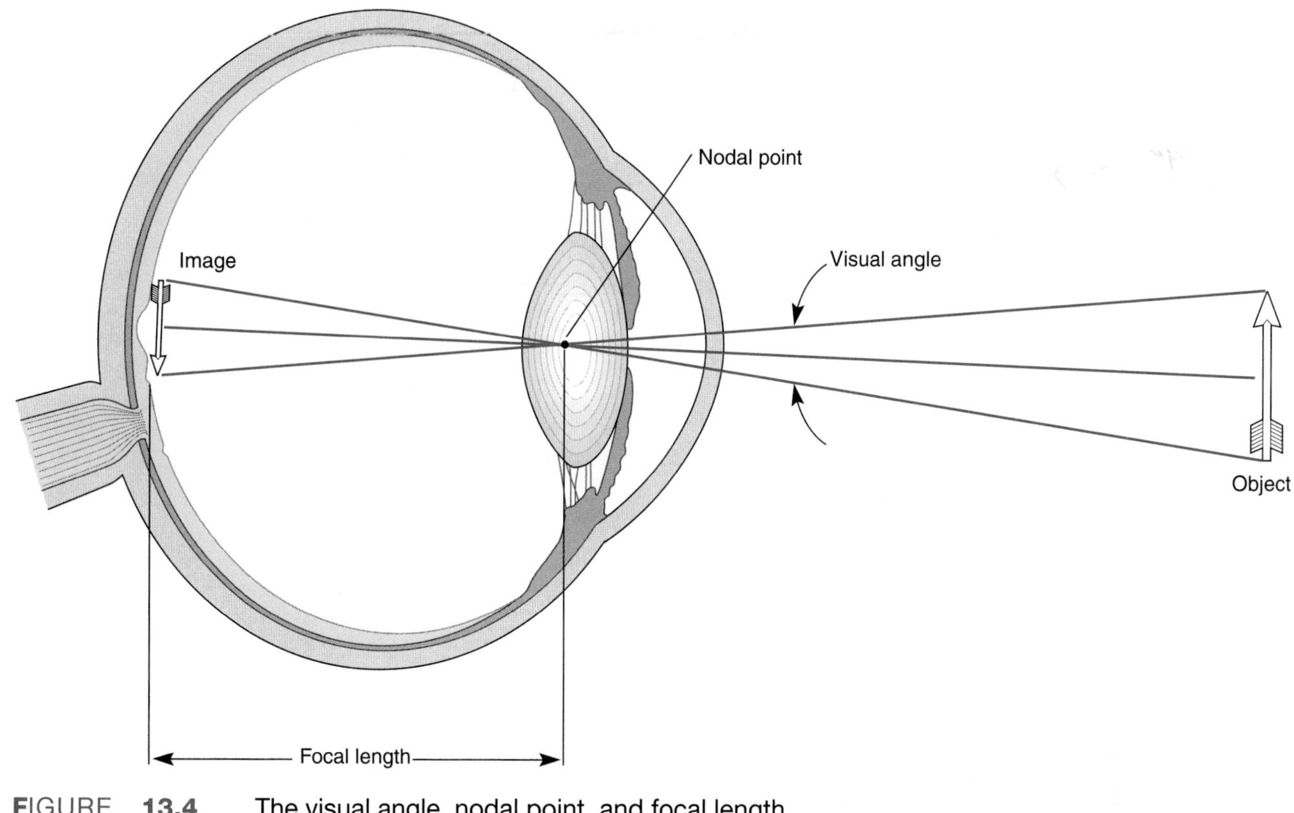

FIGURE **13.4** The visual angle, nodal point, and focal length

your visual acuity is better than normal and is expressed as $V = 20/15 = 1.33$.

Another method can be used with the Snellen test letter chart to determine visual actuity. Stand 25 feet from the chart, focus the test eye (the other eye is covered) on line 8, and slowly move toward the chart until the letters in line 8 become clear and sharp; then, measure the distance in feet from the eye to the chart. Divide the distance by 20 to obtain visual acuity. Record your visual acuity for each eye.

Blind Spot

The blind spot of the retina is the optic disc, about 1.5 mm in diameter, where nerve fibers of the retina converge and exit the eyeball via the optic nerve. There are no rods or cones present in the optic disc; therefore, no image can be formed in this area. Obtain from the laboratory instructor a meterstick designed for the measurement of accommodation. Attached to the meterstick is a brass slide that holds an index card with a cross symbol and a dot symbol printed on it, as shown here.

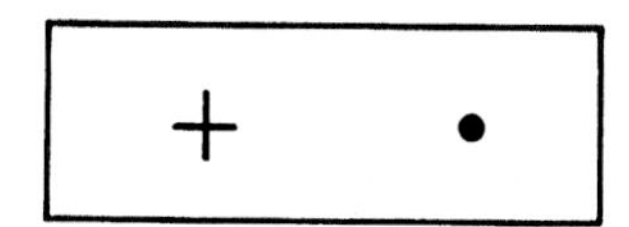

To test for the blind spot, place the end of the meterstick at chin level (figure 13.6). Hold the meterstick at a right angle to your face, and position the card so that the dot is to your right and the cross is directly in front of the right eye at a distance of 50 cm. Cover your left eye. **Always exercise caution when moving objects toward your eyes.** Slide the card *slowly* toward the eye until a point is reached where the cross is seen but the dot is not visible. If you move the slide too quickly, you will pass the blind spot without being aware because the dot reappears after the blind spot is passed. At the point where the blind spot is sensed, record the distance from the card to your eye (meterstick distance to card plus 3 cm, to account for chin-to-eye distance). Calculate the retinal distance from the optic disc to the fovea centralis using the following ratio:

$$\frac{A}{B} = \frac{C}{D} \qquad C = \frac{AD}{B}$$

where:

A = distance between dot and cross on card (10 cm)

B = distance between card and eye (measured in centimeters and equal to meterstick distance to card plus 3 cm, to account for chin-to-eye distance)

C = distance between optic disc and fovea centralis (calculated in centimeters)

D = distance from lens to retina (approximately 2 cm)

Repeat the procedure and calculation for testing the left eye (right eye covered).

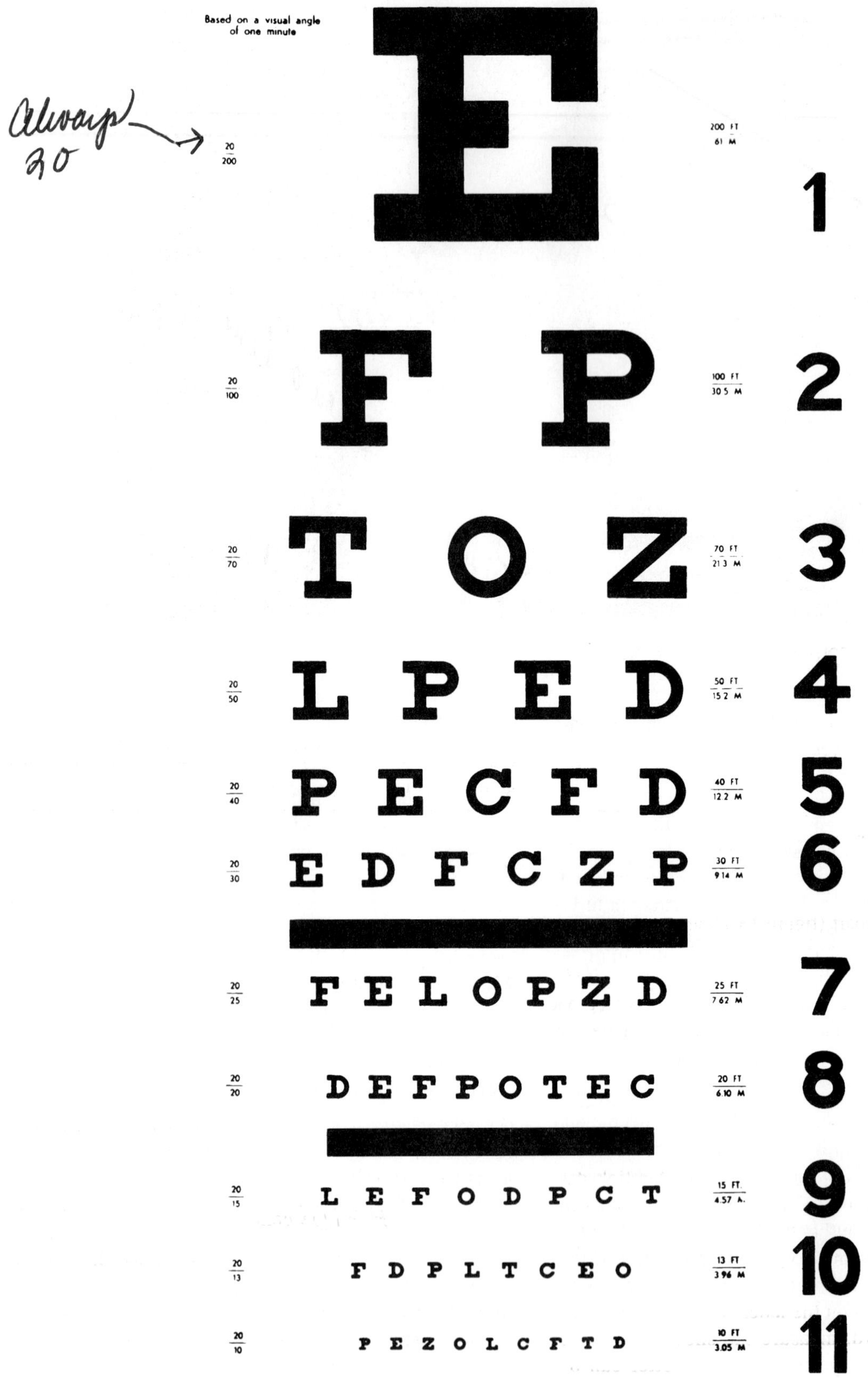

Based on a visual angle of one minute

	Letters		
20/200	E	200 FT / 61 M	1
20/100	F P	100 FT / 30.5 M	2
20/70	T O Z	70 FT / 21.3 M	3
20/50	L P E D	50 FT / 15.2 M	4
20/40	P E C F D	40 FT / 12.2 M	5
20/30	E D F C Z P	30 FT / 9.14 M	6
20/25	F E L O P Z D	25 FT / 7.62 M	7
20/20	D E F P O T E C	20 FT / 6.10 M	8
20/15	L E F O D P C T	15 FT / 4.57 M	9
20/13	F D P L T C E O	13 FT / 3.96 M	10
20/10	P E Z O L C F T D	10 FT / 3.05 M	11

FIGURE 13.5 Snellen test letter chart

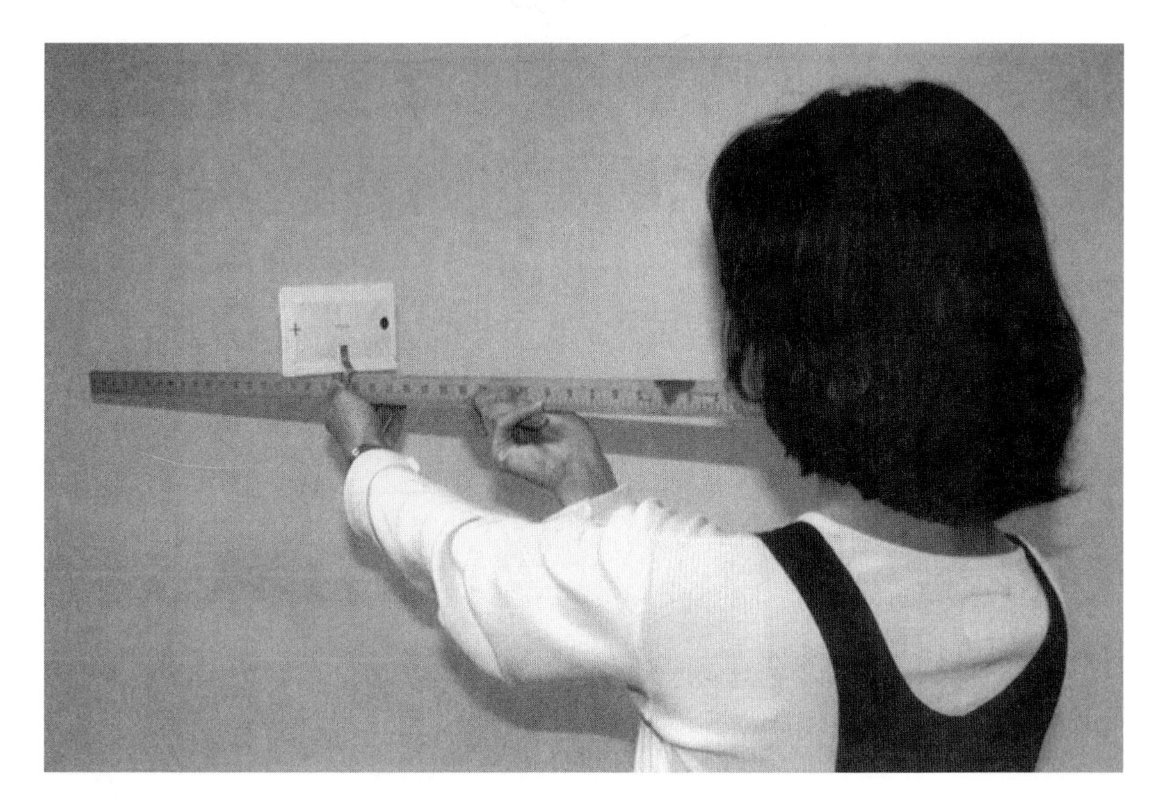

FIGURE 13.6 Determination of the blind spot

Record your calculations in the report. Can the blind spot be sensed when both eyes are open during the test?

Astigmatism

Astigmatism is a form of *ametropia,* abnormal refraction in which a variable degree of refraction exists in different meridians of the eyeball because of abnormal curvature. Using the astigmatism chart (figure 13.7), test each eye for astigmatism. To the normal eye, from 20 feet away from the astigmatic chart all parallel lines in each meridian will appear to have equal contrast. If astigmatism is present, one or more lines will appear darker than all of the others. If figure 13.7 is used in place of a full-sized chart, place the chart about 10 feet from the eye.

Near Point of Accommodation

Obtain from the laboratory instructor a meterstick designed for the measurement of accommodation. Attached to the meterstick is a slide that holds an index card with a sharply defined letter printed on it. Place the end of the meterstick at chin level, and hold the meterstick at a right angle to your face. Focus on the letter with one eye closed and, by sliding the card, measure the shortest distance from the cornea of the eye to where the letter can be sharply seen. Repeat the test for the other eye. Refer to table 13.1 to see how closely the near point of accommodation corresponds to the usual distance for your age. Record your observations.

TABLE 13.1 Near point of accommodation

Age (years)	***Near point (cm)***
10	7
20	10
30	14
40	22
50	40
60	100

Accommodation Reflex

The process of accommodation involves a reflex adjustment of lens diameter and the reflexes of convergence and divergence, as well as a reflex adjustment of pupillary diameter. To observe this process in your lab partner, have the subject focus on the point of a pencil, and then move the pencil toward the subject's eyes. Notice the convergence of both eyeballs and change in the size of the pupils. Move the pencil away from the subject's eyes, and again note a divergence of the eyeballs and change in pupillary diameter. Record your observations.

The convergence and divergence reflexes aid in the judgment of distance, but the chief cue to distance when both eyes are used is the difference in appearance of the object to the two eyes. Place a small cube (1–2 inches on a

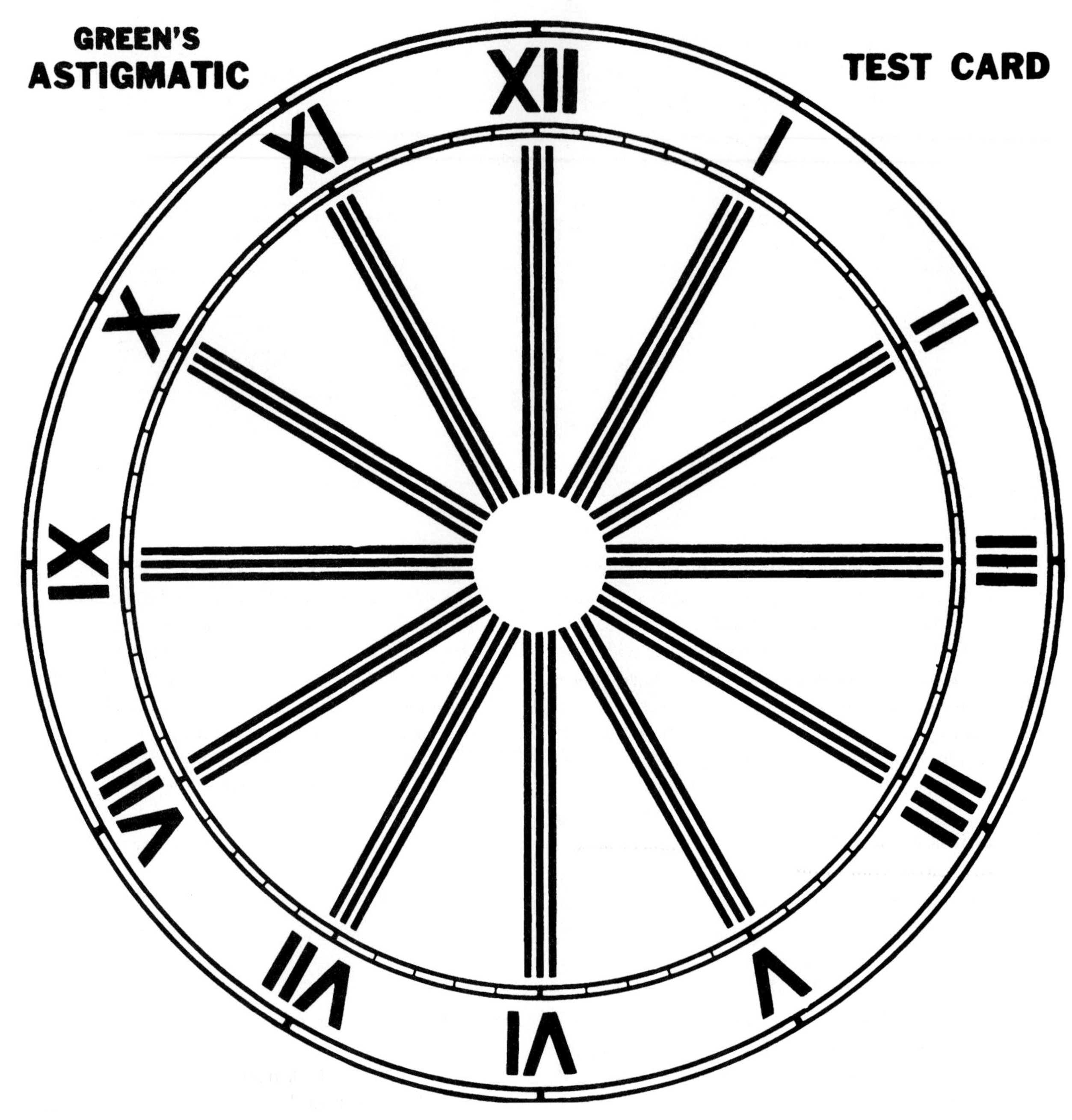

FIGURE 13.7 Green's astigmatic chart

side) about 14 inches in front of your eyes. Cover first one eye and then the other. Sketch the appearance of the cube as seen by each eye. Instruct your laboratory partner to hold a pencil at arm's length from your eye. Close one eye, and attempt to touch the pencil point with the index finger of one hand. The arm movement should start from the side of the body. Repeat the exercise with both eyes open.

Diplopia (Double Vision)

Normal retinal correspondence (NRC) is the occipital cortical integration of similar images projected onto anatomically corresponding areas of each retina into a single perception, producing normal single binocular vision. If there is a disruption of the alignment of the eye, normal binocular vision may produce an annoying diplopia or visual confusion.

Strabismus (cross-eyed) is an abnormal ocular condition defined as the inability to direct both foveas to the same fixation point. There are two kinds of strabismus, paralytic and nonparalytic. *Paralytic strabismus* is caused by failure of one or more of the extraocular muscles to move the eye in concert with the other eye so as to direct their gaze on the same fixation point. The muscle involved may be identified by watching the subject direct the eyes to each of the nine cardinal positions (figure 13.8). Possible causes

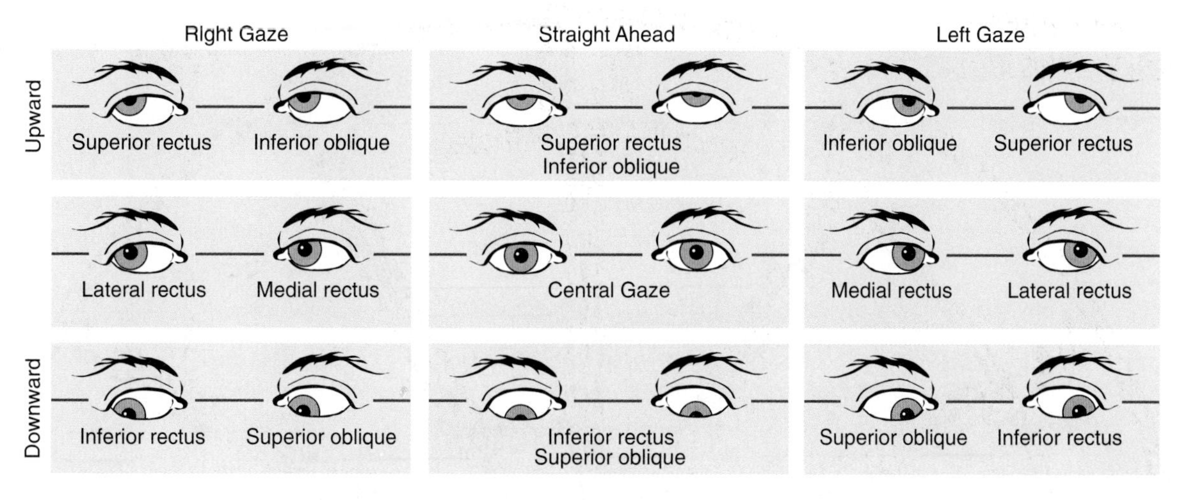

FIGURE 13.8 Nine cardinal directions of gaze

include tumor, infection of the brain, and injury to the cranial nerves, or extraocular muscles, or the eye. *Nonparalytic strabismus* is an inherited defect in the position of the two eyes. The affected person cannot fix the gaze with both eyes but must use one or the other. The eye that is used at a given time is the one that looks straight.

With both eyes open, focus on an object in the distance, and while doing so, gently press on the lateral portion of your left eye. This pressing moves the eye slightly so that the image is no longer focused on the same part of the retina as it is in the right eye. Does the object appear double? Now close your right eye while continuing to maintain slight lateral pressure on the left eye. Does the object now appear as one?

Visual Fields

The visual field of the eye is everything a person can see when the eye is held in a fixed position with the gaze directed forward. The visual field of each eye includes not only that which can be seen clearly but also that which is blurred (peripheral vision). Fields of vision for black/white and colors can be measured using a perimeter (figure 13.9).

Obtain a perimeter, colored perimeter disks, and a perimetric chart from the laboratory instructor. Measure the visual fields of each eye for both black/white and color vision using the following procedure:

1. Have the subject assume a comfortable sitting position with the lower margin of the orbit of the eye touching the center bar of the perimeter. Stand behind the subject while performing the test.
2. Ask the subject to close the untested eye and gaze straight ahead with the tested eye, focusing on the central pivot of the perimeter arc. Place a white disk in the perimeter pointer.
3. With the semicircular carriage horizontal (0–180 degrees), move the pointer along the outer edge of the graduated carriage. Begin at the 90-degree mark and slowly move the pointer around the carriage toward the 10-degree mark. Keep the disk located in the perimeter slot. Record the angle at which the white disk can first be seen by placing a dot on the perimetric chart where the measured angle (the circular lines) intersects the angle denoting position of the semicircular carriage (radial lines).
4. Repeat the test with the semicircular carriage in a vertical position (90–270 degrees) and at various angles. Chart both eyes for black/white fields of vision (figure 13.10).
5. Repeat the tests using the blue, green, and red disks.
6. Attach the perimetric charts to your laboratory report.

Color Vision

The following is an introduction to and instructions for the use of color plates to help determine color blindness. This test was created by Dr. Shinobu Ishihara, Professor Emeritus of the University of Tokyo. Following the instructions, test your laboratory partner for color blindness. Record all observations.

Introduction to Color Vision

The Ishihara series of plates is designed to provide a test that gives a quick and accurate assessment of color vision deficiency of congenital origin. This is the most common form of color vision disturbance.

Most cases of congenital color vision deficiency are characterized by a red-green deficiency, which may be of two types: first, a **protan** type, which may be absolute (protanopia) or partial (protanomalia), and second, a **deutan** type, which may be absolute (deuteranopia) or partial (deuteranomalia).

In *protanopia,* the visible range of the spectrum is shorter at the red end compared with that of the normal,

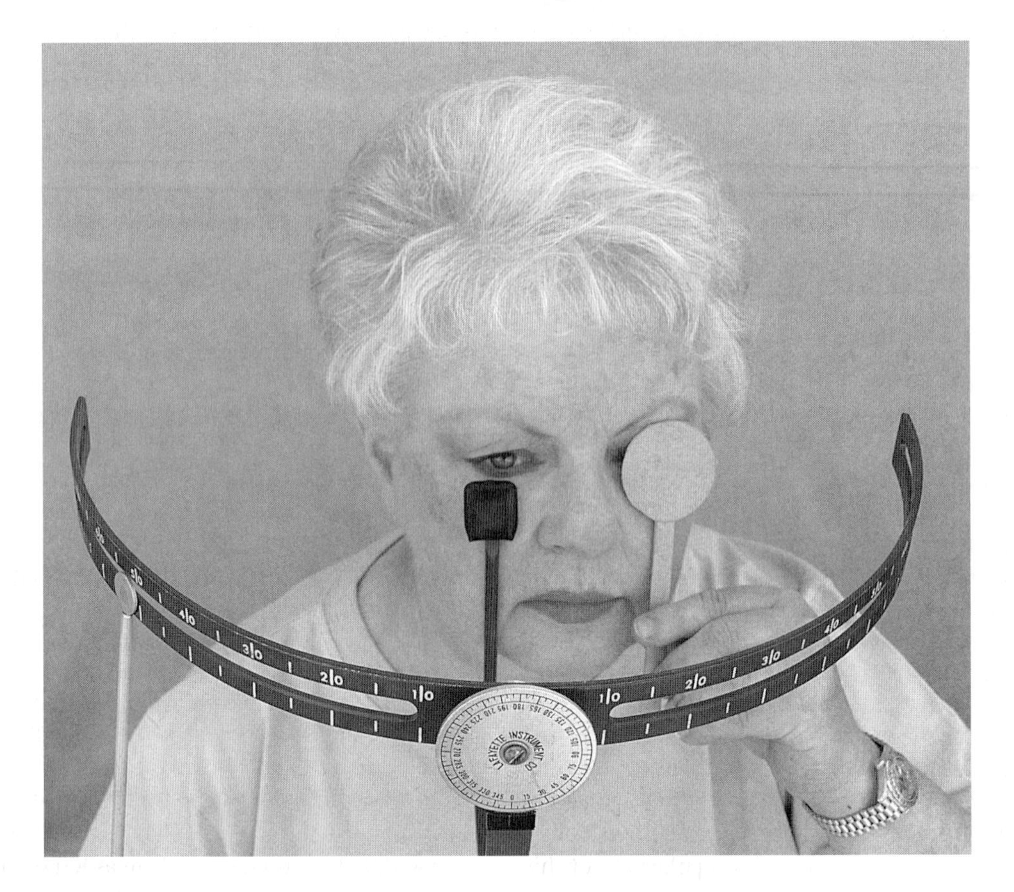

FIGURE 13.9 The perimeter

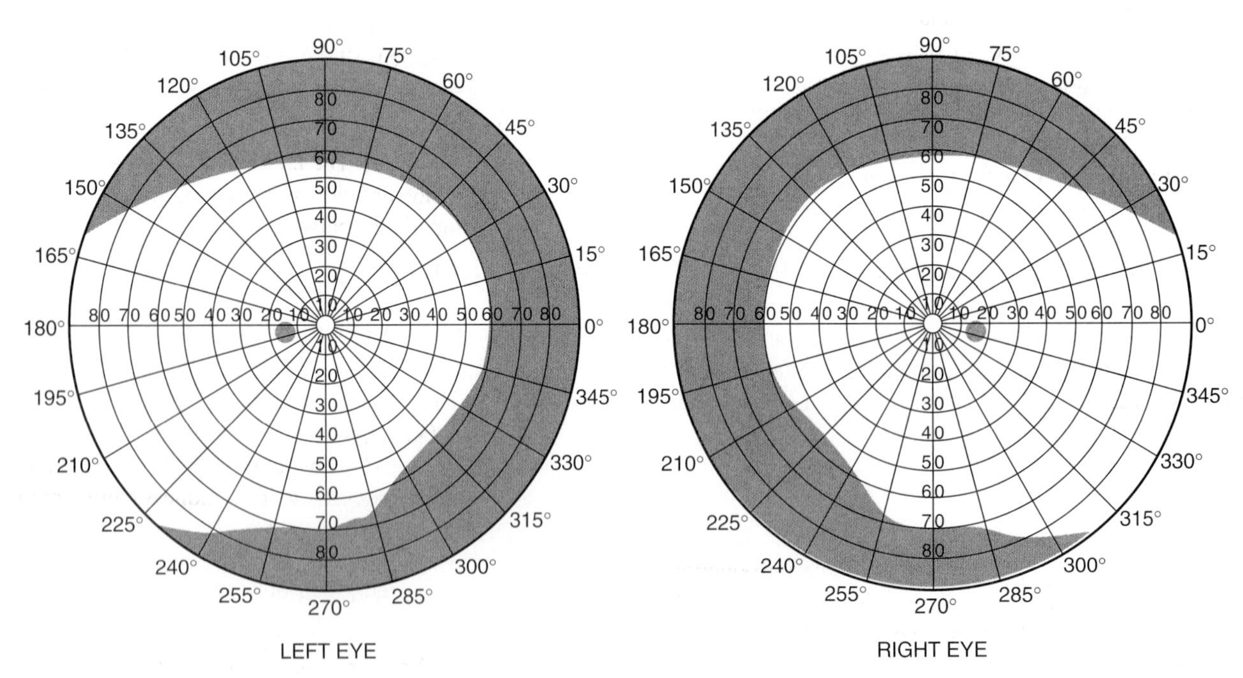

FIGURE 13.10 Perimetric charts of the eyes; record of normal black/white peripheral vision. The white area is normally visible; the darkened area is not normally visible. The circle near the 10 is the blind spot. nose gets in the way
Cheekbone " "

and that part of the spectrum which appears to the normal as blue-green appears to those with protanopia as gray. The whole visible range of the spectrum in protanopia consists of two areas that are separated from each other by this gray part. Each area appears to those with protanopia as one system of color with a different brightness and saturation within each area, the color in one area being different from that of the other. The red with a slight tinge of purple, which is the complementary color of blue-green, appears also as gray.

In *deuteranopia,* that part of the spectrum that appears to the normal person as green appears as gray, and the visible range of the spectrum is divided by this zone into two areas, each of which appears to be of one system of color. The visible range of the spectrum is not contracted, in contrast to protanopia. Purple-red, which is the complementary color of green, also appears as gray.

In *protanomalia* and *deuteranomalia* there is no part of the spectrum that appears as gray, but the part of the spectrum that appears to those with protanopia as gray appears to those with protanomalia as a grayish, indistinct color, and likewise, the gray part of the spectrum seen by the person with deuteranopia appears to those with deuteranomalia as an indistinct color close to gray.

Consequently, one of the peculiarities of red-green deficiencies is that blue and yellow appear to be remarkably clear compared with red and green. The application of this peculiarity to the test for color vision deficiencies is the distinguishing feature of this series.

A very rare congenital color vision deficiency is *total color blindness,* which may be typical or atypical. The subject who suffers from typical total color blindness shows a complete failure to discriminate any color variations and usually has an associated impairment of central vision. In atypical total color blindness, the color sensitivity to red and green, as well as to yellow and blue, is so low that only very clear colors can be perceived. Except for the color sensitivity, there is no abnormality in the visual functions. The plates in the Ishihara test booklets form an easy method of establishing the diagnosis in such cases and in distinguishing them from cases of red-green deficiencies.

Furthermore, a failure in the appreciation of blue and yellow may be termed *tritanomalia* if partial and *tritanopia* if absolute, but if such cases do exist, they are extremely rare. The plates in the booklet are not designed for the diagnosis of such cases.

How to Use the Test

The plates are designed to be appreciated correctly in a room that is lit adequately by daylight. The introduction of direct sunlight or the use of electric light may produce some discrepancy in the results because of an alteration in the appearance of shades of color. When it is convenient only to use electric light, it should be adjusted as far as possible to resemble the effect of natural daylight. The plates are held 75 cm from the subject and tilted so that the plane of the paper is at a right angle to the line of vision. The correct position of each plate is indicated by the number that is printed on the back of the plate. The numerals that are seen on plates 1–17 are stated, and each answer should be given without more than 3 seconds delay. If the subject is unable to read numerals, plates 18–24 are used and the winding lines between the two *x*s are traced with the brush. Each tracing should be completed within 10 seconds.

It is not necessary in all cases to use the whole series of plates. In a large-scale examination, the test is designed merely to separate the color defectives from those with normal color appreciation, and plates 16 and 17 may be omitted. In a short examination, the test may be simplified to an examination of six plates only: no. 1; one of nos. 2, 3; one of nos. 4, 5, 6, 7; one of nos. 8, 9; one of nos. 10, 11, 12, 13; and one of nos. 14, 15.

It may be necessary to vary the order of the plates if it is suspected that there is a deliberate deception on the part of the subject.

Explanation of the Plates

This series of plates is made up of the following 24 plates:

No. 1 Both the normal and those with all sorts of color vision deficiencies read it as 12.

Nos. 2,3 The normal read them as 8 (no. 2) and 29 (no. 3). Those with red-green deficiencies read them as 3 (no. 2) and 70 (no. 3). Those with total color blindness cannot read any numeral.

Nos. 4–7 The normal read them as 5 (no. 4), 3 (no. 5), 15 (no. 6), and 74 (no. 7). Those with red-green deficiencies read them as 2 (no. 4), 5 (no. 5), 17 (no. 6), and 21 (no. 7). Those with total color blindness cannot read any numeral.

Nos. 8,9 The normal read them as 6 (no. 8) and 45 (no. 9). The majority of those with color vision deficiencies cannot read them or read them incorrectly.

Nos. 10–13 The normal read them as 5 (no. 10), 7 (no. 11), 16 (no. 12), and 73 (no. 13). The majority of those with color vision deficiencies cannot read them or read them incorrectly.

Nos. 14,15 The majority of those with red-green deficiencies read them as 5 (no. 14) and 45 (no. 15). The majority of the normal and those with total color blindness cannot read any numeral.

Nos. 16,17 The normal read them as 26 (no. 16) and 42 (no. 17). In protanopia and strong protanomalia, only 6 (no. 16) and 4 (no.17) are read, and in cases of mild deuteranomalia, both numerals on each plate are read, but the 2 (no. 16) and 4 (no. 17) are clearer than the other numerals.

No. 18 In tracing the winding lines between the two *x*s, the normal trace along the purple and red lines. In protanopia and strong protanomalia, only the purple line is traced, and in cases of mild protanomalia, both lines are traced, but the purple line is easier to follow. In deuteranopia and strong deuteranomalia, only the red line is traced, and in cases of mild deuteranomalia, both lines are traced, but the red line is easier to follow.

No. 19 In tracing the winding line between the two *x*s, the majority of those with red-green deficiencies trace along the line, but the majority of the normal and those with total color blindness are unable to follow the line.

No. 20 In tracing the winding line between the two *x*s, the normal trace the bluish green line, but the majority of those with color vision deficiencies are unable to follow the line or follow a line different from the normal line.

No. 21 In tracing the winding line between the two *x*s, the normal trace the orange line, but the majority of those with color vision deficiencies are unable to follow the line or follow a line different from the normal one.

No. 22 In tracing the winding line between the two *x*s, the normal trace the line connecting the bluish green and yellowish green; those with red-green deficiencies trace the line connecting the bluish green and purple; and those with total color blindness cannot trace any line.

No. 23 In tracing the winding line between the two *x*s, the normal trace the line connecting the purple and orange; those with red-green deficiencies trace the line connecting the purple and bluish green; and those with total color blindness and weakness cannot trace any line.

No. 24 Both the normal and those with all sorts of color vision deficiencies can trace the winding line between the two *x*s.

Analysis of the Results

An assessment of the readings of plates 1–15 determines the normality or defectiveness of color vision. If 13 or more plates are read normally, the color vision is regarded as normal. If only 9 or fewer than 9 plates are read normally, the color vision is regarded as deficient. However, in reference to plates 14 and 15, only those who read the numerals 5 and 45 more easily than those on plates 10 and 9 are recorded as abnormal readings.

It is rare to find a person whose recording of normal answers is between 14 and 16 plates. An assessment of such a case requires the use of other color vision tests.

In the assessment of color appreciation by the short method involving 6 plates only, a normal recording of all plates is proof of normal color vision. If there is a discrepancy in any of the recordings, the full series of plates should be used before diagnosing a red-green deficiency.

Ophthalmoscopy

An **ophthalmoscope** is an instrument used for examination of the optic nerve, retina, and blood vessels of the interior of the eye. It is also used for detecting foreign bodies on the cornea, irregularities in the pupil, vitreous opacities, and lens opacities. In addition to detecting abnormalities of the eye, the ophthalmoscope is also useful in the detection or confirmation of systemic diseases such as diabetes mellitus, arteriosclerosis, and hypertension because of associated changes in the fundus (posterior part) of the eye.

The ophthalmoscope is a handheld instrument consisting of a battery-powered light and a set of lenses, one of which will give the clearest view of the fundus when the retina is illuminated. The light is directed by mirrors and prisms into a beam, which is directed through the pupil to illuminate the retina. The intensity of the light beam can be controlled by a rheostat (variable resistor) located on the battery compartment (figure 13.11). The lenses are arranged on a rotating disk. The diopter (D) designation of each lens, as it is rotated into the viewing aperture, may be seen in the

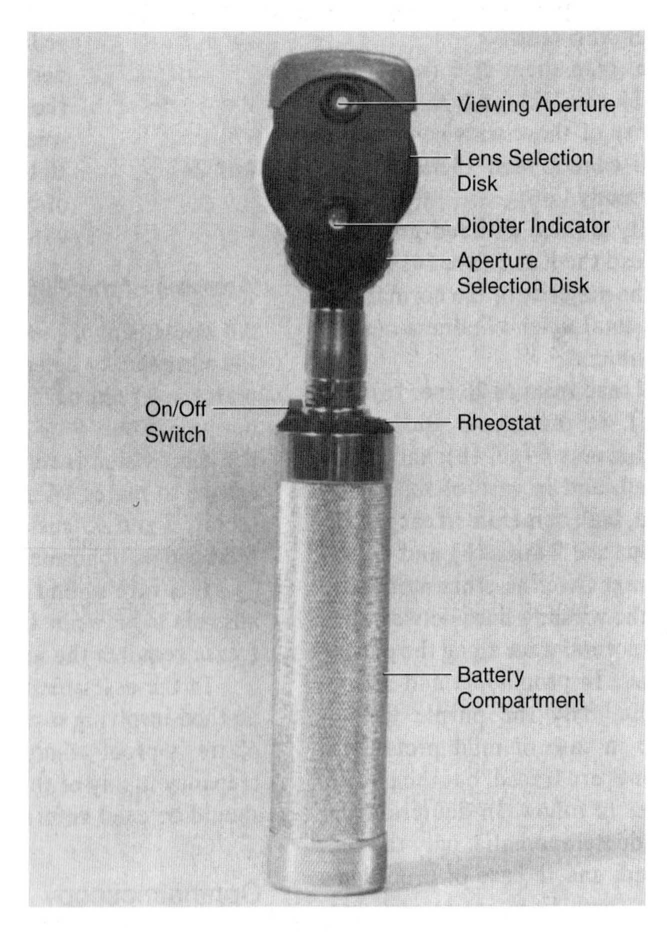

FIGURE **13.11** Ophthalmoscope

small window on the ophthalmoscope head. The 1D lens has a focal length of 1000 mm; the 2D lens, 500 mm; the 3D lens, 250 mm; 4D, 125 mm, and so on. The black numbers represent positive lenses, and the red numbers represent negative lenses. Positive lenses correct for hypermetropia. Negative lenses correct for myopia. Lens selection depends upon the eye condition of both the observer and the subject.

The aperture selection disk is located immediately below the lens selection disk. One of five apertures may be selected: large (circular), small (pinhole), grid, slit, and red free (greenish). Each is designed for a specific use. For example, the grid can be used for estimating the size of fundus lesions and for studying their progress. The red-free filter is useful in studying small retinal blood vessels and small hemorrhagic lesions as well as in detecting small macular changes. With the red-free aperture, the optic disc appears white and the blood vessels are more pronounced.

Operation of the Ophthalmoscope (Welch-Allyn)

Figure 13.12 illustrates the proper method of holding the instrument. The index finger is used to select the lens and aperture. The thumb is used to control light intensity. You will examine your laboratory partner's eyes in a darkened room. Before doing this examination, familiarize yourself with the mechanical operation of the ophthalmoscope.

1. Depress the red button on the rheostat, and rotate the knurled ring clockwise. Direct the light beam down on the tabletop, holding the ophthalmoscope head about 1-1/2 inches from the table surface. Observe the size and shape of the light beam.
2. Rotate the aperture selection disk. Observe the changes in the illuminated field on the table surface.
3. Set the lens selection at zero, and look through the aperture at the illuminated table surface. Rotate the rheostat control, and note how the light intensity changes.
4. Set the aperture to large (circular) and the lens to 15D. Looking through the aperture, focus on a printed word on this page. While continuing to look, rotate the lens selection disk to 12D, 10D, 8D, 6D, and so on, and notice how the focal length changes. The larger the number, the closer you must be to obtain proper focus and vice versa.

Ophthalmoscopic Examination

One problem often encountered by the inexperienced user of the ophthalmoscope is the reflection of light by the subject's cornea back into the examiner's eye, interfering with a clear view of the retina. To minimize this nuisance, direct the light beam toward the edge of the pupil rather than directly through the center. To conduct the ophthalmoscopic examination, use a darkened or semidarkened room and perform the following steps:

1. Place the 0 in the lens selection window, and select the red-free aperture. Turn on the light source.
2. Sit or stand at the subject's right side for examination of the right eye.
3. Take the ophthalmoscope in the right hand, hold it vertically in front of your own eye with the light beam directed toward the subject, and place your right index finger on the serrated edge of the dial in order to change lenses if necessary.
4. Instruct the subject to look straight ahead at some specific object at eye level.
5. Move to a position about 6 inches in front and 25 degrees to the right side of the subject, and direct the light beam into the edge of the pupil. A red "reflex" should appear as you look through the aperture.
6. While the subject holds visual fixation, keep the "reflex" in view and slowly move toward the subject. The optic disc should come into view when you are about 1½–2 inches away from the subject. If the focus is not clear, rotate the lenses into the aperture with your index finger until a clear field is obtained. Adjust the light intensity as necessary.
7. Examine the optic disc for clarity of outline, color, elevation, and condition of the vessels (figure 13.13). The optic disc (optic nerve head) is generally circular

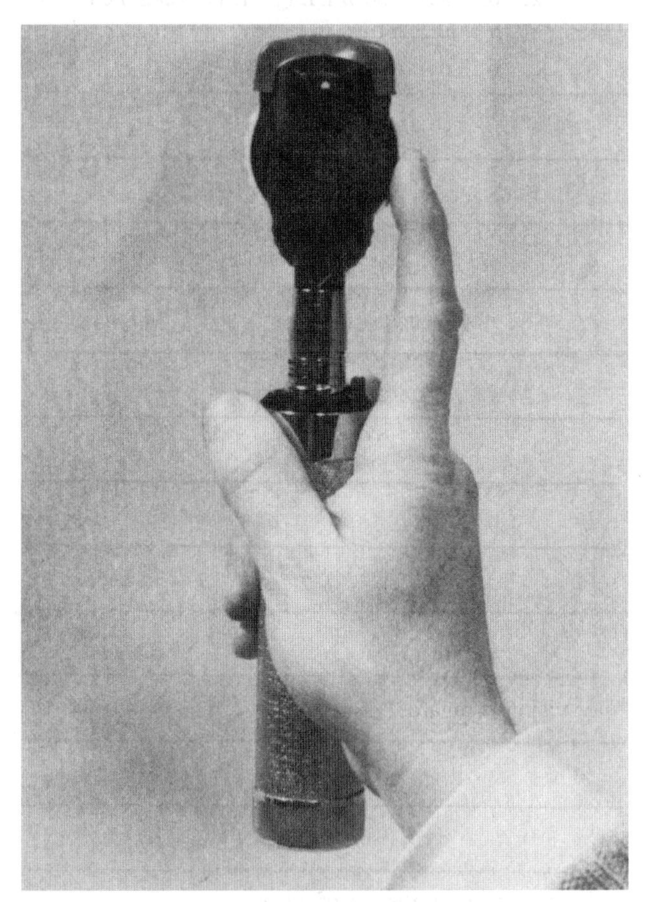

FIGURE 13.12 Correct position for holding the ophthalmoscope

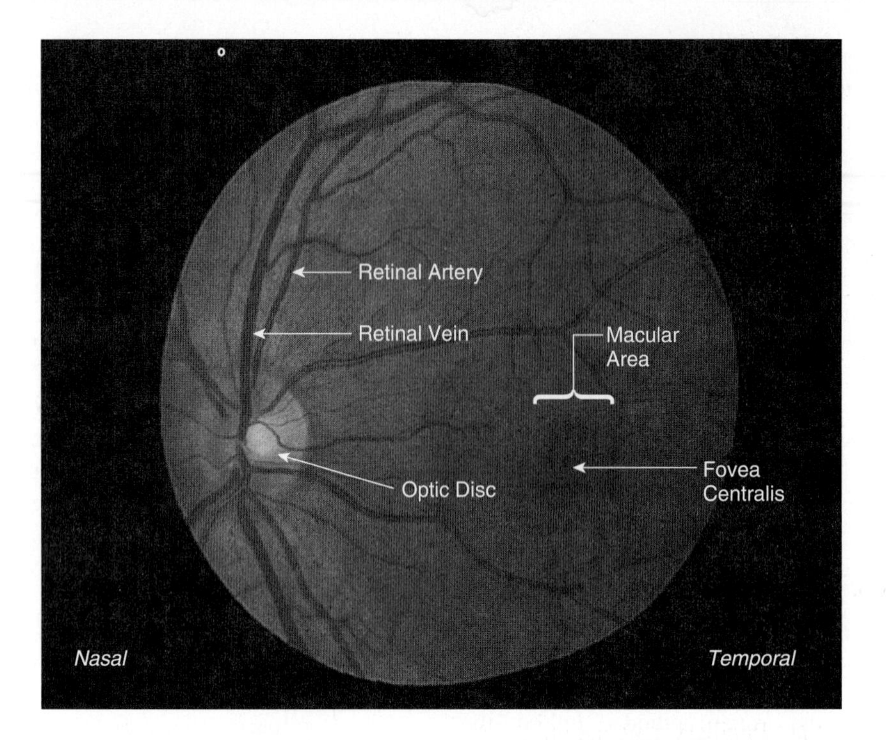

FIGURE 13.13 A normal fundus and optic disc of the left human eye viewed with an ophthalmoscope. The optic disc is generally circular to oval, has a vertical orientation, and is pink. The temporal side is usually lighter pink than the nasal side. The retinal arteries are red and smaller than the retinal veins and often appear to have a faint white stripe (called a central reflex stripe) running along the middle of the blood vessel. Branching is variable, but tortuosity is minimal. The macular area is located about two disc diameters temporal to the optic nerve head and is darker than the surrounding retina. The fovea centralis is at the center of the macula and is more prominent in the young eye, where it appears as a lustrous yellow spot. Light-reflectiveness of the fovea diminishes with age.

to oval with vertical orientation and is pink. The center of the disc may have some depression, which is called the physiologic cup. The depression is normally symmetrical and may involve up to 80% of the nerve head. The normal optic disc does not appear swollen or congested. The border of the optic disc may be sharp and distinct, or it may merge gradually with surrounding retina without any clear-cut edge.

Move the light temporarily, and check the macula (the area lacking blood vessels) for abnormalities. Then go back to the disc, and follow each vessel as far to the periphery as possible. The arteries are red and are smaller than the veins in about 4A:5V ratio. Because of a thicker wall, the arteries will have a shiny central reflex stripe. Branching is variable. Examine the vessels and evaluate them for transparency, the presence of pressure effects such as A-V compression (nicking) where vessels cross each other, the presence of focal narrowing of arterioles, increased tortuosity and widening of venules, and hemorrhages.

8. To examine the extreme periphery, instruct the subject to (a) look up for examination of the superior retina, (b) look down for examination of the inferior retina, (c) look laterally for examination of the temporal retina, and (d) look medially for examination of nasal retina.
9. An examination of the left eye is done in the same manner except that the ophthalmoscope is held in the left hand before the left eye and the examiner is at the subject's left side.
10. Consult the reference material for an interpretation of the images seen.

CHAPTER 14

Hearing and Equilibrium

■ INTRODUCTION

The ear is involved with two distinct special senses. The *auditory* portion of the ear is concerned with the sense of hearing. The *vestibular* portion of the inner ear is concerned with equilibrium and orientation of the body's position in space. Sensory information originating in the receptors in both the auditory and vestibular portions of the ear is relayed to the brain stem by way of the cochlear division and the vestibular division, respectively, of cranial nerve VIII (vestibulocochlear nerve).

Hearing

Sound waves, produced alternately by the compaction and rarefaction of molecules in the air, are conducted through the outer ear (figure 14.1) to the *tympanic membrane* (eardrum), causing the membrane to vibrate. Movement of the tympanic membrane causes the three ossicles of the middle ear—the malleus (hammer), incus (anvil), and stapes (stirrup)—to vibrate in sequence. The plate of the stapes is connected to the *oval window,* a flexible membrane of the cochlea. Vibration of the stapes produces compression waves in the fluid-filled cochlea of the inner ear.

The **cochlea** is a cone-shaped structure wrapped in a coil much like the shape of a snail's shell. Internally, the cochlea is subdivided into three fluid-filled chambers by membranes: the **scala vestibuli,** the **scala media** (cochlear duct), and the **scala tympani** (figure 14.2). The **vestibular membrane** separates the scala vestibuli and scala media. The **basilar membrane** separates the scala media and the scala tympani. The **oval window,** a membrane-covered opening, is located at the beginning of the scala vestibuli, and the **round window** is located at the end of the scala tympani. The scala vestibuli and the scala tympani, confluent at the helicotremma, are filled with a fluid called **perilymph.** The scala media ends at the helicotrema and is filled with **endolymph.**

The **organ of Corti,** containing auditory receptor cells called hair cells, supporting cells, and the tectorial membrane, is attached to the basilar membrane and extends as a continuous sheet of cells from the base to the apex of the cochlea. The organ of Corti contains two morphologically and physiologically distinct populations of hair cells. The **inner hair cell** functions as the primary auditory receptor cell. The **outer hair cell** serves as an effector to modulate and fine-tune auditory stimulation of the inner hair cell. Both inner and outer hair cells are innervated by afferent and efferent fibers connected to the brain stem by way of the auditory division of the vestibulocochlear nerve (C.N. VIII).

The footplate of the stapes, attached to the membrane of the oval window, transfers vibrations of the tympanic membrane to the perilymph in the scala vestibuli and in turn to the perilymph of the scala tympani, causing the flexible basilar membrane to vibrate. Vibration of the basilar membrane displaces the hair cells, bending the hairlike projections, which in turn generate nerve impulses in the afferent fibers connected to the cells (figure 14.3).

Interpretation of the pitch (frequency) of the sound is determined by the location of the maximally stimulated hair cells along the length of the basilar membrane. High-frequency sound waves cause maximal displacement of the basilar membrane and thus maximal stimulation of hair cells near the base of the cochlea, and low-frequency sound causes maximal displacement of the basilar membrane and maximal stimulation of hair cells near the apex of the cochlea. Sound intensity (loudness) is interpreted from the amplitude of the maximal displacement, which in turn determines the frequency of nerve impulses coming from the cochlea.

Impairments of hearing are classified into three categories, according to the part of the auditory apparatus that fails to function properly: (1) conductive hearing loss, (2) sensorineural hearing loss, and (3) central hearing loss.

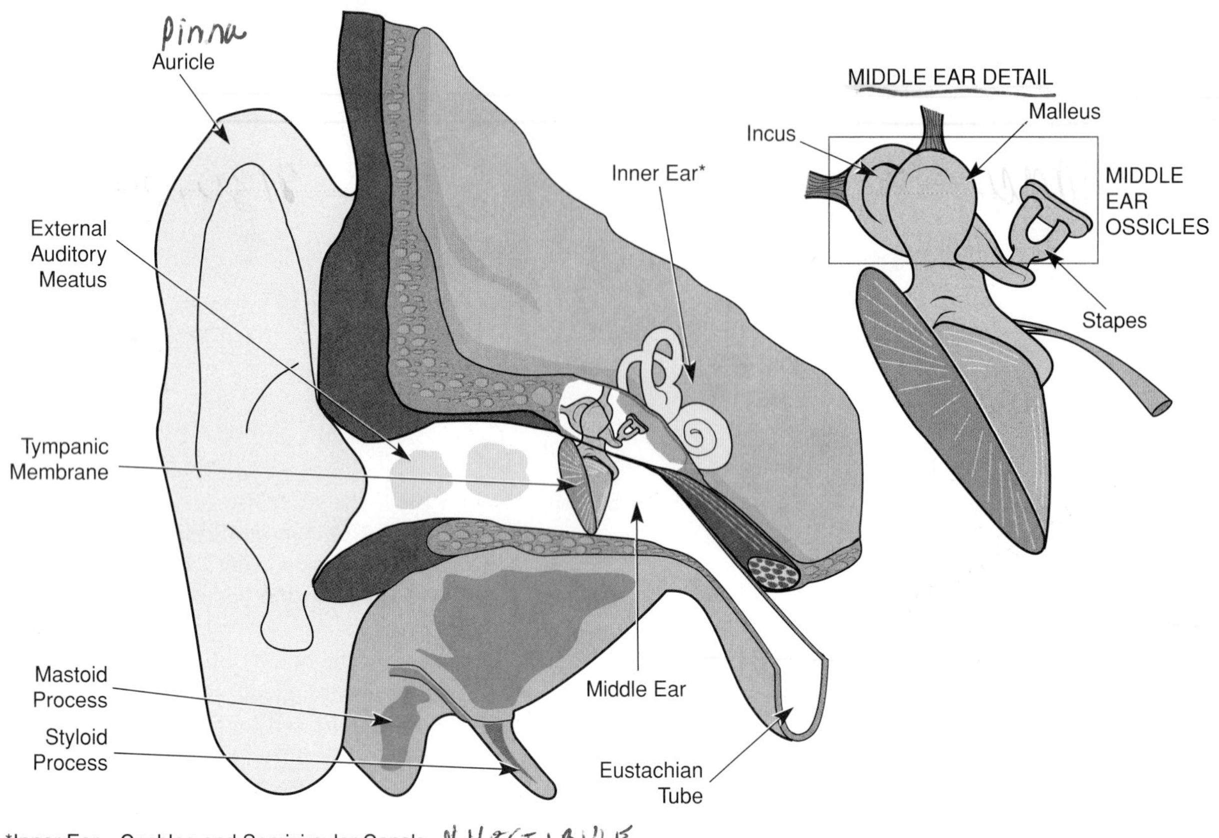

FIGURE 14.1 Frontal view of the human ear

Conductive hearing loss involves inadequate transmission of sound through the outer and middle ear to the inner ear. It may be caused by blockage of the external auditory canal; perforation, inflammation, or scarring of the tympanic membrane; inflammation, pus, and fluid in the middle ear (otitis media); and sclerosis of the ossicles (otosclerosis).

Sensorineural hearing loss is caused by damage to the cochlea, the organ of Corti, or the cochlear nerve fibers of cranial nerve VIII. Other causes include infectious organisms, degenerative bone disease, functional derangement of the organ of Corti due to traumatic sound, and inflammation of the cochlear nerve.

Defects in the auditory tracts of the brain stem or in the auditory portion of the temporal portion of the temporal cerebrum result in **central hearing loss.** Causes include malignancies of the brain, cerebrovascular disease, infections of the central nervous system, cerebral concussion, and hypoxia.

Equilibrium

The nonauditory portion of the inner ear is the *vestibular apparatus,* located adjacent to the cochlea (figure 14.4). It consists of three semicircular canals and the otolith organs (utricle and saccule), all of which are membranous structures enclosed within the bony labyrinth of the temporal bone. They are filled with endolymph and surrounded by perilymph.

The sensory cells of the vestibular system are hair cells located in the **ampullae** of the **semicircular canals** and in the **maculae** of the **utricle** and **saccule.** Each of the three semicircular canals is named after its position and the plane of space it represents: the **anterior (superior), posterior,** and **horizontal (lateral) semicircular canals.** Each canal is situated perpendicular to the other two. Thus, each semicircular canal is positioned in one of the three planes of space. When the horizontal (lateral) canal is oriented in the horizontal plane, the superior canal is in the anterior position. When the head is in its normal resting position, the "horizontal" canal is tilted, anterior end upward, approximately 30 degrees. With this arrangement, each semicircular canal detects the angular (rotational) acceleration in one of three planes in space.

The vestibular apparatus also consists of the otolithic organs, the **utricle** and **saccule.** They detect linear acceleration in the horizontal plane (utricle) and in the vertical plane (saccule) and also sense changes in the position of the head with respect to gravitational force.

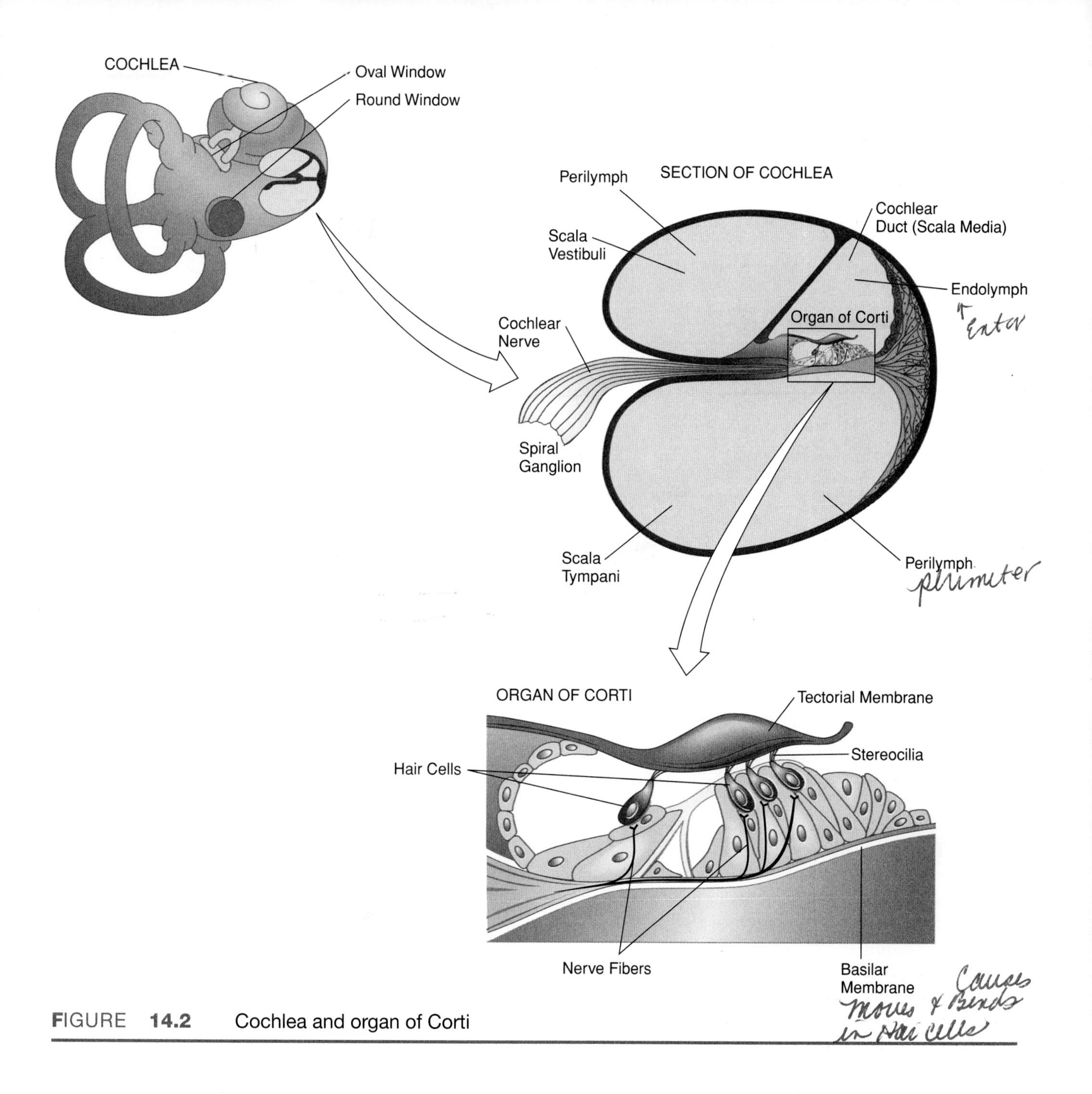

FIGURE 14.2 Cochlea and organ of Corti

The semicircular canals and otolithic organs are filled with the same endolymph fluid that surrounds the hair cells of the cochlea. As does the cochlea, the semicircular canals and otolithic organs have hair cells responsible for the transduction of sensory stimuli. The hairlike projections of the vestibular hair cells are called *stereocilia.* The stereocilia are aligned according to size, with the smallest cilium placed at one end of the group and the largest cilium, called the *kinocilium,* at the opposite end (figure 14.5). The adequate stimulus for the vestibular hair cell is the acceleration of the head. As the head begins to move, the inertia of the endolymph fluid causes the stereocilia to bend. This bending of the stereocilia causes the hair cell either to depolarize or to hyperpolarize. The type of membrane potential change in the hair cell depends on the direction in which the stereocilia bend. If the stereocilia are bent toward the kinocilium, the hair cell will depolarize. If the stereocilia are bent toward the smallest cilium, the hair cell will hyperpolarize.

The vestibular hair cells, like those in the organ of Corti, are not capable of generating action potentials. They do, however, excite or inhibit the **vestibular nerve fibers** that innervate them. As the hair cell depolarizes, it releases a neurotransmitter that excites the innervating vestibular

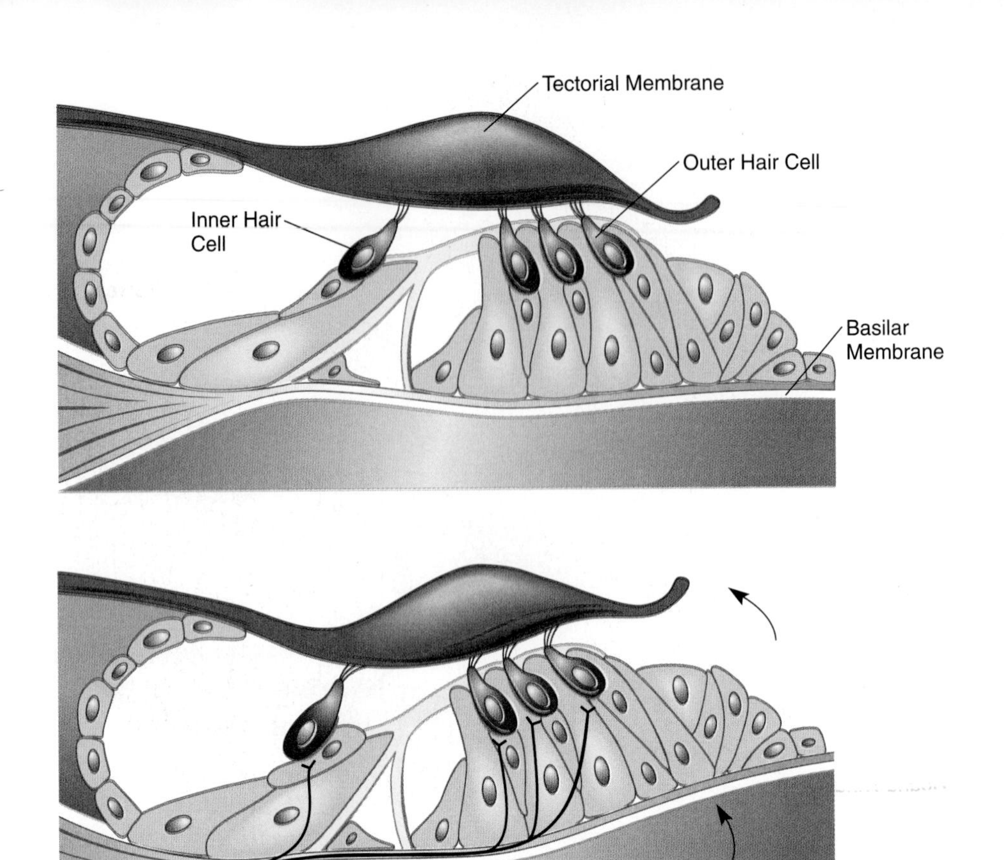

FIGURE **14.3** Distortion of hair cells due to sound wave deflection of basilar membrane

neuron. The hyperpolarization of a hair cell causes the vestibular neuron to produce fewer action potentials. This change may be due to a decrease in the amount of an excitatory transmitter that is tonically being released from the hair cell.

Near the junction of each semicircular canal and the vestibule is an enlarged portion called the *ampulla* (figure 14.6). Each ampulla contains a receptor organ called the *crista,* composed of hair cells and supporting cells embedded in a gelatinous matrix. The stereocilia and the gelatinous matrix form the *cupula,* a flame-shaped process extending into the endolymph of the ampulla. The hair cells are arranged in a polarized manner on both sides of the head. In the ampullae of the horizontal semicircular canals, for example, the stereocilia on each hair cell are positioned so that the kinocilium is nearest to the front of the head, while the smallest cilium is nearest the back of the head. Rotation of the head, termed *angular acceleration,* causes the inertia of the endolymph that fills the semicircular and ampullae to exert a force on the cupula in a direction opposite the rotation of the head. For example, as the head rotates from right to left (counterclockwise), the inertia of the endolymph applies a force on the cupula from left to right (clockwise). The fluid's relative motion causes the stereocilia on the left side of the head to depolarize the hair cells, while the stereocilia on the right side of the head hyperpolarize the hair cells (figure 14.7). Correspondingly, the vestibular nerves on the left side of the head increase their rate of action potential generation, while the nerves on the right side of the head decrease their rate of action potential generation. This information is thus transmitted to inform the brain that the head is rotating counterclockwise. When rotation stops, inertia of the endolymph temporarily deflects the cupulae in the opposite directions, resulting in a false sensation of rotating from left to right (clockwise) if the eyes are closed. The receptors of the semicircular canals respond only to a change in the rotational velocity of head movement (acceleration and deceleration). They are not stimulated by movement of the head at constant velocity.

The information from the vestibular nerves is transmitted to the **vestibular nucleus** located within the brain stem. From there, it is sent down to the spinal cord to act on nerve cells that regulate movement and up to the cerebellum and somatosensory cortex. The cerebellum and somatosensory cortex use this information to coordinate muscular activity in order to maintain balance.

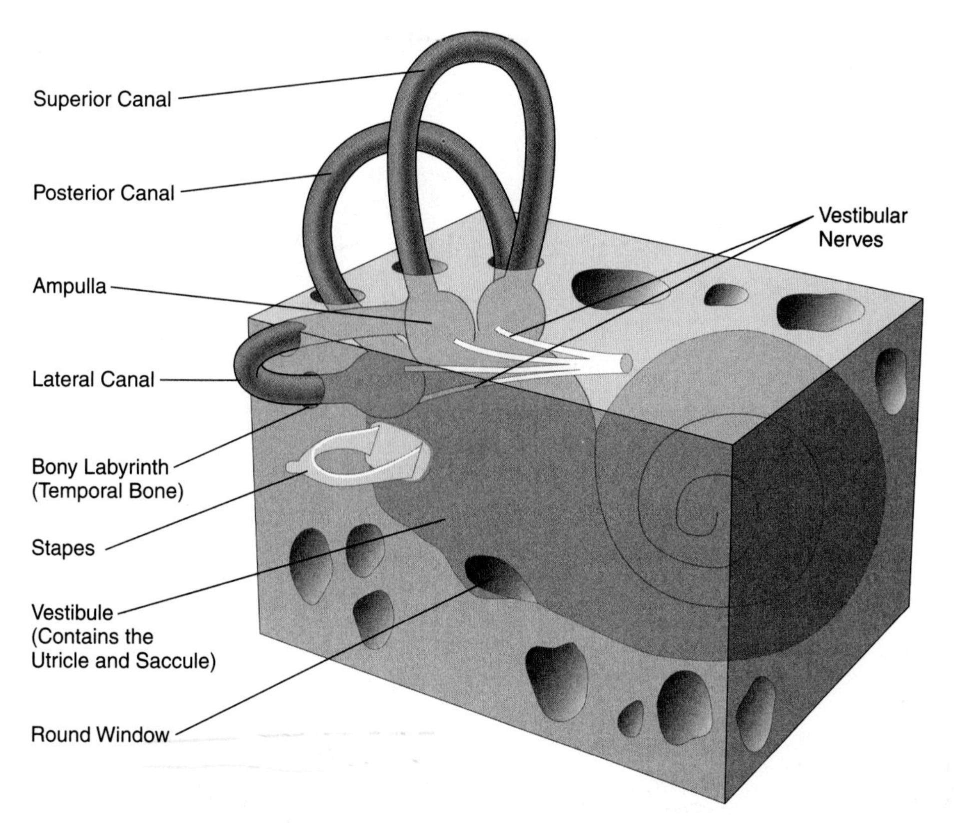

FIGURE **14.4** The vestibular apparatus of the inner ear

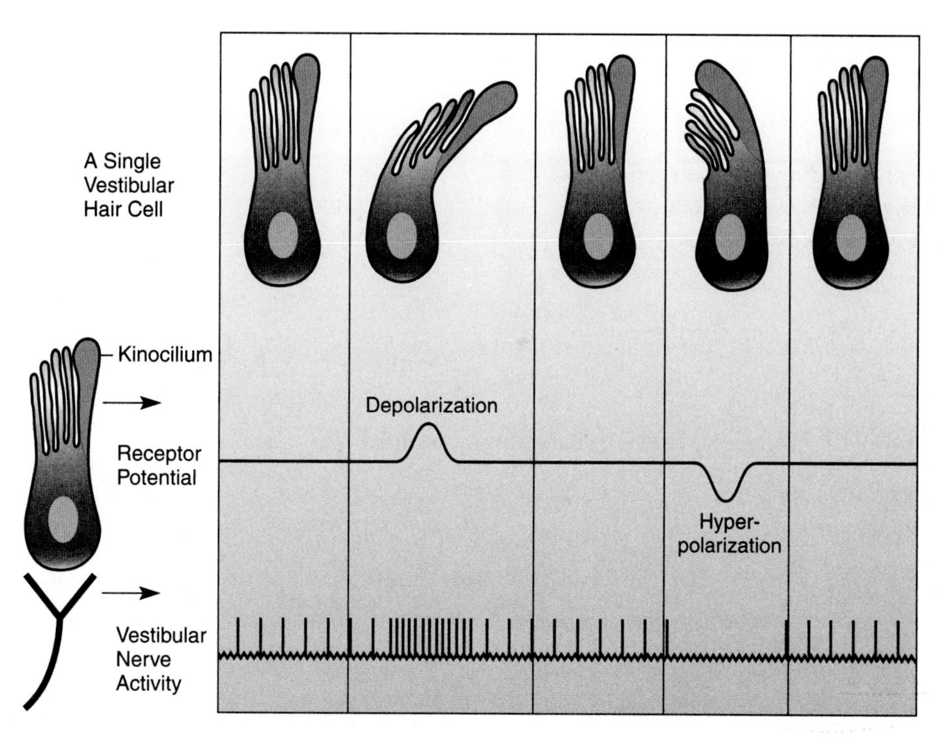

FIGURE **14.5** Depolarization and hyperpolarization of a vestibular hair cell

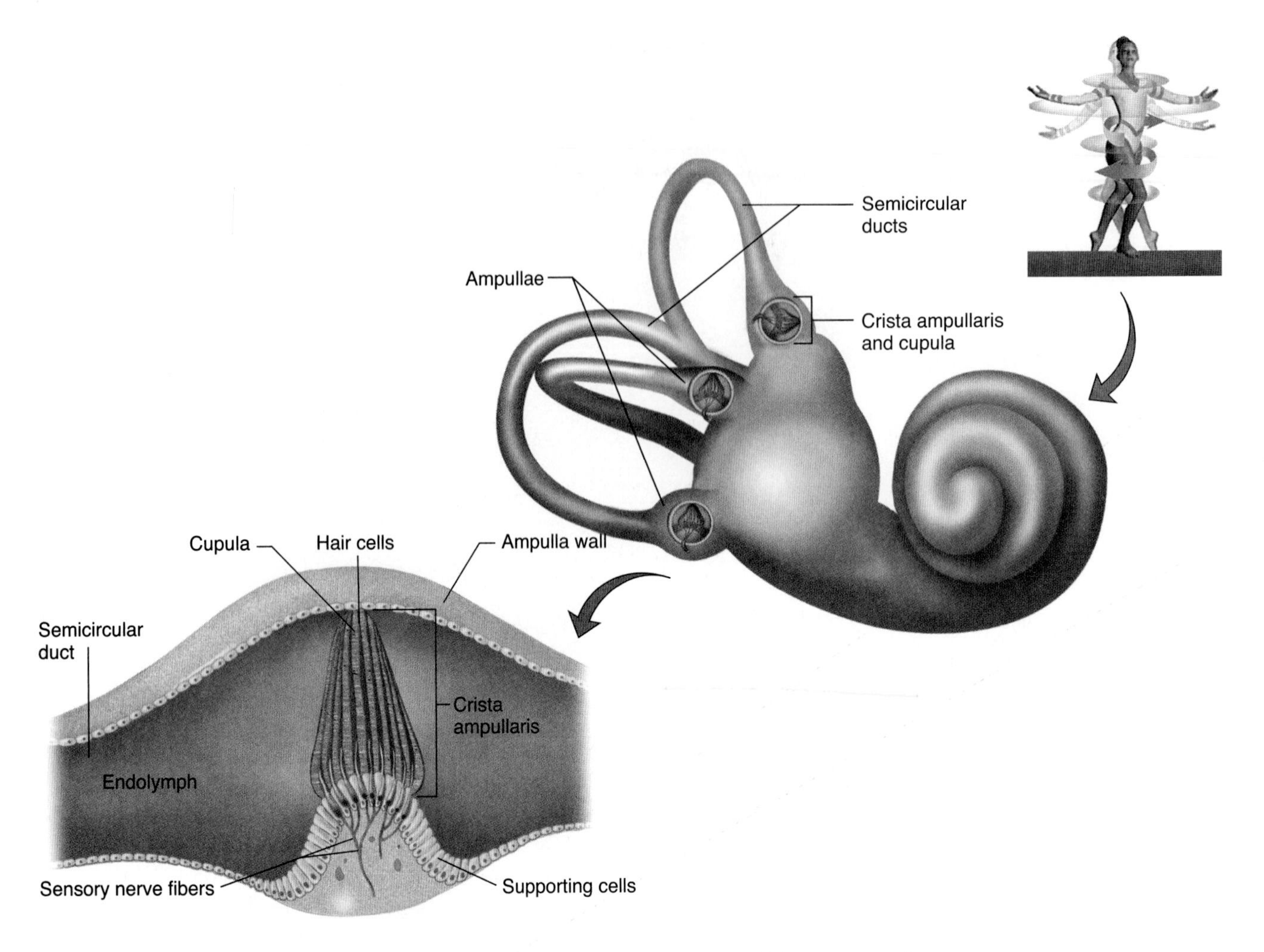

FIGURE 14.6 The ampulla and cupula of the right lateral semicircular canal

The vestibular system also helps the eyes to fixate on a spot in the visual field to provide a stable reference point during acceleration. This process is part of the **vestibulo-ocular reflex** and is achieved by the transmission of vestibular information for the control of eye movements. Outputs from the vestibular nucleus project to the motor nucleus of cranial nerve III (oculomotor nerve) ipsilaterally and cranial nerve VI (abducens nerve) contralaterally. These brain stem nuclei control muscles that cause the eyes to turn inward or outward, respectively. The movement of the head, with eyes open, in a clockwise direction causes a slow counterclockwise movement of both eyes to allow them to fixate on a point of reference. As the head continues to turn and the slow movement brings the eyes to an extreme position, there is a fast movement of the eyes in the direction of rotation. This allows the eyes to fixate on a new point of reference during rotation. The slow and fast movements of the eyes during accelerated angular rotation are called **nystagmus.** The fast-phase movement in the direction of rotation is conventionally chosen to describe the direction of the nystagmus.

The **maculae** are sense organs within the utricle and saccule that respond to change in the position of the head with respect to gravity (figure 14.8). They also respond to linear acceleration of the body. The stereocilia of the hair cells in the utricle and saccule are embedded in an **otolithic membrane** that contains calcium carbonate crystals called **otoliths** (*oto* = ear, *lith* = stone) dispersed in a gelatinous matrix. The otoliths give the otolithic membrane mass, so that when the head is tilted, the gravitational force causes the weighted otolithic membrane to bend the stereocilia. The stereocilia of the maculae are arranged in a complex pattern of orientations. When the head is tilted in any direction, part of the hair cell population depolarizes, part hyperpolarizes, and part is unaffected. Like the function of the stereocilia in the semicircular canals, bending the stereocilia toward the kinocilium depolarizes the hair cell, while bending away from the kinocilium produces a hyperpolarization. The complex signal generated by the total population of hair cells in the maculae on each side of the head provides the brain with information about the orientation of the head with respect to gravity.

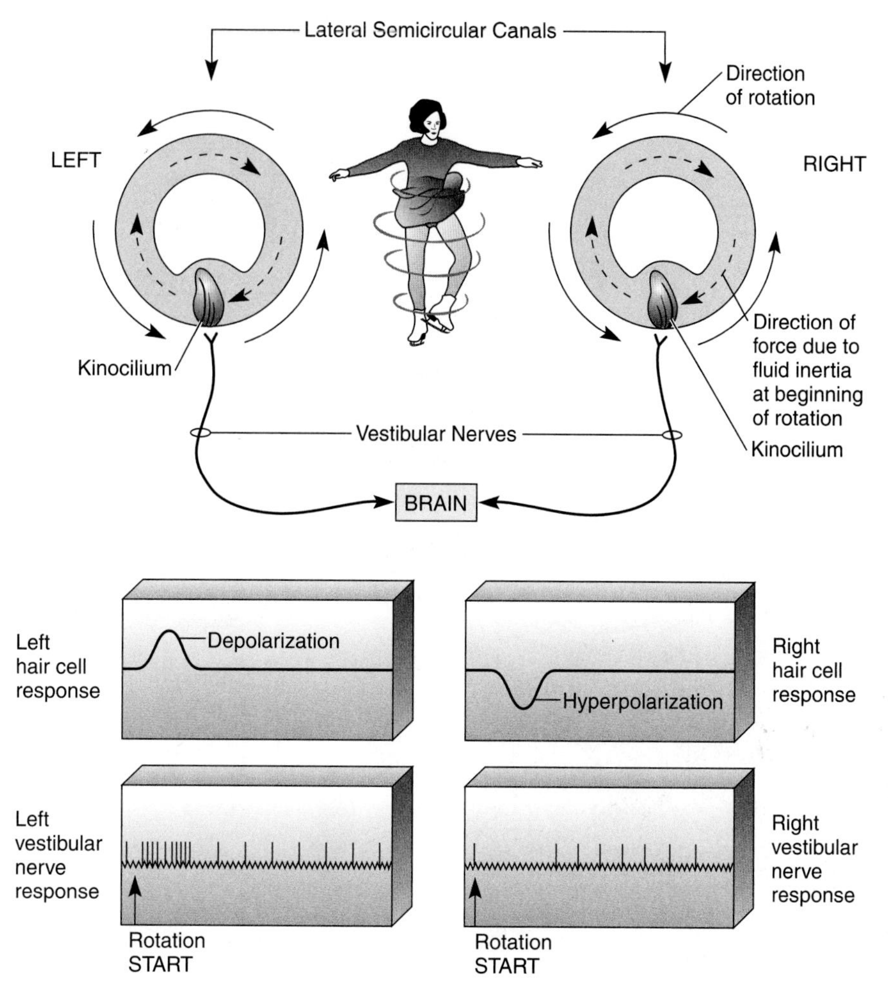

FIGURE 14.7 Vestibular apparatus: dynamic equilibrium

Afferent impulses from the vestibular apparatus supply information to the brain and help make a person aware of his or her position in space. In addition to information coming from the receptors in the muscles, joints, eyes, and skin, vestibular information initiates compensatory reflexes of the eyes and muscles of the neck, trunk, and limbs to keep the body in a proper orientation with respect to gravity. Because of the variety of efferent nerves affected by afferent impulses from the vestibular apparatus, intense or abnormal stimulation of vestibular receptors may result in vertigo (a turning sensation), nystagmus (involuntary, rhythmic, oscillating movements of the eyes), perspiration, vomiting, hypotension, and impairments of body, equilibrium, and balance.

In the experiments that follow, some functions of both the auditory and vestibular portions of the ear will be investigated.

■ EXPERIMENTAL OBJECTIVES

1. To review the functional morphology of the auditory and vestibular apparatus.
2. To assess auditory function relative to the localization of sound and the discrimination of sounds of varying frequency and intensity.
3. To observe the effect of vestibular activity.
4. To gain an understanding of the roles of the auditory and vestibular portions of the ear with respect to the senses of hearing and equilibrium.

Materials

models of the vestibular and auditory portions of the human ear

anatomical charts of the human ear

set of tuning forks (middle C—256 Hz)

Beltone audiometer (or equivalent)

swivel chair

watch

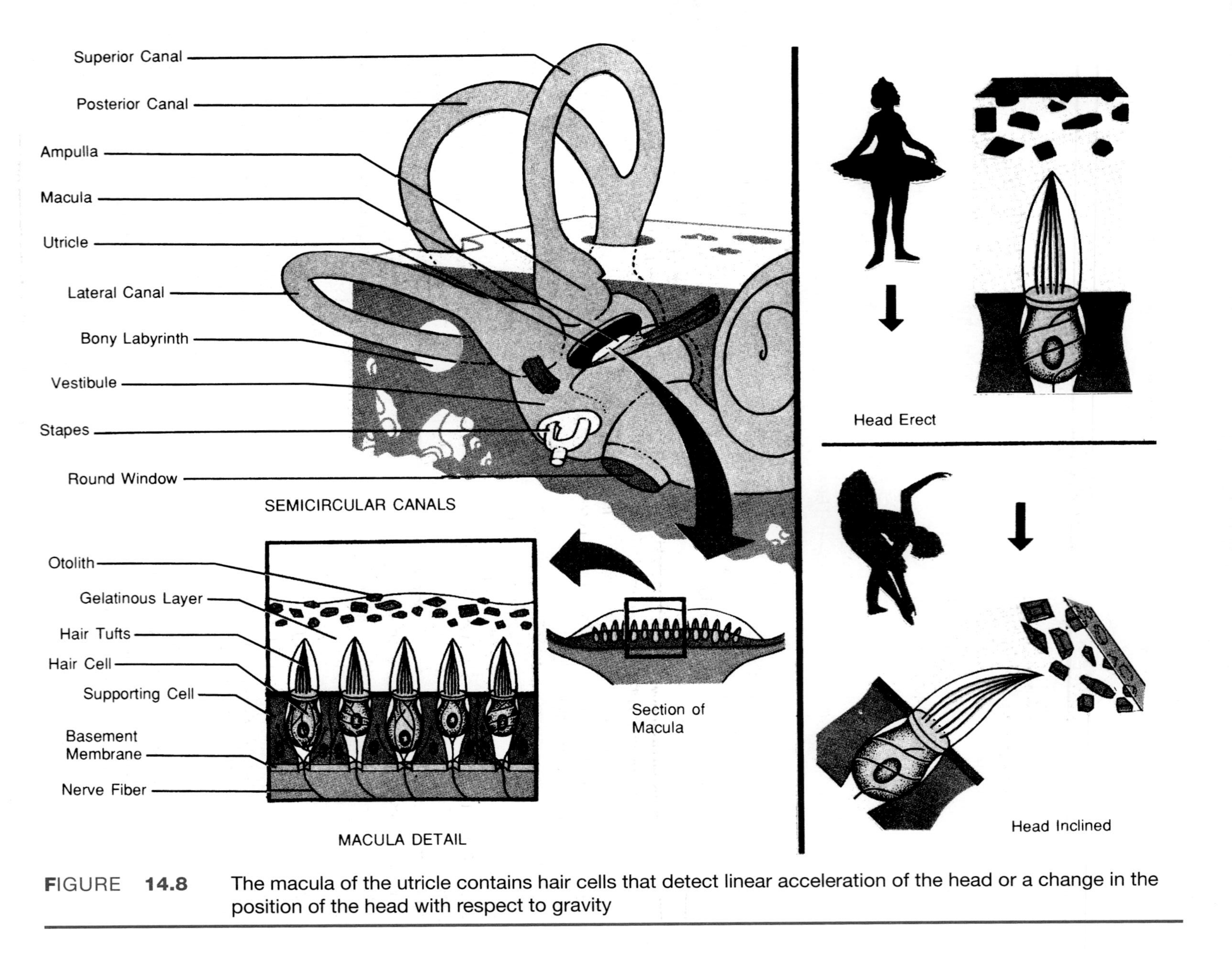

FIGURE 14.8 The macula of the utricle contains hair cells that detect linear acceleration of the head or a change in the position of the head with respect to gravity

■ EXPERIMENTAL METHODS

Auditory Function

Weber's Test and Rinne Test

Simple functional tests using tuning forks have been developed to aid in localizing the site of auditory difficulties. Two tests involving the use of a medium-pitch tuning fork are Weber's test and the Rinne test. The laboratory room must be quiet for each test to be successfully performed.

1. **Weber's test:** Obtain a 256-Hz tuning fork from the laboratory instructor. Strike the tuning fork prong firmly against a semi-hard surface (textbook cover, notebook) to start it vibrating. Place the tip of the handle in the median line of the skull or forehead of the subject. If hearing is normal, the subject will hear the sound equally well in both ears. Perform the test again, asking the subject to block one external auditory canal with a finger before repeating the test. The sound should be louder on the blocked side. Repeat the test blocking the other ear. Similar effects occur in conduction impairments of hearing because the conductive loss masks some of the environmental noise, thus making the cochlea more efficient on the diseased (or blocked) side. In sensorineural impairments, Weber's test results in the sound's being louder in the unaffected ear. Record your observations in the report.
2. **Rinne test:** This test is a comparison of the duration of air conduction with that of bone conduction. Strike the prong of a 256-Hz tuning fork firmly against a semi-hard surface (textbook cover, notebook) to start it vibrating. Place the tip of the tuning fork handle on the mastoid process with the prongs directed toward back of head, being careful not to touch the prongs of the fork against the ear or hand. As soon as the subject no longer hears the sound, move the tuning fork to within 1 inch of the external auditory meatus, and the subject should again be able to hear the sound. Repeat the test for other ear.

   Normally, the sound conducted by air transmission is heard several seconds longer than that conducted by bone transmission because the threshold for air transmission is lower. When this relationship is noted, the Rinne test is designated as positive (+), suggesting no conductive hearing loss. Air conduction that is equal to or less than bone conduction indicates a conductive hearing loss (e.g., as in middle ear damage), and the Rinne test is designated as negative (–). Table 14.1 summarizes results of Weber's test and the Rinne test. Record your test results in the report.

Audiometer

The most precise method of testing auditory acuity is to determine and plot the minimal intensities of sound that are just audible at different frequencies. The device used to test auditory acuity in this manner is called an audiometer (figure 14.9).

TABLE 14.1 Auditory function test results

Hearing	*Weber's*	*Rinne's*
Normal	Midline	AC > BC Both ears
Conductive loss	Lateralizes to affected ear	BC ≥ AC Affected ear
		AC > BC Normal ear
Sensorineural loss	Lateralizes to normal ear	AC > BC Both ears

AC = Air conduction
BC = Bone conduction

A pure tone is presented separately to each ear, and the threshold intensity for hearing is determined. The *reference threshold intensity,* based on norms for a large population with normal hearing, is arbitrarily set at zero decibel (0 db). A *decibel (db)* is a logarithmic unit representing a 25% change in sound intensity. Mathematically, the decibel is defined by the following equation: db = 10 log $I^1 \div I_0$, where I^1 = sound in question and I_0 = reference sound intensity. A common reference level is 10^{-16} watts/cm^2, which is set equal to zero decibel and is approximately the lowest level of sound intensity that can be heard by humans.

If a subject requires a sound intensity 25 db above the reference to hear a sound, the subject has a 25 db hearing loss. A hearing loss of 25–40 decibels causes difficulty in understanding normal speech in the range of frequencies of 500–3000 Hz (1 Hz = 1 cycle per second of sound).

The normal range of frequencies recognized as sound varies approximately between 16 and 20,000 Hz, which is about 10 octaves. Higher frequencies are heard by lower animals (e.g., 35,000 Hz by a dog; 50,000 Hz, cat; 45,000 Hz, grasshopper). The upper range of auditory frequency in the human tends to decrease with age, from 23,000 Hz in children to 15,000 Hz at age 40, to 8000 Hz at age 70–80. Figure 14.10 illustrates a normal audibility curve.

Several models of audiometers are in use for testing human hearing. The model pictured (figure 14.9) and described below is the Beltone Model 119 Audiometer. If your audiometer is different, follow your laboratory instructor's directions for proper use.

1. *Power on-off switch*
2. *Power light*
3. *Tone mode switch:* This switch controls the method of tone presentation. In the "norm off" position the tone is off until the tone presentation switch (7) is pressed. In the "norm on" position the tone is on until the tone presentation switch is pressed and then the tone is interrupted. The pulsed position will present a pulsed tone only when the presentation switch is pressed.

FIGURE **14.9** The Beltone audiometer

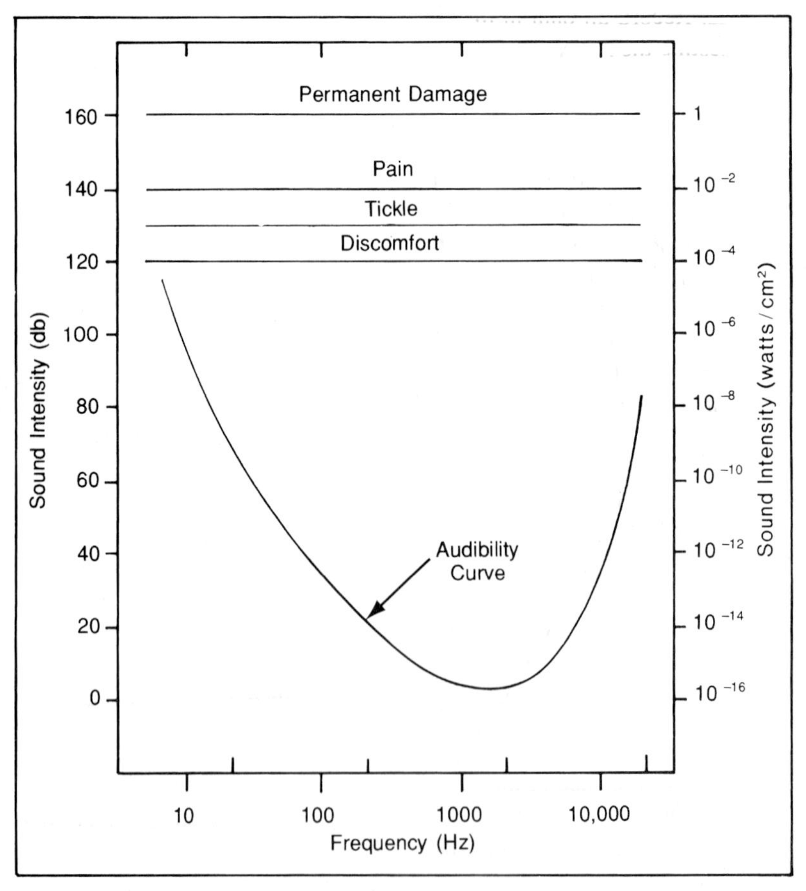

FIGURE **14.10** The human audibility curve

4. *Output selector switch:* Used to select the earphone (red = right ear, blue = left ear) through which the signal is sent.
5. *Patient response light:* Lights when patient presses handheld patient signal switch.
6. *Hearing level control:* The HL dial controls the intensity of the pure tone.
7. *Tone presentation switch*
8. *Tone light:* Will light when pure tone signal is keyed on.
9. *Frequency selector switch:* All 11 audiometric frequencies are available as test stimuli. This dial indicates the maximum calibrated HL setting allowed at the selected frequency. "A" designates air conduction maximum level. "B" indicates bone conduction maximum level.

Select a quiet room for audiometry. It is important that extraneous noise be kept at a minimum. When performing tests with the audiometer, be as quiet as possible. Instruct your laboratory partner to sit opposite the audiometer, so the controls cannot be seen, and to place the earphones correctly about the head. With the intensity control set at 0 db, set the frequency control for 125 Hz and the tone switch for "norm on." Select the right ear for testing, and gradually increase the sound intensity until the subject indicates by raising a finger that sound can be heard. Record the threshold intensity for sound at 250, 500, 750, 1000, 1500, 2000, 3000, 4000, 6000, and 8000 Hz. Record all data in the report. Repeat the procedure for testing the left ear.

Localization of Sound

In addition to decoding the frequency and intensity of sound information, the auditory system determines where sounds originate. This function of the auditory system allows us to orient toward a sound stimulus and respond accordingly. Stereophonic, or binaural, hearing is of practical importance in terms of survival in the environment. When we are crossing a street at a busy intersection, for example, the noise of a speeding car coming around the corner will cause us to stop, look in the direction of the oncoming car, and avoid a collision. The localization of sounds in space is accomplished by comparing differences in the timing and intensity of sounds as they reach each ear. Acoustic waves that originate on the right side of the head would first reach the right ear before traveling on to the left ear. The time delay between the two ears is called the **interaural time difference.** Sounds that originate along the midline would produce no interaural time difference. Sounds that originate off of the midline have been shown to produce interaural differences of 42 μs for every 20-degree shift in space. Some sound waves traveling from the right side of the head to the left ear are absorbed by the head, and others are reflected by objects in the environment into the ear. Absorption and reflection decrease the intensity of sound entering the ear farthest from the sound source. The brain uses the intensity difference to help localize the sound source.

While the subject's eyes are closed, gently strike the prongs of a tuning fork against the heel of your shoe and hold the vibrating tuning fork at various positions about the subject's head (front, back, sides). Ask the subject to locate the source of sound. In which direction is sound localization most accurate? Least accurate? Record your observations.

Vestibular Function

Nystagmus is an involuntary rhythmic movement of the eyes in which the eyes move first in one direction and then back in the opposite direction, then repeat in an oscillating manner. The oscillations may be horizontal (side to side), vertical (up and down), rotary, or mixed. In pendular nystagmus, the rate of eye movement in each direction is the same. In jerking nystagmus, eye movements are faster in one direction.

Sensory information from the semicircular canal is used by the brain to control postural muscles and extrinsic eye muscles in a manner consistent with maintaining balance and body position in space and to provide for a sense of awareness concerning angular acceleration and deceleration when the body rotates about an axis.

The effects of vestibular influence on somatic musculature (in this case, extrinsic eye muscles and muscles of the extremities) can be observed by inducing rotational vestibular nystagmus (Barany's test). The procedure for testing follows the caution message concerning this experiment. Record your observations in the report.

CAUTION

The subject is to grasp the seat of the swivel chair with both hands and remain seated after the rotation period until all feelings of dizziness, vertigo, and, in some cases, mild nausea disappear.

1. Seat the subject in a swivel chair with the subject's head tilted 30 degrees forward. (This tilt places the horizontal semicircular canals parallel to the floor for maximal stimulation during rotation about a vertical axis.) Ask the subject to grasp the chair firmly, so as to prevent falling while being rotated, and to close the eyes.
2. Rotate the subject to the right quickly for about a dozen revolutions. When the subject is first rotated to the right, the cupulae of the semicircular canals will be bent to the left (owing to inertia of the endolymph). If the subject's eyes were open (keep closed), you would see the horizontal nystagmus taking place: the eyes drift slowly to the left and then quickly return to the right. Normally, this movement helps the subject maintain balance by helping the eyes to fix on visual cues. Nystagmus continues until inertia of the endolymph is overcome and the cupulae return to their original position, even though rotation continues.
3. After a dozen rapid revolutions, abruptly stop the chair, ask the subject to open the eyes, and note the

subject's eye movement. When rotation is abruptly stopped, the endolymph continues to move temporarily and the cupula is bent to the right, producing horizontal nystagmus with a slow component to the right and a rapid component to the left.

4. After the subject no longer exhibits postrotational nystagmus, stand in front of the subject, extend both of your hands, and ask the subject to touch both of your index fingers with his or her index fingers. Normally, with the subject's eyes open, there will be no deviation error. With subject and examiner remaining in position, ask the subject to repeat the task with eyes closed. With the normal subject, there will again be no deviation error, because the subject is oriented well enough so as to place the fingers in the correct position. With the abnormal subject, there will be a deviation error to the right or left.
5. Seat the subject in the swivel chair with his or her head tilted forward 30 degrees and again rotate the subject with closed eyes rapidly to the right for a dozen revolutions. Abruptly stop rotation, ask the subject to open the eyes, quickly hold out your extended index fingers, and ask the subject to touch both of your index fingers with his or her index fingers. The normal subject will deviate to the right (pass-point to the right) and also exhibit a tendency to lean to the right for a very brief period of time after rotation ceases.
6. Repeat steps 2–5, but rotate the subject to the left instead of to the right.

Destructive lesions of the labyrinth or of the vestibular nerve are characterized in Barany's test by failure to exhibit nystagmus and pass-pointing, as well as falling.

CHAPTER 15

Electroencephalography I: Relaxation and Brain Rhythms

■ INTRODUCTION

The brain is encased by the **cranium,** bones of the skull that immediately cover and protect brain surfaces. A thin cover of skin, called the scalp, covers most of the cranium. The largest part of the brain, located immediately beneath the cranium, is the **cerebrum** (figure 15.1). The cerebrum is divided into hemispheres, and each hemisphere is divided into frontal, parietal, temporal, and occipital lobes. The outer cell layers of the cerebrum form the **cerebral cortex,** the "gray matter" of the brain often referred to in popular literature. The cerebral cortex contains billions of nerve cells (neurons), many of which are functionally connected to each other and connected to other parts of the brain.

Functions of the cerebral cortex include abstract thought, reasoning, memory, voluntary and involuntary control of skeletal muscle, and the recognition and differentiation of somatic, visceral, and special sensory stimuli. Specific regions of the cerebral cortex (figure 15.2) process or generate various kinds of information. For example, the **frontal lobe** generates nerve signals that voluntarily control skeletal muscle contractions such as in walking or riding a bicycle. The **occipital lobe** processes visual (sight) information, and the **temporal lobe** processes auditory (hearing)

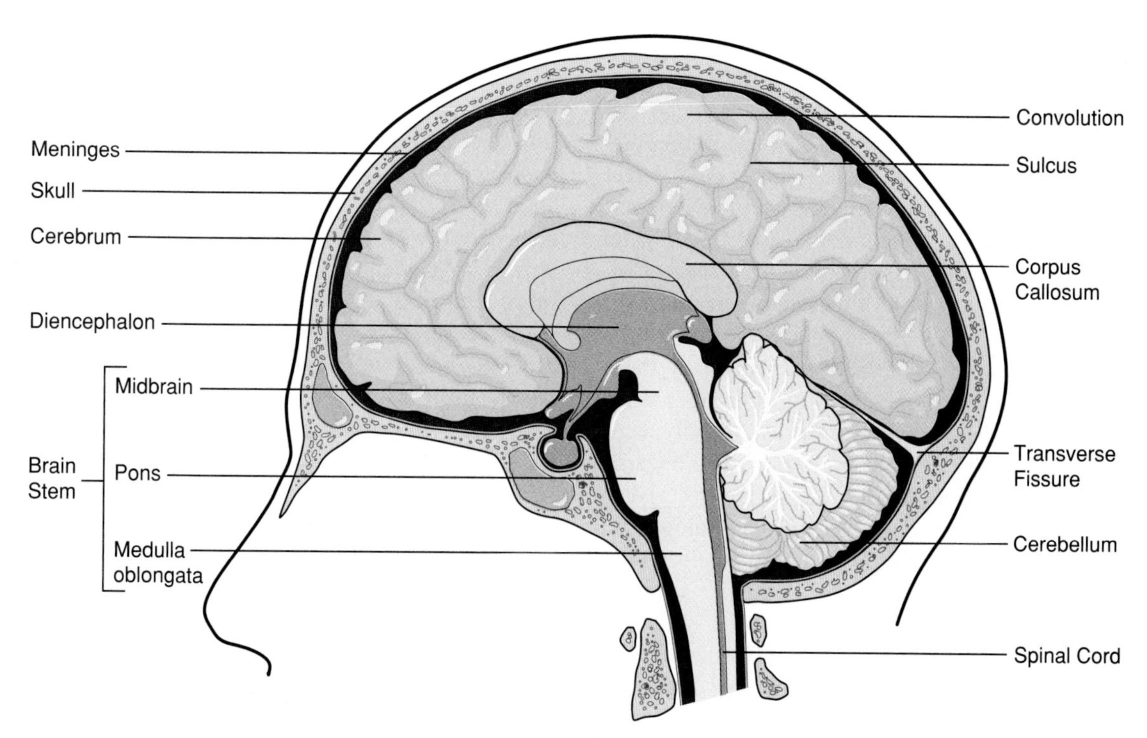

FIGURE **15.1** The major portions of the brain: the cerebrum, cerebellum, and brain stem

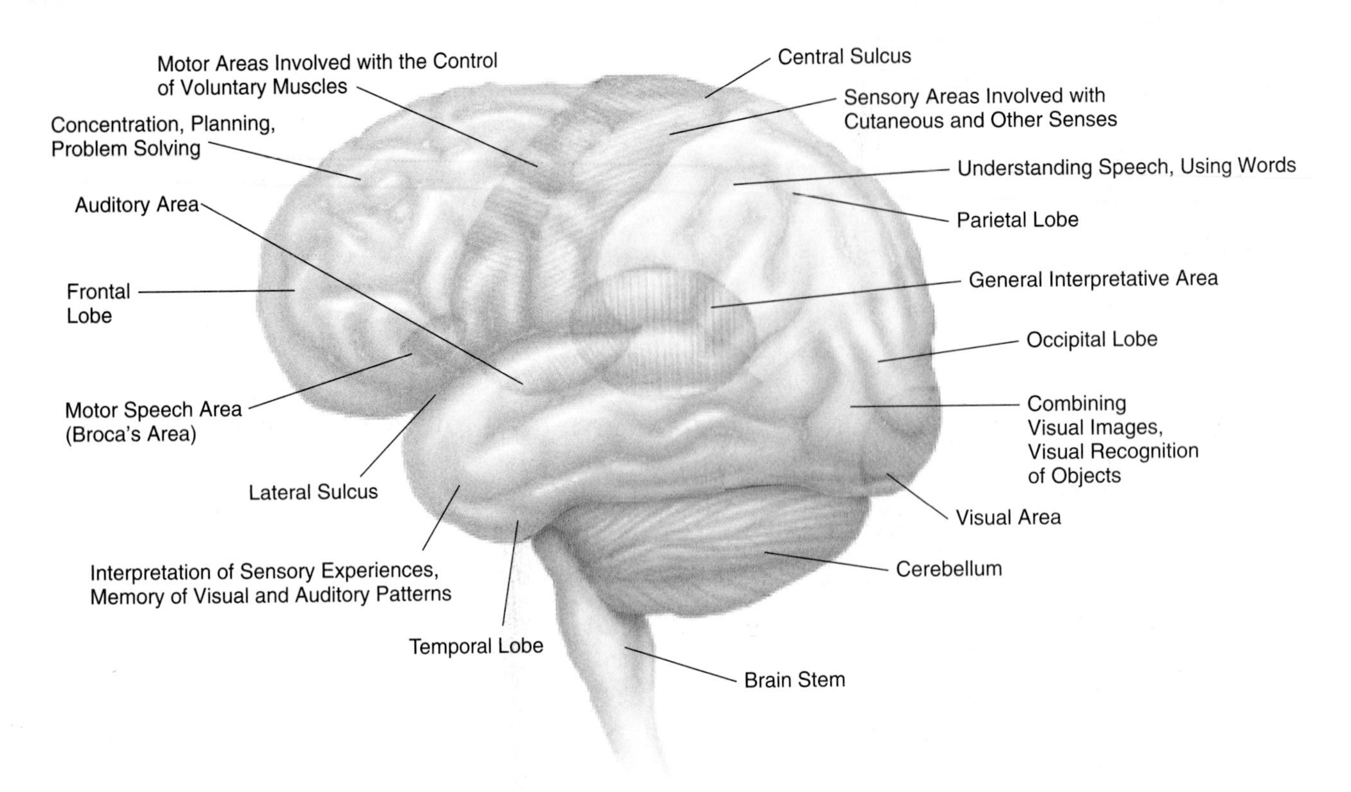

FIGURE 15.2 Some motor, sensory, and association areas of the cerebral cortex

information. Cutaneous pain and temperature information and other somatosensory information are processed in the **parietal lobe.** Electrical activity in the form of nerve impulses being sent and received to and from cortical neurons is always present, even during sleep or other states when the level of consciousness is reduced. In a legal sense as well as a medical or biological sense, absence of electrical activity in the human cerebral cortex signifies death.

In 1929 an Austrian physician named Hans Burger discovered that electrodes placed on the scalp could detect various patterns of electrical activity and that this detected electrical activity was not due simply to artifacts of scalp musculature. Burger recorded the patterns of electrical activity and called the record an **electroencephalogram** (*electro* = electric, *encephelo* = brain, *gram* = record). Soon after Burger's discovery, scientists began to study these "brain waves," and the detection, amplification, recording, and interpretation of the patterns of electrical activity associated with functioning of the cerebral cortex became known as **electroencephalography.** The hardware used to record such patterns is called an **electroencephalograph,** and the record obtained from its use is called an **electroencephalogram,** or **EEG.**

Today, the EEG is still medically useful for recording brain function. In medical and basic research, the correlation of particular brain waves with sleep phases, emotional states, psychological profiles, and types of mental activities is ongoing.

EEG signals are recorded as a series of complex waveforms. Basic knowledge regarding waveform terminology and analysis is useful and therefore is reviewed here to assist you as you record, examine, and analyze an EEG.

Two fundamental characteristics of a regular, repeating waveform are its amplitude and its frequency.

Amplitude refers to the "height" or "depth" of a waveform as measured from a reference point called the **baseline.** Amplitude values above the baseline are considered *positive* (+), and this part of the waveform appears as a "hill" or "peak." Amplitude values below the baseline are considered *negative* (–), and this part of the waveform appears as a "trough" or "valley" (figure 15.3). The amplitude of an electrical waveform may be measured in *volts* (V), *millivolts* (mV), or *microvolts* (μV).

Frequency refers to the number of times a waveform repeats itself in a given interval of time, such as 1 second or 1 minute. A waveform repeating itself 60 times in 1 second has a frequency of 60 **cycles per second** (cps) or 60 **hertz** (Hz). One Hertz equals one cycle per second.

Frequency is measured by counting the number of peaks or troughs (but not both) within 1 second (cps) or 1 minute (cpm).

A waveform with a constant interval of time between peaks is called **periodic.** A waveform with variable intervals of time between peaks is termed **nonperiodic.** The waveforms shown in figure 15.3 are periodic sine waves.

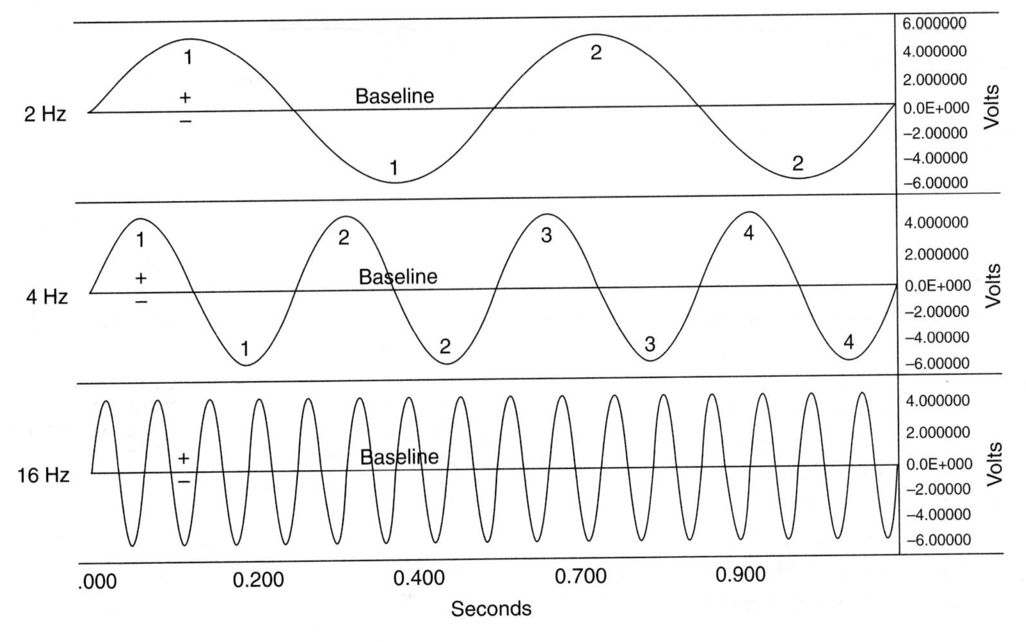

FIGURE 15.3 Three waveforms of equal amplitude but different frequency

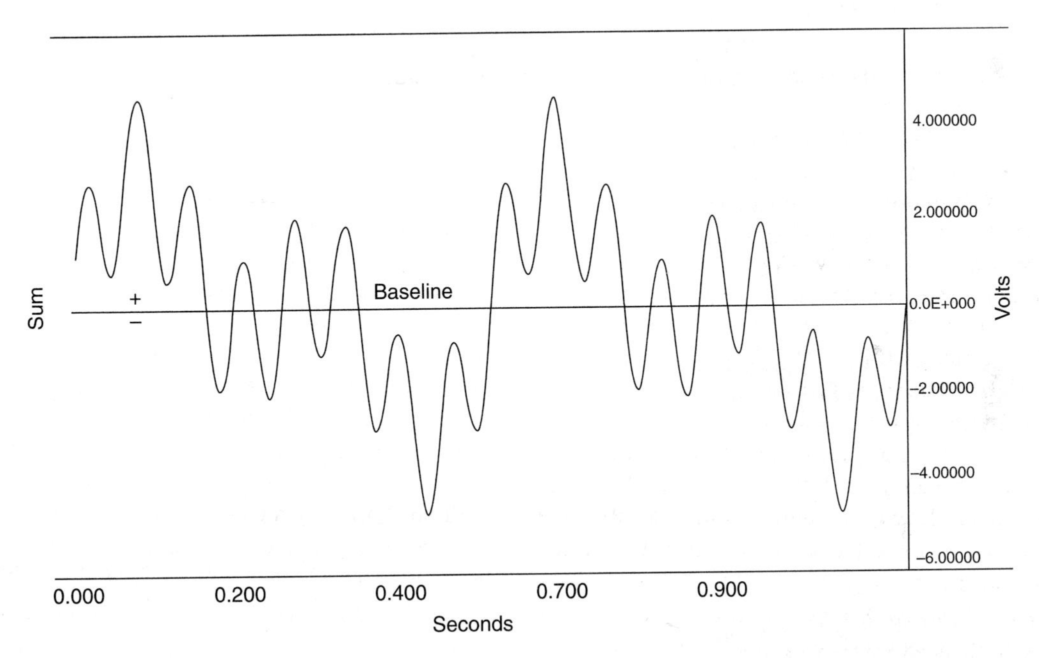

FIGURE 15.4 Complex waveform representing the sum of 2-Hz, 4-Hz, and 16-Hz waveforms

Waveforms may have identical amplitudes but differ in frequencies, as shown in figure 15.3 or they may have identical frequencies but differ in amplitudes. A **complex waveform** results when two or more waveforms, each with different amplitudes and frequencies, are added together. Figure 15.4 shows the complex waveform that represents the sum of the 2-Hz, 4-Hz, and 16-Hz sine waves shown in figure 15.3. If we were to mathematically remove (by waveform subtraction) the 2-Hz and 4-Hz waveforms, we would be left with a 16-Hz waveform as shown in figure 15.3. Similarly, when an EEG is analyzed, dominant waveforms or rhythms are separated from the rest of the record so as to make the examination easier.

Four simple periodic rhythms recorded in the EEG are **alpha, beta, delta** and **theta** (table 15.1) These rhythms are identified by frequency (Hz, or cps) and amplitude. The amplitudes recorded by scalp electrodes are in the range of microvolts (μV, or 1/1,000,000 of a volt). The amplitude

TABLE 15.1 Typical frequencies and amplitudes of synchronized brain waves		
Rhythm	***Typical frequencies (Hz)***	***Typical amplitude (μV)***
Alpha	8–13	20–200
Beta	13–30	5–10
Delta	1–5	20–200
Theta	4–8	10

measurements shown in table 15.1 are those values reported for clinical settings; in a classroom setting, the amplitudes may be much lower.

Alpha

In general, the alpha rhythm is the prominent EEG wave pattern of an adult who is *awake but relaxed with eyes closed.* Each region of the brain has a characteristic alpha rhythm, but alpha waves of the greatest amplitude are recorded from the occipital and parietal regions of the cerebral cortex. Results from various studies indicate that:

1. Females tend to have higher mean frequencies of alpha waves than males.
2. Alpha wave amplitudes are likely to be higher in "outgoing" subjects.
3. Alpha wave amplitudes vary with the subject's attention to mental tasks performed with the eyes closed.

In general, amplitudes of alpha waves diminish when subjects open their eyes and are attentive to external stimuli, although some subjects trained in relaxation techniques can maintain high alpha amplitudes even with their eyes open.

Beta

When the eyes are open and the individual becomes alert and attentive to external stimuli or exerts conscious mental effort such as when performing mental computations, the alpha rhythm is replaced by the lower, faster beta rhythm. This transformation is known as *desynchronization* of the alpha rhythm and represents *arousal of the cortex to a state of alertness.* Paradoxically, beta rhythms also occur during deep sleep—that is, rapid eye movement (REM) sleep—when the eyes rapidly move back and forth beneath closed eyelids. The beta rhythm is best recorded from precentral regions of the frontal cortex. Figure 15.5 illustrates various patterns of the EEG associated with the awake state. Notice that the amplitudes of beta waves tend to be lower than those of alpha waves. This difference does not mean that there is less electrical activity; rather, it means that the "positive" and "negative" waveforms are starting to offset one another so that the sum of the electrical activity is less.

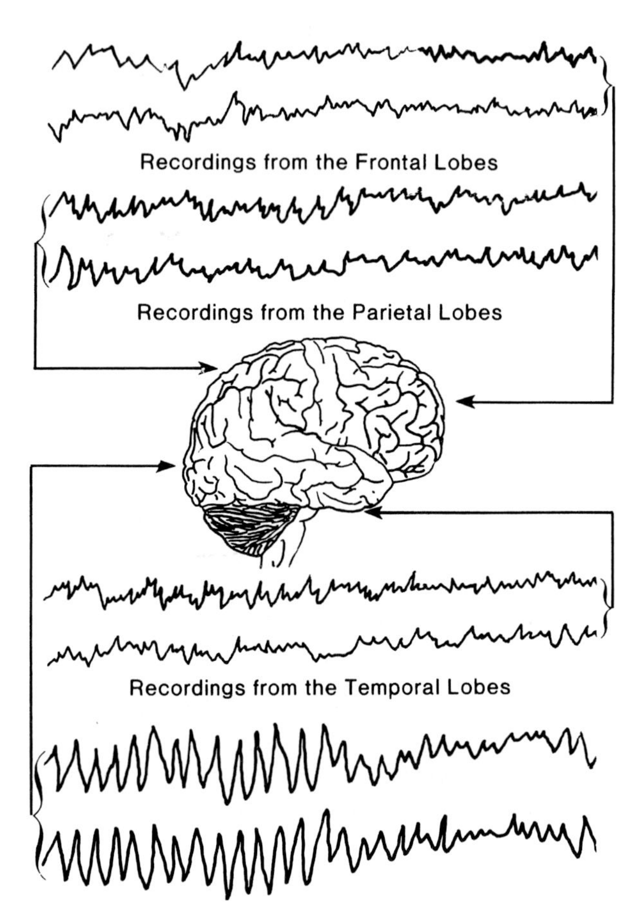

FIGURE 15.5 EEG patterns associated with each lobe of the cerebrum in the awake state

Delta and Theta

Delta and theta rhythms are low-frequency EEG patterns associated with *drowsiness and sleep* in the normal adult. As a person becomes drowsy, the alpha rhythm is gradually replaced by the lower-frequency theta rhythm. As sleep deepens, the slow-wave delta rhythm becomes dominant. Periodically during slow-wave sleep, the delta rhythm is interrupted by episodes of *paradoxical sleep* during which the subject appears to be asleep but has an EEG pattern similar to the beta rhythm of an alert individual.

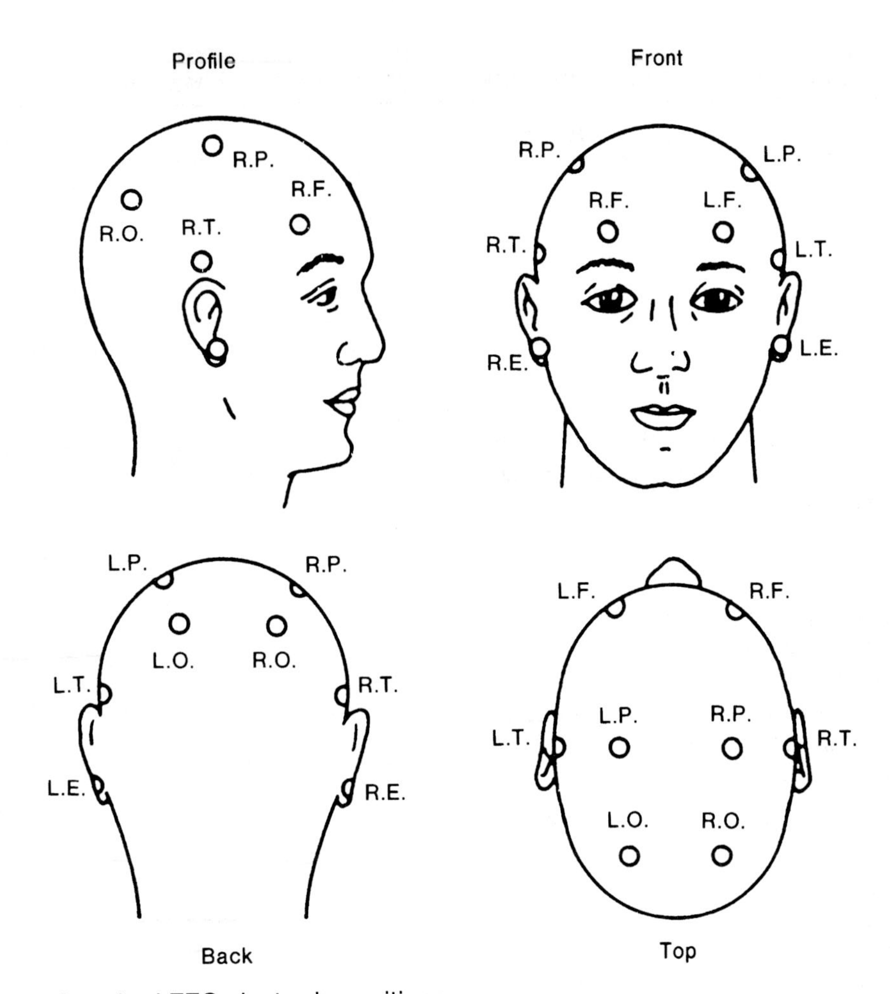

FIGURE 15.6 Standard EEG electrode positions

During episodes of paradoxical sleep, twitching of muscles in the face and limbs and rapid eye movements behind closed lids occur. As a result, paradoxical sleep is also referred to as rapid eye movement (REM) sleep.

Electrode Positions

Electrode positions have been named according to the brain region below that area of the scalp: **frontal, central** (sulcus), **parietal, temporal,** and **occipital.** There are two basic methods of electrode placement when recording the EEG. In a *monopolar recording,* an active electrode is placed over the cortical region of interest and a "reference" electrode is attached to the earlobe or a more distant part of the body. In a *bipolar recording,* the voltage difference between two electrodes placed over the cortical region of interest is measured with respect to a third "reference" electrode. Bipolar recording of the EEG is more localized. Figure 15.6 illustrates standard and commonly employed electrode positions.

Analysis of the EEG involves determination of the dominant frequency or rhythm, measurement of the amplitudes of different frequencies; calculation of the percentage of time that each frequency is present; and consideration of waveform, synchronism, and topographical distribution. Detailed analysis of the EEG is beyond the scope of this experiment. Instead, you will record an EEG using the bipolar method and perform some simple experiments and observations.

■ EXPERIMENTAL OBJECTIVES

1. To record a frontal lobe EEG from an awake, resting subject with eyes open and eyes closed.
2. To identify and examine alpha, beta, delta, and theta components of the EEG complex.
3. To observe the effect of opening the eyes, mental arithmetic, and hyperventilation on the alpha rhythm.

■ EXPERIMENTAL METHODS

Lafayette Minigraph

Materials

Minigraph model 76107 or 76107VS

model 76322 time/marker channel + remote marker button

model 76406MG basic amplifier channel

model 76422 biopotential preamplifier (or model 76402 MG)

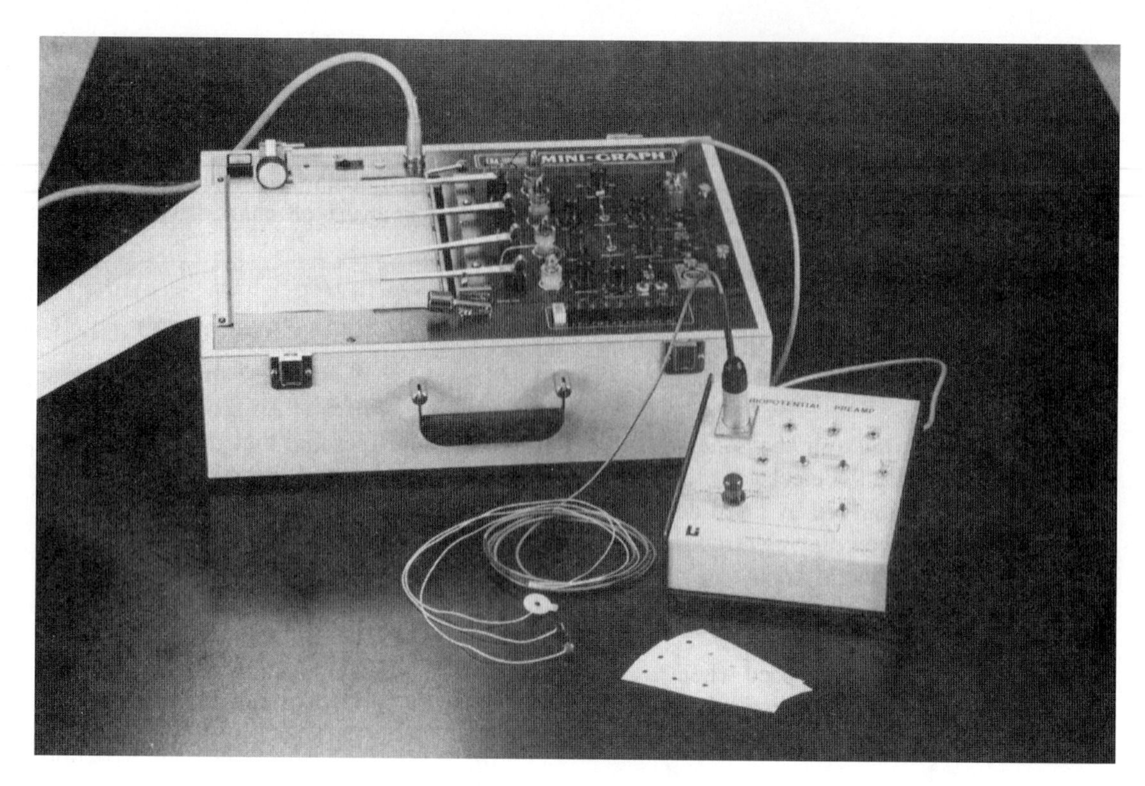

FIGURE 15.7 The Minigraph setup for electroencephalography

model 76623M miniature electrodes

model 76716M adhesive collars

model 76621 biogel

model 76724 skin cleaner

Preparation of the Recorder

1. Prepare the Minigraph for two-channel recording. Check to make sure there is sufficient recording paper and the inking system is working properly. Refer to chapter 2, if necessary, and to figure 15.7.
2. Attach the EEG electrode cable to the biopotential preamplifier or optional 76402MG amplifier via the 4-pin connector. Connect the preamplifier to the channel amplifier via the 9-pin amphenol connector.
3. Set the timer to mark off 1-second intervals.
4. Set the following controls on the preamplifier: mode—EEG; standby—cal/run on standby-cal; filter—in.
5. Set the amplifier polarity switch to (+), the gain switch to 1000, and the sensitivity to 1.
   Optional: Set the 76402 sensitivity to ×10 and 1 and the cal/run switch on cal.
6. Turn on mainframe power and center the amplifier pens using the pen-position control.
7. Periodically depress the 50 μV calibrate button on the preamplifier or amplifier and advance the sensitivity control on the amplifier until a pen deflection of 1 cm is obtained.
8. Request that the subject assume a comfortable resting supine position on a cot or laboratory table near the Minigraph. The subject should not be in contact with nearby metal objects (faucets, pipes, etc.) and should remove facial jewelry.
9. Using cotton or a paper towel soaked in alcohol (or skin cleaner) cleanse the skin on the lateral aspects of the forehead about 1 inch above the corner of the eyebrow. Also, cleanse the skin on one earlobe.
10. Apply the self-sticking paper washers to the raised plastic area around the electrode plates. Squeeze electrolyte gel onto the metal electrode plates. Use a paper towel to smooth the gel so that it completely fills the well between the metal plates and the surrounding plastic.
11. Remove the protective covering over the adhesive area of the washers and apply the electrodes to the skin. The ground electrode, marked by a red dot, is applied to an earlobe. The recording electrodes are applied to the right and left frontal areas (see figure 15.6) about 1 inch above the lateral margin of the eyebrow.
12. A paper speed of 10 mm/s will be used when recording the EEG. If a continuously variable speed recorder is to be used, a paper speed of 10 mm/s may be approximated by adjusting the speed control until two 1-second time marks are separated by 10 mm (one large square) on the recording paper.

Experimental Procedure

1. Allow the subject to rest, eyes closed, for approximately 5–10 minutes before beginning to record the EEG. The room should be reasonably quiet so as to allow the subject to become mentally relaxed.

2. Flip the standby/cal/run switch to "run" position and turn on the chart drive power. Record the frontal lobe EEG. The rhythm should resemble the alpha pattern in figure 15.5.
3. Request the subject to open his or her eyes. Note and mark the change in the EEG. The alpha rhythm should become desynchronized when the eyes are opened. After observing the change in the EEG, return the preamplifier or biopotential amplifier switch to "standby/cal" position and turn off the chart drive power.
4. Request the subject to close the eyes and relax. After a 5-minute waiting period, turn on the chart drive power and flip the standby/cal switch to "run" position. Record the alpha rhythm and then ask the subject to perform mental arithmetic, such as going through the multiplication table (i.e., $2 \times 2 = 4$, $2 \times 3 = 6$, $2 \times 4 = 8$, etc.). Note and mark the change in the EEG. After observing the change in the EEG, return the preamplifier switch to "standby/cal" position and turn off the chart drive power.
5. Request the subject to cease mental arithmetic and again relax with eyes closed. After a 5-minute waiting period, turn on the chart drive power and flip the standby/cal switch to "run" position. Record the alpha rhythm and then ask the subject to breathe deeply and more rapidly for about 20 seconds. Note and mark the change in the EEG.
6. Return the preamplifier switch to "standby/cal" position and turn off chart drive and mainframe power.

Electroencephalography I: Relaxation and Brain Rhythms

Name: ______________________ Date: ______________________

Lab Section: ______________________

1. Data: Append a copy of the experimental record to this report.

   Determine the average amplitude and frequency of the EEG patterns associated with:

   a. Awake, eyes closed: amplitude ______ frequency ______

   b. Awake, eyes open: amplitude ______ frequency ______

   c. Performing mental arithmetic, eyes closed: amplitude ________ frequency ________

   d. Hyperventilation, eyes closed: amplitude ______ frequency ______

2. Define the following:

   a. Alpha rhythm ______________________

   b. Beta rhythm ______________________

   c. Delta rhythm ______________________

   d. Theta rhythm ______________________

3. What is paradoxical sleep? ______________________

4. List and define two characteristics of regular, periodic waveforms. ______________________

5. Examine the alpha and beta waveforms for change between the "eyes closed" state and the "eyes open" state.

   a. Does desynchronization of the alpha rhythm occur when the eyes are open?

   ____

   b. Does the beta rhythm become more pronounced in the eyes-open state?

   ____

CHAPTER 16

Electroencephalography II: Occipital Lobe Alpha Rhythms

INTRODUCTION

As noted in chapter 15, the cerebrum is divided into **hemispheres,** and each hemisphere is divided into frontal, parietal, temporal, and occipital lobes (figure 16.1). Each lobe has functions that are unique, but each lobe also shares functions with other lobes and, indeed, with other parts of the brain. For example, as a child, we may see *(occipital lobe)* a flame and touch *(frontal lobe)* it to see what it is like, experiencing heat and pain *(parietal lobe),* and remembering *(temporal lobe)* not to repeat the experience. These functions and many others, such as reasoning and abstract thought, occur in the cerebral cortex, the thin covering of gray matter forming the surface of the cerebrum.

Electrical activity in the cerebral cortex is continuous from formation of the cerebrum in utero to death. As demonstrated in chapter 15, electrical activity of the cerebral cortex can be detected and recorded using scalp electrodes and an electroencephalograph. The record obtained, called an electroencephalogram (EEG), is complex and variable among subjects, although under certain conditions, the EEG exhibits simpler, rhythmic activity.

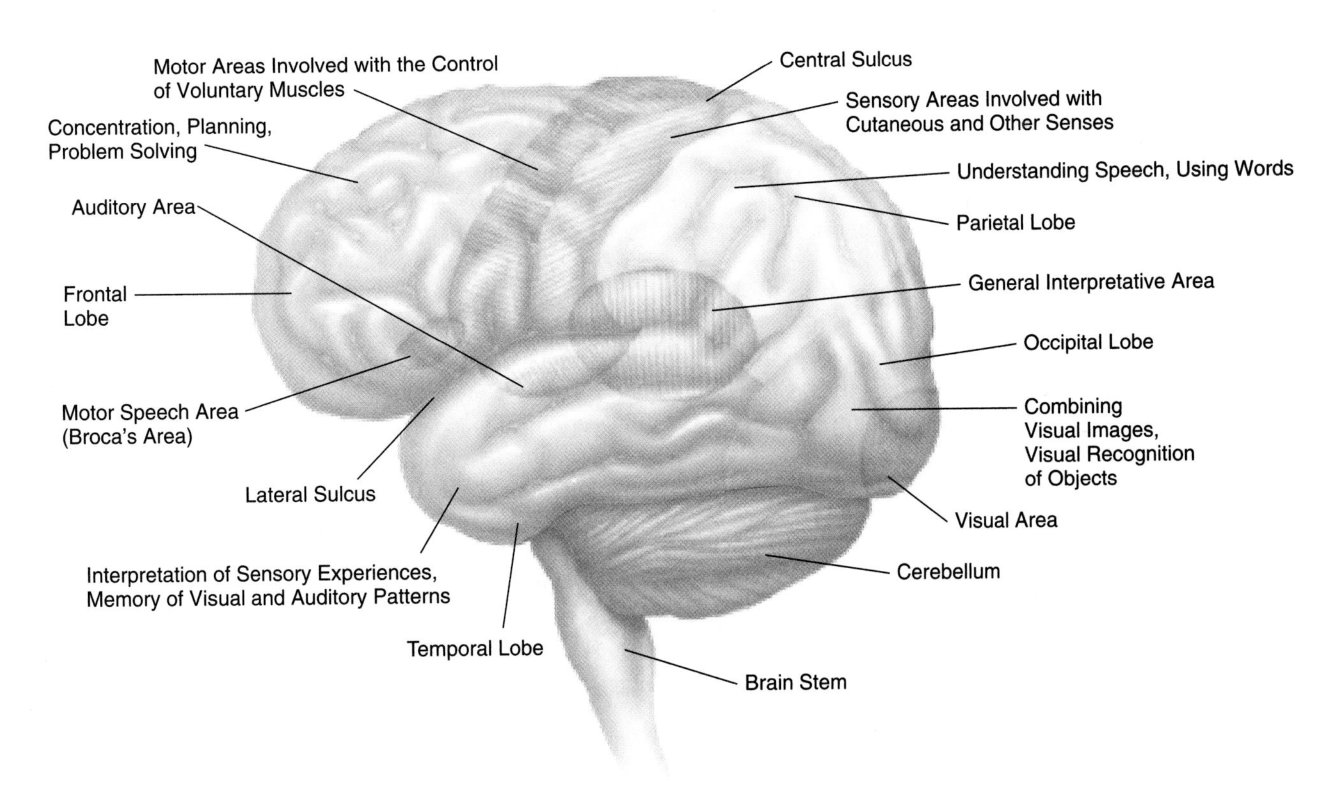

FIGURE 16.1 Some motor, sensory, and association areas of the cerebral cortex

Your EEG changes as you grow. The development of EEG is rapid with newborns. As neural development proceeds, the EEG recorded from the posterior regions of the brain of an infant of 3–4 months begins to resemble EEGs recorded from the posterior region of adults. The difference is that the 3–4-month-old infants have EEGs in the frequency range of 3–4 Hz, whereas adults tend to have average frequencies of 10 Hz. By the time the infant is 1 year old, the posterior region EEG is approximately 6 Hz; by 3 years it is 8 Hz; and by puberty (13–14 years old) the average frequency is similar to that of adults at 10 Hz.

Recall from chapter 15 that one of the simpler patterns is the alpha rhythm. The alpha rhythm is characterized by a frequency of 8–13 Hz and amplitudes of 20–200 μV. Each region of the brain has a characteristic frequency of alpha rhythm. Alpha waves of the greatest amplitude tend to be recorded from the occipital and parietal regions of the cerebral cortex.

The EEG is variable, depending on the mental state of an individual and the frequency and amplitude of alpha rhythms within an individual change. In general, the alpha rhythm is the prominent EEG wave pattern of a relaxed, inattentive state in an adult with eyes closed; however, specific conditions can influence the alpha rhythm, for example:

1. *Hyperventilation:* Breathing abnormally quickly and deeply causes carbon dioxide levels of the blood and cerebrospinal fluid to fall and pH levels to rise. These effects increase electrical activity of cortical nerve cells, often increasing amplitude of the alpha waves.
2. *Gender:* Females tend to have higher mean frequencies of alpha waves than males, although the differences are small.
3. *Memory:* Frequency may affect the speed of "remembering" and may be approximately 1 Hz higher during memory tests for high-scoring subjects than for subjects who scored lower.
4. *Personality:* Amplitudes tend to be higher in subjects who are extroverted.
5. *Mental stress:* Amplitudes vary with the difficulty of mental tasks performed with the eyes closed.
6. *Visual attention:* Amplitudes of alpha waves diminish when subjects open their eyes and are attentive to external stimuli. Thus, instead of getting the wavelike synchronized pattern of alpha waves, *desynchronization* occurs.
7. *Time of day:* Amplitudes increase when subjects are less alert and tend to be higher from 1:30 to 4:30 P.M.

In this chapter you will record the EEG and alpha rhythm under several conditions.

Experimental Objectives

1. To record an EEG from an awake, resting subject under the following conditions: relaxed with eyes closed; performing mental arithmetic with eyes closed; hyperventilating (breathing quickly and deeply) with eyes closed; relaxed with eyes open.
2. To examine differences in the level of alpha rhythm activity during mental arithmetic and hyperventilation compared with the control condition of eyes closed and relaxed.

■ EXPERIMENTAL METHODS

Lafayette Minigraph

Materials

Minigraph model 76107 or 76107VS

model 76322 time/marker channel + remote marker button

model 76406MG basic amplifier channel

model 76422 biopotential preamplifier (or model 76402 MG)

model 76623M miniature electrodes

model 76716M adhesive collars

model 76621 biogel

model 76724 skin cleaner

Preparation of the Recorder

1. Prepare the Minigraph for two-channel recording. Check to make sure there is sufficient recording paper and the inking system is working properly. Refer to chapter 2, if necessary, and to figure 16.2.
2. Attach the EEG electrode cable to the biopotential preamplifier or optional 76402MG amplifier via the 4-pin connector. Connect the preamplifier to the channel amplifier via the 9-pin amphenol connector.
3. Set the timer to mark off 1-second intervals.
4. Set the following controls on the preamplifier: mode—EEG; standby—cal/run on standby-cal; filter—in.
5. Set the amplifier polarity switch to (+), the gain switch to 1000, and the sensitivity to 1.
   Optional: Set the 76402 sensitivity to ×10 and 1 and the cal/run switch on cal.
6. Turn on mainframe power and center the amplifier pens using the pen-position control.
7. Periodically depress the 50 μV calibrate button on the preamplifier or amplifier and advance the sensitivity control on the amplifier until a pen deflection of 1 cm is obtained.
8. Select a subject for electroencephalography. Have the subject assume a relaxing position. A supine position with the head resting comfortably but tilted to one side is recommended. The best recordings occur when the subject is relaxed throughout the session. Read the following sections carefully *before* attaching electrodes to the subject. Electrode adhesion to the scalp is crucial for obtaining a meaningful EEG recording.

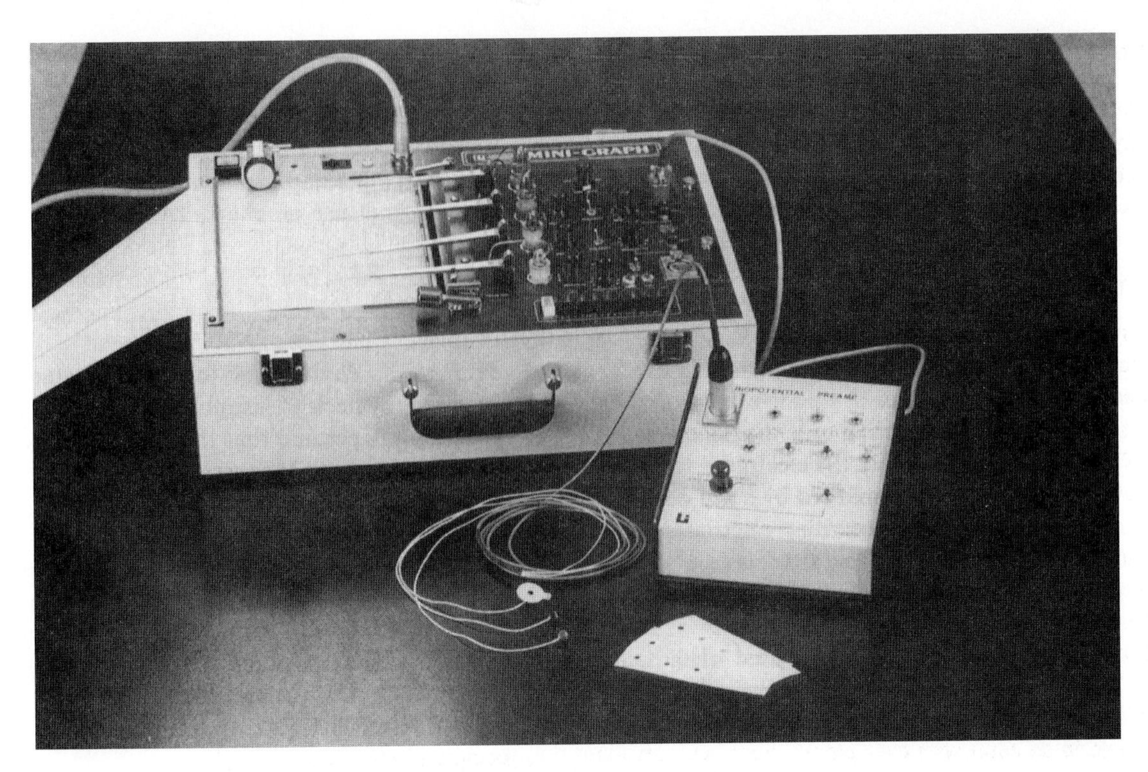

FIGURE 16.2 The Minigraph setup for electroencephalography

Hints for Obtaining Optimal Data

(a) As much as possible, move the hair away from the electrode adhesion area. Otherwise, the hair will pull the electrodes up, away from the scalp.
(b) Apply pressure to the electrodes for about 1 minute after the initial placement.
(c) Subject should try to remain still because blinking and other movement will affect the recording of all four rhythms.
(d) Despite your best efforts, electrode adhesion may not be strong enough to record data; if it is not strong enough, try another subject or different electrode placement.

Guidelines for Electrode Placement

(a) The placement of the scalp electrodes can vary (within limits) depending on your instructor's or the subject's preference. A suggested electrode placement is shown in figure 16.3.
(b) Keep the electrodes on one side (right or left) of the head.
(c) The third electrode is the *ground* electrode and is connected to the earlobe. Although the adhesive collar is larger than the earlobe, it can be folded under the ear for proper adhesion. Alternatively, the ground electrode can be placed on the facial skin behind the earlobe.

9. Apply the self-sticking paper washers to the raised plastic area around the electrode plates. Squeeze electrolyte gel onto the metal electrode plates. Use a paper towel to smooth the gel so that it completely fills the well between the metal plates and the surrounding plastic.
10. Remove the protective covering over the adhesive area of the washers and apply the electrodes to the skin. The ground electrode, marked by a red dot, is applied to an earlobe.

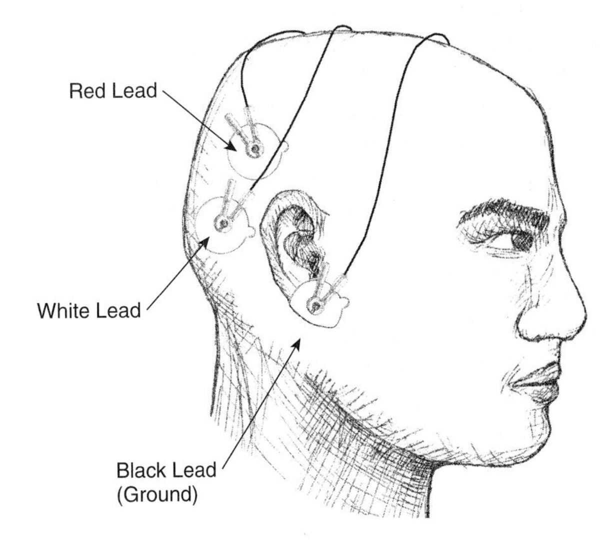

FIGURE 16.3 Electrode placement for recording occipital lobe EEG

11. A paper speed of 10 mm/s will be used when recording the EEG. If a continuously variable speed recorder is to be used, a paper speed of 10 mm/s may be approximated by adjusting the speed control until two 1-second time marks are separated by 10 mm (one large square) on the recording paper.

Experimental Procedure

1. Allow the subject to rest, eyes closed, for approximately 5–10 minutes before beginning to record the EEG. The room should be reasonably quiet so as to allow the subject to become mentally relaxed.
2. Flip the standby/cal/run switch to "run" position and turn on the chart drive power. Record the occipital lobe EEG. The rhythm should resemble the alpha pattern in figure 16.4. Record the occipital EEG (eyes closed) for approximately 30 seconds.
3. Request the subject to open his or her eyes. Note and mark the changes in the EEG. The alpha rhythm should become desynchronized when the eyes are opened. Mark the record "eyes open" and continue recording for an additional 20 seconds. The subject should try not to blink during this recording segment. After observing the change in the EEG, return the preamplifier or biopotential amplifier switch to "standby/cal" position and turn off the chart drive power.
4. Request the subject to close the eyes and relax. After a 5-minute waiting period, turn on the chart drive power and flip the standby/cal switch to "run" position. Record the alpha rhythm and then ask the subject to perform mental arithmetic, such as going through the multiplication table (i.e., 2 × 2 = 4, 2 × 3 = 6, 2 × 4 = 8, etc.). Note and mark the change in the EEG. After observing the change in the EEG, return the preamplifier switch to "standby/cal" position and turn off the chart drive power.
5. Request the subject to cease mental arithmetic and again relax with eyes closed. After a 5-minute waiting period, turn on the chart drive power and flip the standby/cal switch to "run" position. Record the alpha rhythm and then ask the subject to breathe deeply and more rapidly for about 20 seconds. Note and mark the change in the EEG.
6. Ask the subject to open eyes and continue to record the occipital EEG for 30 seconds.
7. Return the preamplifier switch to "standby/cal" position and turn off chart drive and mainframe power.

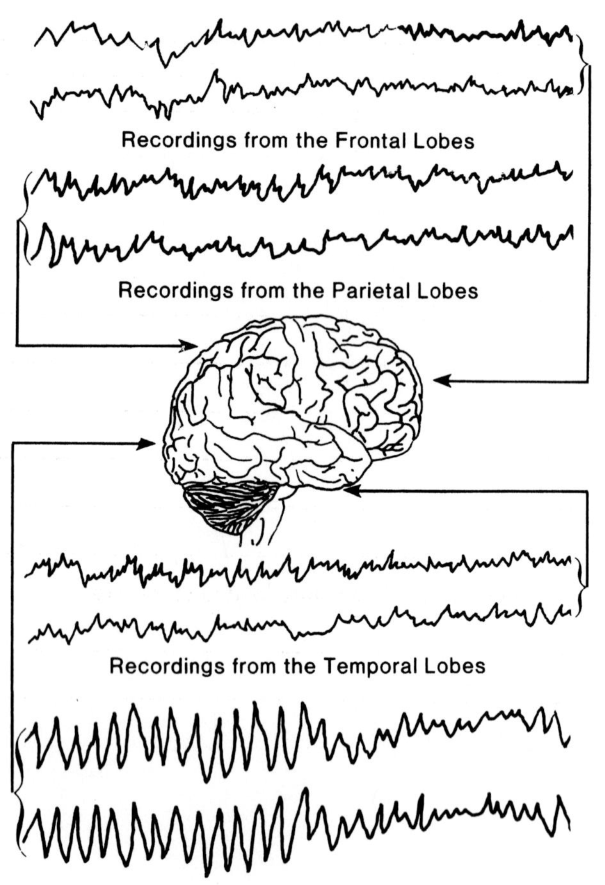

FIGURE **16.4** EEG patterns associated with each lobe of the cerebrum in the awake state

Electroencephalography II: Occipital Lobe Alpha Rhythms

REPORT 16

Name: ______________________________ Date: ______________________________

Lab Section: ______________________________

1. Data and calculations:

   a. Subject __________ Gender __________ Age __________ Height __________ Weight __________

   b. Complete table 16.1 with amplitudes of the recorded data in the control and experimental conditions.

TABLE 16.1

Condition	Average amplitude; raw EEG	Average amplitude, alpha
Eyes closed (control) first 30 seconds		
Eyes closed, performing mental math		
Eyes closed, recovering from hyperventilation		
Eyes open		

2. Refer to table 16.1, column 1. When was the general amplitude of the EEG highest? ______________________________

3. Refer to table 16.1, column 2. When were the alpha wave levels highest? ______________________________

4. Refer to table 16.1. How do your results compare with the information presented in the introduction to this lab experiment? ______________________________

5. Did the subject need to concentrate during the math problems? Yes _________ No _________

   How would the level of concentration required affect the data? ______________________________

   ____

   ____

6. Which conditions produced the lowest alpha activity? ______________________________

   ____

   ____

Blood Cells and Blood Types

CHAPTER 17

■ INTRODUCTION

Whole blood consists of *formed elements*—erythrocytes (red blood cells), leukocytes (white blood cells), and platelets (thrombocytes)—suspended in a complex amber fluid, the plasma.

Platelets, or thrombocytes, are not blood cells, they are cytoplasmic fragments split from large cells called megakaryocytes in the bone marrow (color plate 1). They are small, anucleate bodies about 2–4 micrometers (μm) in diameter and consist of a dark-staining granular portion (chromomere) surrounded by a light-staining cytoplasmic portion (hyalomere). Platelets are normally present in numbers ranging from 150,000 to 350,000 per microliter of blood and play several important roles in hemostasis.

The **leukocytes** (*leuko* = white, *cyte* = cell) are not actually white; they appear translucent or colorless in an unstained preparation because they do not contain a colored pigment as do erythrocytes. In contrast to erythrocytes, all leukocytes contain a nucleus. Leukocytes may be divided into two groups (color plate 1) according to the appearance of their cytoplasm: (1) granulocytes, which contain cytoplasm distinguished by the presence of large granules, and (2) agranulocytes, which contain cytoplasm that appears relatively homogeneous, or free of large granules.

Granulocytes consist of neutrophils, eosinophils, and basophils. These cells are produced in the bone marrow and range in size from 10 to 14 μm in diameter. Their cytoplasm contains large granules, believed to be mostly lysosomes, and their nuclei appear irregular and lobular; hence, they are frequently called polymorphonuclear leukocytes.

Agranulocytes consist of lymphocytes and monocytes. Some are produced in lymphatic tissue (lymph nodes, spleen) and others in bone marrow. They range in size from 8 to 20 μm in diameter. The cytoplasm appears relatively homogeneous, and the large, numerous granules present in the granulocytes are absent. The nucleus is generally spherical and occasionally indented, giving rise to a horseshoe shape. The amount of cytoplasm is reduced in an agranulocyte because of its relatively large nucleus.

Leukocytes combat foreign substances that enter the body. In performing this primary function, they may (1) phagocytize (engulf and destroy) material, (2) take up toxins, (3) release enzymes and other important substances, and (4) produce antibodies.

Erythrocytes (*erythro* = red, *cyte* = cell) are biconcave, disk-shaped cells, with a diameter of approximately 8.5 μm and a thickness of 2.5 μm (near the edge of the cell). The shape of the cell provides a very large surface area for diffusion coupled with a very small intracellular volume. The biconcave discoid shape of the erythrocyte is ideal because it allows the red cell to rapidly exchange gases (O_2 and CO_2) with the plasma. Gas exchange is important because the primary function of the red cell is to transport oxygen and carbon dioxide in the blood.

The mature circulating erythrocyte does not have a nucleus; it is lost during a developmental stage in the bone marrow (color plate 1). About 35% of an erythrocyte is composed of a respiratory pigment, hemoglobin, which when bound with oxygen imparts a red color to the cell. The biconcave shape is due to the tension of the cell membrane and interaction with intracellular contents. Abnormally shaped erythrocytes (e.g., sickle cells and microcytes) are frequently due to the presence of abnormal hemoglobin molecules or to an abnormal amount of hemoglobin.

Erythrocytes are pliable cells, bending and twisting as they pass through the vasculature. Because of this pliability, and because they lack a nucleus, red cells live an average of only 125 days in circulation. The normal adult, therefore, completely replaces all red blood cells once every 4 months. Because the average number of circulating erythrocytes is some 5 million cells per microliter of blood, the adult bone marrow normally forms 2.5 million per second—hundreds of billions of red cells daily—to replace those being destroyed at the same rate.

Functionally, the erythrocytes (1) participate in the transport and release of oxygen and carbon dioxide, (2) participate in the buffer system of the blood, and (3) participate, in a minor way, in coagulation of the blood.

Despite the fact that all human blood contains erythrocytes, leukocytes, and thrombocytes, not all blood is the same. When some types of blood are mixed, the result is an aggregation of red blood cells into clumps (agglutination), which may travel and become lodged in the vasculature, causing malfunction of the kidneys, brain, heart, and skeletal muscle. Incompatibilities of blood are due to the presence of antigens on the membrane surface of red blood cells. **Antigens** are chemical substances that, when introduced into an animal, will stimulate the animal's immune system to produce antibodies, which, in turn, react with the antigen in a way that destroys the antigen's harmful influence. When red blood cell antigens and antibodies react, the result is *agglutination* (figure 17.1), a clumping of the red cells that represents a defensive attempt to minimize dispersion of the antigen throughout the body. Therefore, erythrocyte antigens are called **agglutinogens**, and the corresponding antibodies are called **agglutinins**. The presence or absence of erythrocyte agglutinogens is determined genetically; that is, blood type is inherited.

More than 300 types of erythrocyte agglutinogens have been identified. Nearly all of them are weakly antigenic; that is, they do not provoke a strong immune response when given to a person who does not normally possess it. Such a recipient would neither have nor develop large amounts of corresponding agglutinin. However, a few erythrocyte agglutinogens are strongly antigenic, causing the recipient to reject, by agglutination, the donated erythrocytes. Three strong agglutinogens that may be present in erythrocytes are the A, B, and D agglutinogens. The corresponding agglutinins, normally absent if the agglutinogen is present, are designated as agglutinin a (anti-A), agglutinin b (anti-B), and agglutinin d (anti-D).

The presence or absence of agglutinogens and agglutinins allows the blood to be classified into types. Related types form a blood group. We will examine only the blood types of the ABO and D (Rh) blood groups.

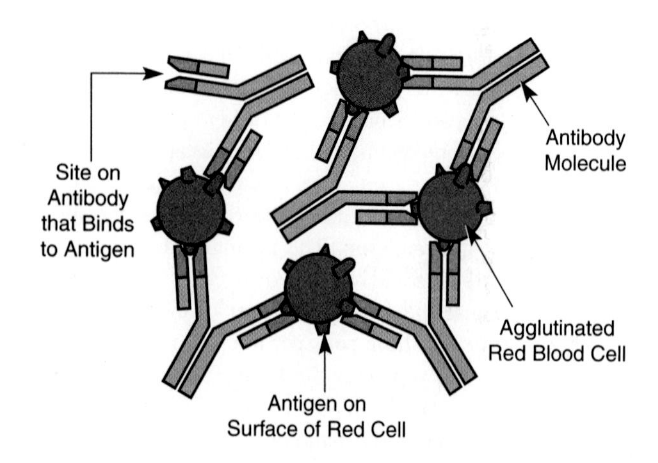

FIGURE **17.1** Antigen–antibody binding: red cell agglutination

ABO Blood Group

Type A blood contains the A agglutinogen on its red cell membranes. The plasma contains agglutinin b.

Type B blood contains the B agglutinogen on its red cell membranes. The plasma contains agglutinin a.

Type AB blood contains both the A and B agglutinogens on its red cell membranes, but the plasma lacks both agglutinins a and b.

Type O blood contains neither the A nor the B agglutinogens on its red cell membranes, but the plasma contains both agglutinins a and b.

The blood types of the ABO group and their constituents are listed in table 17.1.

When type B blood is transfused into a type B recipient, no reaction occurs because the recipient's body does not recognize the blood as being foreign. When type B is transfused into a type A recipient, two reactions occur:

1. The agglutinin a in the donor's plasma reacts with the A agglutinogen on the recipient's red cells **(minor agglutination).**
2. Agglutinin b in the recipient's plasma reacts with the B agglutinogen on the donor's red cells **(major agglutination).**

The first reaction is of little consequence because the agglutinin a in the donor's plasma (serum) becomes greatly diluted on entering the recipient's blood and thus becomes inconsequential. Agglutination of recipient cells occurs, but the agglutination is minor. The second reaction, however, is of major consequence because larger quantities of recipient agglutinin b (compared with much smaller quantities of donor agglutinin a) are available to react with the B agglutinogens of the donor's cells. When large numbers (about 5 million per microliter of blood) of donor red cells enter the recipient's body, widespread agglutination occurs and a transfusion reaction develops in the recipient as the donor's cells are agglutinated by the recipient's agglutinins. A **transfusion reaction** is characterized by general discomfort, anxiety, difficulty in breathing, flushing of the face, pain in the chest and neck, and other variable symptoms leading to

TABLE **17.1** ABO blood types: constituents and population distribution

Blood type	*RBC agglutinogen (antigen)*	*Plasma agglutinin (antibody)*	*Population distribution (white)*
A	A	b	41%
B	B	a	10%
AB	AB	Neither a nor b	4%
O	Neither A nor B	a,b	45%

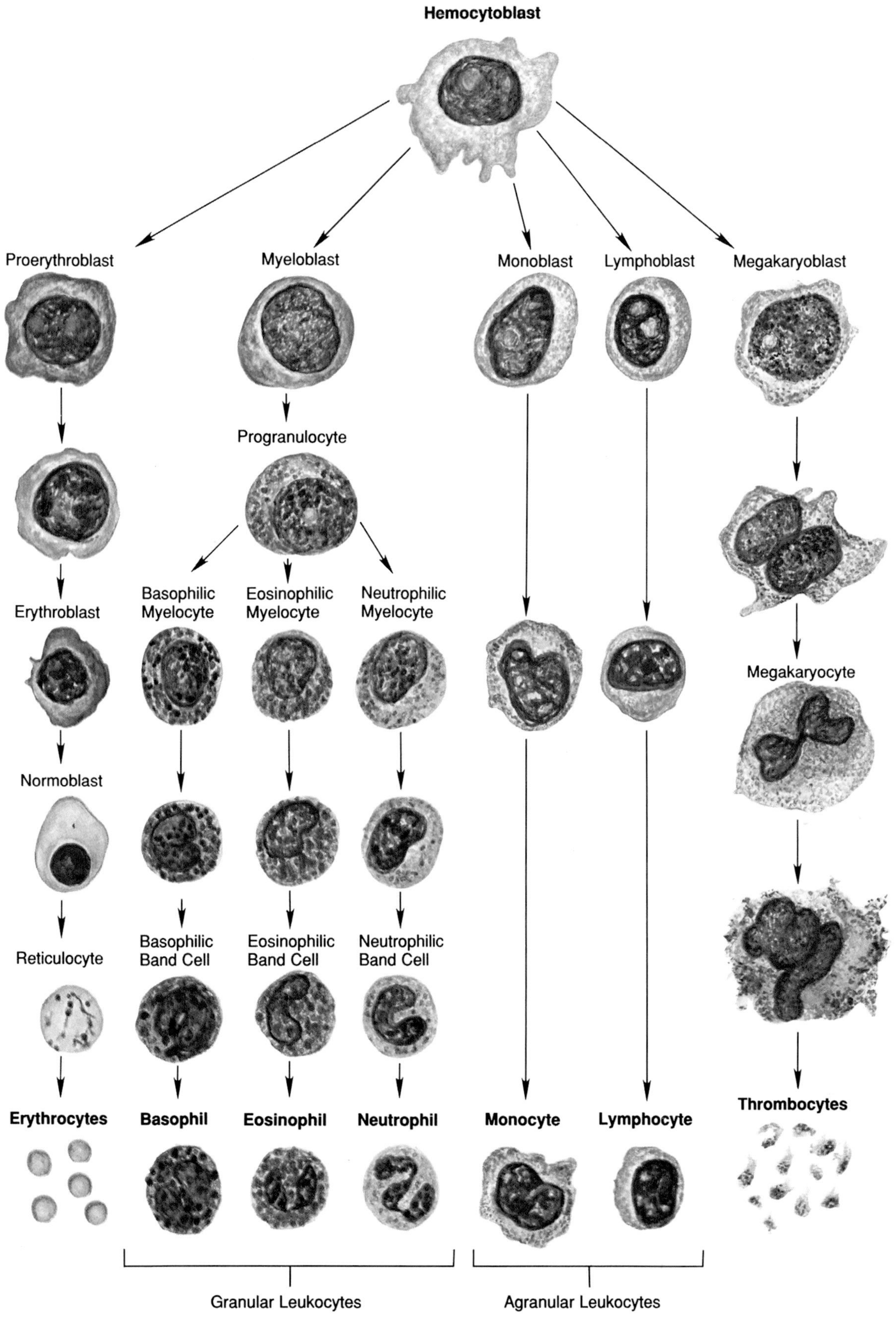

COLORPLATE 1 Cellular components of human blood

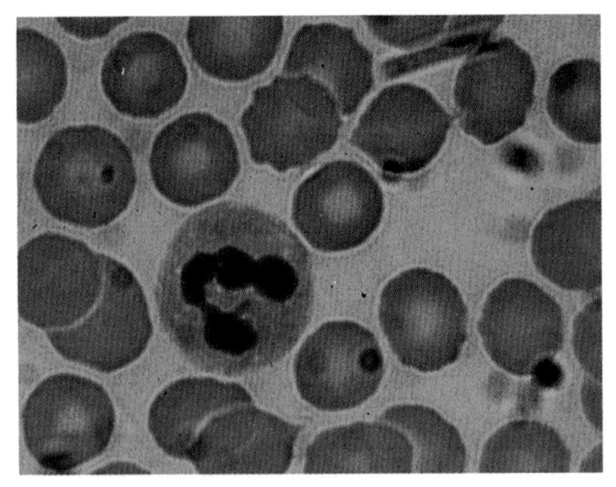
Neutrophils

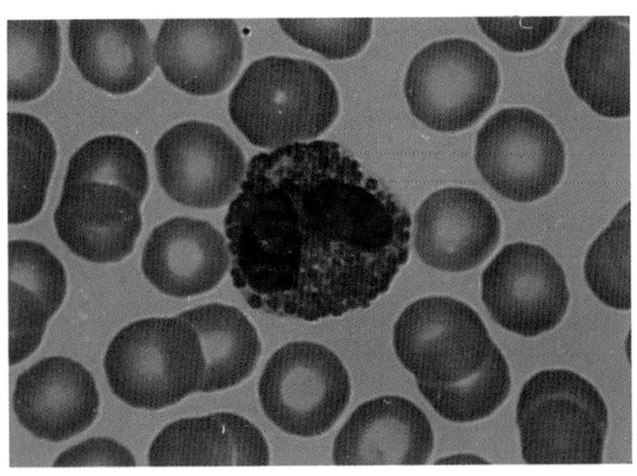
Eosinophils

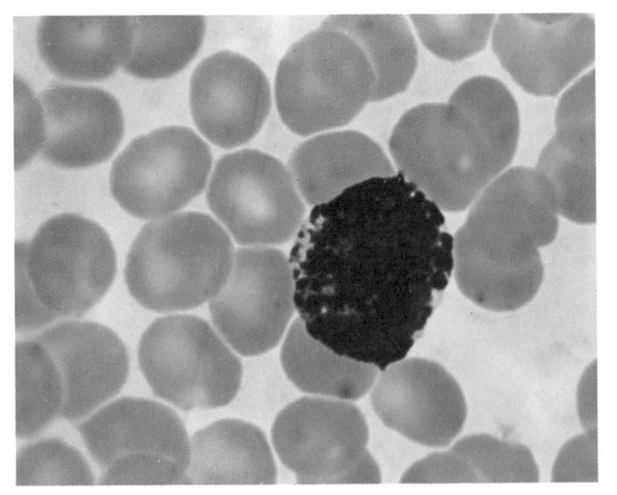
Basophils

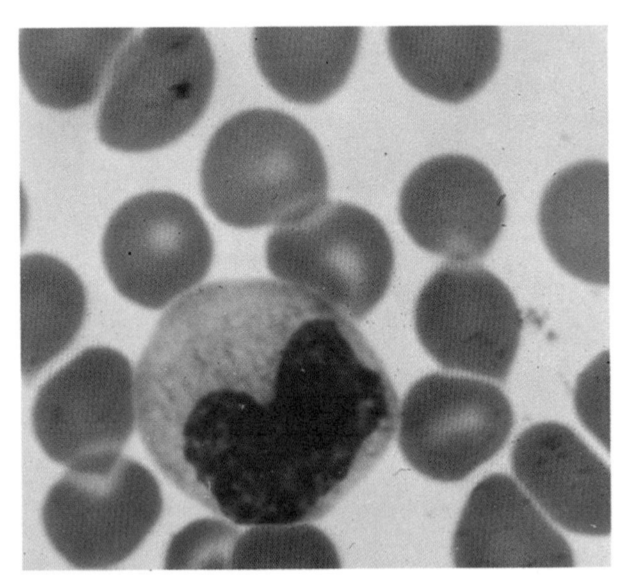
Monocytes

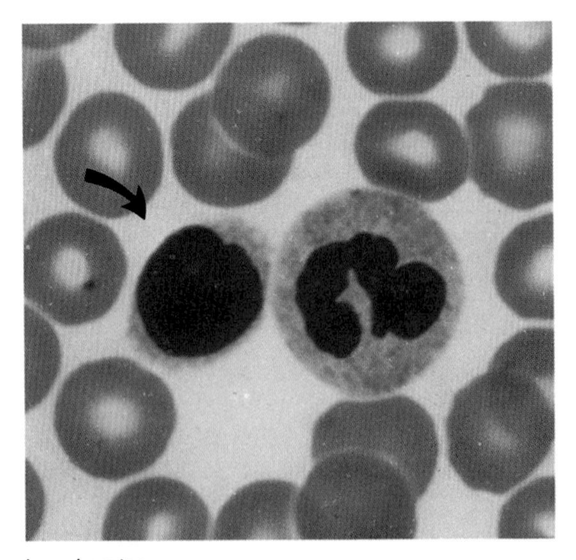
Lymphocytes

COLORPLATE 2 Cellular components of human blood

shock—as evidenced by rapid feeble pulse, cold clammy skin, fall in blood pressure, nausea, and vomiting. These acute symptoms usually occur within the first 2 hours. Depending on the amount of blood administered, more severe complications involving kidney, heart, lung, liver, and brain damage may develop, leading to death several days later.

Rh Blood Group

The Rhesus blood group, first described in 1940, is named after the monkey in which the antigens were discovered. As with the A and B agglutinogens, the presence or absence of the Rh agglutinogen is determined genetically. Most of the Rh agglutinogens are weakly antigenic and thus of little clinical importance; however, one of the Rh agglutinogens, called **agglutinogen D,** is strongly antigenic. The D agglutinogen is commonly called the **Rh factor.**

If the Rh factor is present on the red cell, the blood is said to be **positive;** if it is absent, the blood is said to be **negative.** Agglutinins to the Rh factor are not normally present in the plasma of Rh-negative blood. However, on exposure to the Rh agglutinogen (through transfusion or through mixing of fetal and maternal blood during childbirth), Rh-negative persons will produce agglutinins against the Rh agglutinogen, so that the next time negative blood is exposed to positive blood, the positive cells will be destroyed. Although an Rh-negative recipient transfused with Rh-positive blood, or an Rh-negative mother exposed at the birth of her Rh-positive fetus may be temporarily desensitized to the Rh agglutinogen by being given RhoGam (a preparation of Rh agglutinins), as a general rule Rh-positive blood should never be transfused into an Rh-negative recipient. However, assuming compatibility in the ABO and other blood groups, Rh-negative blood may be given to an Rh-positive recipient. Blood of the type A^+, for example, may not be given to an A^- recipient, but type A^- blood could be given to an A^+ recipient.

■ EXPERIMENTAL OBJECTIVES

1. To become familiar with erythrocytes and leukocytes and the major functions of each type.
2. To become familiar with the ABO–Rh system of blood typing and one of the clinical procedures for typing human blood.

Materials

compound light microscope
microscope slides
human blood smear (Wright stain)
coverslips
Wright stain
Wright-Giesma stain
Hype-Wipes (bleach-soaked disposable napkins)
staining dishes
distilled water
warming box
immersion oil
lancets (sterile, prepackaged, single use, manual or automatic) such as Single-Let (Bayer) or EZ-Lets II (Palco Labs, Inc.)
alcohol preps
gauze pads
finger bandages
disposable examination gloves (optional)
charts of human blood
Ward's 36W0019 simulated ABO–Rh blood typing kit
anti-A and anti-B sera (human)
anti-D serum

■ EXPERIMENTAL METHODS

CAUTION

In this and subsequent experiments on blood, you may be asked to obtain a drop of blood from your fingertip. If you have any history of a blood-clotting disorder or if you are being treated with anticoagulant medicine, inform your laboratory instructor and you will be excused from obtaining samples of your blood. Do not use samples from another student. Avoid contact with the blood of other persons. Handle only your own blood, and do so in a manner that minimizes its exposure to other students.

Properly dispose of all sharps (lancets, needles, blades, etc.) by discarding them in appropriately marked hazardous waste containers as soon as possible after use. Nondisposable items that come in contact with your blood are to be washed with a detergent and tap water, then rinsed successively with tap water, bleach solution, distilled water, and then acetone.

Disinfect your laboratory table or laboratory bench top by wiping it clean with a bleach solution before obtaining a fingertip blood sample and again at the end of the experiment.

Blood Smear Preparation

You may elect, with approval of your laboratory instructor, to prepare and examine a microscope slide of your own blood, instead of using a commercially prepared slide of human blood. If a blood smear is to be prepared and examined, follow directions carefully and observe all safety precautions.

Obtain two clean microscope slides from the laboratory instructor, and place them on the laboratory bench.

Technique for Obtaining Fingertip Blood

NOTE: Read all steps before proceeding. Ask questions if you do not understand the directions.

1. Clean the palmar surface of the third or fourth finger with a sterile gauze pad soaked with 70% alcohol or a sterile disposable alcohol prep. Save the pad.
2. Allow the skin to dry. Do not blow on it to make it dry faster.
3. Obtain an EZ-Lets II sterile, single-use capillary blood sampling device from your laboratory instructor. Your laboratory instructor may require use of another capillary blood sampling device; if so, follow the instructor's directions carefully regarding procedures for its use.
   (a) Push the yellow stem into the body of the device (figure 17.2a). **Do not use the device if the yellow stem is missing.**
   (b) Hold the device securely with one hand while twisting the yellow stem with the other hand (figure 17.2b). Twist the stem completely off.
   (c) Place the opening against the cleansed, dry surface of the fingertip and firmly depress the end of the green lever (figure 17.2c). The needle automatically retracts after use.
4. Allow a good-sized drop to form before using the blood sample. Do not "milk" the finger or squeeze it to increase flow, as doing so will alter the composition of the sample. Do not allow blood to drip from your finger. Place the sample on microscope slides as directed in the various experiments outlined in this chapter.
5. Discard the lancet in the appropriately marked hazardous waste container for sharps. If another attempt at obtaining a fingertip sample is necessary, always use a new sterile lancet.

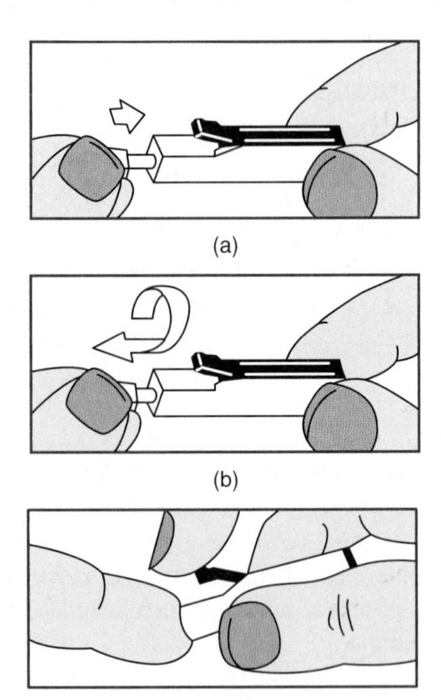

FIGURE **17.2** Procedure for obtaining a sample of fingertip capillary blood

6. After obtaining the sample, compress the gauze pad over the cut until the bleeding ceases or cover the wound with a finger bandage. Discard the gauze pad, alcohol prep, and bandage in the appropriately marked hazardous waste container.

Place one drop of blood on one end of a clean slide. Position the remaining clean slide at a 45-degree angle to the first slide (figure 17.3) so that the angled slide touches the drop of blood on the first slide. Quickly move the second slide across the top of the first slide, dragging the drop of blood into a thin, uniform layer. Allow the blood smear to dry thoroughly (10 minutes).

Stain the blood smear with Wright stain or Wright-Giesma stain.

Wright and Wright-Giesma Stain

1. Dip slide in fixative solution *5 times, 1 second each time.* Allow excess to drain.
2. Dip slide in solution I *5 times, 1 second each time.* Allow excess to drain.
3. Dip slide in solution II *5 times, 1 second each time.* Allow excess to drain.
4. Rinse slide with distilled or deionized water.
5. Allow to dry.

Identification of Blood Cells

Place the slide of stained human blood on the microscope stage. With the condenser raised to maximum height and the iris diaphragm adjusted to admit maximum light, focus on the blood cells using the high-power objective. Identify, if possible, the following cellular components of human blood, using the laboratory charts and color plate 2 for reference.

1. *Erythrocytes* are the most numerous of the cellular components of blood. They appear as biconcave disks of approximately 8.5 μm in diameter and are the smallest cells of the blood. They are nearly colorless and from a lateral view appear dumbbell-shaped. Occasionally erythrocytes may be seen stacked together like coins (rouleaux formation). Their cytoplasm is nearly transparent, and no nucleus is present.
2. Granulocytes of three types, the neutrophil, the eosinophil, and the basophil, are found in human blood. They are produced primarily in the bone marrow and are named because of the appearance of granules in the cytoplasm. They usually contain lobulated nuclei (from three to five lobes) and are 10–14 μm in diameter.
   (a) *Neutrophils* (heterophiles) are the most numerous type of leukocyte. They are very motile and actively phagocytic cells, playing a key role in the body's defense against bacterial invasion. They pass with ease through the capillary wall (diapedesis) and are attracted to and move toward the chemical substances liberated by damaged

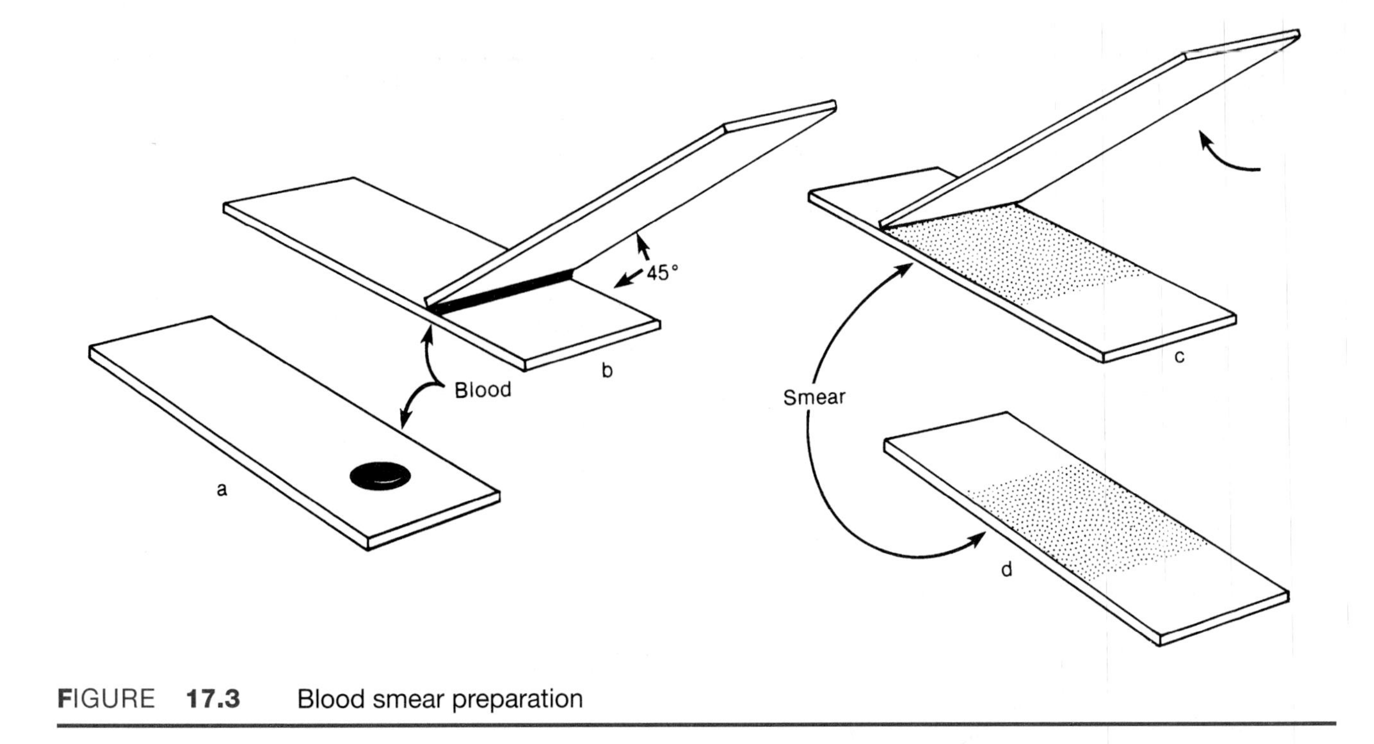

FIGURE 17.3 Blood smear preparation

cells and bacteria (chemotaxis). Three types of neutrophils are found in blood:

(1) "Juvenile" neutrophils are immature neutrophils, constituting about 0.5% of all leukocytes. Their nuclei are nonlobulated and relatively spherical and stain dark blue. Their cytoplasm stains a lighter shade of violet and contains small violet-staining granules.

(2) Neutrophilic band cells account for about 3% to 5% of all leukocytes. Their nuclei are lobulated (generally two lobes) and stain dark blue. Their cytoplasm stains a lighter shade of violet and contains small violet-staining granules.

(3) "Segmented" neutrophils are the most numerous type of leukocyte in the blood, constituting about 60% of all leukocytes. Their nuclei are lobulated (generally three or more lobes) and stain dark blue. Their cytoplasm stains a lighter shade of violet and contains small violet-staining granules.

(b) *Eosinophils* constitute approximately 1% to 5% of all leukocytes. They are less motile than the neutrophils and not as active in phagocytosis. Their cytoplasmic granules contain enzymes such as oxidases, peroxidases, and phosphatases, indicating that the primary function of the eosinophil is the detoxification of foreign proteins and other substances. The eosinophil plays an important role in the destruction of antigen–antibody complexes. Although relatively rare in the blood, the eosinophil is abundant in other connective tissues. The nucleus of the eosinophil is lobulated (generally three or more lobes) and stains dark blue. The cytoplasm is distinguished by the presence of numerous large red-orange–staining granules.

(c) *Basophils* are rare, constituting 0% to 1.0% of all leukocytes. They are found predominantly outside of the blood in connective tissues, where they are known as mast cells. They rarely display motility and phagocytosis. Many of their cytoplasmic granules contain anticoagulants (e.g., heparin) and vasodilators (e.g., histamine). A primary function of the basophil is to release histamine in areas of tissue damage to increase blood flow, attract neutrophils, and facilitate the repair of tissue. The nucleus is nonlobulated and generally spherical, staining a dark blue. The cytoplasm is marked by the presence of large dark-violet staining.

3. *Agranulocytes* of two types are found in human blood. Some are produced in lymphoid tissue (spleen, lymph nodes) and others in bone marrow. They derive their name from the absence of granules in their cytoplasm. They contain nonlobulated nuclei and are, therefore, occasionally called monocunuclear leukocytes. The cytoplasm appears homogeneous and stains light blue. The agranulocytes are 8–20 μm in diameter.

(a) *Lymphocytes* constitute approximately 20% to 40% of all leukocytes. They are actively motile cells but rarely display phagocytosis. Some lymphocytes function in the body's defense, synthesizing and releasing antibody molecules; others function as phagocytes. They are small (10 μm) and are easily

TABLE 17.2 Hemagglutination test

Blood type	Agglutination observed with anti-A serum	anti-B serum
A	Yes	No
B	No	Yes
AB	Yes	Yes
O	No	No

TABLE 17.3 Confirmation of blood types

Test cells A	B	O	Serum antibody
+	–	–	a
–	+	–	b
+	+	–	a + b
–	–	–	Neither a nor b

(+) = Agglutination
(–) = No agglutination

identified because the nucleus occupies most of the cytoplasmic space. The nucleus stains dark blue and usually is slightly indented.

(b) *Monocytes* constitute about 1% to 6% of all leukocytes; they are the largest leukocytes at 15–20 μm in diameter. They are active in movement and in phagocytosis and play important roles in destroying bacteria and cleaning up cellular debris in areas of tissue damage. Their nuclei are large, stain a lighter shade of blue than those of the lymphocytes, and are frequently indented, giving them a horseshoe appearance. Their cytoplasm stains light blue and appears homogeneous.

4. *Platelets* (thrombocytes) are actually not whole cells but rather fragments of a large cell called a megakaryocyte. Thrombocytes are usually quickly destroyed when the blood is drawn, but it may be possible to see clumps of them on the stained slide (small purple dots). Compare each cell type you saw on the stained slide with colored drawings in your text or on the wall charts.

Blood Typing

In the sections that follow, principles of blood typing can be learned through experimentation with real human blood or, if preferred, simulated human blood (Ward's 36B0019). Ward's simulated blood is made with nonbiological components that look like real red and white blood cells. These microcomponents are similar in proportion and size to those found in real human blood. The simulated red cells agglutinate in a manner similar to real red cells when mixed with appropriate artificial antisera (figure 17.4, table 17.2).

Blood Typing (ABO–Rh): Simulated Blood

The laboratory instructor has prepared four unknown simulated blood samples marked 1, 2, 3, and 4. Students will work in groups of four, each student determining the type of one of the four unknowns. Record the results of your determination and those of the others in your group in the laboratory report.

1. Decide who in your group will test sample 1, who will test sample 2, and so on.
2. Obtain a specially prepared slide containing three wells marked A, B, and Rh and three toothpicks.
3. Place three drops of the unknown sample into each of the three wells.
4. Place three drops of simulated anti-A serum into the A well, mix with a clean toothpick, and observe agglutination (+) or nonagglutination (–). Record.
5. Place three drops of simulated anti-B serum into the B well, mix with a clean toothpick, and observe agglutination (+) or nonagglutination (–). Record.
6. Place three drops of simulated anti-Rh serum into the Rh well, mix with a clean toothpick, and observe agglutination (+) or nonagglutination (–). Record.

Blood Typing (ABO): Human Blood

Use a wax marking pencil to draw two large circles on the surface of a clean, plain microscope slide. Follow the technique for obtaining a fingertip sample of blood (given earlier in this chapter) and place a small drop of blood in the center of each wax circle. Immediately ask the laboratory instructor to add one drop of anti-A serum to the blood in the first circle and one drop of anti-B serum to the blood in the second circle. Mix the contents of each circle with separate toothpicks, place the slide against a white background (e.g., paper), and gently tilt the slide back and forth, being careful not to mix the two groups together.

Look for clumping (agglutination) of blood cells, which should occur within 1 minute after mixing (figure 17.4). Do not read after 2 minutes.

Interpret the results according to table 17.2 and record the data.

Confirmation of blood type may be obtained by the testing of agglutinin (antibody) in the sera of the blood whose type is to be confirmed. Such a procedure is called crossmatching. A small quantity of sera to be tested is placed on a microscope slide and mixed with red cells of a known type. The presence or absence of red cell agglutination indicates the type of agglutinin (antibody) present in the unknown sera, thus confirming the type of agglutinogen present on the unknown red cells (table 17.3).

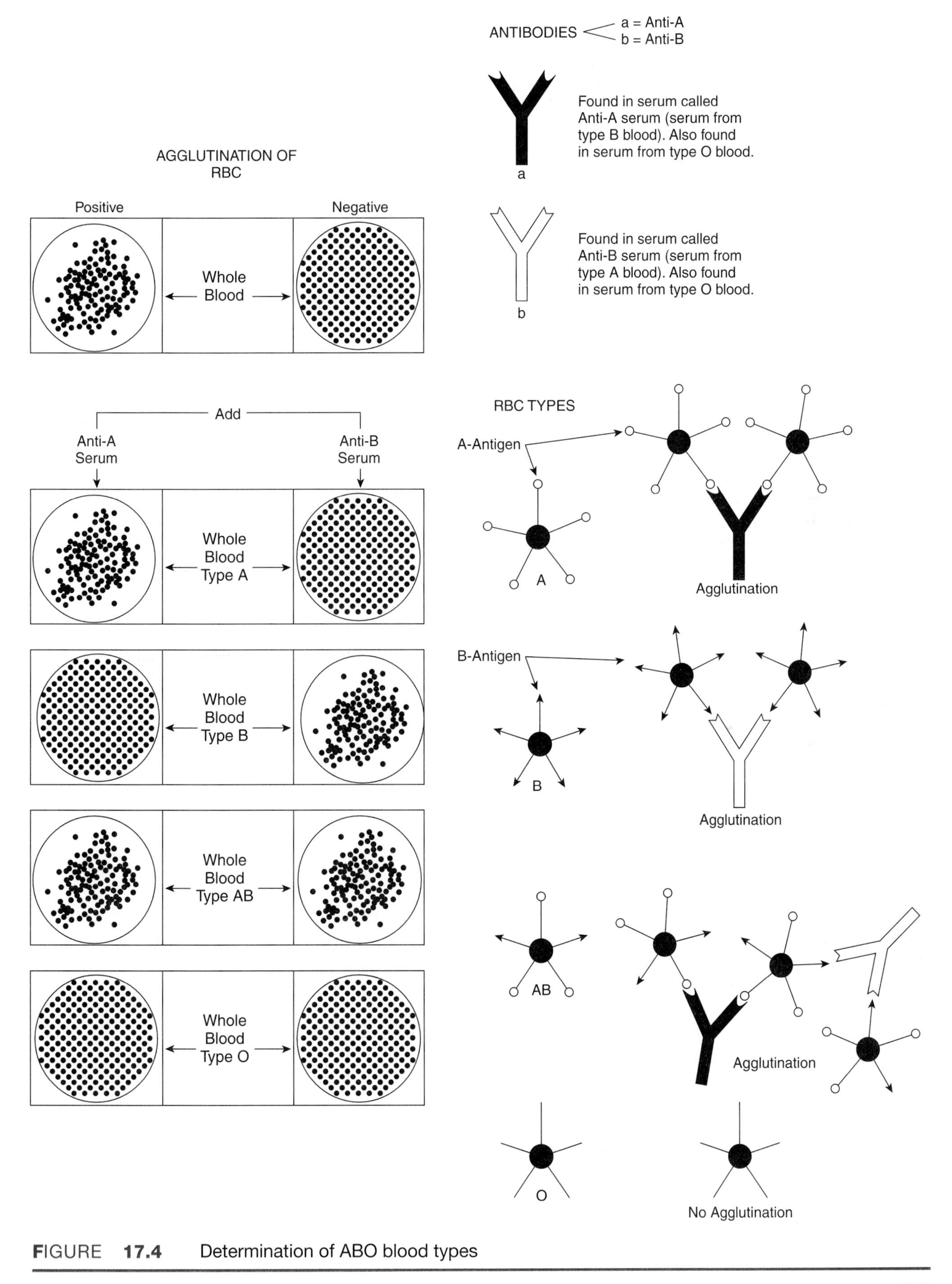

FIGURE 17.4 Determination of ABO blood types

Rh Factor: Human Blood

Use a wax marking pencil to draw a large circle on the surface of a clean, plain microscope slide. Follow the technique for obtaining a fingertip sample of blood (given earlier in this chapter) and place 2 small drops of blood in the center of the wax circle. Immediately ask the laboratory instructor to add one drop of anti-D serum to the blood in the circle.

Mix the blood and serum with a toothpick and place the slide against a white background (e.g., paper or an illuminated warming box), and gently tilt the slide back and forth. Look for agglutination of red cells. Do not read after 2 minutes. Record the data.

If agglutination occurs, the erythrocytes possess the Rh agglutinogen, and the blood is said to be Rh-positive (Rh^+). If agglutination does not occur, the erythrocytes do not possess the Rh agglutinogen, and the blood is said to be Rh-negative (Rh^-).

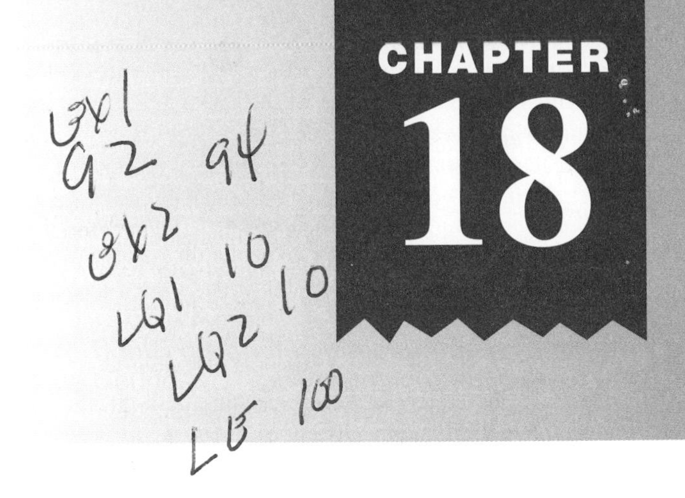

Red Blood Cell Count

■ INTRODUCTION

Erythrocytes, or red blood cells, originate in the red bone marrow of the adult and in the yolk sac, liver, spleen, and red bone marrow of the fetus. The mature normal erythrocyte is a biconcave, disk-shaped cell with a diameter of approximately 8.5 micrometers and thickness of 1 micrometer (μm) through the center of the cell. It lacks a nucleus, and its cytoplasm consists of a meshlike framework composed of proteins and lipids within which hemoglobin is bound. Despite the lack of a nucleus, erythrocytes normally live approximately 125 days in circulation.

Human blood contains approximately 5 million red blood cells per microliter of blood. Normal ranges are 5.4 ± 0.8 million per microliter for males and 4.8 ± 0.6 million per microliter for females.

A decrease in the number of circulating erythrocytes below normal range constitutes **erythrocytopenia,** commonly seen in many types of anemia.

An increase in the number of circulating erythrocytes above normal range constitutes **polycythemia.** Polycythemia occurs in three forms: relative polycythemia, polycythemia vera, and secondary polycythemia. *Relative polycythemia* (pseudopolycythemia) results from an increase in the concentration of erythrocytes without an increase in the total red cell mass, as may occur in dehydration, severe burn, and shock. *Polycythemia vera* (true polycythemia, or erythremia) is an increase in the concentration of erythrocytes, accompanied by an increase in total red cell mass due to hyperactivity of the bone marrow. In most cases, the cause is unknown, but in others, the polycythemia may be symptomatic of malignancies in the bone marrow, kidney, or brain. Polycythemia vera is not associated with any form of arterial hypoxia. *Secondary polycythemia* (erythrocytosis) occurs as a physiologic response to chronic arterial hypoxia. The hypoxia may be associated with living in a low-oxygen environment (high altitude) or may occur in emphysema, pulmonary fibrosis, carbon-monoxide poisoning, and other states in which oxygenation of the blood is reduced.

■ EXPERIMENTAL OBJECTIVES

1. To become familiar with one clinical procedure for counting blood cells.
2. To perform and record one's own erythrocyte count.
3. To become familiar with normal values of red blood cell count.
4. To acquire an appreciation for the value of performing a red blood cell count in the diagnosis of disease.

Materials

Clay-Adams hemacytometer (improved Neubauer)

Becton-Dickinson 5851 Unopette

Hayem's solution

acetone

alcohol

distilled water

alcohol preps

finger bandages

compound light microscope

microscope lens paper

gauze pad

sterile, prepackaged, single-use lancets such as Single-Let (Bayer) or EZ-Lets II (Palco Labs, Inc.)

disposable examination gloves (optional)

Hype-Wipes (bleach-soaked disposable napkins)

■ EXPERIMENTAL METHODS

CAUTION

In this and subsequent experiments on blood, you may be asked to obtain a drop of blood from your fingertip. If you have any history of a blood-clotting disorder or if you are being treated with anticoagulant medicine, inform your laboratory instructor and you will be excused from obtaining samples of your blood. Do not use samples from another student. Avoid contact with the blood of other persons. Handle only your own blood, and do so in a manner that minimizes its exposure to other students.

Properly dispose of all sharps (lancets, needles, blades, etc.) by discarding them in appropriately marked hazardous waste containers as soon as possible after use. Nondisposable items that come in contact with your blood are to be washed with a detergent and tap water, then rinsed successively with tap water, bleach solution, distilled water, and then acetone.

Disinfect your laboratory table or laboratory bench top by wiping it clean with a bleach solution before obtaining a fingertip blood sample and again at the end of the experiment.

Technique for Obtaining Fingertip Blood

NOTE: Read all steps before proceeding. Ask questions if you do not understand the directions.

1. Clean the palmar surface of the third or fourth finger with a sterile gauze pad soaked with 70% alcohol or a sterile disposable alcohol prep. Save the pad.
2. Allow the skin to dry. Do not blow on it to make it dry faster.
3. Obtain an EZ-Lets II sterile, single-use capillary blood sampling device from your laboratory instructor. Your laboratory instructor may require use of another capillary blood sampling device; if so, follow the instructor's directions carefully regarding procedures for its use.
   (a) Push the yellow stem into the body of the device (figure 18.1a). **Do not use the device if the yellow stem is missing.**
   (b) Hold the device securely with one hand while twisting the yellow stem with the other hand (figure 18.1b). Twist the stem completely off.

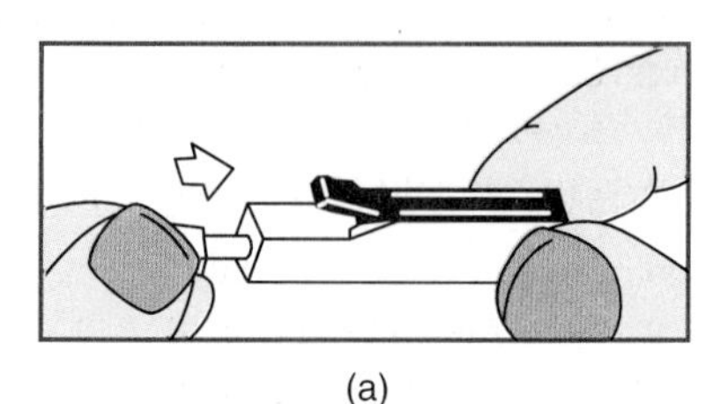

(a)

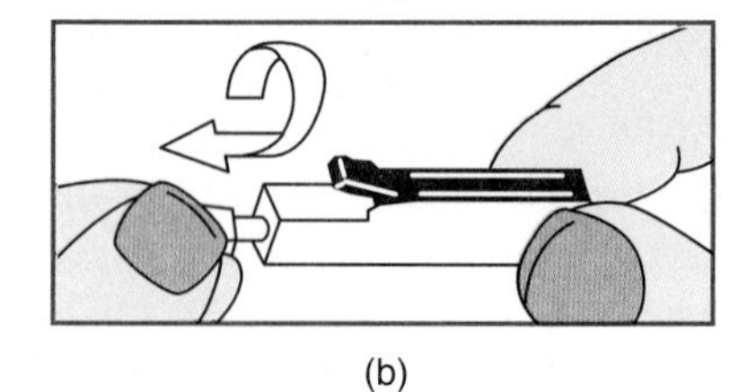

(b)

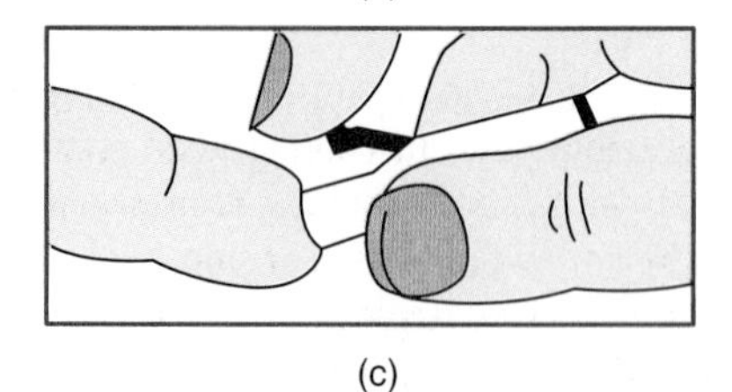

(c)

FIGURE 18.1 Procedure for obtaining a sample of fingertip capillary blood

   (c) Place the opening against the cleansed, dry surface of the fingertip and firmly depress the end of the green lever (figure 18.1c). The needle automatically retracts after use.
4. Allow a good-sized drop to form before using the blood sample. Do not "milk" the finger or squeeze it to increase flow, as doing so will alter the composition of the sample. Do not allow blood to drip from your finger. Place the sample where it belongs as directed in the experiments that follow.
5. Discard the lancet in the appropriately marked hazardous waste container for sharps. If another attempt at obtaining a fingertip blood sample is necessary, always use a new sterile lancet.
6. After obtaining the sample, compress the gauze pad over the cut until the bleeding ceases or cover the wound with a finger bandage. Discard the gauze pad, alcohol prep, and bandage in the appropriately marked hazardous waste container.

Counting Slide

Obtain a hemacytometer (figure 18.2) from the laboratory instructor. The hemacytometer consists of a special glass slide, coverslip, and two special pipettes. In the center of the counting slide are three grooves arranged in the shape of the letter H. The two polished platforms within the H lie exactly 0.1 mm below the surface of the slide. When the special coverslip is in place, a chamber 0.1 mm deep is formed. The scale, engraved on each platform, is a ruled square, the smallest division of which is 1/20 mm × 1/20 mm (1/400 mm^2). After becoming familiar with the appearance of the counting slide, proceed with the dilution technique.

Place the counting slide without the coverslip on the microscope stage. Using first the low-power and then the high-power objective, locate and identify the subdivisions with the aid of figure 18.3.

Dilution Technique Using Hemacytometer Pipettes

Within the hemacytometer case are two special pipettes. The larger one with the red plastic float in the bulb is used for counting red blood cells. The smaller one with the white plastic float is used for counting white blood cells. The red-cell pipette is graduated to dilute blood 200 times when the sample is drawn to the 0.5 mark on the stem and Hayem's solution is drawn and mixed with the sample to the 101 mark.

1. Review the technique for obtaining fingertip blood given previously in this chapter; then proceed to obtain your sample. Wipe off the first drop, and allow a good-sized drop to form.
2. Dip the red-cell pipette tip into the drop, and fill to the 0.5 mark exactly. Be careful to keep the tip constantly in the blood; otherwise, air will enter the column and

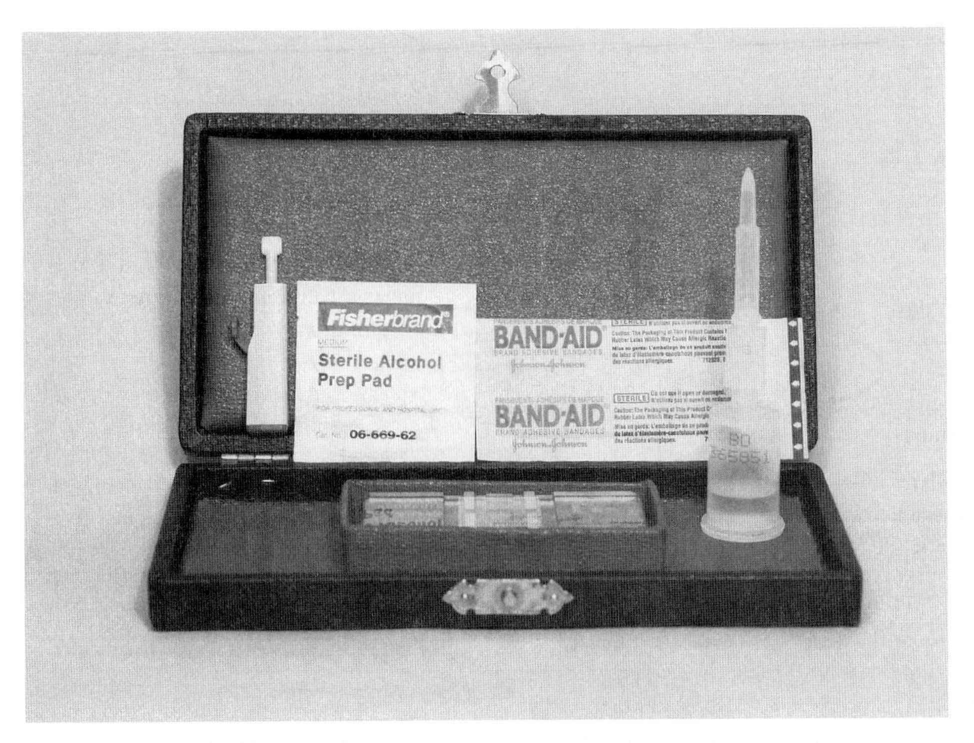

FIGURE **18.2** Hemacytometer and supplies for obtaining blood specimen

the sample will be inaccurate. If air does enter, clean and dry the pipette and start again.

3. Immediately wipe the excess blood from the outside of the tip. Quickly insert the pipette in the Hayem's solution, and fill it to the 101 mark exactly. (The sample is now diluted 200 times.) Close the tips of the pipette with your thumb and forefinger, and shake the contents for 2 minutes by alternately pronating and supinating the hand (pivoting the hand at the wrist) (i.e., moving the pipette as though it were a teeter-totter). **It is very important not to shake in the direction of the longitudinal axis. Doing so would be a serious mistake because the red cells would tend to settle toward one end of the pipette, resulting in dilution error.**
4. With the hemacytometer and the coverslip in place on the microscope stage, discharge 4–5 drops of the dilution on a paper napkin, and place the next small drop at the edge between the counting chamber and the coverslip. The counting chamber will fill by capillary action. Do not attempt to adjust the coverslip or in any way disturb the chamber after it has been filled.
5. Allow about 3 minutes for the blood cells to settle on the ruled squares. Immediately after discharging blood into the hemacytometer, rinse the pipette with bleach solution, distilled water, and then acetone. Place the pipette back in the hemacytometer case. Procedures to be followed in counting the red blood cells are outlined later.

Dilution Technique Using the Unopette

A more convenient method for obtaining and diluting the blood sample before counting involves the use of the Becton-Dickinson 5851 Unopette. The Unopette consists of a plastic reservoir containing isotonic diluent and a pipette with a shield (figure 18.4). Obtain a Unopette from your laboratory instructor and proceed as follows:

1. Place the reservoir on a firm, flat surface. Grasping the reservoir in one hand and the pipette assembly with the other, apply firm, even pressure, and push the tip of the pipette shield through the diaphragm in the neck of the reservoir (figure 18.5). This action punctures the diaphragm and opens the reservoir chamber containing diluent.
2. Remove the pipette assembly from the reservoir and disengage the shield from the pipette with a twist action. Leave the pipette loose in the shield until you are ready to use it.
3. Review the technique for obtaining fingertip blood given previously in this chapter; then proceed to obtain your sample. Wipe off the first drop, and allow a good-sized drop to form. In the next step, the capillary pipette bore must completely fill with blood, so a good-sized drop is required.
4. Holding the pipette almost horizontally, touch the tip of the capillary pipette to the blood and keep it there until the pipette fills (figure 18.6). Capillary action fills the pipette, and blood collection stops automatically

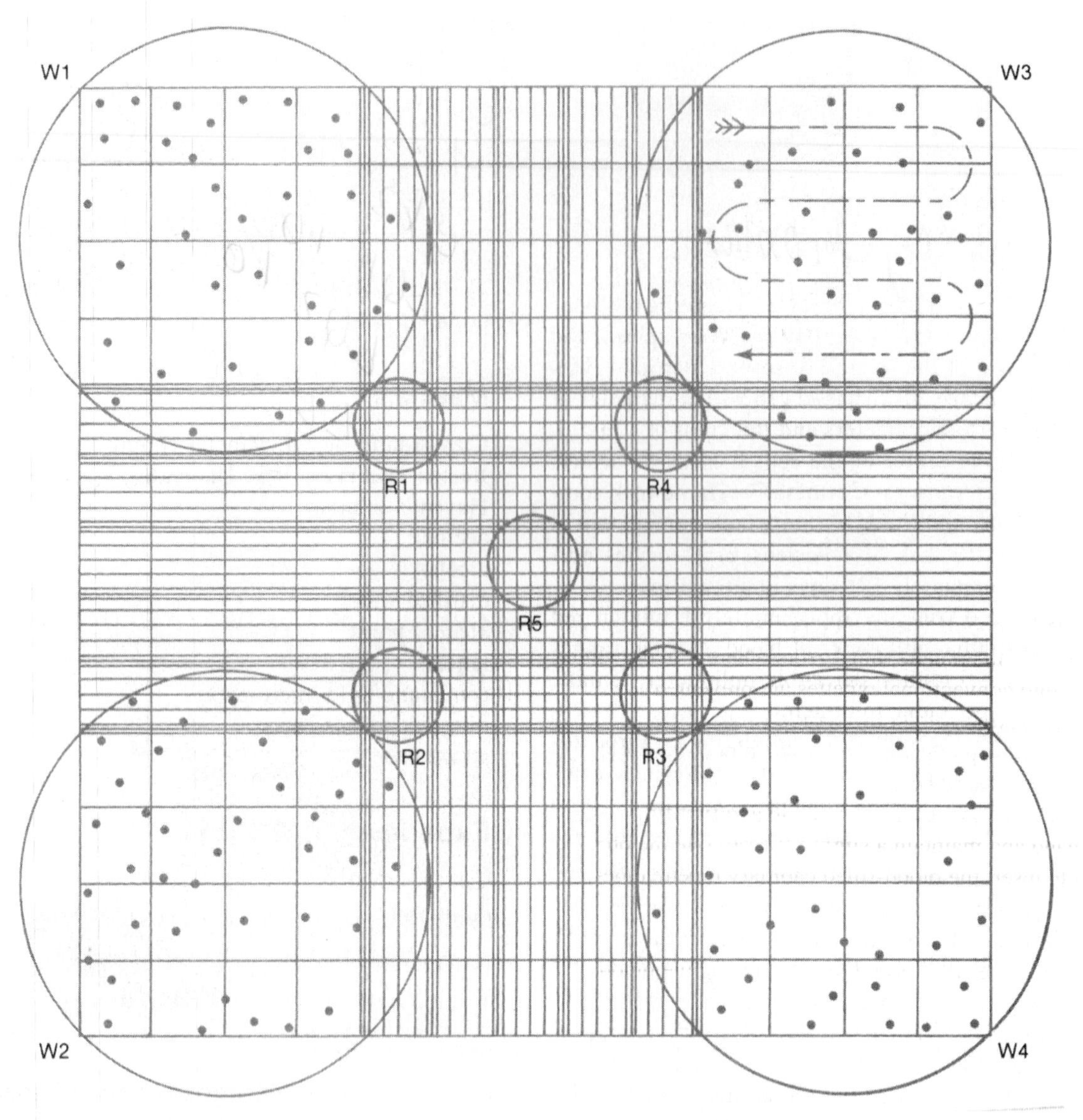

FIGURE **18.3** Improved Neubauer counting surface

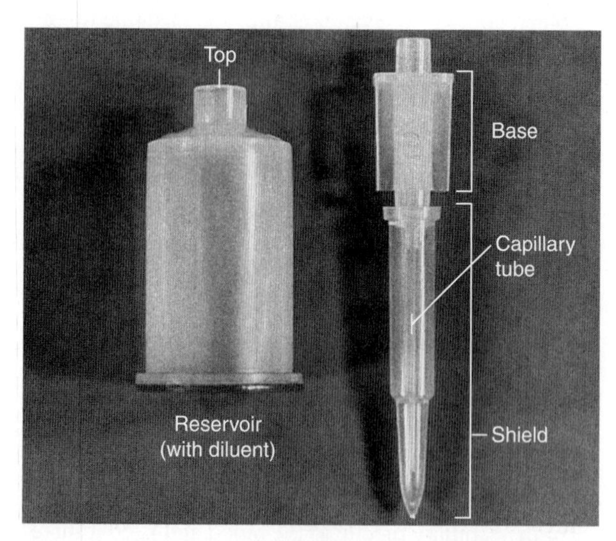

FIGURE **18.4** The Becton-Dickinson Unopette microcollection system

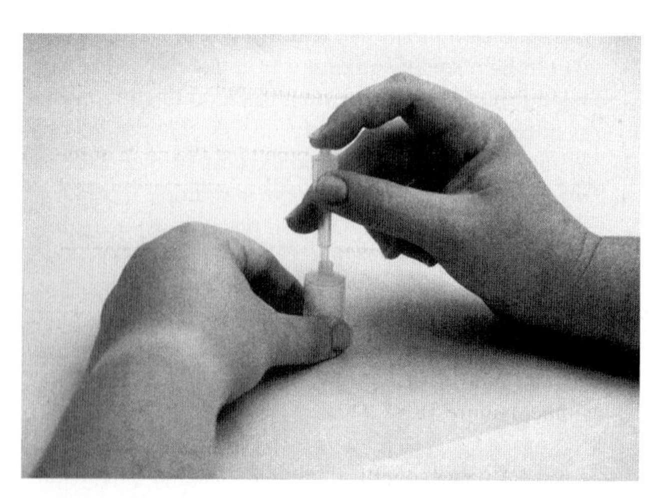

FIGURE **18.5** Use the pipette shield to puncture the reservoir diaphragm

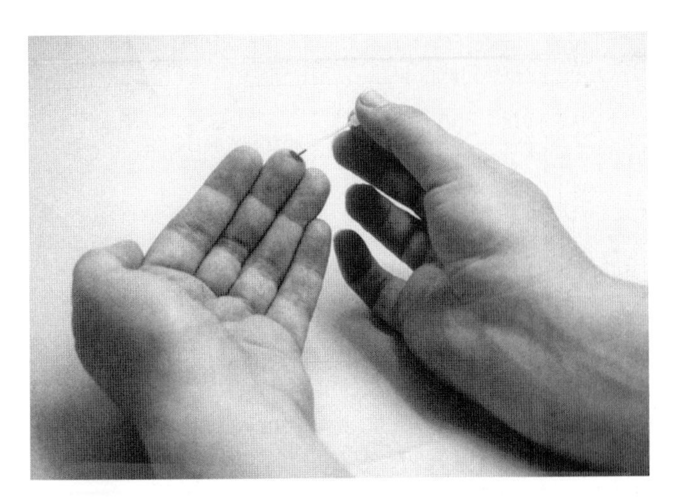

FIGURE 18.6 Fill the pipette bore with fingertip blood

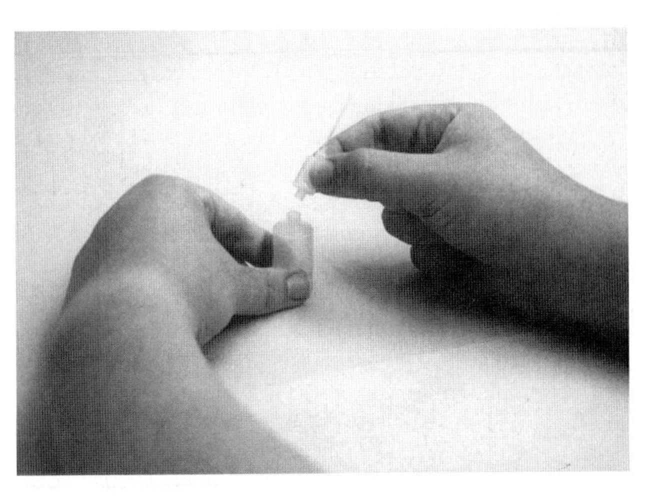

FIGURE 18.7 Convert to a dropper assembly

when the blood reaches the end of the capillary bore in the neck of the pipette; thus errors inherent in drawing blood into conventional pipettes are prevented.

5. Wipe any excess blood from the outside of the capillary pipette, making certain no blood is removed from the capillary bore.
6. Grasp the reservoir with the thumb and first finger of one hand and maintain a slight squeeze. Use the other hand to insert the blood-filled capillary pipette tube-first snugly into the reservoir. Remove your finger from the end of the pipette, then release the finger pressure on the reservoir, thereby sucking the blood out of the pipette and into the diluent.
7. Squeeze the reservoir gently two or three times to rinse the capillary bore; force the diluent up into but not out of the overflow chamber, and release the pressure each time to return the mixture to the reservoir.
8. Place your index finger over the upper opening of the pipette and gently invert the reservoir a few times to mix the blood and diluent thoroughly. The blood sample is now diluted 200-fold.
9. To convert to a dropper assembly, remove the pipette from the reservoir and invert it. Compress the reservoir and seat the large end of the pipette at the neck of the reservoir (figure 18.7). Release the compression on the reservoir.
10. Place the clean, dry hemacytometer slide on a napkin. Position the hemacytometer coverslip over the counting chambers. Discharge 2 drops of the dilution on the paper napkin and place the next small drop at the edge between the coverslip and the counting chamber (figure 18.8). The counting chamber will fill by capillary action. **Do not force fluid under the coverslip.** Doing so will cause the coverslip to rise, introducing a serious volume error when cells are counted. Do not attempt to adjust the coverslip or in any way disturb the chamber after it has been filled.

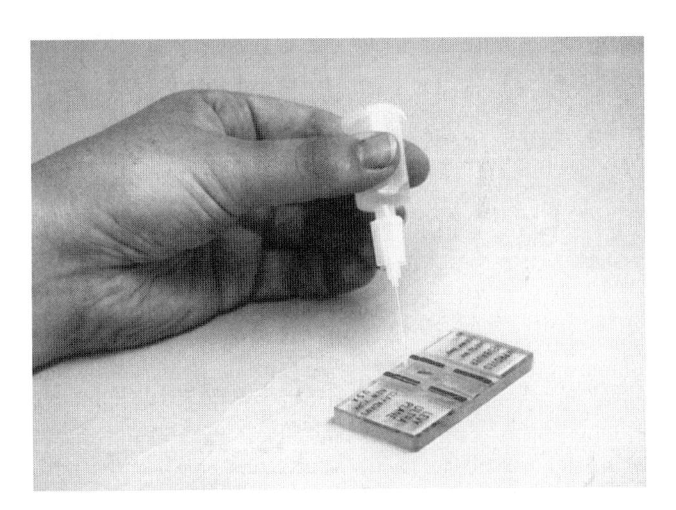

FIGURE 18.8 Charge the hemacytometer with a diluted blood sample

11. Gently transfer the charged hemacytometer to the microscope stage and allow about 3 minutes for the blood cells to settle on the ruled squares.

Counting Red Cells

1. Ensure that the microscope condensor is fully raised and the low-power objective is aligned for viewing the specimen. Reduce the light intensity, while attempting to focus on the ruled squares, until the cells become visible. Remember, these cells are not stained for visibility; therefore, the light intensity must be reduced to increase contrast and make the cells easier to see. After the cells and grid lines are in focus, switch to the high-power objective to perform the cell count.
2. Count all of the cells in the five R squares (see figure 18.3). Each R square is subdivided into 16

smaller squares, 4 vertical and 4 horizontal. The volume beneath each small square is: $1/20 \text{ mm} \times 1/20 \text{ mm} \times 1/10 \text{ mm} = 1/4000 \text{ mm}^3 = 1/4000\ \mu\text{L}$. Count all of the cells in each of the 16 squares and record the count in each of the corresponding squares in the laboratory report. To avoid counting a cell twice, count all of the cells that touch the line at the left but none that touch the line at the right of each square. Similarly, count all of the cells that touch the line at the top of each square but none that touch at the bottom line.

3. To calculate the number of red blood cells (RBC) in 1 microliter (μL) of blood, the following formula is used:

$$\text{RBC}/\mu\text{L of blood} = \frac{\text{Number of cells counted} \times \text{dilution} \times 4000}{\text{Number of small squares counted}}$$

For example, if a total of 500 cells were counted in a total of 80 small squares, then the RBC count would be:

$$\frac{500 \times 200 \times 4000}{80} = 5 \text{ million RBC}/\mu\text{L}$$

As you can see from the above formula, if cells in all 80 small squares of five R squares are counted, and 4000 is a constant, then $4000 \div 80 = 50$ and the previous formula simplifies to:

$$\text{RBC}/\mu\text{L of blood} = \text{Number of cells counted in 5 R squares} \times 10{,}000$$

4. Record your cell counts and your calculation in the data section of your report.

Hemoglobin Content, Hematocrit, and Red Cell Indices

CHAPTER 19

■ INTRODUCTION

Erythrocytes contain a respiratory pigment, **hemoglobin,** which imparts a characteristic color to the cells. Hemoglobin readily associates and dissociates with oxygen and carbon dioxide and is responsible for the red blood cell's ability to transport these gases.

The hemoglobin molecule is composed of a *protein portion (globin)* to which are attached four molecules of iron-containing *heme.* When oxygen is transported by the hemoglobin molecule, it is bound to the heme portion. Each hemoglobin molecule can transport four molecules of oxygen. Carbon dioxide molecules may be transported in combination with the globin portion of the molecule (carbamino hemoglobin). Hemoglobin can, therefore, transport oxygen and carbon dioxide simultaneously. When hemoglobin is saturated with oxygen (one molecule of hemoglobin transporting four molecules of oxygen), it imparts a bright red color to the blood. When hemoglobin gives up its oxygen (reduced hemoglobin), its color changes to a dark red. The primary function of hemoglobin is to enable the red blood cell to transport oxygen from the lungs to the body tissues and carbon dioxide from the body tissues to the lungs. Hemoglobin also plays an important role as a chemical buffer in the body's defense against changes in hydrogen ion concentration (pH).

Normal blood contains approximately 15 g of hemoglobin per deciliter of whole blood. Normal adult ranges are 14–18 g/dL for males and 12–16 g/dL for females. At birth, hemoglobin concentration is about 24 g/dL and decreases to approximately 11.5 g/dL at 1–2 years of age for both sexes. Normal adult levels are reached by 16 years of age.

Each gram of hemoglobin can transport 1.34 mL of oxygen. The oxygen-carrying capacity of normal blood is approximately 20 mL/dL (20 mL of oxygen per deciliter of whole blood). Normal adult ranges are 18–25 mL O_2/dL for males and 16–22 mL O_2/dL for females.

A reduction in the oxygen-carrying capacity of the blood below 13.5 mL/dL defines **anemia.** Anemias may result from marked deficiencies of circulating red blood cells or reductions below 10 g/dL in the hemoglobin content of the blood or both. Because of the reduced oxygen-carrying capacity of their blood, anemic persons may experience physical weakness, shortness of breath, heart palpitation, and difficulty in performing mental work. Anemias are most frequently detected by hemoglobin determination and by the hematocrit.

When heparinized whole blood is centrifuged in a glass tube, the heavier elements of the blood, such as the red and white blood cells, settle to the bottom of the tube, and the lighter elements contained in the plasma remain near the top. The percentage of the total blood volume in the tube occupied by the packed red blood cells is known as the **hematocrit.** For example, if 20 mL of blood is centrifuged and the packed red cell volume is 10 mL, the hematocrit is 50%. The broadest acceptable ranges for the normal hematocrit are 42%–52% for adult males and 37%–47% for adult females.

After the centrifugation of blood, a certain amount of plasma always remains trapped among the red cells. Studies indicate that approximately 4% of the plasma remains in the packed red cell layer. To obtain a corrected hematocrit, simply multiply the value of the apparent hematocrit by 0.96.

Hematocrit values increase in various forms of polycythemia, in some cases approaching 70%, whereas in various forms of anemia hematocrit values may drop to 25%.

With respect to morphology, red cells are described in specific terms. A *normocytic* erythrocyte is normal in size and shape; a *normochromic* erythrocyte is normal in color (and thus in hemoglobin content). A *microcytic* erythrocyte is smaller than normal; a *macrocytic* erythrocyte is larger than normal. A *hypochromic* erythrocyte is paler than normal (indicating insufficient hemoglobin); a *hyperchromic* erythrocyte is deeper in color than normal. *Anisocytosis* is a variation in size, and *poikilocytosis* is a variation in shape. The optimum situation is for erythrocytes to be normocytic and normochromic.

Red cell indices are used to detect abnormalities in erythrocyte size, shape, and color. Three commonly used indices are the mean corpuscular volume (MCV), the mean corpuscular hemoglobin (MCH), and the mean corpuscular hemoglobin concentration (MCHC).

The *MCV index* is a quantitative assessment of the red cell volume reported in picoliters (pL) per red blood cell. It is computed using the following formula:

$$\text{MCV} = \frac{\text{Hematocrit ratio} \times 1000}{\text{RBC (millions per } \mu\text{L)}}$$

For example, if the measured hematocrit ratio is 0.45 and the RBC count is 5.4 million per μL, MCV = 83.3 pL.

The normal adult value of MCV is 87 pL ± 5 pL. A low MCV indicates microcytosis, and a high MCV indicates macrocytosis.

The *MCH index* is used to determine the concentration of hemoglobin within the erythrocytes. It is reported in picograms (pg) per red blood cell and is computed as follows:

$$\text{MCH} = \frac{\text{g/dL hemoglobin} \times 10}{\text{RBC (millions per } \mu\text{L)}}$$

The normal adult value of MCH is 29 pg ± 5 pg. A low MCH indicates microcytosis, hypochromia, or both. A high MCH suggests hyperchromia.

The *MCHC* is an index of the proportion of hemoglobin measured (weight/volume) per average red cell in a sample of blood. It is reported as a percentage and is calculated using the following formula:

$$\text{MCHC} = \frac{\text{g/dL hemoglobin}}{\text{Hematocrit ratio}}$$

The normal adult value for MCHC is 34% ± 2%. A low MCHC means hypochromia, but a high MCHC indicates a loss of RBC volume without a proportionate loss of hemoglobin (microcytosis). Table 19.1 summarizes normal red cell values for adult males and females.

TABLE **19.1** Normal red cell values (adult)

RBC measurement	***Male***	***Female***
Red cell count (million μL)	5.4 ± 0.8	4.8 ± 0.6
Hemoglobin (g/dL)	16.0 ± 2.0	14.0 ± 2.0
Hematocrit (%)(apparent)	47.0 ± 5.0	42.0 ± 5.0
MCV (pL)	87 ± 5	87 ± 5
MCH (pg)	29 ± 2	29 ± 2
MCHC (%)	34 ± 2	34 ± 2

EXPERIMENTAL OBJECTIVES

1. To estimate the amount of hemoglobin found in normal blood.
2. To compare the accuracy of calorimetric methods of hemoglobin estimation.
3. To estimate the percentage of red cells found in normal blood by use of microhematocrit.
4. To compute MCV, MCH, and MCHC by the use of hemoglobin content, hematocrit, and red cell count.

Materials

microhematocrit centrifuge

heparinized capillary tubes

sealing wax (Sealease, Critoseal)

hematocrit scale

alcohol preps

Kimwipes

absorbent cotton

Tallquist hemoglobin scale + grade 909 test papers (Ward's 36-2068)

AO hemoglobinometer

STAT-Site M hemoglobinometer

STAT-Site M test cards

STAT-Site M hemaglobin controls

disposable examination gloves

finger bandages

distilled water

dropper

sterile, prepackaged, single-use lancets such as Single-Let (Bayer) or EZ-Lets II (Palco Labs, Inc.)

Hype-Wipes (bleach-soaked disposable napkins)

EXPERIMENTAL METHODS

CAUTION

In this and subsequent experiments on blood, you may be asked to obtain a drop of blood from your fingertip. If you have any history of a blood-clotting disorder or if you are being treated with anticoagulant medicine, inform your laboratory instructor and you will be excused from obtaining samples of your blood. Do not use samples from another student. Avoid contact with the bloods of other persons. Handle only your own blood, and do so in a manner that minimizes its exposure to other students.

Properly dispose of all sharps (lancets, needles, blades, etc.) by discarding them in appropriately marked hazardous waste containers as soon as possible after use. Nondisposable items that come in contact with your blood are to be washed with a detergent and tap water, then rinsed successively with bleach solution, distilled water, and then acetone.

Disinfect your laboratory table or laboratory bench top by wiping it clean with a bleach solution before obtaining a fingertip blood sample and again at the end of the experiment.

Technique for Obtaining Fingertip Blood

NOTE: Read all steps before proceeding. Ask questions if you do not understand the directions.

1. Clean the palmar surface of the third or fourth finger with a sterile gauze pad soaked with 70% alcohol or a sterile disposable alcohol prep. Save the pad.
2. Allow the skin to dry. Do not blow on it to make it dry faster.
3. Obtain an EZ-Lets II sterile, single-use capillary blood sampling device from your laboratory instructor. Your laboratory instructor may require use of another capillary blood sampling device; if so, follow the instructor's directions carefully regarding procedures for its use.
   (a) Push the yellow stem into the body of the device (figure 19.1a). **Do not use the device if the yellow stem is missing.**
   (b) Hold the device securely with one hand while twisting the yellow stem with the other hand (figure 19.1b). Twist the stem completely off.
   (c) Place the opening against the cleansed, dry surface of the fingertip and firmly depress the end of the green lever (figure 19.1c). The needle automatically retracts after use.

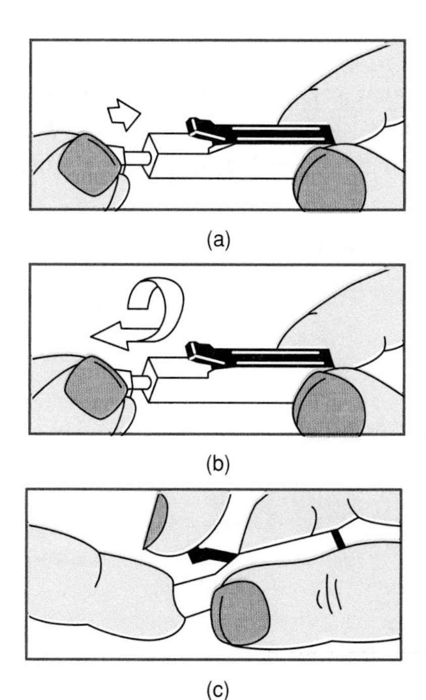

FIGURE **19.1** Procedure for obtaining a sample of fingertip capillary blood

4. Allow a good-sized drop to form before using the blood sample. Do not "milk" the finger or squeeze it to increase flow, as doing so will alter the composition of the sample. Do not allow blood to drip from your finger. Place the sample where it belongs, as directed in the experiments that follow.
5. Discard the lancet in the appropriately marked hazardous waste container for sharps. If another attempt at obtaining a fingertip sample is necessary, always use a new sterile lancet.
6. After obtaining the sample, compress the gauze pad over the cut until the bleeding ceases or cover the wound with a finger bandage. Discard the gauze pad, alcohol prep, and bandage in the appropriately marked hazardous waste container.

Tallquist Test for Hemoglobin

Obtain a Tallquist booklet from the laboratory instructor. Follow the procedure for obtaining fingertip blood. Observing the directions given in the booklet (figure 19.2), place a drop of blood on the special filter paper in the booklet, and match the color with the color scale as directed. The standard for the Tallquist method of hemoglobin estimation is 15.8 g = 100% normal. What do the results represent in terms of grams of hemoglobin per deciliter of blood? Record the results.

The Tallquist method of hemoglobin estimation is a direct method based on the natural color of blood, which is, in turn, determined by the amount of hemoglobin present. The error in this type of determination is approximately 20%; therefore, it is not generally used by clinical laboratories to determine hemoglobin content of the blood.

AO Hemoglobinometer

The AO hemoglobinometer is a simple handheld photometer that permits a visual determination of blood hemoglobin concentration by comparing the light absorption of a hemolyzed blood sample with the light absorption of a glass standard. The inherent error in this method is approximately 2%.

Obtain an AO hemoglobinometer (figure 19.3) from the laboratory instructor and proceed as follows to determine your blood's hemoglobin concentration:

1. Disassemble the blood chamber (figure 19.4) by pulling the two pieces of glass from the metal clip. Note that one piece of glass has an H-shaped moat cut into it. This piece will receive the blood. The other piece of glass has two flat surfaces and serves as a cover plate.
2. Clean both pieces of glass with alcohol and Kimwipes. Handle by edges to keep clean.
3. Reassemble the glass plates in the clip so that the grooves on the moat plate face the cover plate. The

Actual Anemia				Suggestive Anemia		Normal	
Men and Women below 70%				Men: 70% to 85% Women: 70% to 80%		Men: above 85% Women: above 80%	
30%	40%	50%	60%	70%	80%	90%	100%
4.7g	6.3g	7.8g	9.4g	10.9g	12.5g	14.1g	15.6g

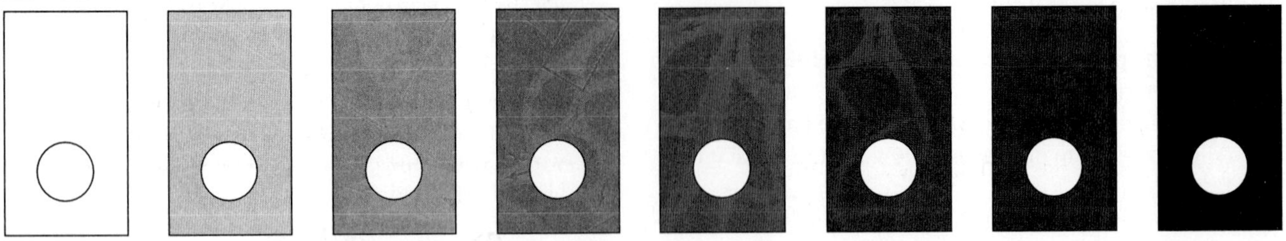

FIGURE **19.2** Hemoglobin scale (after Tallquist): After the proper puncture of fingertip or ear lobe, a perforated paper swatch should be placed so that it absorbs the blood evenly and thoroughly. Any excess blood should be blotted with another paper swatch. Before the blood starts to dry, the comparison should be made under natural light by placing the specimen under the color comparison chart so that it appears at the apertures. The approximate hemoglobin content of the blood can be read from the figures alongside the color that is the closest match to the specimen.

moat plate should be inserted only halfway to provide an exposed surface to receive the drop of blood.

4. Follow the technique for obtaining fingertip blood given previously in this chapter.
5. Place a drop of blood on the exposed surface of the moat plate, as shown in figure 19.5.
6. Hemolyze the blood on the plate by mixing the blood with the pointed end of a hemolysis applicator, as shown in figure 19.6. It will take 30–45 seconds for all the red blood cells to rupture. Complete hemolysis has occurred when the blood loses its cloudy appearance and becomes a transparent red liquid.
7. Push the moat plate in flush with the cover plate and insert the sample into the side of the instrument, as shown in figure 19.7.
8. Place the eyepiece to your eye with the left hand in such a manner that the left thumb rests on the light-switch button on the bottom of the hemoglobinometer (figure 19.8).
9. While pressing the light button with the left thumb, move the slide button on the side of the instrument back and forth with the right index finger until the two halves of the split field match. The index mark on the

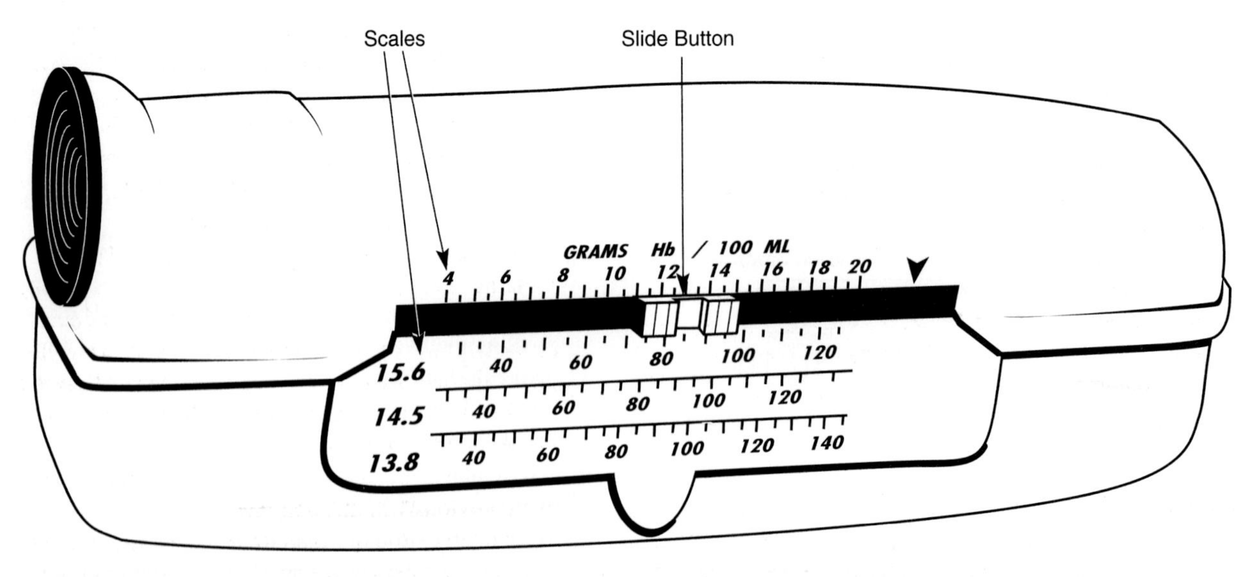

FIGURE **19.3** Hemoglobinometer

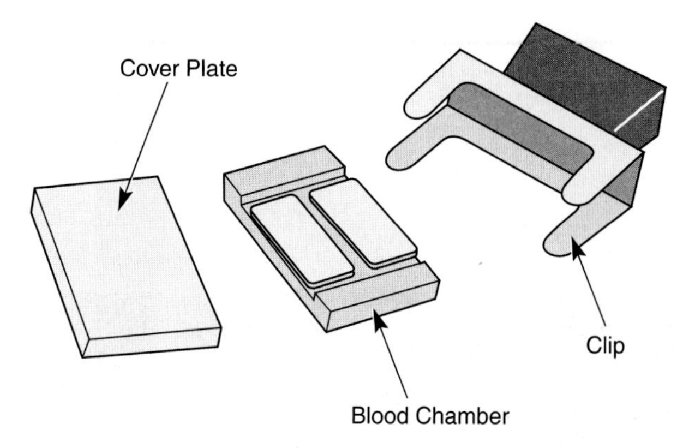

FIGURE 19.4 Blood chamber assembly

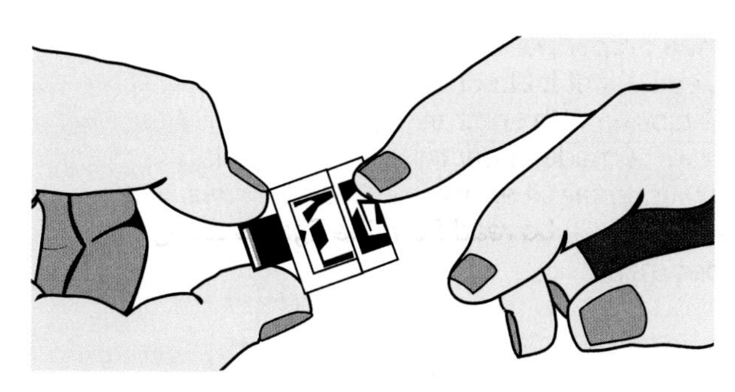
FIGURE 19.5 Charging the blood chamber

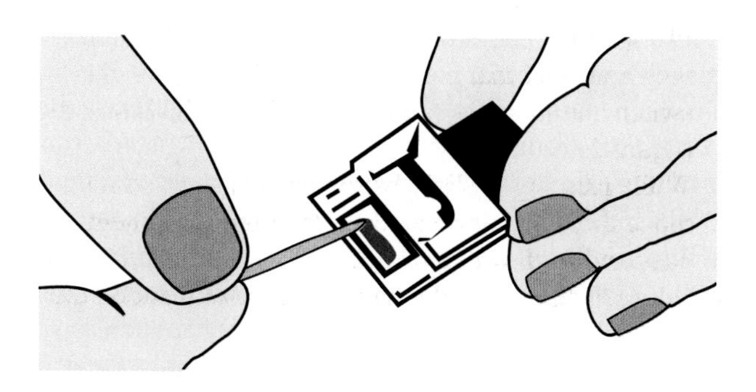
FIGURE 19.6 Hemolyzing the blood

slide knob indicates the grams of hemoglobin per 100 mL of blood. Read the percent hemoglobin on the 15.6 scale. Enter the data in the report.

10. Remove, disassemble, and rinse the blood chamber with bleach solution, distilled water, and then acetone. Return it to its container. Discard the hemolysis applicator in the appropriately marked hazardous waste container.

STAT-Site M Hemoglobinometer

The STAT-Site M hemoglobinometer shown in figure 19.9 is a reflectance photometer which operates by illuminating an area containing the blood sample and measuring the amount of light that area reflects. The wavelength of the reflected light is dependent on the wavelength of the illuminating light and the color or hemoglobin concentration of the blood sample. The hemoglobinometer contains a microprocessor that converts reflected light data to hemoglobin concentration and displays the result on a liquid crystal screen.

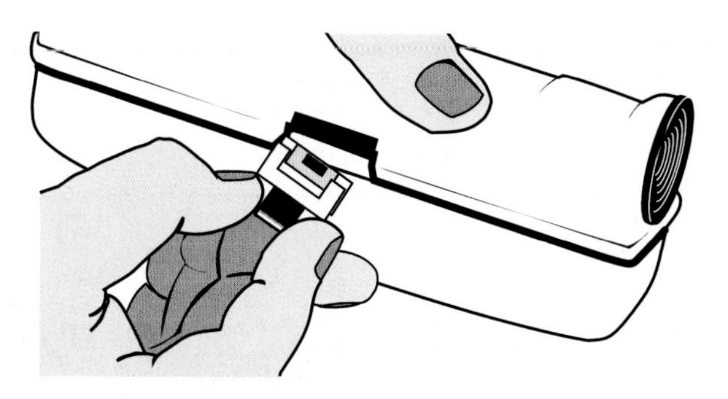
FIGURE 19.7 Inserting the charged chamber into the hemoglobinometer

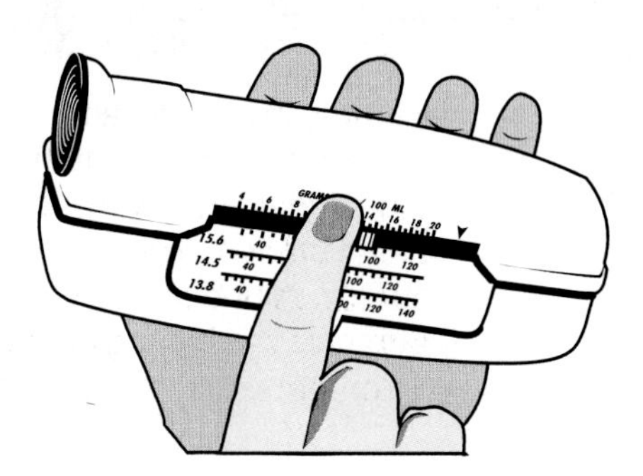
FIGURE 19.8 Determining hemoglobin content

Obtain a STAT-Site M hemoglobinometer and test card from your laboratory instructor and become familiar with the meter components as shown in figure 19.10. Proceed as follows to determine your blood's hemoglobin concentration:

1. Note the code number on the test card container (Figure 19.9). Turn on the meter and ensure the code number displayed matches the container's code number. If the numbers do not match, ask the laboratory instructor to code the meter to obtain a match. The numbers must match in order for the hemoglobin measurement to be accurate.
2. When you see the test card symbol (figure 19.9) blinking, insert the test card into the meter by sliding it under the guide tabs and pushing it firmly to the back of the test card holder. You will feel and hear the test card lock into place. A flashing drop of blood symbol will be displayed, indicating you may now apply a blood sample for analysis.
3. Follow the technique for obtaining fingertip blood given previously in this chapter.
4. Touch a large drop of blood to the center hole in the test card as shown in figure 19.11. The countdown to test result will begin.

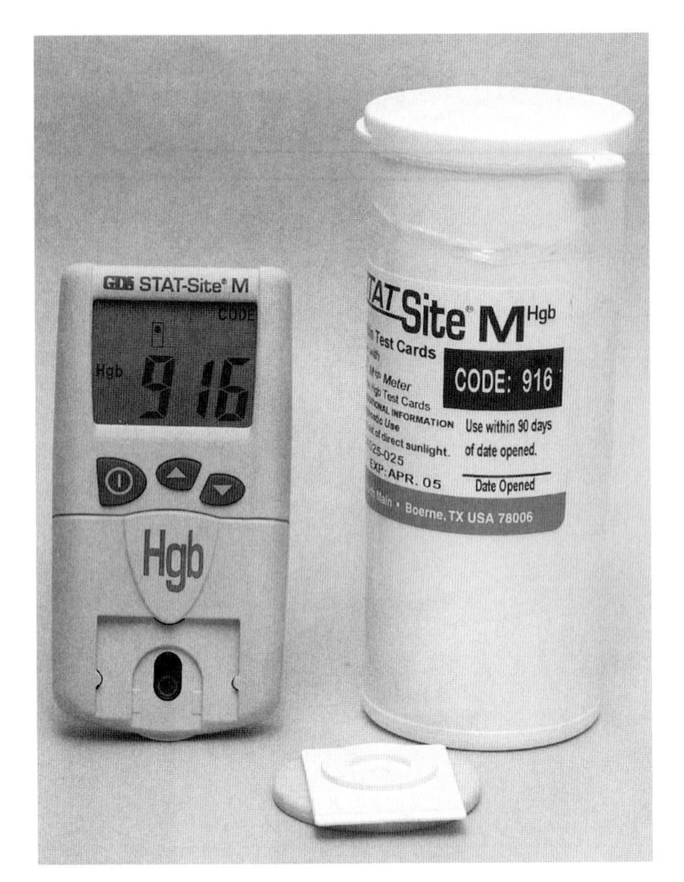

FIGURE **19.9** The STAT-Site M hemoglobinometer, test cards, and test card holder

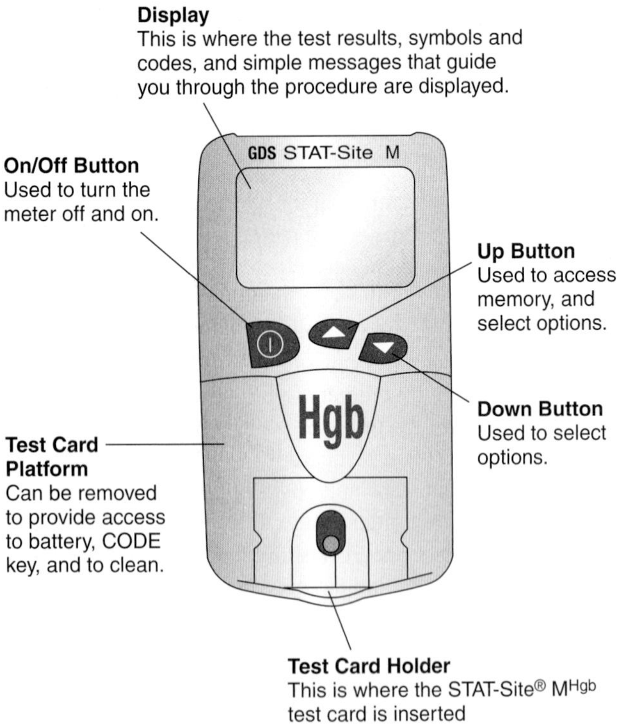

Display
This is where the test results, symbols and codes, and simple messages that guide you through the procedure are displayed.

On/Off Button
Used to turn the meter off and on.

GDS STAT-Site M

Up Button
Used to access memory, and select options.

Hgb

Down Button
Used to select options.

Test Card Platform
Can be removed to provide access to battery, CODE key, and to clean.

Test Card Holder
This is where the STAT-Site® M^{Hgb} test card is inserted

FIGURE **19.10** Meter components

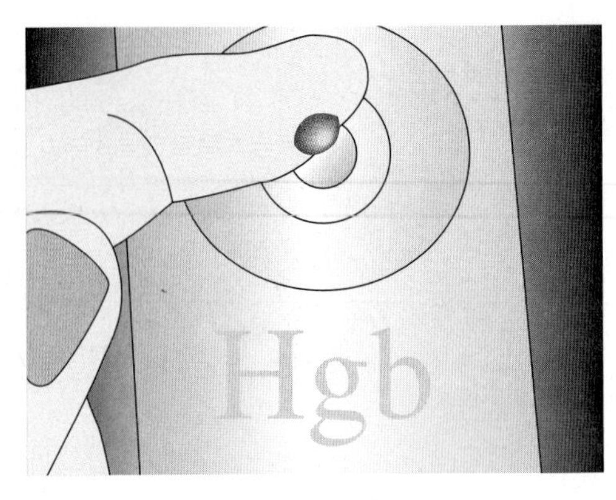

FIGURE **19.11** Applying a blood sample to the test card

5. In less than 2 minutes, the test result will be displayed as hemoglobin concentration expressed in grams per deciliter. Record the hemoglobin concentration in the report.
6. The meter will automatically turn off after 2 minutes. Carefully remove the test card so as not to contaminate the meter. Discard the test card in the appropriately marked hazardous waste container. To retrieve the result of the last test, turn on the meter and push the up button.

The Microhematocrit

Obtain a heparinized capillary tube, a sterile lancet, and an alcohol prep from the laboratory instructor. Obtain a sample of fingertip blood, according to the technique given previously in this chapter, and allow a large drop to form on the fingertip for about 3 seconds. Holding the capillary tube parallel to the plane of the finger and at a right angle to the fingertip, place the end of the capillary tube in the drop of blood and allow it to fill to at least three-fourths capacity by capillary action. Holding the finger over the other end of the capillary tube, withdraw it from the finger surface, and then seal the blood-filled end with wax.

Place the sealed capillary tube in one of the numbered slots of the microhematocrit centrifuge (figure 19.12), making certain that the sealed end is placed at the outer circumference of the centrifuge plate. Remember the slot number. When all slots in the centrifuge are filled, the laboratory instructor will centrifuge the blood. After centrifugation, remove the capillary tube and determine the hematocrit ratio using the hematocrit scale (figure 19.13).

Place the centrifuged capillary tube on the microhematocrit scale so that the bottom of the red-cell layer (at the plugged end) rests on the zero line and the top of the clear plasma layer rests on the 100% line. Note the line that intersects the top of the red-cell layer, and read its percentage value on the scale. This percentage value is the uncorrected hematocrit ratio.

Record the results in the report and on the chalkboard. Compute the class average for both male and female hematocrit ratios.

Red Cell Indices

Use the value determined in chapter 18 for your red blood cell count and the values determined in this chapter for hemoglobin content and hematocrit ratio to compute the following red cell indices: mean corpuscular volume (MCV), mean corpuscular hemoglobin (MCH), and mean corpuscular hemoglobin concentration (MCHC). Use the formulas given in the introduction to this chapter and record your computations in the report. Compare your determinations with normal values given in table 19.1.

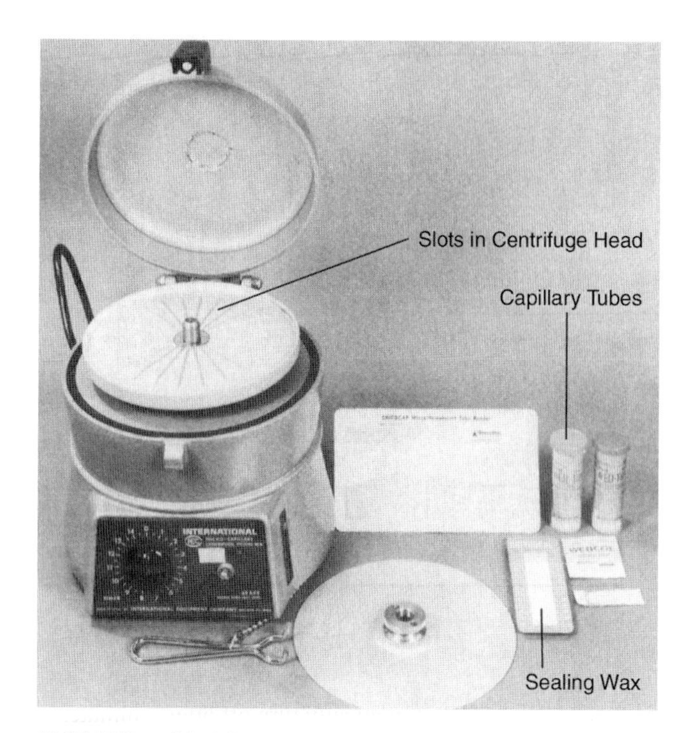

FIGURE 19.12 Microhematocrit centrifuge and supplies

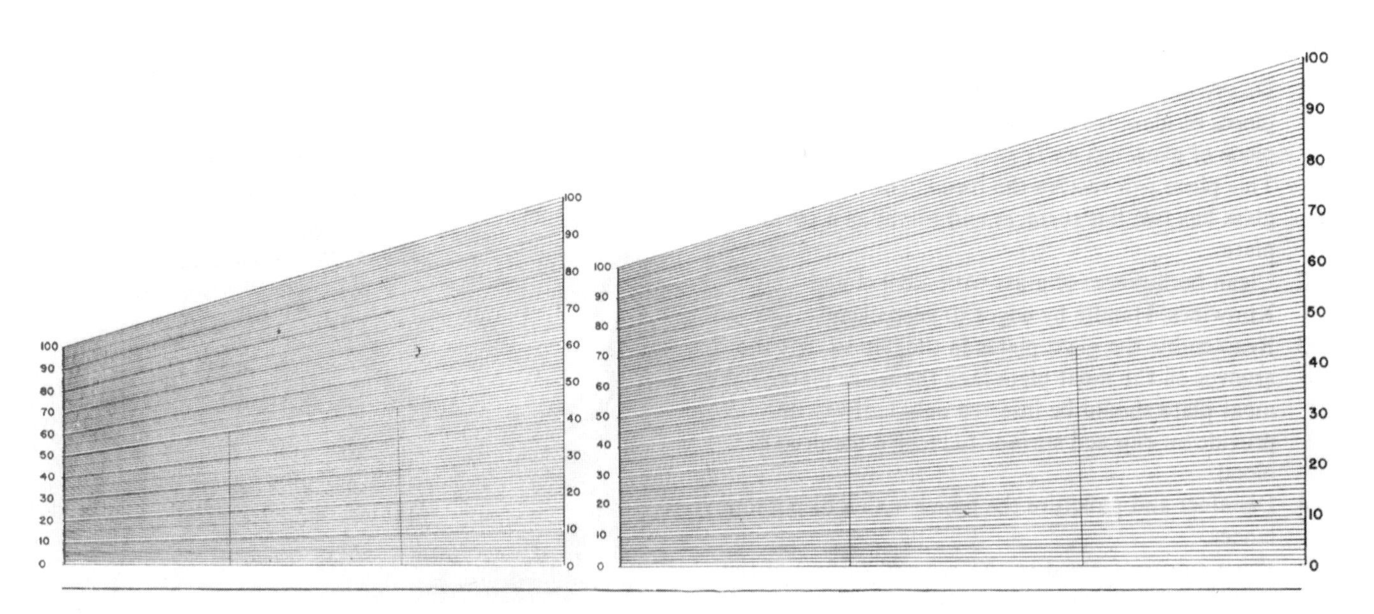

FIGURE 19.13 Microhematocrit scale

Hemostasis: The Platelet Count, Bleeding Time, and Coagulation Time

CHAPTER 20

■ INTRODUCTION

The term **hemostasis** refers to the processes that restrict the loss of blood from the vasculature. When a blood vessel ruptures, the amount of blood lost is reduced by (1) local vasoconstriction, (2) the formation of a platelet aggregate, (3) coagulation of blood and the formation of a blood clot, and (4) clot retraction and dissolution.

Immediately after a blood vessel is cut or ruptures, it constricts to reduce hemorrhage (figure 20.1). The *vasoconstriction* appears to be due to (1) local spasms of smooth muscle in the vascular wall, (2) sympathetic reflexes, and (3) the release of vasoconstrictor substances, such as serotonin, by the blood platelets, which accumulate at the site of vascular damage. In addition to reducing blood loss, vasoconstriction promotes the aggregation of platelets and the accumulation of procoagulants at the area of damage.

The *platelet aggregate* that forms at the site of vascular damage is a temporary plug that helps prevent further hemorrhage. When platelets come into contact with the torn, rough edges of the vascular wall, they adhere to the exposed connective tissue, degranulate, and release several substances, including adenosine diphosphate (ADP), into the

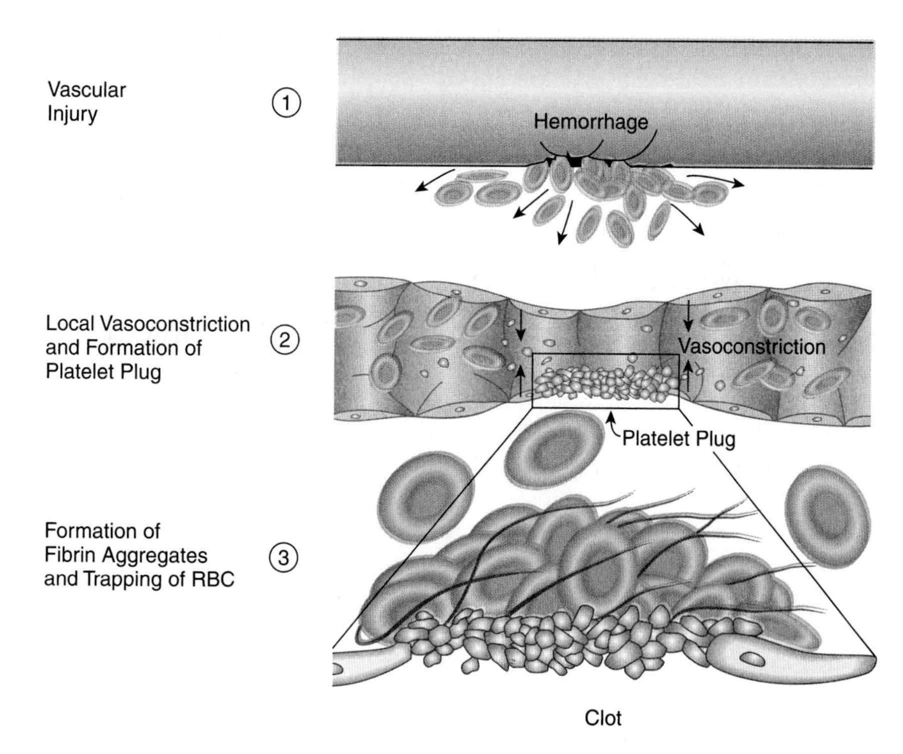

FIGURE 20.1 Formation of a clot

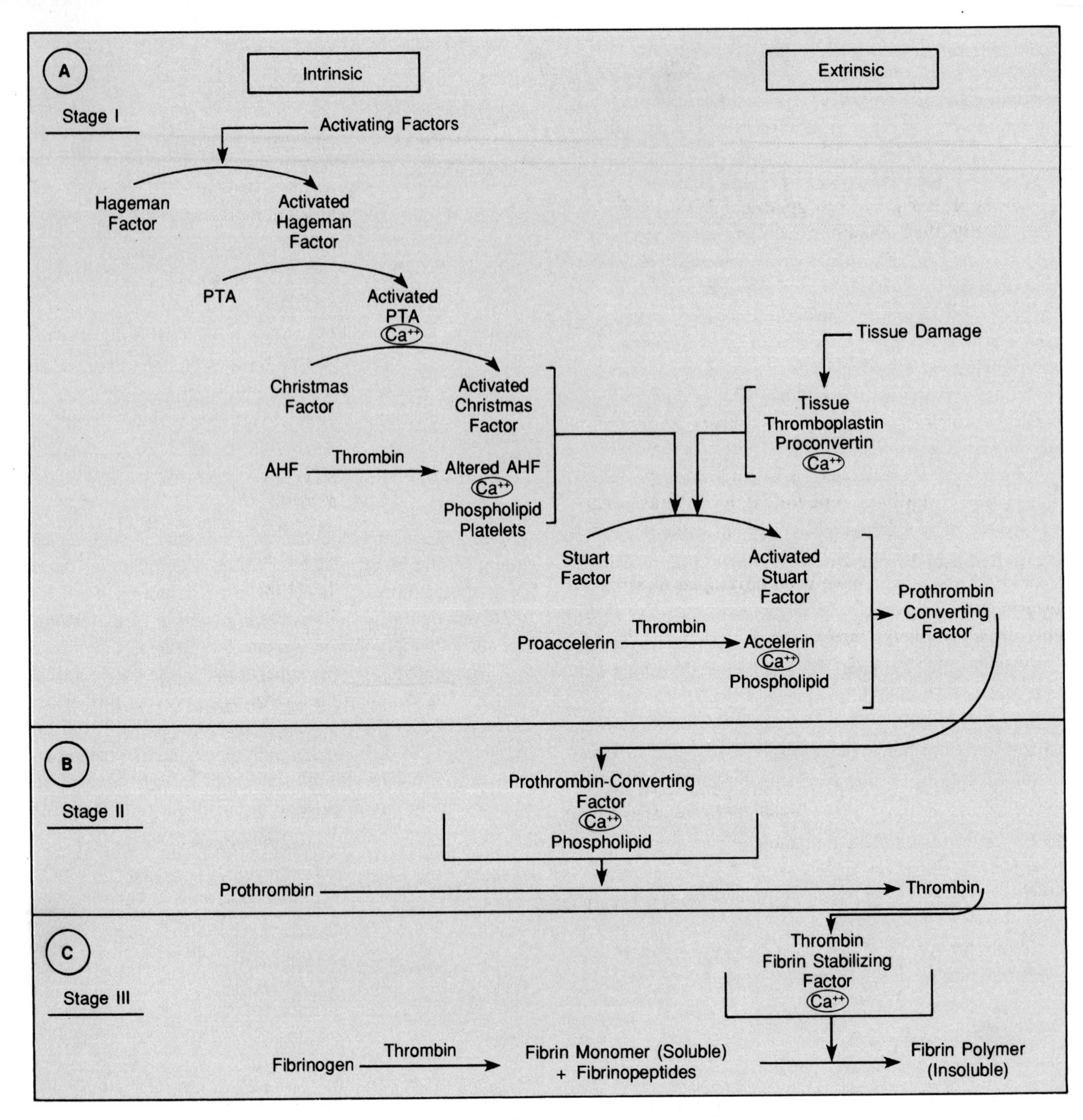

FIGURE **20.2** Cascade theory of blood coagulation

surrounding fluid. The release of ADP attracts more platelets and causes them to swell and become sticky; they adhere in increasing numbers to the damaged site and form a plug (a platelet aggregate). Activated platelets also produce thromboxane A_2, a powerful vasoconstrictor and platelet aggregator. In addition, the spherical platelets extend pseudopodia (footlike extensions of their cytoplasm) to nearby exposed collagen, anchoring the platelet plug and establishing a framework upon which clot formation can proceed.

Coagulation refers to the process in which an insoluble protein called fibrin forms and causes the blood to lose its fluid consistency and become a semisolid mass, similar in consistency to gelatin. Coagulation is a complex sequence of chemical events, which may be outlined as occurring in three stages, or phases: Stage I refers to the sequence of chemical events that produce an enzyme, the prothrombin-converting factor. Stage II involves the conversion of prothrombin to thrombin. *Prothrombin* is a plasma protein manufactured by the liver and normally present in the plasma. During stage II, prothrombin is enzymatically split by the prothrombin-converting factor, in the presence of calcium ions, into an inert fragment and *thrombin,* an active enzyme. Stage III involves the conversion of *fibrinogen,* a soluble plasma protein manufactured by the liver, into *fibrin,* an insoluble protein polymer. Figure 20.2 outlines the chemical events in each of the three stages of coagulation.

It should be noted that the "cascade" or "waterfall" sequence of events in stage I, as illustrated in figure 20.2, is a theoretical explanation consistent with available evidence. There are other equally plausible theories, but it is beyond the scope of this book to discuss them. There are 13 known factors, along with the blood platelets (thrombocytes), that are responsible for proper coagulation. These are listed in table 20.1 with their synonyms and location. If any of these clotting factors or platelets are missing or deficient, the coagulating process will be impaired.

After a clot forms, it retracts over a period of several hours. Retraction, caused by a platelet factor contractile protein, serves to draw wound surfaces together and open the vessel if it has been occluded by the clot, thereby increasing blood flow and promoting tissue repair and clot dissolution.

Clots are gradually dissolved after hemorrhage has been checked and tissue repair is well under way. Plasmin, an enzyme normally absent unless a clot has formed, breaks down fibrin into soluble fragments, thereby gradually dissolving the clot (figure 20.3).

Many clinical tests have been developed to assess hemostasis and its disorders. In the experiments that follow, some of the relatively simple screening tests for disorders of hemostasis will be performed.

TABLE 20.1 Coagulation factors

International committee designation	*Synonyms*	*Location*
Factor I	Fibrinogen	Plasma
Factor II	Prothrombin	Plasma
Factor III	Tissue thromboplastin	Tissue cells
Factor IV	Calcium ion	Plasma
Factor V	Proaccelerin	Plasma
	Prothrombin accelerator	
	Accelerator globulin	
	Labile factor	
Factor VI	(Obsolete)	
Factor VII	Serum prothrombin conversion accelerator (SPCA)	Plasma
	Proconvertin	
	Autoprothrombin I	
	Stable factor	
Factor VIII	Antihemophilic factor (AHF)	Plasma
	Platelet cofactor I	
	Thromboplastinogen	
	Antihemophilic factor A	
Factor IX	Plasma thromboplastin component (PTC)	Plasma
	Christmas factor	
	Platelet cofactor II	
	Antihemophilic factor B	
	Autoprothrombin II	
Factor X	Stuart-Power factor	Plasma
	Stuart factor	
	Autoprothrombin III	
Factor XI	Plasma thromboplastin antecedent (PTA)	Plasma
Factor XII	Hageman factor	Plasma
	Contact factor	
Factor XIII	Fibrin-stabilizing factor	Plasma
	Plasma transglutaminase	
	Laki-Lorand factor	
Platelet factors	PF 1,2,3,4	Platelets

■ EXPERIMENTAL OBJECTIVES

1. To become familiar with the hemacytometer as an instrument used to count formed elements of the blood.
2. To perform a platelet count and compare it with normal adult values.
3. To determine normal bleeding time as assessed by the Duke method.
4. To measure coagulation time and compare it with normal adult values.

Materials

compound light microscope

Clay-Adams hemacytometer (improved Neubauer)

alcohol preps

sterile, prepackaged, single-use lancets such as Single-Let (Bayer) or EZ-Lets II (Palco Labs, Inc.)

finger bandages

disposable examination gloves (optional)

Hype-Wipes (bleach-soaked disposable napkins)

gauze pads

filter paper disks

watch (with second hand)

capillary tubes (nonheparinized)

clean microscope slides

straight pins

large petri dish

acetone

Rees and Ecker fluid (3.8 g sodium citrate + 0.2 mL formalin + 0.1 g brilliant cresyl blue, to which is added distilled water to bring the total volume to 100 mL; filter before using)

Becton-Dickinson 5855 Unopette

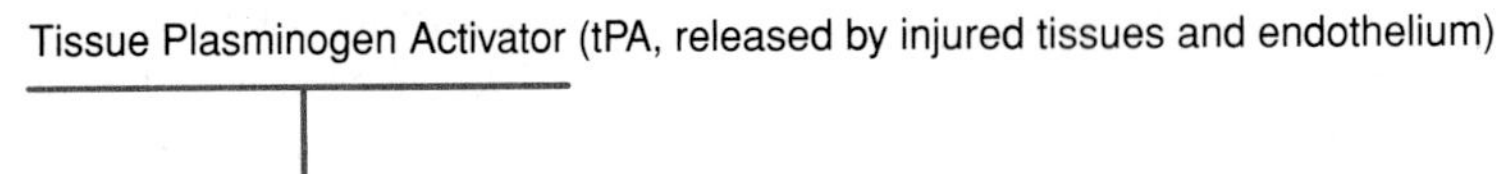

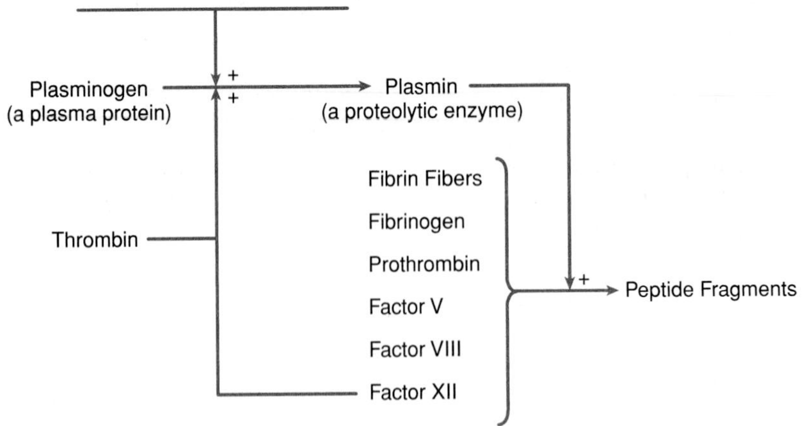

FIGURE 20.3 Dissolution of a blood clot by digesting proteins that have formed the clot

EXPERIMENTAL METHODS

CAUTION

In this and subsequent experiments on blood, you may be asked to obtain a drop of blood from your fingertip. If you have any history of a blood-clotting disorder or if you are being treated with anticoagulant medicine, inform your laboratory instructor and you will be excused from obtaining samples of your blood. Do not use samples from another student. Avoid contact with the bloods of other persons. Handle only your own blood, and do so in a manner that minimizes its exposure to other students.

Properly dispose of all sharps (lancets, needles, blades, etc.) by discarding them in appropriately marked hazardous waste containers as soon as possible after use. Nondisposable items that come in contact with your blood are to be washed with a detergent and tap water, then rinsed successively with tap water, bleach solution, distilled water, and then acetone.

Disinfect your laboratory table or laboratory bench top by wiping it clean with a bleach solution before obtaining a fingertip blood sample, and again at the end of the experiment.

Technique for Obtaining Fingertip Blood

NOTE: Read all steps before proceeding. Ask questions if you do not understand the directions.

1. Clean the palmar surface of the third or fourth finger with a sterile gauze pad soaked with 70% alcohol or a sterile disposable alcohol prep. Save the pad.
2. Allow the skin to dry. Do not blow on it to make it dry faster.
3. Obtain an EZ-Lets II sterile, single-use capillary blood sampling device from your laboratory instructor. Your laboratory instructor may require use of another capillary blood sampling device; if so, follow the instructor's directions carefully regarding procedures for its use.
   (a) Push the yellow stem into the body of the device (figure 20.4a). **Do not use the device if the yellow stem is missing.**
   (b) Hold the device securely with one hand while twisting the yellow stem with the other hand (figure 20.4b). Twist the stem completely off.
   (c) Place the opening against the cleansed, dry surface of the fingertip and firmly depress the end of the green lever (figure 20.4c). The needle automatically retracts after use.

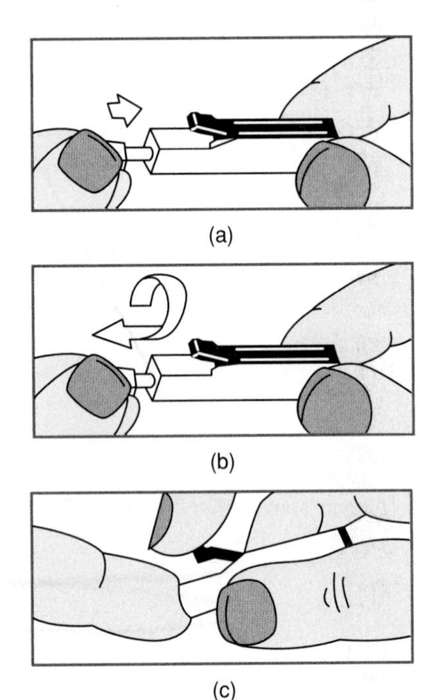

FIGURE 20.4 Procedure for obtaining a sample of fingertip capillary blood

4. Allow a good-sized drop to form before using the blood sample. Do not "milk" the finger or squeeze it to increase flow, as doing so will alter the composition of the sample. Do not allow blood to drip from your finger. Place the sample where it belongs, as described in the experiments that follow.
5. Discard the lancet in the appropriately marked hazardous waste container for sharps. If another attempt at obtaining a fingertip sample is necessary, always use a new sterile lancet.
6. After obtaining the sample, compress the gauze pad over the cut until the bleeding ceases, or cover the wound with a finger bandage. Discard the gauze pad, alcohol prep, and bandage in the appropriately marked hazardous waste container.

Platelet Count

Defects in blood coagulation may result from a deficiency of platelets or from qualitative defects in the platelets. Defects may be hereditary or acquired. The normal adult number of circulating platelets is 150,000 to 350,000 platelets per microliter of whole blood.

An increase in the number of circulating platelets is called *thrombocytosis* or *thrombocythemia.* Pathologically, it may be associated with malignancies and with polycythemia vera. It may also occur under normal physiologic circumstances such as menstruation, pregnancy, and during severe exercise.

A decrease in the number of circulating platelets is called *thrombocytopenia.* Usually, the decrease does not become critical until it descends below 50,000 μL, at which point bleeding may result from minor trauma (e.g., skin bruises easily). If the platelet count is 20,000 μL or less, bleeding may occur spontaneously. Thrombocytopenia may be due to hypoactivity of the bone marrow, resulting in a decrease in megakaryocytes; excessive destruction of platelets; or a disturbance in platelet formation.

Hemacytometer (Counting Slide)

Obtain a hemacytometer (figure 20.5) from the laboratory instructor. The hemacytometer consists of a special glass slide, coverslip, and two special pipettes. In the center of the counting slide are three grooves arranged in the shape of the letter H. The two polished platforms within the H lie exactly 0.1 mm below the surface of the slide. When the special coverslip is in place, a chamber 0.1 mm deep is formed. The scale, engraved on each platform, is a ruled square, the smallest division of which is 1/20 mm × 1/20 mm (1/400 mm^2) in area. Larger squares, each containing 16 small squares, are marked off by heavier lines. The volume of each small cube is 1/4000 μL. These small cubes are used for counting platelets and for counting red blood cells. Place the counting slide, without the coverslip, on the microscope stage, and using first the low-power and then the high-power objective, locate and identify the subdivisions with the aid of figure 20.6.

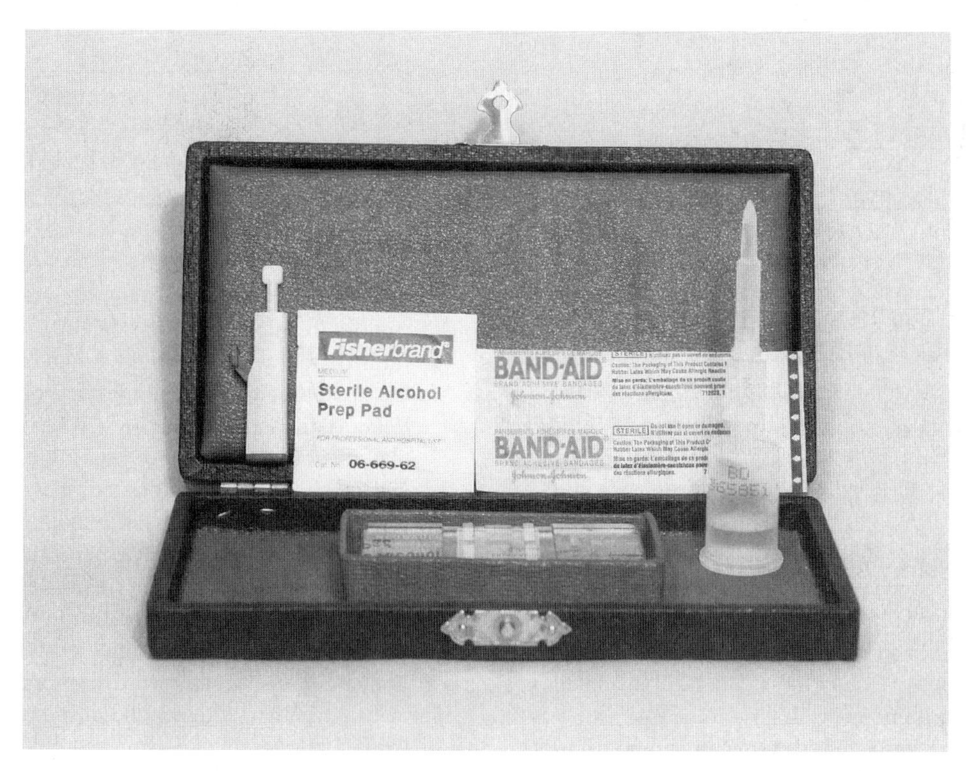

FIGURE **20.5** Hemacytometer and supplies for obtaining blood specimen

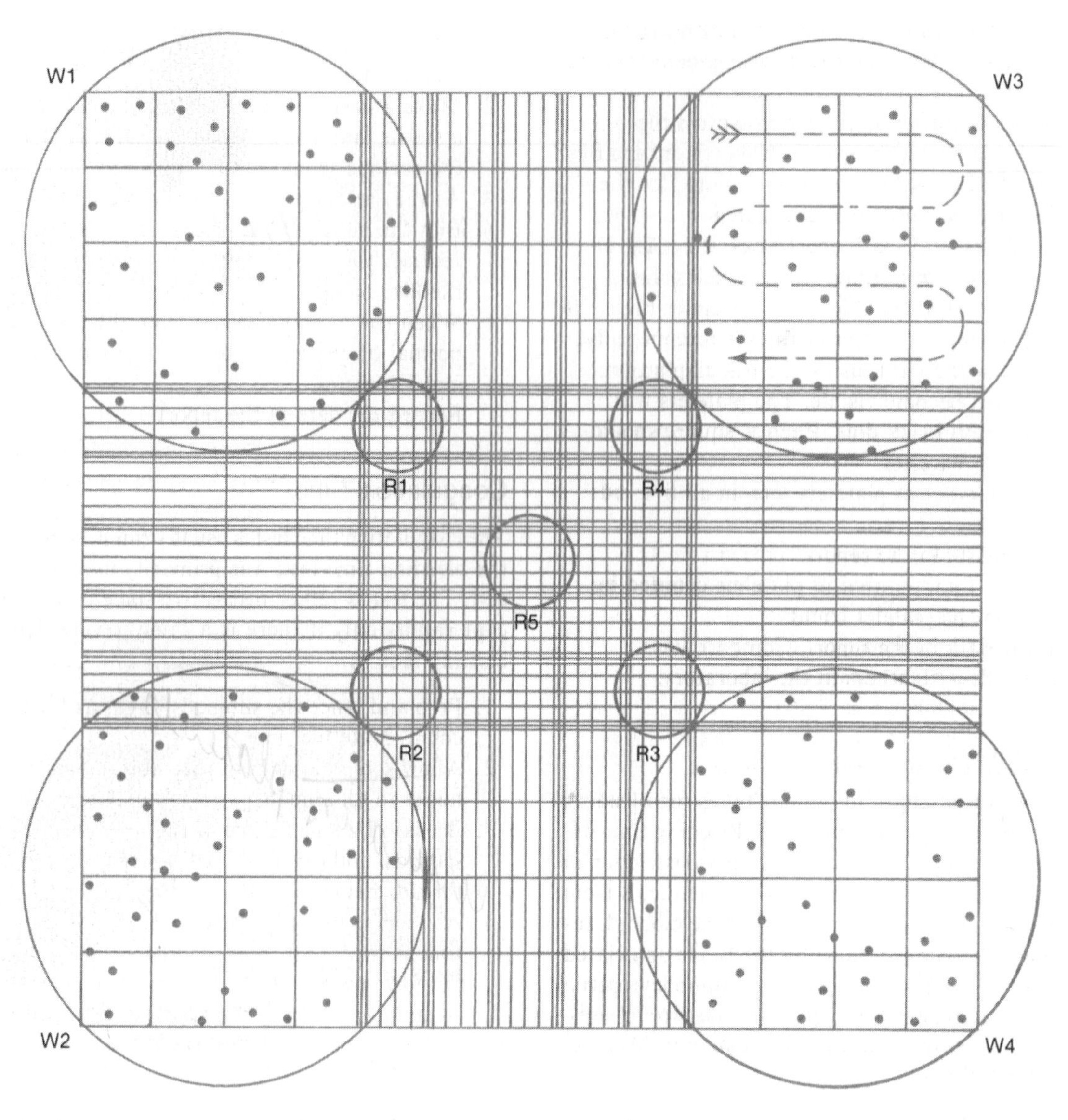

FIGURE 20.6 Improved Neubauer counting surface

Dilution Technique Using Hemacytometer Pipette

Within the hemacytometer case are two special pipettes. The larger one with the red plastic float in the bulb is used for counting platelets or for counting red blood cells. The red cell pipette is graduated to dilute blood 200 times when the sample is drawn to the 0.5 mark on the stem and the diluting solution is drawn and mixed with the sample to the 101 mark.

1. Follow the technique for obtaining fingertip blood outlined previously in this chapter; then proceed to obtain your sample. Wipe off the first drop, and allow a good-sized drop to form.
2. Dip the red cell pipette tip into the drop, and fill to the 0.5 mark exactly. Be careful to keep the tip constantly in the blood; otherwise, air will enter the column and the sample will be inaccurate. If air does enter, clean and dry the pipette and start again.
3. Immediately wipe the excess blood from the outside of the tip. Quickly insert the pipette into the Rees and Ecker fluid, and fill it to the 101 mark exactly. Close the tips of the pipette with your thumb and forefinger, and shake the contents for 2 minutes by alternately pronating and supinating the hand (i.e., moving the pipette as though it were a teeter-totter). **It is very important not to shake in the direction of the longitudinal axis. Doing so would be a serious error because the formed elements tend to settle toward one end of the pipette, resulting in dilution error.**
4. With the hemacytometer and the coverslip in place on the laboratory table, discharge 4–5 drops of the dilution on a paper napkin and place the next small drop at the edge between the counting chamber and the

coverslip. The counting chamber will fill by capillary action. Do not attempt to adjust the coverslip after the chamber has been filled.

5. Allow the charged hemacytometer to stand 15 minutes (cover with a petri dish; also place wet gauze under the dish next to the counting chamber). Immediately rinse the pipette with tap water, bleach solution, distilled water and then acetone. Place the pipette back in the hemacytometer case. Failure to clean the pipette immediately after discharging the sample will result in clogging of the pipette by dried blood, making it very difficult to clean.
6. After waiting 15 minutes, place the charged hemacytometer on the stage of the microscope, and adjust the microscope for viewing the R squares (figure 20.6) under high power. With the condenser of the microscope fully raised, adjust the light, using the iris diaphragm, until the platelets become visible. The platelets will appear as small black dots. Each R square (16 small squares) should contain a total of 4–8 platelets.
7. Count the number of platelets seen in all five R squares. Record the data in the report.
8. Multiply the total number of platelets counted in the five R squares by 10,000 to find the platelet count. Record the count in the report. Compare your platelet count with the counts of others in the laboratory.

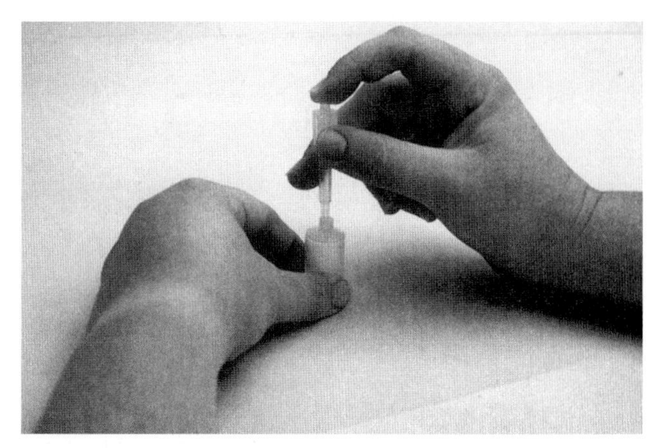

FIGURE 20.7 Use the pipette shield to puncture the reservoir diaphragm

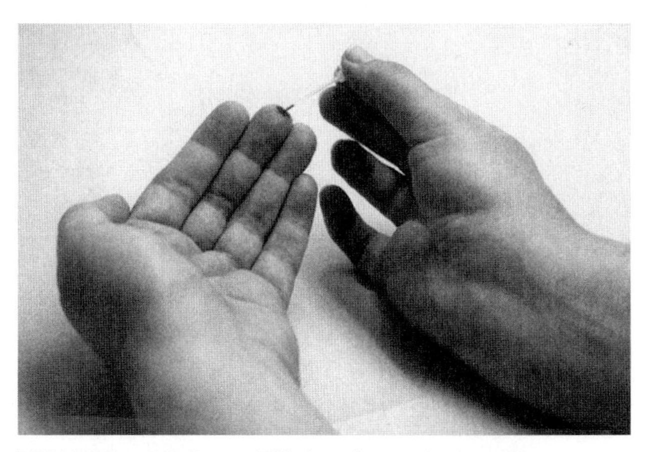

FIGURE 20.8 Fill the pipette bore with fingertip blood

Dilution Technique Using the Unopette

A more convenient method for obtaining and diluting the blood sample before counting involves the use of the Becton-Dickinson 5855 Unopette. The Unopette consists of a plastic reservoir containing a diluent and a pipette with a shield. Obtain a Unopette from your laboratory instructor and proceed as follows:

1. Place the reservoir on a firm, flat surface. Grasping the reservoir in one hand and the pipette assembly with the other, apply firm, even pressure, and push the tip of the pipette shield through the diaphragm in the neck of the reservoir (figure 20.7). This action punctures the diaphragm and opens the reservoir chamber containing diluent.
2. Remove the pipette assembly from the reservoir and disengage the shield from the pipette with a twist action. Leave the pipette loose in the shield until you are ready to use it.
3. Review the technique for obtaining fingertip blood given previously in this chapter; then proceed to obtain your sample. Wipe off the first drop, and allow a good-sized drop to form. In the next step, the capillary pipette bore must completely fill with blood, so a good-sized drop is required.
4. Holding the pipette almost horizontally, touch the tip of the capillary pipette to the blood and keep it there until the pipette fills (figure 20.8). Capillary action fills the pipette, and blood collection stops automatically when the blood reaches the end of the capillary bore in the neck of the pipette; thus errors inherent in drawing blood into conventional pipettes are prevented.
5. Wipe any excess blood from the outside of the capillary pipette, making certain no blood is removed from the capillary bore.
6. Grasp the reservoir with the thumb and first finger of one hand and maintain a slight squeeze. Use the other hand to insert the blood-filled capillary pipette tube-first snugly into the reservoir. Remove your finger from the end of the pipette, then release the finger pressure on the reservoir, thereby sucking the blood out of the pipette and into the diluent.
7. Squeeze the reservoir gently two or three times to rinse the capillary bore; force the diluent up into but not out of the overflow chamber, and release the pressure each time to return the mixture to the reservoir.
8. Place your index finger over the upper opening of the pipette and gently invert the reservoir a few times to mix the blood and diluent thoroughly. The blood sample is now diluted 100-fold.

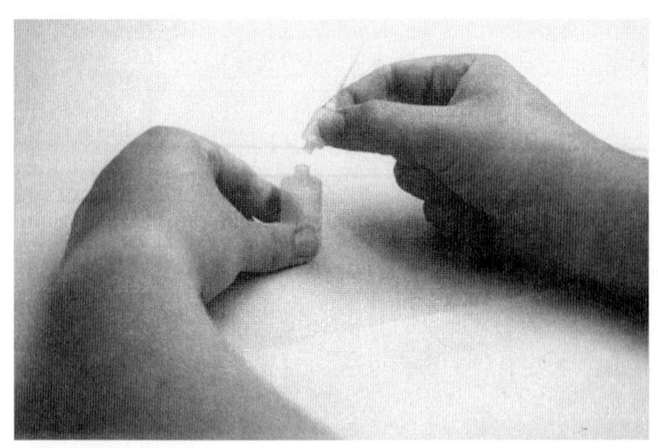

FIGURE **20.9** Convert to a dropper assembly

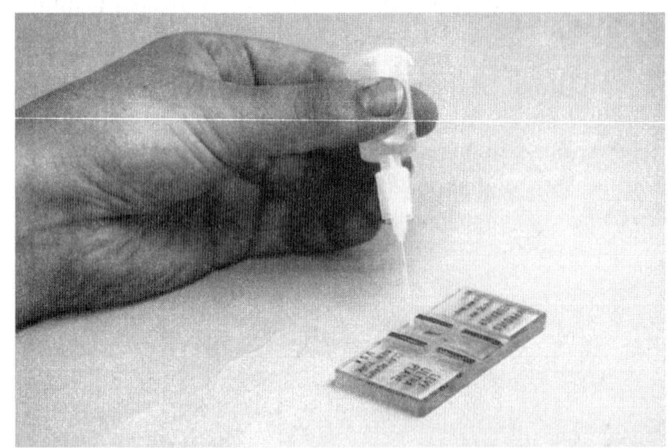

FIGURE **20.10** Charge the hemacytometer with a diluted blood sample

9. To convert to a dropper assembly, remove the pipette from the reservoir and invert it. Compress the reservoir and seat the large end of the pipette at the neck of the reservoir (figure 20.9). Release the compression on the reservoir.
10. Place the clean, dry hemacytometer slide on a napkin. Position the hemacytometer coverslip over the counting chambers. Discharge 2 drops of the dilution on the paper napkin and place the next small drop at the edge between the coverslip and the counting chamber (figure 20.10). The counting chamber will fill by capillary action. **Do not force fluid under the coverslip.** Doing so will cause the coverslip to rise, introducing a serious volume error when cells are counted. Do not attempt to adjust the coverslip or in any way disturb the chamber after it has been filled.
11. Gently transfer the charged hemacytometer to the microscope stage and allow about 3 minutes for the platelets to settle on the ruled squares.

Counting Platelets

1. Ensure that the microscope condensor is fully raised and the low-power objective is aligned for viewing the specimen. Reduce the light intensity, while attempting to focus on the ruled squares, until the grid lines of the R squares become visible. Remember, the platelets are not stained for visibility; therefore, the light intensity must be reduced to increase contrast and make them easier to see. After the grid lines are in focus, switch to the high-power objective and adjust the light to perform the platelet count. The platelets will appear as small black dots. Each R square (16 small squares) should contain a total of 4–8 platelets.
2. Count all of the platelets in the five R squares (see figure 20.6). Because each R square is subdivided into 16 smaller squares, 4 vertical and 4 horizontal, many of the small squares will not contain platelets. Record the data in the report.
3. Multiply the total number of platelets counted by 5000 to find the platelet count in one microliter (μL) of blood. For example, if R1 = 9, R2 = 11, R3 = 8, R4 = 15, and R5 = 14, then the platelet count is $57 \times 5000 =$ 285,000 platelets/μL.
4. Record your calculations in the report, and compare your platelet count with others in the laboratory.

Bleeding Time

Bleeding time is the length of time it takes for bleeding to stop after a minor skin puncture. Bleeding time depends on local vasoconstrictor reflexes, vasoconstriction in response to the release of vasoconstrictor chemicals from connective tissue cells, and coagulation of the blood. Usually, abnormally long bleeding times are associated with disorders of coagulation, such as thrombocytopenia and hypoprothrombinemia (inadequate amount of prothrombin). There are several methods of assessing bleeding time. The following method, called the Duke method, is most common.

1. Put on latex or vinyl examination gloves.
2. Cleanse the subject's earlobe with an alcohol prep. Allow it to dry.
3. Make a deep puncture in the subject's earlobe with a sterile lancet. The blood should flow freely, and the earlobe should not be squeezed.
4. As the earlobe is punctured, note the time.
5. Blot the blood with a piece of filter paper every 30 seconds. Do not touch the surface of the lobe with the filter paper.
6. When the bleeding ceases, note the time. The normal bleeding time by this method is between 2 and 5 minutes. Record the data in the report.
7. Discard the lancet in the appropriately marked hazardous waste container for sharps. Discard the filter paper and alcohol prep in the appropriately marked hazardous waste container.

Coagulation Time

The coagulation time test is usually run in conjunction with the bleeding time test; it is primarily used as a screening test. The test is rather insensitive because it renders significant results only if there is a fairly severe deficiency of clotting factors. One of two methods can be used to determine coagulation time: the capillary tube method or the microscope slide method.

Capillary Tube Method

1. Obtain a fingertip sample as for the platelet count. Note the time.
2. Fill a nonheparinized capillary tube from your fingertip blood sample.
3. Break off a small piece of the tube every 30 seconds and carefully look for the presence of fibrin threads between the two pieces. Place all pieces of glass on a paper napkin for easy disposal.
4. As soon as fibrin threads appear, note the time. The time is measured from the instant of puncture to the formation of fibrin threads. The normal coagulation time by this method is 4–7 minutes.
5. Discard the lancet, capillary tubc, and alcohol prep in appropriately marked hazardous waste containers.

Record the data in the report, and compare your results with those of others in the laboratory.

Microscope Slide Method

1. Use a wax pencil to draw a circle on a clean microscope slide.
2. Obtain a fingertip sample of blood as for the platelet count. Note the time.
3. Place a drop of blood within the wax circle.
4. Draw a pin through the blood every 30 seconds, slightly elevating the tip above the blood to look for fibrin threads. As soon as fibrin threads appear, note the time.
5. Discard the lancet, microscope slide, pin, and alcohol prep in appropriately marked hazardous waste containers.

Record the data in the report, and compare your results with those of others in the laboratory.

CHAPTER 21

White Blood Cell Count

■ INTRODUCTION

The **leukocytes,** or white blood cells, range in the adults of both sexes between 5000 and 10,000 cells per microliter of blood. The circulating leukocytes are either granulocytic (neutrophils, basophils, or eosinophils) or agranulocytic (lymphocytes and monocytes). The life span of leukocytes varies according to their function: The range extends from neutrophils, which generally survive a few hours to a few days, to the lymphocytes, which may survive for more than a year.

The granulocytes are formed in the bone marrow of the adult, as are the monocytes. The lymphocytes are formed in lymphatic tissues of the spleen, thymus, and lymph glands and in the bone marrow.

An increase in the number of leukocytes above the upper limit of the normal range (10,000 leukocytes per μL) is called *leukocytosis* (literally, "a condition of too many white blood cells"). An increase in the count above 50,000 per μL indicates possible leukemia, a malignant proliferation of leukocytes. Any count above 100,000 per μL is almost certain to be due to leukemia. In most cases, leukocytosis is a favorable sign indicating the stimulation of the body's defense against foreign material (bacteria, parasites, toxins).

A decrease in the number of circulating leukocytes below the lower limit of the normal range (5000 per μL) constitutes *leukopenia* (literally, "too few leukocytes"). Leukopenia may also be caused by certain infections as well as depression of the bone marrow due to radiation, poisoning, or, according to recent reports, alcoholism.

When white blood cells are counted regardless of type, the count is known as a *combined white blood cell count.* In some cases it is diagnostically useful to count white blood cells according to type. This is known as the differential white blood cell count.

In the *differential white blood cell count,* leukocyte types are commonly expressed as a percentage of the combined white blood cell count. Normal values for the differential percentage are indicated in table 21.1.

TABLE **21.1** Normal values for differential WBC count

Leukocyte type	***Percentage***	***Number/μL***
"Juvenile" neutrophils	0–1	0–100
Neutrophilic band cells	3–5	100–500
"Segmented" neutrophils	50–70	2500–7000
Eosinophils	1–5	50–500
Basophils	0–1	0–100
Monocytes	1–6	100–600
Lymphocytes	20–40	1000–4000

An increase in the number of young neutrophils ("juvenile" and neutrophilic band cells) is seen in many infections in which neutrophil production is rapid and immature forms of neutrophils are released in larger numbers.

An increase in the number of segmented neutrophils, or the number of nuclear lobes of the neutrophils is seen in pernicious anemia and other related disorders.

In the *absolute differential white blood cell count,* leukocyte types are counted and expressed as an absolute number in 1.0 μL of blood, for example, 300 eosinophils/μL. More commonly, leukocyte types are counted and expressed as a percentage of the total number of leukocytes counted in a single blood sample. In the *percentage differential white blood cell count,* if the total number of leukocytes counted is 200, and 120 are segmented neutrophils, the segmented neutrophil count would be expressed as 60%. Normal values for the absolute differential count are indicated in table 21.1.

The value and interpretation of the differential white blood cell count require a knowledge of the origin and production of these cells. Basic information concerning leukocyte function was presented in chapter 17. For a more detailed discussion of leukocyte physiology, consult your textbook. Some of the more common conditions associated with abnormal white blood cell counts and examples of causative factors are listed here.

1. *Neutropenia* is a decrease in the number of neutrophils, which occurs in aplasia of the marrow, depression of the marrow (due to cytotoxic drugs, X-radiation, etc.), and from certain viral infections.
2. *Neutrophilia* is an increase in the number of neutrophils, which occurs normally during the first few days of an infant's life, as well as late in pregnancy, after strenuous exercise, or after prolonged vomiting. It occurs in pneumonia, smallpox, appendicitis, meningitis, and other chronic infections, as well as in myelogenous leukemia, intestinal obstruction, acute hemorrhage, coronary thrombosis, uremia, and other states.
3. *Eosinophilia* is an increase in the number of eosinophils, which occurs in parasitic infections (e.g., trichinosis, tapeworm); in such allergic disorders as a serum sickness, asthma, and hay fever; in malignant blood diseases, such as Hodgkin's disease; and in some diseases of the skin (e.g., psoriasis).
4. *Basophilia* is an increase in the number of basophils, which occurs in inflammatory diseases of connective and epithelial tissues (e.g., sinusitis), malignant blood diseases, and pernicious anemia.
5. *Monocytosis* is an increase in the number of monocytes, which occurs in monocytic leukemia, tuberculosis, infectious mononucleosis, typhoid fever, malaria, carbon tetrachloride poisoning, and bacterial endocarditis.
6. *Lymphocytosis* is an increase in the number of lymphocytes, which occurs in lymphatic leukemia, mumps, tuberculosis, whooping cough, syphilis, and other malignant and infectious diseases.

■ EXPERIMENTAL OBJECTIVES

1. To become familiar with one clinical procedure for counting white blood cells.
2. To perform and record one's own combined white blood cell count.
3. To become familiar with normal values for the combined and the differential white blood cell counts.
4. To acquire an appreciation for the value of performing a combined and a differential white blood cell count in the diagnosis of disease.

Materials

Clay-Adams hemacytometer (improved Neubauer)

Becton-Dickinson 5853 Unopette

Turk's solution (diluting fluid)

acetone

alcohol

distilled water

staining dish

Wright stain

alcohol preps

sterile, prepackaged, single-use lancets such as Single-Let (Bayer) or EZ-Lets II (Palco Labs, Inc.)

compound light microscope

microscope lens paper

finger bandages

disposable examination gloves (optional)

Hype-Wipes (bleach-soaked disposable napkins)

gauze pad

immersion oil

microscope slides

■ EXPERIMENTAL METHODS

CAUTION

In this and subsequent experiments on blood, you may be asked to obtain a drop of blood from your fingertip. If you have any history of a blood-clotting disorder or if you are being treated with anticoagulant medicine, inform your laboratory instructor and you will be excused from obtaining samples of your blood. Do not use samples from another student. Avoid contact with the bloods of other persons. Handle only your own blood and do so in a manner that minimizes its exposure to other students.

Properly dispose of all sharps (lancets, needles, blades, etc.) by discarding them in appropriately marked hazardous waste containers as soon as possible after use. Nondisposable items that come in contact with your blood are to be washed with a detergent and tap water, then rinsed successively with tap water, bleach solution, distilled water, and then acetone.

Disinfect your laboratory table or laboratory bench top by wiping it clean with a bleach solution before obtaining a fingertip blood sample, and again at the end of the experiment.

Technique for Obtaining Fingertip Blood

NOTE: Read all steps before proceeding. Ask questions if you do not understand the directions.

1. Clean the palmar surface of the third or fourth finger with a sterile gauze pad soaked with 70% alcohol or a sterile disposable alcohol prep. Save the pad.
2. Allow the skin to dry. Do not blow on it to make it dry faster.
3. Obtain an EZ-Lets II sterile, single-use capillary blood sampling device from your laboratory instructor. Your laboratory instructor may require use of another capillary blood sampling device; if so, follow the

instructor's directions carefully regarding procedures for its use.

(a) Push the yellow stem into the body of the device (figure 21.1a). Do not use the device if the yellow stem is missing.

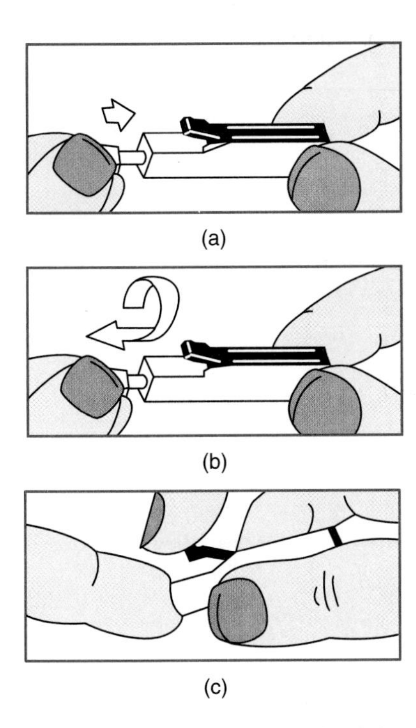

FIGURE 21.1 Procedure for obtaining a sample of fingertip capillary blood

(b) Hold the device securely with one hand while twisting the yellow stem with the other hand (figure 21.1b). Twist the stem completely off.

(c) Place the opening against the cleansed, dry surface of the fingertip and firmly depress the end of the green lever (figure 21.1c). The needle automatically retracts after use.

4. Allow a good-sized drop to form before using the blood sample. Do not "milk" the finger or squeeze it to increase flow, as doing so will alter the composition of the sample. Do not allow blood to drip from your finger. Place the sample where it belongs, as described in the experiments that follow.
5. Discard the lancet in the appropriately marked hazardous waste container for sharps. If another attempt at obtaining a fingertip sample is necessary, always use a new sterile lancet.
6. After obtaining the sample, compress the gauze pad over the cut until the bleeding ceases. Discard the gauze pad or alcohol prep in the appropriately marked hazardous waste container.

Counting Slide

Obtain a hemacytometer (figure 21.2) from the laboratory instructor. The hemacytometer consists of a special glass slide, coverslip, and two special pipettes. In the center of the counting slide are three grooves arranged in the shape of the letter H. The two polished platforms within the H lie exactly 0.1 mm below the surface of the slide. When the special coverslip is in place, a chamber 0.1 mm deep is formed. A ruled square is engraved on each platform, the smallest division of which is 1/20 mm × 1/20 mm (1/400 mm^2) in area. Larger

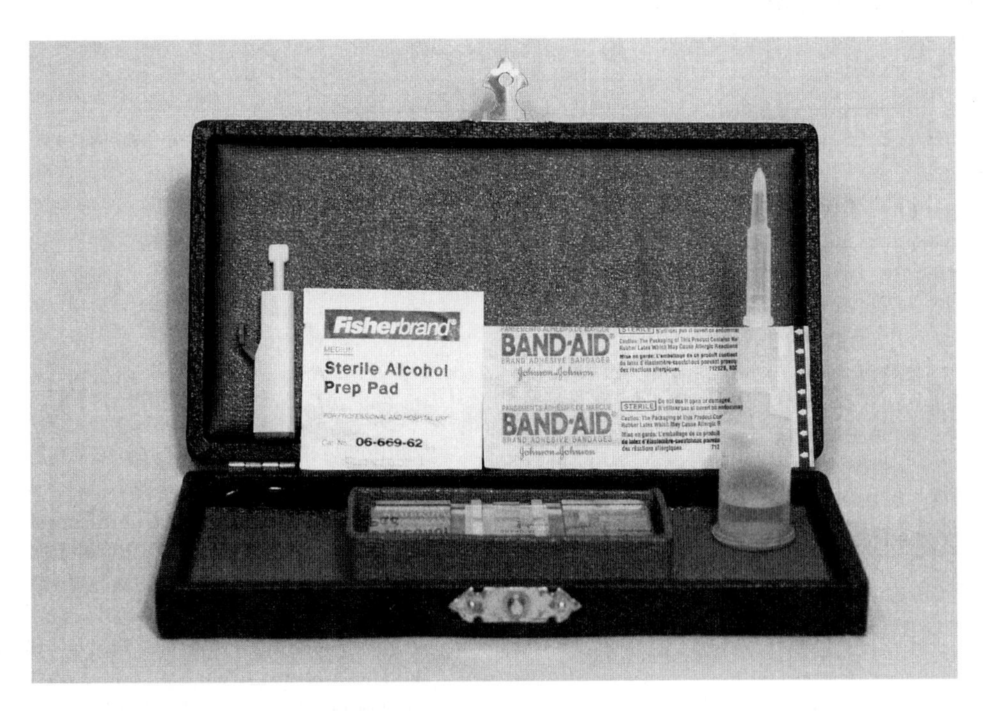

FIGURE 21.2 Hemacytometer and supplies for obtaining blood specimen

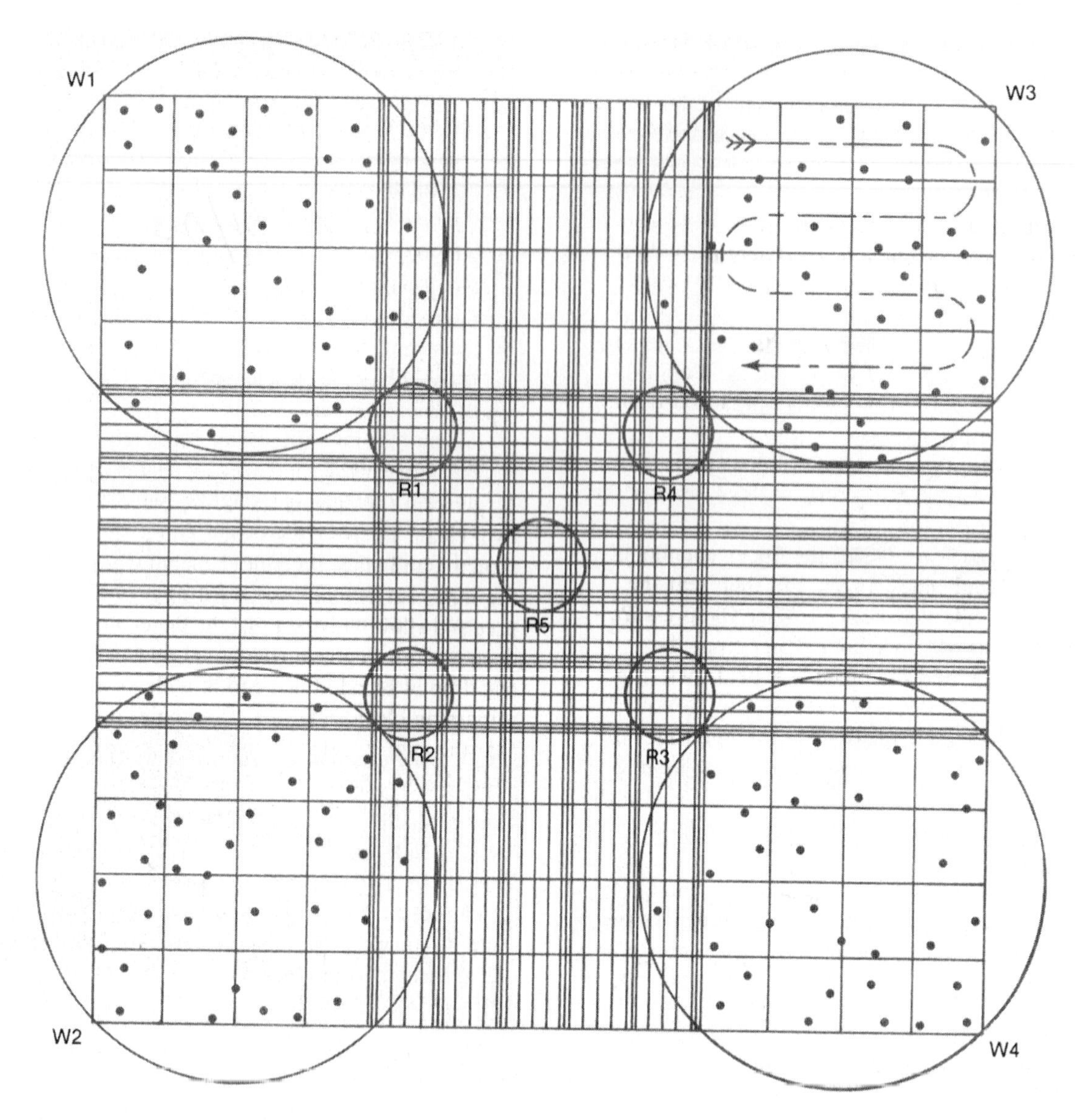

FIGURE 21.3 Improved Neubauer counting surface

squares, each containing 16 small squares, are marked off by heavier lines. The volume of each small cube is 1/4000 µL. These cubes are used for counting red blood cells.

At each of the four corners of an improved Neubauer pattern is a 1-mm^2 area subdivided into 16 large squares (figure 21.3). These four areas are used for counting white blood cells.

Place the counting slide, without the coverslip, on the microscope stage. Using first the low-power and then the high-power objective, locate and identify the subdivisions with the aid of figure 21.3. After becoming familiar with the counting slide, proceed to the sections on dilution techniques.

Dilution Technique Using Hemacytometer Pipettes

Within the hemacytometer case are two special pipettes. The larger one with the red plastic float in the bulb is used for counting red blood cells. The smaller one with the white plastic float is used for counting white blood cells. The white blood cell pipette is graduated to dilute blood 20 times when the sample is drawn to the 0.5 mark on the stem and Turk's solution is drawn and mixed with the sample to the 101 mark.

1. Follow the technique for obtaining fingertip blood outlined previously in this chapter; then proceed to obtain your sample. Wipe off the first drop, and allow a good-sized drop to form.
2. Dip the white cell pipette tip into the drop, and fill to the 0.5 mark exactly. Be careful to keep the tip constantly in the blood; otherwise, air will enter the column and the sample will be inaccurate. If air does enter, clean and dry the pipette and start again.
3. Immediately wipe the excess blood from the outside of the tip. Quickly insert the pipette into the Turk's solution,

and fill to the 101 mark exactly. (The sample is now diluted 20 times.) Close the tips of the pipette with your thumb and forefinger, and shake the contents for 2 minutes by alternately pronating and supinating the hand (i.e., moving the pipette as though it were a teeter-totter). **It is very important not to shake in the direction of the longitudinal axis. Doing so would be a serious error, because the white cells tend to settle toward one end of the pipette, resulting in dilution error.**

4. With the hemacytometer and the coverslip in place on the microscope stage, discharge 2 drops of the dilution on a paper napkin, and place the next small drop at the edge between the counting chamber and the coverslip. The counting chamber will fill by capillary action. Do not attempt to adjust the coverslip or in any way disturb the chamber after it has been filled.
5. Allow about 3 minutes for the blood cells to settle on the ruled squares.

Immediately after discharging blood into the hemacytometer, rinse the pipette with tap water, bleach solution, distilled water, and then acetone. Place the pipette back in the hemacytometer case. Procedures to be followed in counting the white blood cells are outlined later.

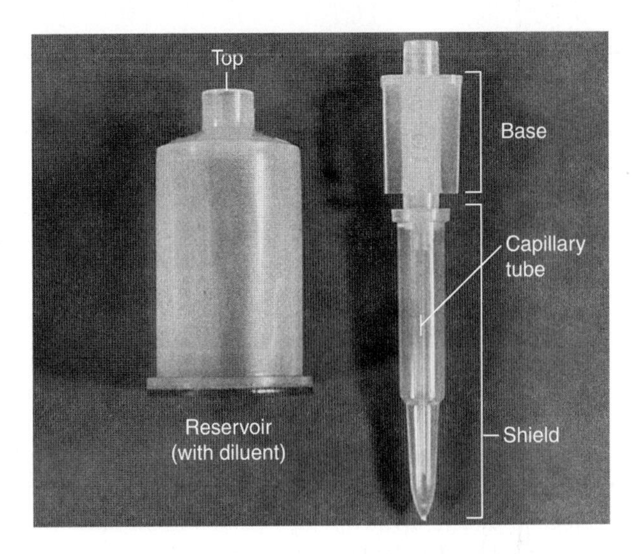

FIGURE 21.4 The Becton-Dickinson Unopette microcollection system

Dilution Technique Using the Unopette

A more convenient method for obtaining and diluting the blood sample before counting involves the use of the Becton-Dickinson 5853 Unopette. The Unopette consists of a plastic reservoir containing isotonic diluent and a pipette with a shield (figure 21.4). Obtain a Unopette from your laboratory instructor and proceed as follows:

1. Place the reservoir on a firm, flat surface. Grasping the reservoir in one hand and the pipette assembly with the other, apply firm, even pressure, and push the tip of the pipette shield through the diaphragm in the neck of the reservoir (figure 21.5). This action punctures the diaphragm and opens the reservoir chamber containing diluent.
2. Remove the pipette assembly from the reservoir and disengage the shield from the pipette with a twist action. Leave the pipette loose in the shield until you are ready to use it.
3. Review the technique for obtaining fingertip blood given previously in this chapter; then proceed to obtain your sample. Wipe off the first drop, and allow a good-sized drop to form. In the next step, the capillary pipette bore must completely fill with blood, so a good-sized drop is required.
4. Holding the pipette almost horizontally, touch the tip of the capillary pipette to the blood and keep it there until the pipette fills (figure 21.6). Capillary action fills the pipette, and blood collection stops automatically when the blood reaches the end of the capillary bore in the neck of the pipette, thus errors inherent in drawing blood into conventional pipettes are prevented.

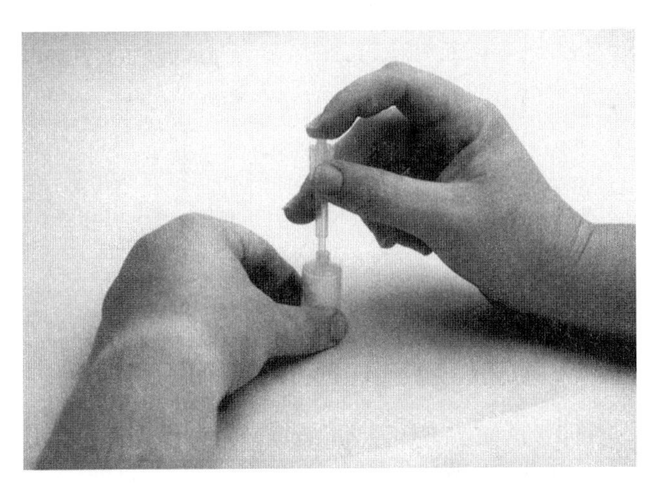

FIGURE 21.5 Use the pipette shield to puncture the reservoir diaphragm

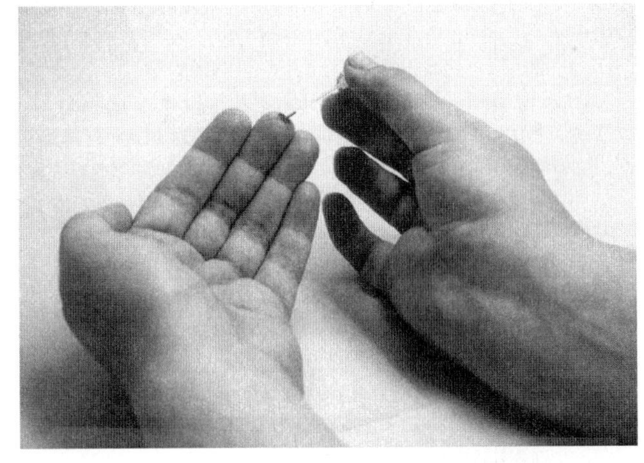

FIGURE 21.6 Fill the pipette bore with fingertip blood

5. Wipe any excess blood from the outside of the capillary pipette, making certain no blood is removed from the capillary bore.
6. Grasp the reservoir with the thumb and first finger of one hand and maintain a slight squeeze. Use the other hand to insert the blood-filled capillary pipette tube-first snugly into the reservoir. Remove your finger from the end of the pipette, then release the finger pressure on the reservoir, thereby sucking the blood out of the pipette and into the diluent.
7. Squeeze the reservoir gently two or three times to rinse the capillary bore; force the diluent up into but not out of the overflow chamber, and release the pressure each time to return the mixture to the reservoir.
8. Place your index finger over the upper opening of the pipette, and gently invert the reservoir a few times to mix the blood and diluent thoroughly. The blood sample is now diluted 100-fold.
9. To convert to a dropper assembly, remove the pipette from the reservoir and invert it. Compress the reservoir and seat the large end of the pipette at the neck of the reservoir (figure 21.7). Release the compression on the reservoir.
10. Place the clean, dry hemacytometer slide on a napkin. Position the hemacytometer coverslip over the counting chambers. Discharge 2 drops of the dilution on the paper napkin and place the next small drop at the edge between the coverslip and the counting chamber (figure 21.8). The counting chamber will fill by capillary action. **Do not force fluid under the coverslip.** Doing so will cause the coverslip to rise, introducing a serious volume error when cells are counted. Do not attempt to adjust the coverslip or in any way disturb the chamber after it has been filled.
11. Gently transfer the charged hemacytometer to the microscope stage and allow about 3 minutes for the blood cells to settle on the ruled squares.

Counting Combined White Blood Cells

1. Ensure that the microscope condensor is fully raised and the low-power objective is aligned for viewing the specimen. Reduce the light intensity, while attempting to focus on the ruled squares, until the cells become visible. Remember, these cells are not stained for visibility; therefore, the light intensity must be reduced to increase contrast and make the cells easier to see. After the cells and grid lines are in focus, switch to the high-power objective to perform the cell count.
2. Count all of the cells in the four W squares (see figure 21.3). Each W square is subdivided into 16 smaller squares, 4 vertical and 4 horizontal. Count all of the cells in each of the 16 squares and record the count in each of the corresponding squares in the laboratory report. To avoid counting a cell twice, count all of the cells that touch the line at the left but none that touch the line at the right of each square. Similarly, count all of the cells that touch the line at the top of each square but none that touch at the bottom line.
3. To calculate the number of white blood cells (WBC) in 1 microliter (μL) of blood, the following formula is used:

$$\text{WBC}/\mu\text{L of blood} = \frac{\text{Number of cells counted} \times \text{Dilution} \times 10}{\text{Number of small squares counted}}$$

For example, if a total of 500 cells were counted in a total of 64 small squares, then the WBC count would be:

$$\frac{500 \times 100 \times 10}{64} = 7812 \text{ WBC}/\mu\text{L}$$

As you can see from the above formula, if cells in all 64 small squares of four W squares are counted, and

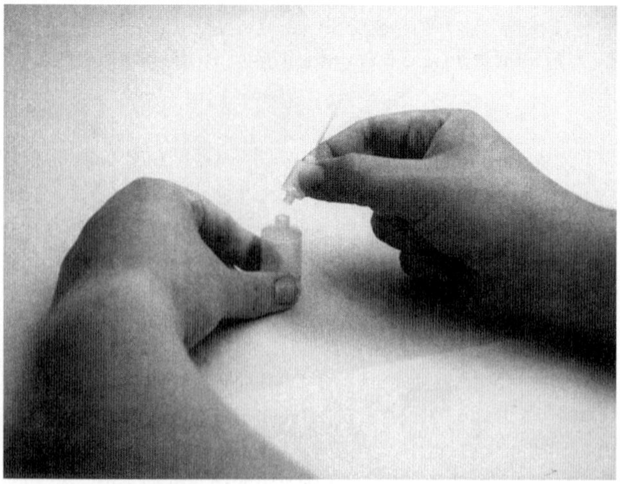

FIGURE **21.7** Convert to a dropper assembly

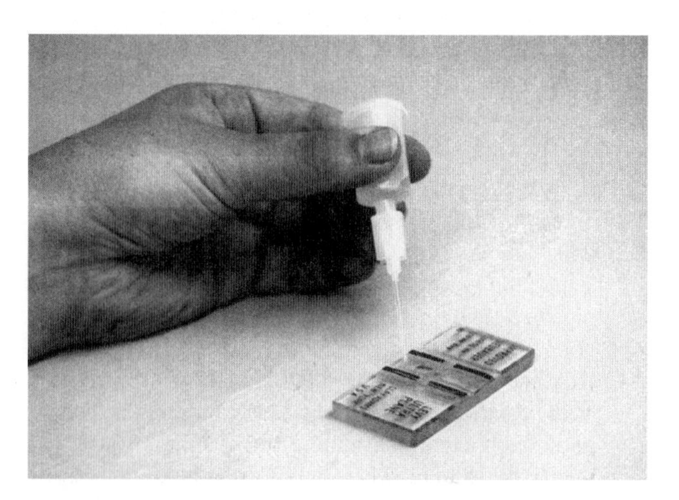

FIGURE **21.8** Charge the hemacytometer with a diluted blood sample

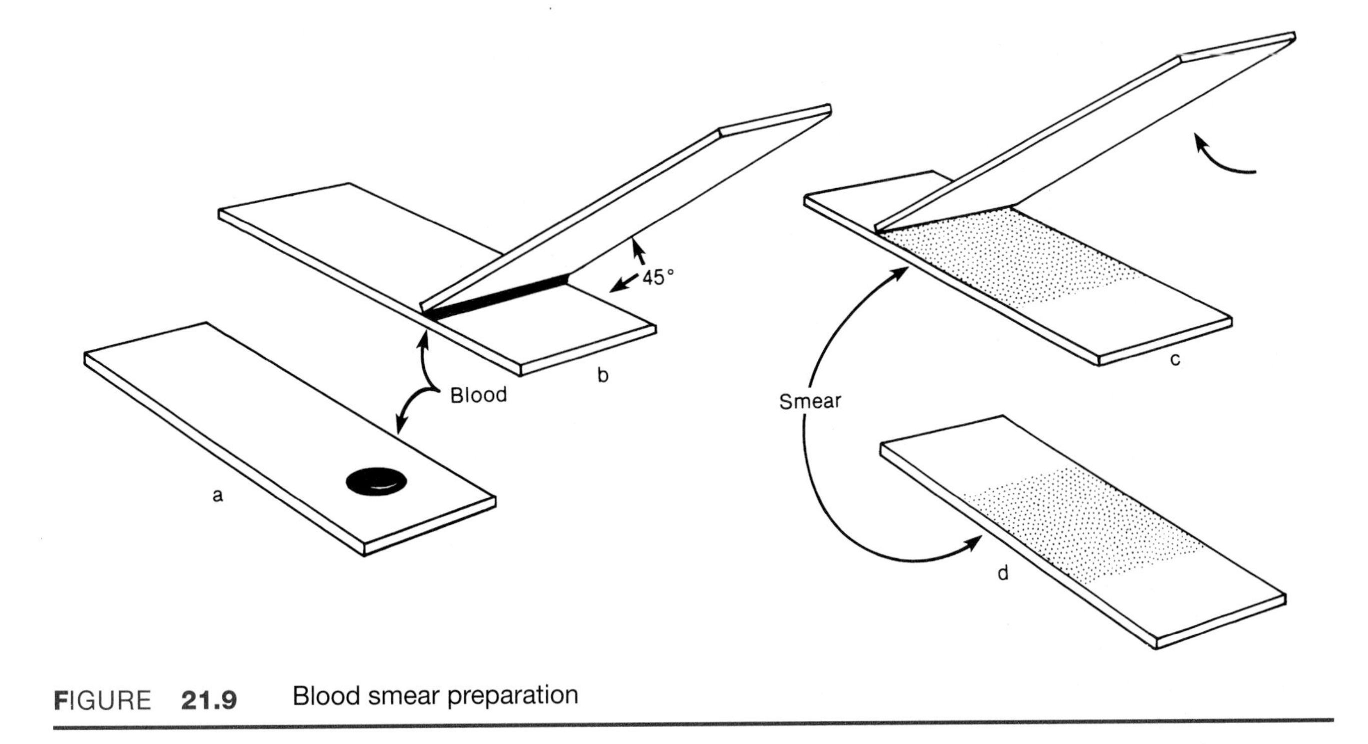

FIGURE 21.9 Blood smear preparation

1000 is a constant, then 1000 ÷ 64 = 15.625 and the formula simplifies to:

WBC/μL of blood = Number of cells counted in 4 W squares × 15.625

4. Record your cell counts and your calculation in the data section of your report.

Differential Count

To perform the differential white blood cell count, obtain two clean microscope slides from the laboratory instructor and place them on the laboratory bench. Follow the technique for obtaining fingertip blood outlined previously in this chapter; then proceed to obtain your sample. Place one drop of blood at one end of a clean slide. Position the remaining clean slide at a 45-degree angle to the first slide, so that the angled slide touches the drop of blood on the first slide (figure 21.9). Quickly move the second slide across the top of the first slide, dragging the drop of blood into a thin, uniform layer. Allow the blood smear to dry thoroughly.

Wright Stain

After drying, lower the slide into a staining dish containing Wright stain (Quik-Stain, a special preparation of Wright stain, may be used) and leave it in the stain for 5 seconds. Remove the slide and lower it into a second dish containing distilled water, letting it remain for 10 seconds. Remove the slide from the distilled water, gently rinse with tap water, wipe the bottom of the slide, and allow the slide to air-dry.

Counting Cells

Place the slide of stained blood on the microscope stage after it has completely dried. With the condensor raised to its maximum height and the iris diaphragm adjusted to admit maximum light, focus on the blood cells through the high-power objective. Refer to chapter 17 for a description of each white blood cell type. Using the mechanical stage, begin at the lower left-hand portion of the smear and move the field of view to the right, then up, then to the left, then up, and so on, counting white blood cells as the field of view changes (figure 21.10). Each time a white blood cell comes into view, identify it and record it by making a hash-mark (l) in the laboratory report table. Count a total of 100 white blood cells, then compute the percentage differential white blood cell counts and record the values in the report table.

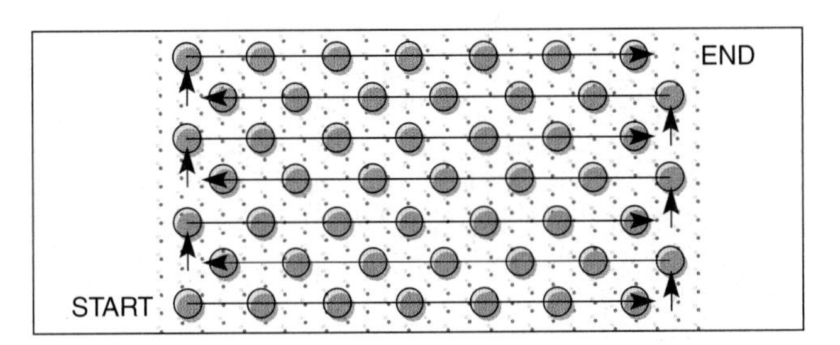

FIGURE 21.10 Directional changes in the field of view when performing the differential white blood cell count

Microcirculation

INTRODUCTION

In vertebrates and other animals with a closed circulatory system, blood circulates in the following pattern (figure 22.1):

heart → arteries → arterioles → metarterioles → capillaries → venules → veins → heart

The function of the blood is to provide a homeostatic environment for the cells of the organism. In doing so, not only must the blood be circulated by heart contractions, but essential nutrients and gaseous materials must also be allowed to leave the blood and enter the interstitial fluids in exchange for the waste products of cellular metabolism. The exchanges occur across the wall of the capillary, one of several vessel types that constitute the microcirculation.

A **microcirculatory unit** (figure 22.2) consists of an *arteriole* and its major branches, the *metarterioles.* The metarterioles lead to the *true capillaries* through which blood flow is controlled by precapillary sphincters—rings of smooth muscle in the metarteriole wall at its junction with a capillary. The capillaries then join to form small *venules,* which drain the blood from the capillary network into larger collecting venules.

In some cases, blood may pass directly from the metarterioles to the venules by way of arteriovenous shunts, called *AV (arteriovenous) capillaries.* These should not be confused with true capillaries that form capillary beds and across the walls of which exchanges occur between the blood and interstitial fluid.

True capillaries are the most numerous type of blood vessel of the vascular system. They have an average diameter of 9 µm, so red blood cells (8 µm in diameter) must pass through the capillary in single file. Capillary walls consist of a single layer of flat endothelial cells arranged in the form of a tubular mosaic (figure 22.2). Materials of the plasma (e.g., water, electrolytes, nutrient molecules, gases) pass through the capillary wall by going either through the endothelial cells or between them. Because of the vast number of capillaries, their combined diameter is extremely

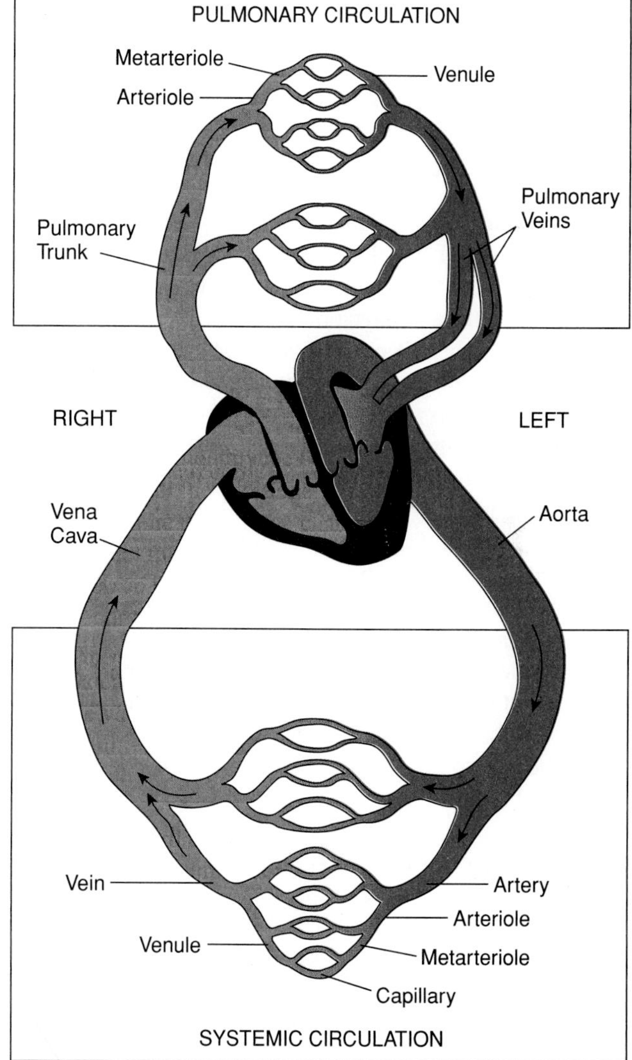

FIGURE 22.1 An outline of the human cardiovascular system. Arrows denote the direction of blood flow through the heart and blood vessel types.

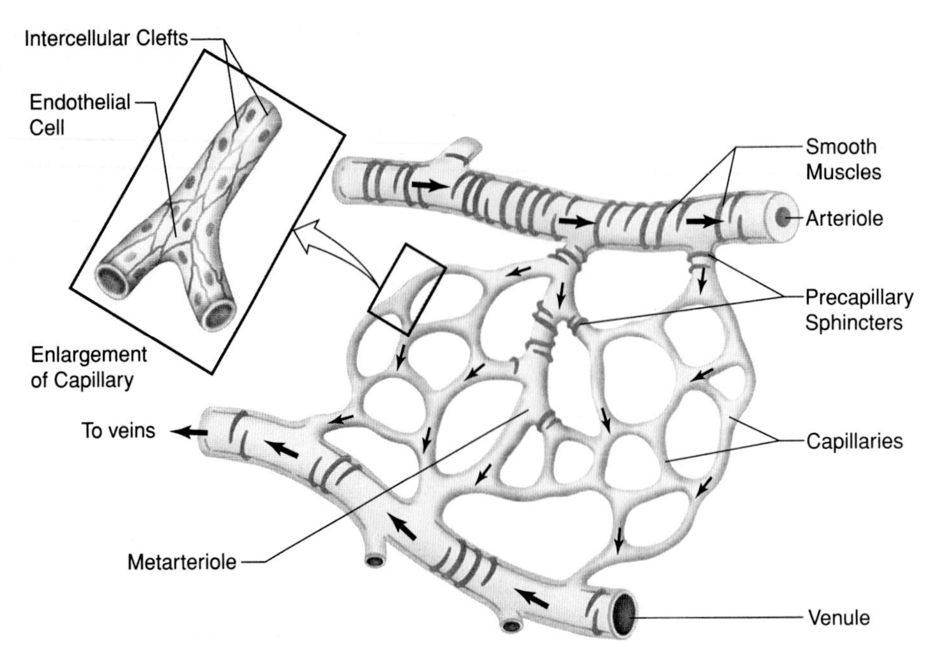

FIGURE 22.2 Microcirculatory unit. Note the absence of smooth muscle in the capillaries.

large in comparison to the rest of the vasculature, and, hence, the flow of blood is slowest through the capillaries; this slower flow facilitates the exchange of materials between the plasma and interstitial fluid.

Not all the body's capillaries are open to permit a flow at any given moment. Because the total cross-sectional area of the capillaries is so large, if all of the capillary networks would open simultaneously systemic blood pressure would be drastically lowered, blood flow would rapidly decrease, and the organism would die of circulatory insufficiency. Capillary flow is, therefore, intermittent; it is governed, in part, by the needs of the body.

The perfusion of capillary beds is controlled by either increasing arteriole diameter *(vasodilation)* or decreasing arteriole diameter *(vasoconstriction)* and by the relaxation or constriction of the precapillary sphincter. The smooth muscle of the arteriole and the precapillary sphincter are under the autonomic control of the vasomotor centers in the medulla oblongata.

The regional distribution (e.g., skin versus gut) of capillary blood is controlled via the autonomic nervous system and its adjustments of arteriole and precapillary sphincter diameters. On a hot summer day, for example, the perfusion of capillary networks in the skin is increased to allow for a greater dissipation of body heat, while the perfusion of the capillary networks elsewhere is diminished.

Locally, the perfusion of capillary beds is influenced by the presence of metabolites (e.g., carbon dioxide) and vasoactive substances such as histamine and hormones. An increase in the concentration of histamine, for example, in the interstitial fluid results in vasodilation of arterioles and precapillary sphincters, thereby increasing the perfusion of local tissue.

Histamine is released by mast cells of the connective tissue in response to the presence of an irritating chemical. Histamine release is part of a protective mechanism to increase blood flow to traumatized tissue to prevent further injury and promote healing. Histamine release in respiratory tissues (e.g., nasal passages) may promote local blood flow to the extent of causing congestion, which is why antihistamines and vasoconstrictors are contained in decongestant sprays. Massive systemic releases of histamine, such as may occur in severe allergic reactions, may result in extensive peripheral vasodilation, a precipitous fall in blood pressure and flow, and death.

In the experiments that follow, capillaries in the web of the foot of the frog will be used for observation of microcirculation, although similar experiments can be performed using the capillary networks in the intestinal mesentery, urinary bladder, wings of insects, and other animals.

■ EXPERIMENTAL OBJECTIVES

1. To observe and identify anatomic components of a microcirculatory unit.
2. To observe normal patterns of blood flow through the microcirculatory unit.
3. To observe the influence of vasoconstrictors, vasodilators, and electric stimulation on blood vessel diameter and blood flow.

Materials

grass frog
corkboard
compound light microscope
stimulator
hand electrode set
latex or vinyl examination gloves
anesthetic pan
straight pins
2% MS-222 anesthetic
1.8 mg/mL acetylcholine in frog Ringer's solution
1:1000 norepinephrine in frog Ringer's solution
11 mg/100 mL histamine in frog Ringer's solution
oil of wintergreen (methyl salicylate)
dropper bottles
frog Ringer's solution

■ EXPERIMENTAL METHODS

Anesthesia

Put on latex or vinyl examination gloves. MS-222, an anesthetic commonly used in veterinary medicine, is water soluble and readily absorbed by amphibian skin. Place enough 2% MS-222 in a large glass anesthesia pan to cover the feet and legs of a frog. Put the frog in the pan, cover, and wait until the frog becomes unresponsive (about 2–4 minutes). At the 2-minute mark, remove the frog and test for the level of anesthesia using the corneal and withdrawal reflexes. Absence of reflex activity signifies an appropriate level of anesthesia. Do not overanesthetize, as doing so will severely depress blood circulation.

If the animal is properly anesthetized (corneal and withdrawal reflexes are absent; see chapter 10), wash off the excess anesthetic from the skin by temporarily immersing the frog in a beaker of tap water and then proceed with surgery. Maintain a proper level of anesthesia by periodically checking for the return of reflex activity; if necessary, cover the frog's skin with a napkin soaked in MS-222.

Preparation for Observation

1. The web of the foot of the frog may be used to view capillary circulation, although the presence of pigment cells in the connective tissue may partially obscure the vasculature (figure 22.3). Place the frog with its dorsal side up on the corkboard, and secure the animal to the board by wrapping the body in a laboratory napkin soaked with frog Ringer's solution and tying the animal to the board (figure 22.3). Extend one leg of the frog, and pin the web of the foot over the circular opening in the corkboard. Keep the webbing moistened with frog Ringer's solution. Secure the frog corkboard to the stage of the microscope using a large rubber band.
2. Using the low-power objective of the microscope, examine the capillary circulation. Periodically moisten the preparation with a drop of saline to prevent desiccation. Keep the frog's skin moist by covering it with a napkin soaked in frog Ringer's solution.

Observations

1. Study the circulation and identify the arterioles, metarterioles, capillaries, and venules. How do you distinguish between the various vessel types? Sketch your observations in the report, and indicate the direction of blood flow. Compare the blood vessel diameters, and make rough estimates of the diameters on the basis of the size of the frog erythrocyte (15–20 μm in diameter, 5 μm thick). Compare the relative thicknesses of the blood vessel walls. Do some of the vessels pulsate? Make notes on the relative velocity of the blood flow in the arterioles, capillaries, and venules. Does the blood flow at the same velocity in all capillaries? Does the blood always flow in the same direction through the capillaries? Do the blood cells travel at the same velocity at the margin of an arteriole as they do in the center? Why? Do you observe any other blood cell types besides the red blood cells (erythrocytes)?
2. After making the preceding observations and answering the questions, apply electrodes directly to the surface of the mesentery or the web of the foot, and deliver a repetitive train of stimuli of 60 pulses per second at 2–5 V for approximately 1 minute, while observing the tissue under the microscope. Do you observe vasoconstriction? Where? Do the capillaries constrict? Why or why not? Does capillary blood flow increase or decrease during the period of stimulation?
3. After electric stimulation, allow the vessel diameter and blood flow to return to normal values. Apply 1 or 2 drops of the norepinephrine solution directly on the vessels. Norepinephrine induces vasoconstriction of the arterioles. As soon as vasoconstriction is apparent, wash the surface with Ringer's solution and allow the vessel diameter to return to normal.
4. Place 1 or 2 drops of the acetylcholine solution directly on the blood vessels. Do you observe vasodilation? Which vessels dilate? Again wash the surface with Ringer's solution.
5. Place 1 or 2 drops of the histamine solution directly on the blood vessels. Histamine is a very powerful vasodilator. Which vessels dilate? Are there changes in the blood flow? Of what practical

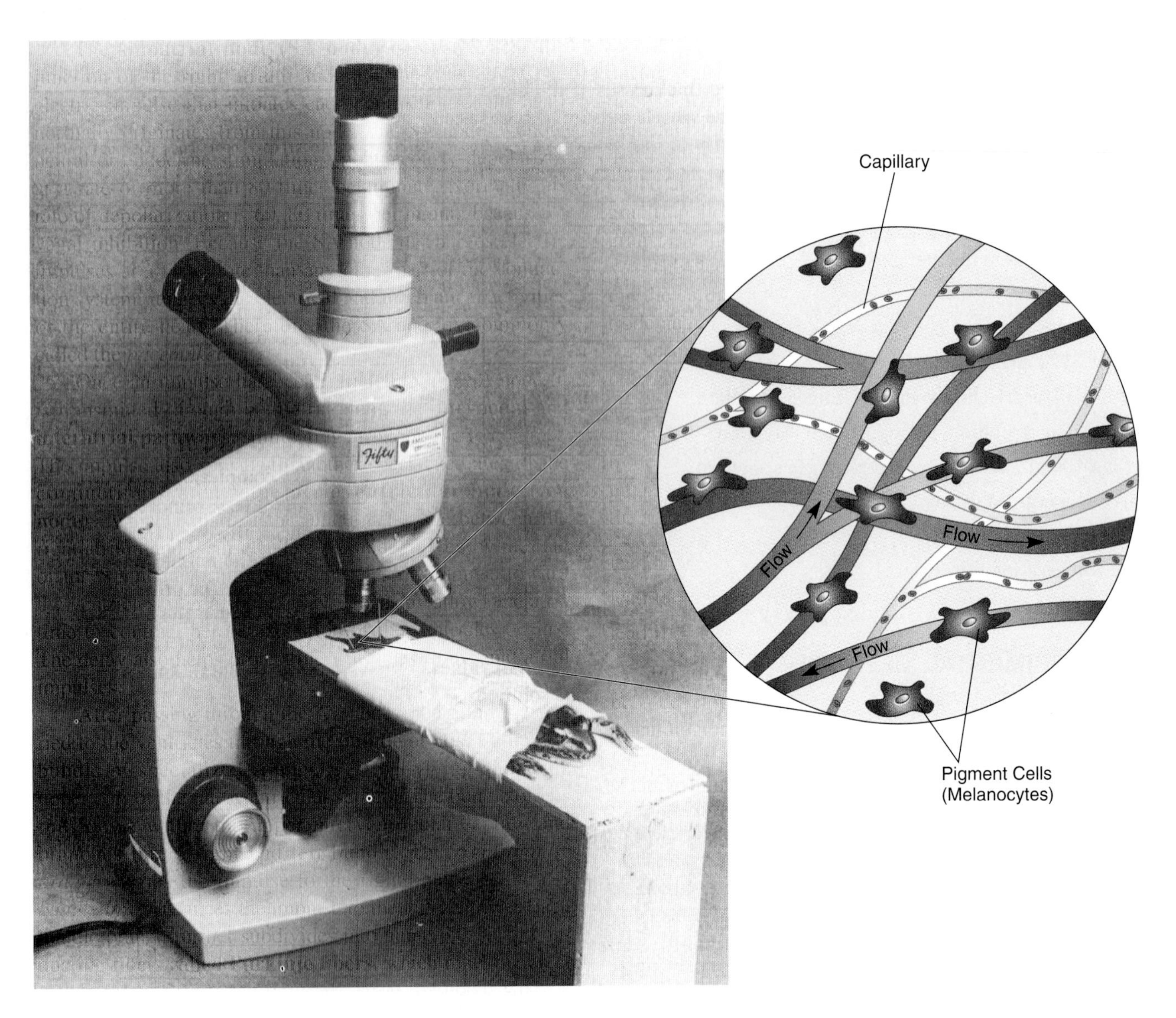

FIGURE 22.3 Circulation in the frog foot web. Note the individual blood cells in the capillary moving very slowly, or not at all, and the fast laminar flow in the larger blood vessels.

importance are antihistamine compounds? Wash the surface with Ringer's solution, and allow the vessel diameter and flow to return to normal.

6. Place a drop of methyl salicylate directly on the blood vessels. Record changes in the vessel diameter. Why is this substance, under the name of "oil of wintergreen," sometimes used in liniments?
7. On completing your observations of the frog foot web, remove the animal from the cork and return the animal to the holding tank. Clean the microscope, corkboard, and instruments, and return them to their proper location.

Microcirculation

REPORT 22

Name: ______________________ Date: ______________________

Lab Section: ______________________

1. Sketch your observation of the microcirculation, labeling the vessel types and indicating the direction of blood flow.

2. When observing microcirculation, how did you distinguish between an arteriole and a venule?

____

____

____

3. Indicate which of the following are vasoconstrictors and which are vasodilators with respect to the peripheral (foot web) circulation.

Substance	Constriction	Dilation
Norepinephrine		
Acetylcholine		
Histamine		
Methyl salicylate		

4. What is the function of the arteriovenous shunt? ______________________

____

5. Why are antihistamines sometimes administered to persons experiencing anaphylactic shock? ______________________

____

____

____

6. Explain the nasal congestion associated with the common cold. ______________________

____

____

7. Why is methyl salicylate (oil of wintergreen) sometimes used in liniments? ________________

____

____

____

8. Explain the difference between an arteriovenous capillary and a true capillary. ________________

____

____

____

9. The velocity of blood flow is slowest in the capillaries and fastest in the large arteries and veins, despite the fact that blood pressure is higher in the capillary than in the vein. Why? ________________

____

____

____

10. Why is it desirable to have a lower rate of blood flow through the capillary networks? ________________

____

____

____

CHAPTER 23

The Cardiac Cycle

INTRODUCTION

The primary function of the heart is to pump blood through the systemic and pulmonary vasculature. For the heart to move blood efficiently, it must receive an adequate supply of blood from the systemic and pulmonary veins, ensure a unidirectional flow of blood through the cardiac chambers, and effectively pump the blood out through the systemic and pulmonary arterial systems. The sequence of electrical and mechanical events of the heart associated with receiving blood from the venous systems and pumping it out into the arterial systems during one heartbeat is known as the **cardiac cycle.**

The heart possesses four chambers: two *atria,* which receive blood from the venous systems, and two *ventricles,* which receive blood from the atria and pump it out into the arterial systems (figure 23.1). In the mammalian heart, the atrium and ventricle on the right side are separated from their counterparts on the left side by *septa* (interatrial and interventricular). The right atrium receives blood from the systemic venous system via the superior and inferior venae cavae and delivers it to the right ventricle. The right ventricle pumps blood to the lungs by way of the pulmonary trunk and arteries. The left atrium receives blood from the pulmonary veins and delivers it to the left ventricle. The left ventricle pumps blood to all body tissues via the aorta and subsequent branches of the systemic arterial system.

Blood flow through the vasculature is unidirectional; that is, blood flows from the heart into (in sequence) the arteries, arterioles, capillaries, venules, veins, and then back into the heart. The unidirectional nature of blood flow within the heart is ensured by the presence of four intracardiac valves (figure 23.2).

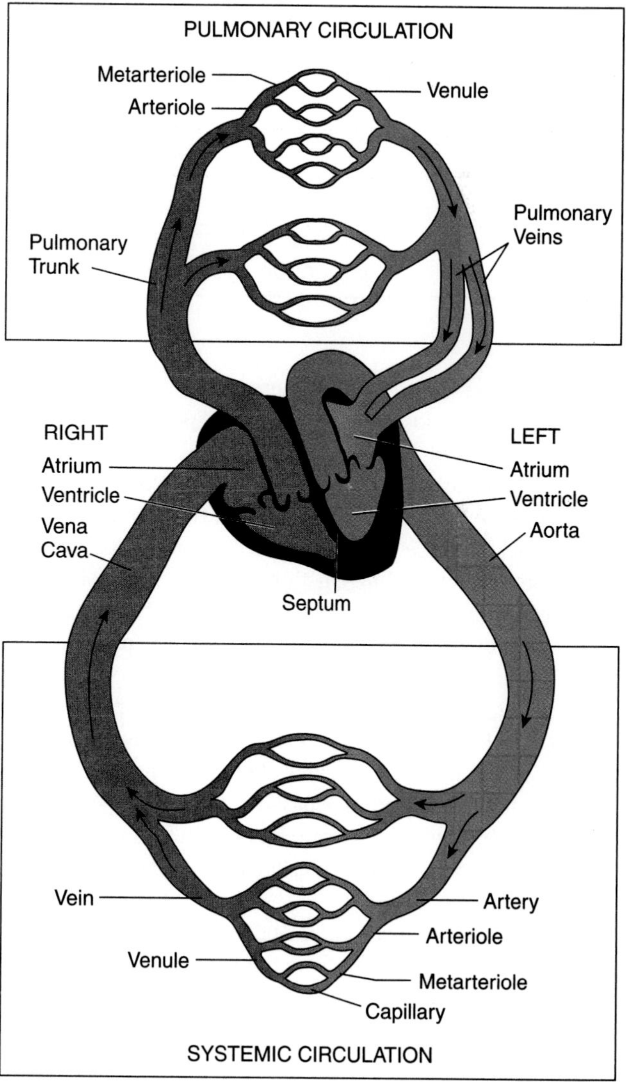

FIGURE 23.1 Chambers of the mammalian heart and the pulmonary and systemic circulation

1. The *right atrioventricular valve* (tricuspid) opens into the right ventricle, allowing blood to flow from the right atrium into the right ventricle but not in the reverse direction. Valve action is passive; the valve

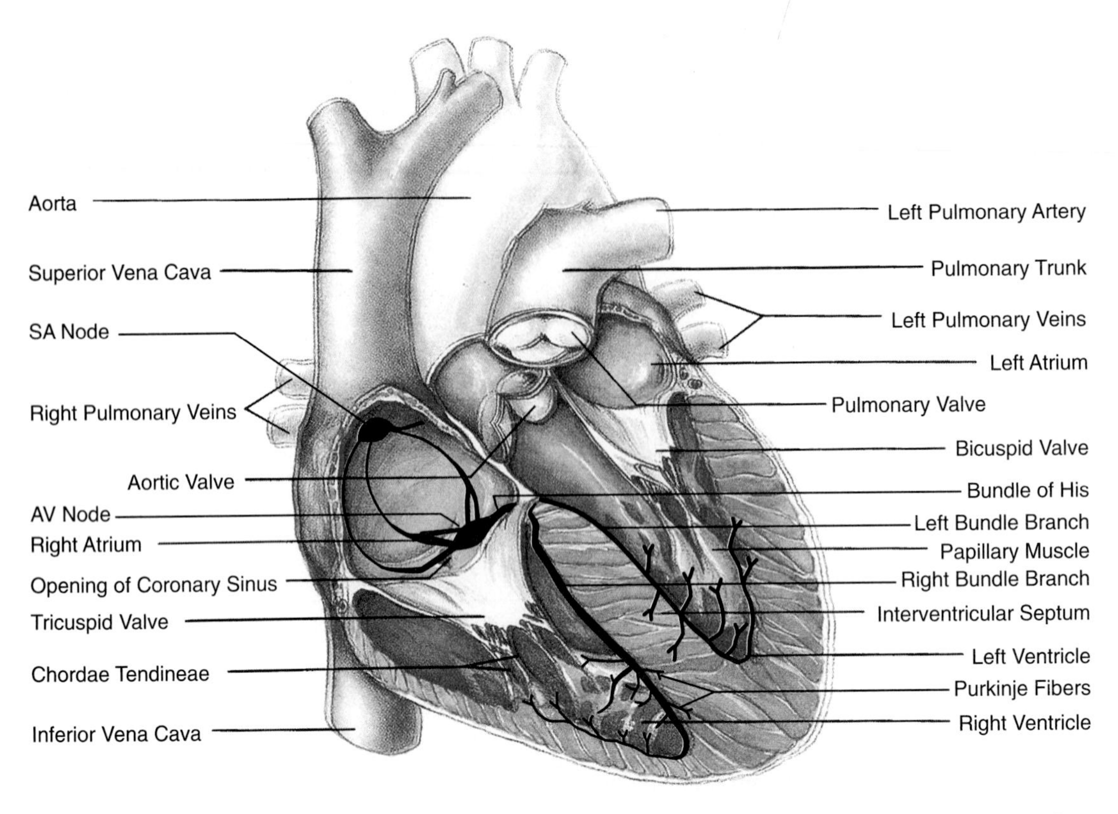

FIGURE 23.2 Posterior view of the internal structure of the human heart with emphasis on the conduction system (Black).

opens when ventricular pressure falls below atrial pressure and closes when ventricular pressure exceeds atrial pressure.

2. The *pulmonary semilunar valve* opens into the pulmonary trunk, allowing blood to flow from the right ventricle into the pulmonary trunk but not in the reverse direction. Valve action is passive; the valve opens when ventricular pressure exceeds pulmonary trunk pressure and closes when ventricular pressure falls below pulmonary trunk pressure.
3. The *left atrioventricular valve* (mitral, bicuspid) opens into the left ventricle, allowing blood to flow from the left atrium into the left ventricle but not in the reverse direction. Valve action is passive; the valve opens when ventricular pressure falls below atrial pressure and closes when ventricular pressure exceeds atrial pressure.
4. The *aortic semilunar valve* opens into the aorta, allowing blood to flow from the left ventricle into the aorta but not in the reverse direction. Valve action is passive; the valve opens when ventricular pressure exceeds aortic pressure and closes when ventricular pressure falls below aortic pressure.

The changes in pressure within the chambers of the heart are caused by the contraction and relaxation of cardiac muscle in the walls. The contraction of cardiac muscle is known as *systole,* and the relaxation of cardiac muscle is called *diastole.* The sequence of cardiac muscle contraction and relaxation during one cardiac cycle is atrial systole, ventricular diastole, atrial diastole, ventricular systole. The relationships between systole and diastole of the cardiac chambers and the opening and closing of cardiac valves is illustrated in figures 23.3 and 23.4.

The alternating sequence of atrial contraction, ventricular relaxation, atrial relaxation, and ventricular contraction is coordinated by the pacemaker system of the heart. The pacemaker system consists of the sinoatrial node (SA node), atrioventricular node (AV node), atrioventricular bundle (bundle of His), atrioventricular bundle branches (right and left), and the Purkinje network. The components of the pacemaker system consist of modified cardiac cells, which have become specialized in the generation and conduction of electric impulses.

The cardiac cycle is initiated by the firing of the *SA node,* the primary pacemaker of the heart (figure 23.2). As the impulse spreads through atrial muscle, the atria contract, the impulse reaches the *AV node,* and after a short delay (to allow the atria time to complete contraction), the AV node fires and relays the electric impulse, by way of the AV bundle, bundle branches, and the Purkinje network, to ventricular muscle. As the ventricles undergo systole, the atria are already in diastole awaiting the initiation of the next cardiac cycle. The electrical events associated with the cardiac cycle are explained in greater detail in chapter 26.

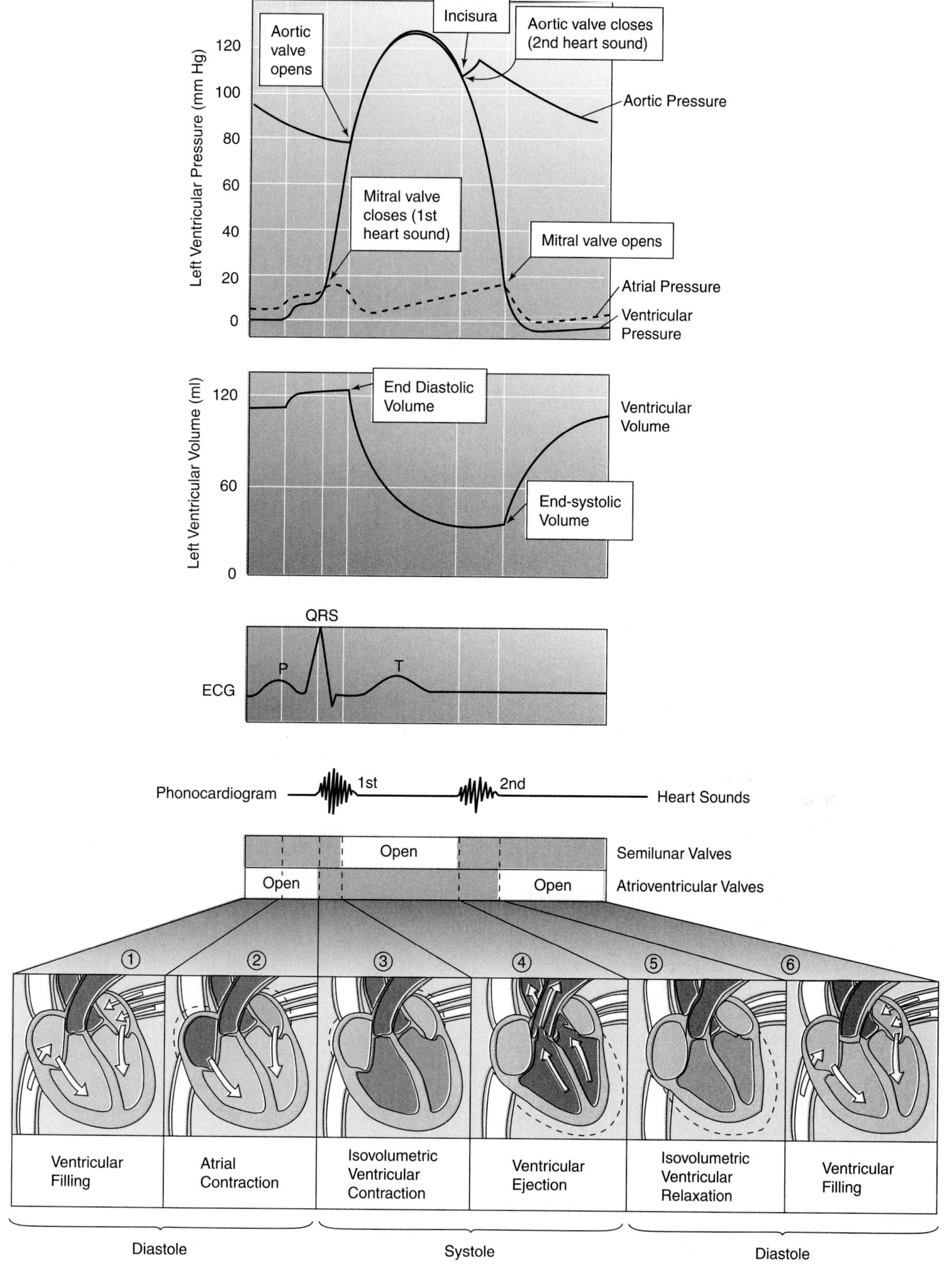

FIGURE 23.3 The mechanical and electrical events of one human cardiac cycle

From *Human Physiology* 4th edition by Rhoades/Pflanzer. © 2003. Reprinted with permission of Brooks/Cole, a Division of Thomson Learning: www.thomsonrights.com. Fax 800 730 2215

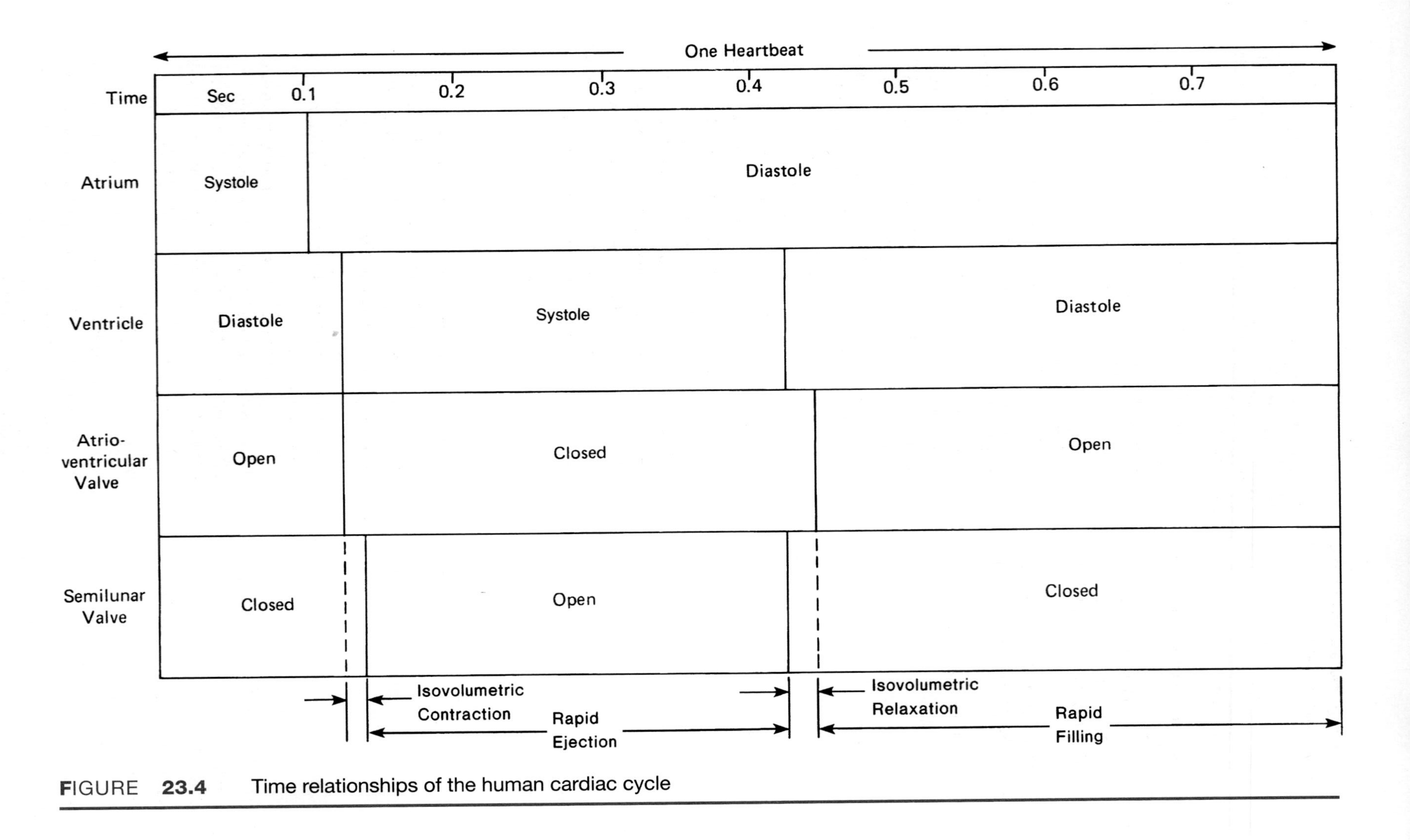

FIGURE **23.4** Time relationships of the human cardiac cycle

In the experiments that follow, some aspects of the cardiac cycle and properties of cardiac muscle will be investigated.

■ EXPERIMENTAL OBJECTIVES

1. To observe and record the mechanical events occurring during the normal cardiac cycle.
2. To observe and record the cardiac phenomena of refraction, extrasystole, and compensatory pause.
3. To observe and record the effects of vagal stimulation on the cardiac cycle.
4. To observe and record the effects of acetylcholine and epinephrine on heart rate and the strength of contraction.
5. To observe and record the effects of atrioventricular block.

■ EXPERIMENTAL METHODS

Lafayette Minigraph

Materials

turt le or bullfrog
brass hook
glass plate
dropper
turtle board or frog board
⅜-inch VSR electric drill
2½-inch hole saw with pilot bit
dissecting kit
glass probe
thread
twine
50-mL beaker
frog Ringer's solution
modeling clay
flat-base stand
double clamps
1:1000 epinephrine in frog Ringer's solution
1:1000 acetylcholine in frog Ringer's solution
Minigraph model 76107 and 76107VS
model 76322 time/marker channel + remote marker button
model 76406MG basic amplifier channel
model 76613 semi-isotonic force transducer
model 76613-T tension adjuster
model 76802 transducer stand
model 76634 electrodes
model 82415 square-wave stimulator

Preparation of the Recorder

Prepare the Minigraph for two-channel recording. Check to make sure there is sufficient recording paper and the inking system is working properly. Refer to chapter 2, if necessary, and to figure 23.5.

1. Attach the 76613 force transducer to the 76613-T tension adjuster and then secure the apparatus to a ring stand or flat-base stand.
2. Turn on mainframe power and center the basic amplifier pen using the pen-position control.
3. Connect the transducer to the balance control box via the standard ¼-inch phone jack and connect the balance control box to the amplifier via the 9-pin amphenol connector. Swing the top four leaves of the transducer to the side, exposing the thinnest leaf spring.
4. Connect the square-wave stimulator to a suitable AC outlet and connect the signal report terminals to the time/event marker. Set the timer to mark off 1-second intervals. Connect the model 76634 hand electrode to the output terminals of the stimulator.
5. Set the basic amplifier polarity switch to (+), the gain switch to 1000, and the sensitivity to 1. Recenter the pens using the balance control box. Whenever the gain and sensitivity controls are adjusted the balance control box must be used to recenter the pen.
6. Turn on mainframe and chart drive power. Check for proper operation of the transducer–amplifier channel by gently deflecting the leaf spring upward. The amplifier pen should move upward and return to center when the leaf spring returns to its original position. Turn off mainframe power.

Preparation of the Animal

1. A large grass frog, large bullfrog, or medium-sized freshwater turtle may be selected for experimentation. Pith the animal according to procedures outlined in appendix C or properly anesthetize the animal according to directions of your laboratory instructor. In either case, the animal must be rendered completely insensitive to pain.
2. Following the pithing procedure or anesthesia, secure the animal to the frog board or turtle board, as in figure 23.5, and expose the pericardial sac and heart. If using a turtle, also expose the vagus nerve on one side of the neck. Surgical procedures are outlined in appendix D.
3. After examining the appearance of the heart beating within the pericardial sac, expose the heart by making a longitudinal incision and a horizontal incision, with scissors, through the pericardial sac to its margins. Be careful not to injure the heart or the major blood vessels. Note the appearance of the atria and the ventricle (the turtle heart has two atria and one

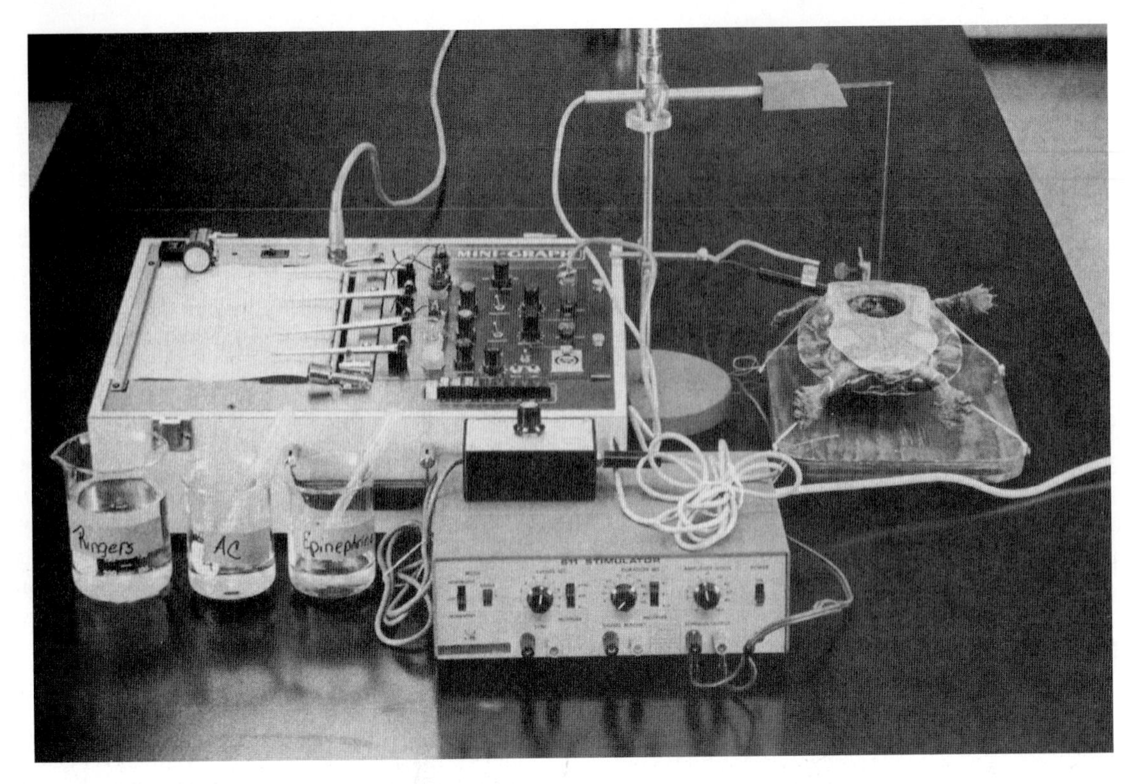

FIGURE **23.5** Minigraph setup for cardiac muscle experiments

ventricle) as they fill with and eject blood. Does atrial systole occur during ventricular diastole?

4. Using a dropper, carefully remove a drop of pericardial fluid and feel its consistency between your finger and thumb. What is the function of pericardial fluid? Keep the animal heart moist by occasionally dripping frog Ringer's solution (at room temperature) on it. Avoid touching the heart with your hands or metal instruments.
5. The apex (tip) of the ventricle is attached to the pericardium by the *frenulum,* a fine white ligament. Pass a long thread (18 inches) around the frenulum and tie tightly. Cut the frenulum distal to the knot.
6. Position the animal board beneath the force transducer so that the exposed heart is directly beneath the hole in the leaf spring. Secure the thread attached to the frenulum to the hole in the thinnest leaf spring.
7. Adjust the position of the tension adjuster on the flat-base stand so that the thread is taut between the leaf spring and the apex of the ventricle.
8. Turn on mainframe power. Gradually increase the sensitivity control and/or height of the tension adjuster until a 2-cm pen deflection occurs with each ventricular contraction. Readjust the baseline as necessary using the balance control box.
9. A paper speed of 10 mm/s will be used to record the cardiac cycle. If a continuously variable speed recorder is to be used, a paper speed of 10 mm/s may be approximated by adjusting the speed control until two 1-second time marks are separated by 10 m on the record.
10. Turn off mainframe power.

■ EXPERIMENTAL PROCEDURE

Cardiac Cycle

Record 10 cardiac cycles and label the following divisions of the cardiac cycle if they can be observed on the record: atrial systole and diastole, ventricular systole and diastole. Although atrial contraction and relaxation are not being directly recorded, components of these events may be superimposed on the ventricular contraction–relaxation record (figure 23.6). Measure the length of time of one cardiac cycle, the amount of time spent in systole, and the amount of time spent in diastole. Calculate the recorded heart rate. Record the data in the report.

Absolute Refractory Period

The *absolute refractory period* of electrically excitable cells (nerve and muscle cells) is defined as the period during which the cell will fail to respond to any stimulus. The absolute refractory period allows the cell time to "recharge" and be able to respond rapidly to a second stimulus. The absolute refractory period of cardiac muscle is relatively long. Because of the long refractory period, the contraction

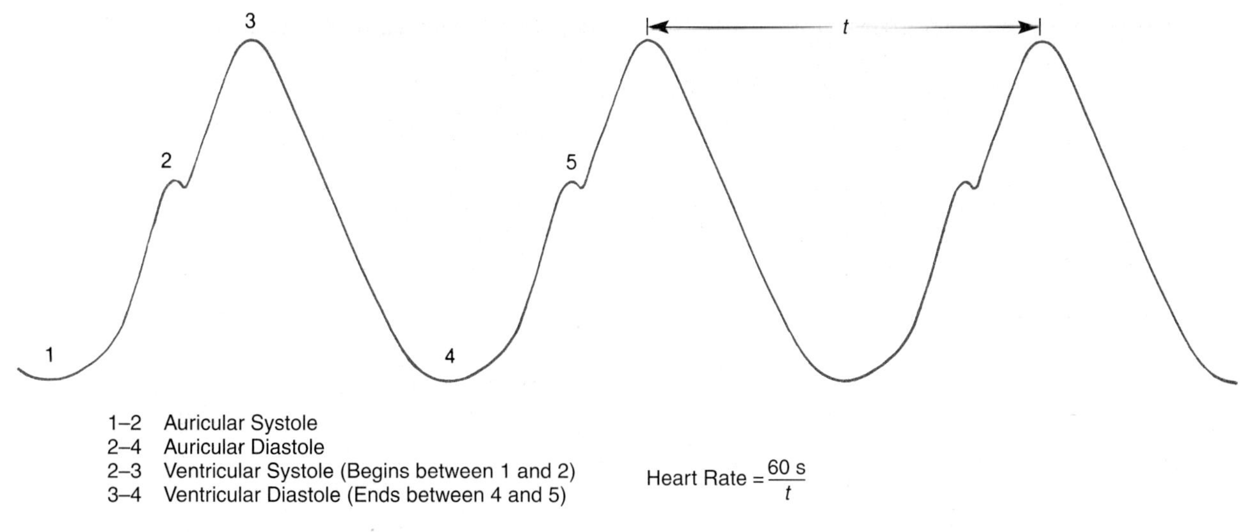

FIGURE 23.6 Components of the turtle heart cardiac cycle

of cardiac muscle is complete before the tissue again becomes electrically excitable.

If the ventricle is stimulated by means of an electrode immediately before it is normally stimulated through discharge of the AV node, the ventricle will respond by a contraction (extrasystole), then skip a normal beat (compensatory pause) as the AV node discharges, and resume normal rhythm with the next pacemaker discharge. During the compensatory pause, the ventricle is refractory and, therefore, fails to respond to the pacemaker stimulus. Arrange the hand electrode position to deliver a single electric stimulus of 50 V, 6 milliseconds in duration, to the ventricular muscle. Make sure the electrode position will not interfere mechanically with contraction.

Record a normal tracing of the cardiac cycle for 10 seconds. Continue recording and deliver a single shock to the ventricle during several stages of the cardiac cycle, such as during ventricular systole and diastole.

Does the ventricle respond to a stimulus applied during ventricular systole? Does the ventricle respond to a stimulus applied during ventricular diastole? Was a compensatory pause after extrasystole observed? Record your observations in the report.

All or None Phenomenon

Histologically, cardiac muscle is arranged in the form of a syncytium, with all branching fibers interconnected. This arrangement differs from skeletal muscle, in which fibers are not connected to each other. Because of the syncytial arrangement of cardiac muscle cells, the entire muscle responds "all or none" (i.e., maximally) to a threshold stimulus, as opposed to skeletal muscle, which does not. Because of both the syncytial structure of cardiac muscle and its long refractory period, cardiac muscle cannot be tetanized, nor does it exhibit the phenomenon of summation.

Record a normal tracing of the cardiac cycle for 10 seconds and stimulate the ventricle with repetitive 50-V, 6-millisecond stimuli at a frequency of one stimulus per second. Gradually increase the frequency of stimulation by adjusting the frequency control on the electronic stimulator. Does tetany of ventricular muscle occur?

Repeat the experiment, this time delivering single stimuli of increasing strength to the ventricle. Begin with a stimulus strength of 0.2 mV. Can summation of cardiac muscle be produced? Record your observations in the report.

Vagal Stimulation

The initiation of the cardiac cycle through spontaneous depolarization of the sinoatrial node is intrinsic to the heart. Depolarization of the pacemaker does not depend on stimulation of the heart by peripheral nerves. If all of the extrinsic nerves of the heart (cardioaccelerator and cardioinhibitory) are severed, the heart will continue to beat, although its rate will be somewhat increased.

Extrinsic cardiac nerves control the rate, not the genesis of the cardiac cycle. Innervation is by way of the autonomic nervous system, with sympathetic nerves (cardioaccelerator) originating from thoracic spinal nerves and parasympathetic nerves (cardioinhibitory) originating in the medulla oblongata. They are distributed to the heart via cranial nerve X (vagus). Both sets of extrinsic nerves control heart rate by releasing chemical transmitters in the region of the SA node.

Sympathetic fibers release epinephrine (adrenaline), which increases heart rate by increasing the rate of SA node depolarization. Parasympathetic fibers slow down heart rate by releasing acetylcholine, which decreases the rate of SA node depolarization.

Observe and record the effects of vagal stimulation on the heart rate by using the following procedure:

1. Using the electrode holder clamped to the flat-base stand, position both tips of the electrode directly on the exposed vagus nerve by gently lifting the nerve by the thread and placing it over the curved tips of the electrode. Be sure to keep the nerve moist with frog Ringer's solution.
2. Record a normal tracing of the cardiac cycle for 10 seconds using moderate paper speed.
3. As the recording continues, stimulate the vagus nerve with repetitive stimuli of 5–10 V, 0.66 milliseconds in duration, at a frequency of 0.1 pulse per second. Gradually increase the frequency of stimulation until a decrease in heart rate is observed.
4. Discontinue vagal stimulation and record the cardiac cycle for an additional 10 seconds.

What influence did vagal stimulation have on the heart rate? What influence did vagal stimulation have on the strength of the contraction? Calculate the heart rate and measure the height of the contraction before, during, and after vagal stimulation. Record the data.

Repeat the experiment, this time increasing the frequency of stimulation until the heart stops beating. Continue to record and maintain vagal stimulation until the heart begins to beat despite vagal stimulation. This is known as *vagal escape.* Under laboratory conditions, why does vagal escape occur? Does vagal escape ever occur under physiologic conditions?

Effects of Epinephrine and Acetylcholine

The effects of sympathetic and parasympathetic neurotransmitters on the cardiac cycle may be observed by dropping epinephrine or acetylcholine directly on the heart:

1. Record a normal heartbeat for 15 seconds and then apply 1:1000 epinephrine solution in frog Ringer's solution, drop by drop, directly on the heart.
2. Continue to record until an increase in the heart rate is noted.
3. Flush the heart with frog Ringer's solution until the heart rate returns to the original rate.
4. Repeat the experiment and use 1:1000 acetylcholine instead of epinephrine.

What are the effects of acetylcholine and epinephrine on the strength of contraction? Calculate the heart rates and the strength (height) of the contractions before, during, and after the application of epinephrine and acetylcholine. Record the data in the report.

Heart Block

In the normal heart, the ventricular beat is generated by the firing of the AV node, which, like the SA node, possesses the ability to depolarize spontaneously. However, the rate of spontaneous AV node depolarization is slower than that of the SA node; therefore, the rate of AV node depolarization and contraction of the ventricle is determined by the firing rate of the SA node.

If the AV node or the AV bundle is damaged, the ventricle may fail to respond to every depolarization of the SA node, resulting in *heart block.* In complete heart block, the ventricular beats are totally unrelated to atrial beats. In incomplete heart block, the ventricle responds to every second or third atrial beat (i.e., 2:1 or 3:1 heart block).

To observe heart block in the animal follow these directions:

1. Cut the thread from the heart and place the turtle (still attached to the board) on an area of the laboratory table free of equipment, where it can be observed by all members of the laboratory group.
2. Observe the rhythmic coordinated alternate contractions of the atria and ventricle, noting that each time the atria contract, ventricular contraction soon follows.
3. Loop a piece of thread around the junction between the atria and the ventricle and tie a single loose knot in the thread. Gradually tighten the knot, constricting the heart between the atria and ventricle. Do not constrict the heart too severely, only enough to observe synchrony between atrial and ventricular beats.
4. After observing heart block, loosen the thread. Does the ventricular beat soon become synchronous with the atrial beat?

Remove the heart from the animal by severing all connections with a small pair of scissors. Place the isolated heart on a glass plate and moisten it with frog Ringer's solution.

Observe the contractions of each chamber. Are the atria and ventricle in synchrony? Determine the heart rate. Is the rate slower or faster than when the heart was in situ? Why?

Sever the ventricle from the atria and observe the contractile rate of each. Are they identical? Why or why not?

Record your observations in the report.

The Cardiac Cycle

REPORT 23

Name: ______________________________ Date: ______________

Lab Section: ______________________________

1. Data:

   a. Resting cardiac: Cycle duration _____ Heart rate (BPM) _____

      Systole duration _____ Diastole duration _____

   b. Vagal stimulation:

	Before	During	After
Heart rate	______	______	______
Amplitude	______	______	______

   c. Epinephrine:

	Before	During	After
Heart rate	_______	_______	_______
Amplitude	_______	_______	_______

   d. Acetylcholine:

	Before	During	After
Heart rate	_______	_______	_______
Amplitude	_______	_______	_______

2. What is the function of pericardial fluid? ______________________________

____

____

____

3. Define the following:

   a. Absolute refractory period ______________________________

   ____

   b. Extrasystole ______________________________

   ____

   c. Compensatory pause ______________________________

   ____

4. What was the effect of stimulating the ventricle with a single shock during ventricular diastole? During ventricular systole? Why? ______________________________

____

____

____

5. Do human hearts ever exhibit extrasystole? What are the possible causes of extrasystole? ____________________

6. Can cardiac muscle exhibit (normally or under experimental conditions) summation or complete tetanus as does skeletal muscle? Why not? ____________________

7. What advantage is it for the ventricle to have a long refractory period? ____________________

8. Does the human heart spend more time in diastole or more time in systole (assume a heart rate of 72 beats per minute)? Of what benefit is this? ____________________

9. Which of the following valves are open and which are closed during systole of the ventricle?
   a. Pulmonary semilunar ____________________
   b. Aortic semilunar ____________________
   c. Mitral ____________________
   d. Tricuspid ____________________

10. Why is adrenaline used in some cases of heart stoppage in humans? ____________________

11. Define 3:1 heart block. What are the possible causes of heart block? ____________________

12. What conclusions may be drawn from the observation of the excised turtle heart? ____________________

CHAPTER 24

Heart Sounds, Pulse Rate, and Systemic Blood Pressure

■ INTRODUCTION

The human cardiovascular system consists of the heart and blood vessels arranged to form a double circulation: the systemic circulation and the pulmonary circulation. The circulatory pattern resembles a figure 8 with the heart located at the center (figure 24.1).

The primary functions of the heart are to receive blood from the pulmonary veins and pump it into the systemic arteries and to receive blood from the systemic veins and pump it into the pulmonary arteries. The sequence of electrical and mechanical events of the heart associated with receiving blood from the venous systems and pumping it out into the arterial systems during one heartbeat is known as the cardiac cycle.

The normal flow of blood through the heart and blood vessels is unidirectional and is as follows: *left ventricle–systemic arterial vessels–systemic capillaries–systemic venous vessels–right atrium–right ventricle–pulmonary arterial vessels–pulmonary capillaries–pulmonary venous vessels–left atrium–left ventricle.* Blood flowing through the left side of the heart is kept separate from blood flowing through the right side of the heart by the *septa* (walls) between the atria and between the ventricles. The unidirectional flow of blood through the chambers on each side of the heart is ensured by an *atrioventricular valve* and a *semilunar valve.* On the left side of the heart, the atrioventricular valve is called the *mitral valve,* and the semilunar valve is called the *aortic valve.* On the right side of the heart, the atrioventricular valve is called the *tricuspid valve,* and the semilunar valve is called the *pulmonary valve* (figure 24.2).

The atrioventricular valve opens when the pressure in the ventricle falls below the pressure in the atrium during relaxation of the ventricle *(ventricular diastole).* The open valve allows blood to flow from the atrium into the ventricle. The atrioventricular valve closes when the pressure in the ventricle rises above the pressure in the atrium during

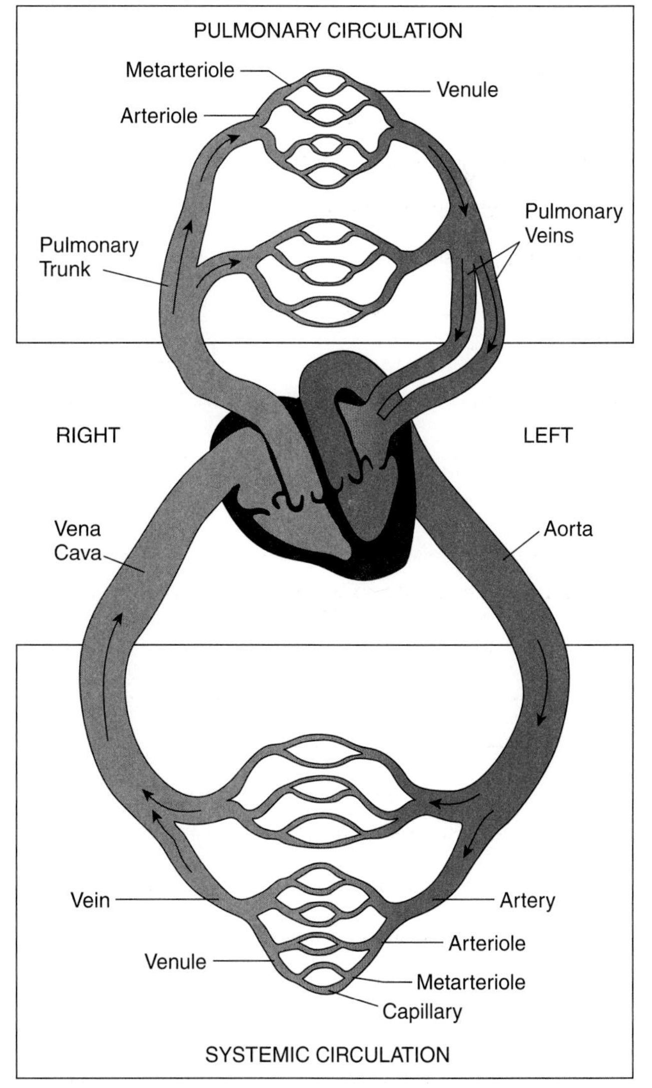

FIGURE 24.1 An outline of the cardiovascular system

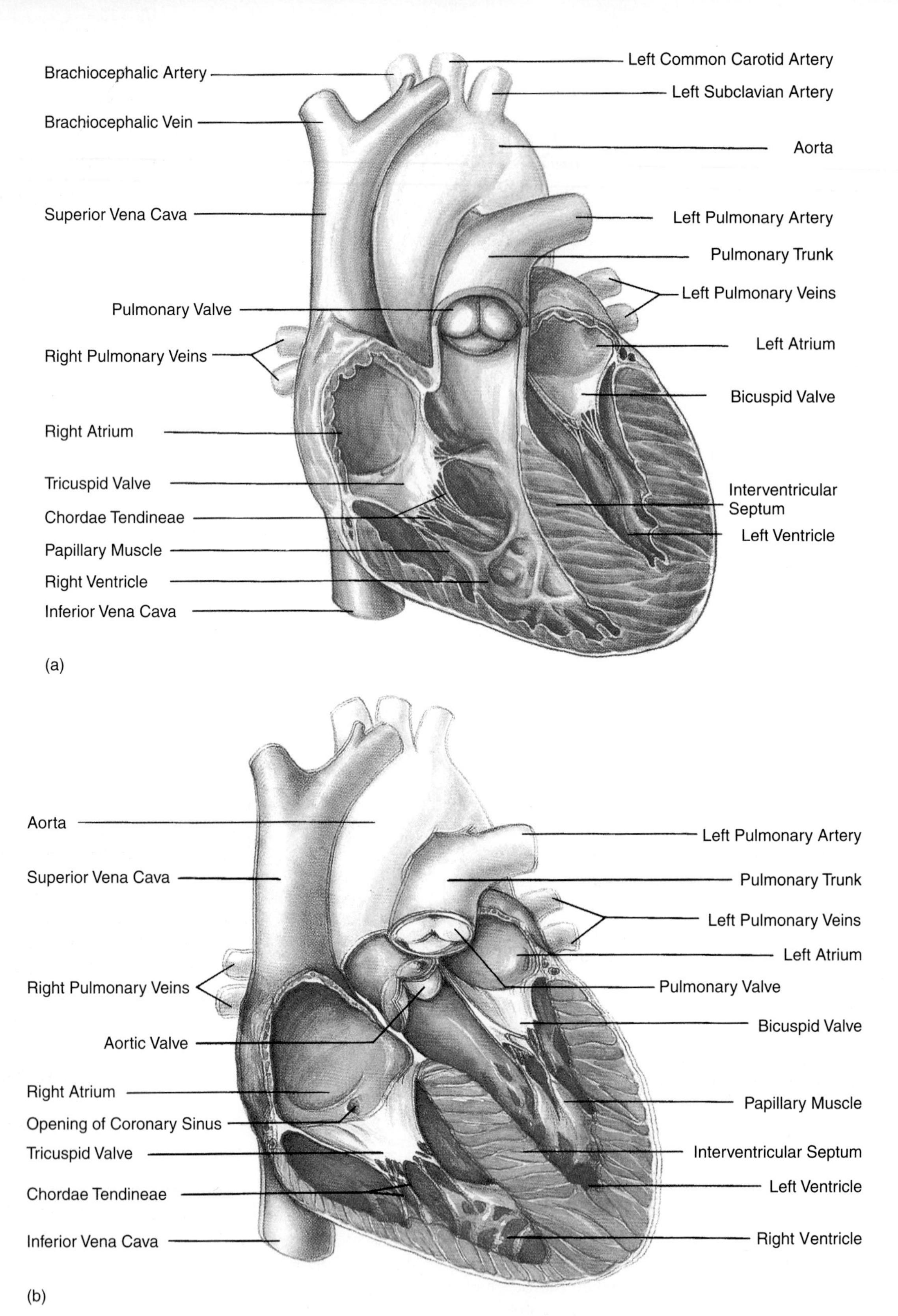

FIGURE 24.2 (a) Human heart, coronal plane, anterior view of posterior section showing origin of the pulmonary trunk from right ventricle. (b) Human heart, coronal plane, anterior view of posterior section showing origin of ascending aorta from left ventricle.

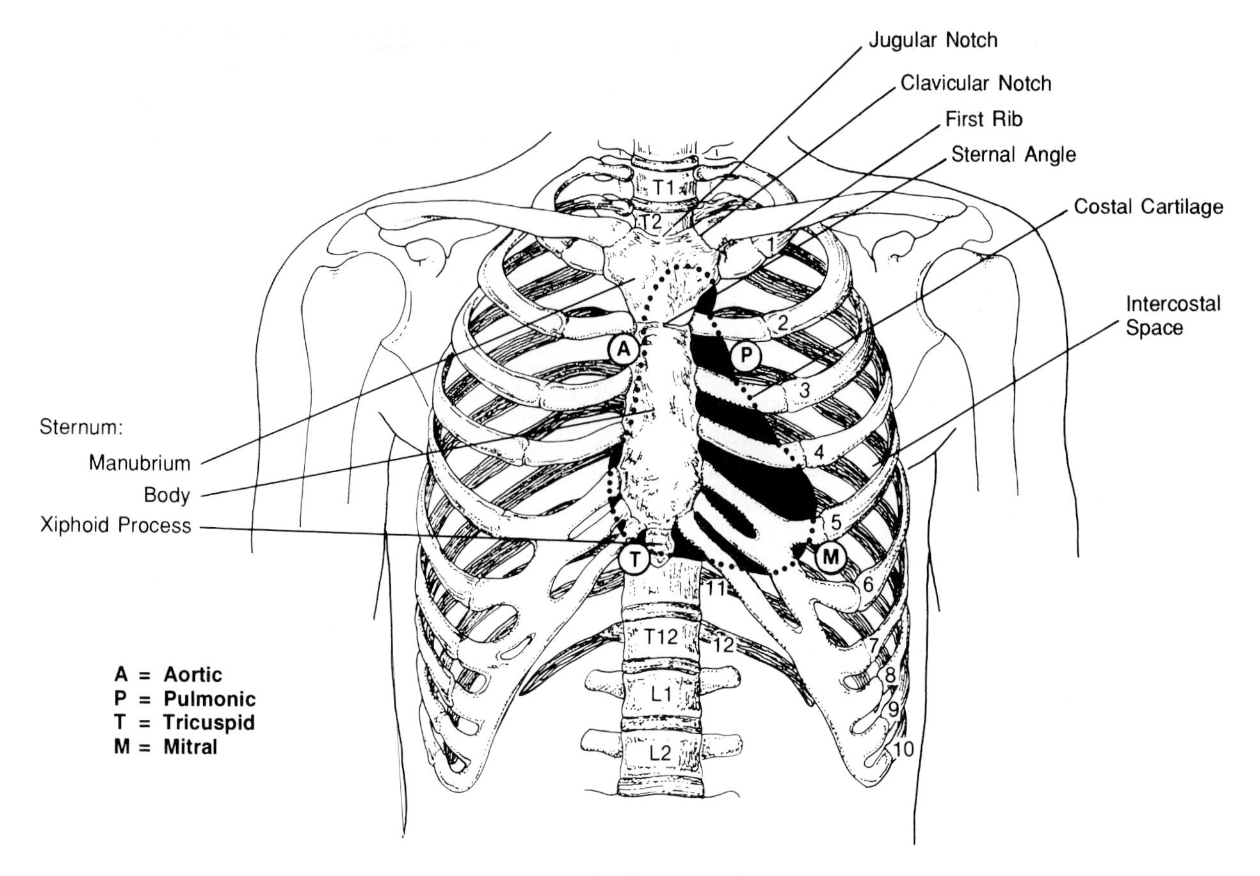

FIGURE 24.3 Ausculatory areas for individual valve sounds

contraction of the ventricle *(ventricular systole).* The closed valve prevents blood from flowing back into the atrium. Right and left atrioventricular valves open and close simultaneously.

The semilunar valve opens when the pressure in the ventricle rises above the pressure in the artery (pulmonary trunk or aorta) during ventricular systole. The open valve allows blood to flow out of the ventricle into the artery. The semilunar valve closes when the pressure in the ventricle falls below the pressure in the artery during ventricular diastole. The closed valve prevents blood from flowing back into the ventricle. Aortic and pulmonary semilunar valves open and close simultaneously.

From the preceding discussion it follows that the atrioventricular valves are open when the semilunar valves are closed, and the semilunar valves are open when the atrioventricular valves are closed. As a consequence, blood flows in progression from vein to atrium to ventricle to artery.

The opening and closing of the four cardiac valves produce sounds that may be heard over the anterior surface of the chest. Figure 24.3 depicts areas of the chest where these sounds are heard best with the aid of a stethoscope.

The four major heart sounds associated with the opening and closing of the valves and the intracardiac flow of blood are as follows:

The *first heart sound* occurs during ventricular systole and is caused by closure of the atrioventricular valves and opening of the semilunar valves. This sound is the "lub" of the characteristic "lub-dub" heard with each heartbeat when the stethoscope is placed on the chest over the heart.

The *second heart sound* occurs during ventricular diastole and is caused by closure of the semilunar valves and opening of the atrioventricular valves. This sound is the "dub."

The first and second heart sounds are sharp and distinct, easily heard by the untrained ear. The third and fourth heart sounds are muffled and less distinct and require more careful listening.

The *third heart sound* is caused by the turbulence associated with rapid filling of the ventricles shortly after the opening of the atrioventricular valves.

The *fourth heart sound* is caused by the turbulence associated with the passage of blood from the atria into the ventricles during atrial systole. This sound is heard immediately before the ventricles begin to contract and force the atrioventricular valves to close.

Damage to the cardiac valves, which results, for example, from disease processes that scar valve tissue (such as bacterial endocarditis and rheumatic fever), produces abnormal heart sounds known as *murmurs.* Murmurs may be associated with valvular insufficiency or incomplete closure (allowing a backward flow) and with narrowing (stenosis) of the valvular orifice. Murmurs may be heard during ventricular systole (systolic murmurs) or during ventricular diastole (diastolic murmurs). Systolic murmurs may involve aortic stenosis or AV valve insufficiency, whereas diastolic murmurs generally involve semilunar insufficiency. Murmurs may also be produced by atrial septal defects, ventricular septal defects, and vascular abnormalities, which create an excessive turbulence on the blood flow. Occasionally, murmurs that are nonpathologic in nature (functional murmurs) may be detected, such as the murmur heard when blood is rapidly ejected from the left ventricle into the aorta (caused by a turbulence in the blood flow).

At the beginning of left ventricular ejection, the blood flows faster going into the aorta than it does leaving through the arterioles. As a result, the blood ejected by the ventricle is at first opposed by the blood remaining in the aorta from the previous ventricular ejection and tension in the wall of the aorta is increased as the vessel distends to accommodate more blood. The aortic wall then elastically recoils, forcing blood into the adjacent segment of the aorta, where the sequence of distension and elastic recoil is repeated. In this manner, a pulse of pressure moves rapidly down the aorta and is conducted along the systemic arterial system at a velocity determined by the elasticity of the vascular walls and the pressure of the blood.

Using your third and fourth fingers, you may palpate the *arterial pulse* and determine the pulse rate on the body surface where large arteries are superficially located or where they pass over underlying bone. Figure 24.4 illustrates the pressure points where the arterial pulse may be palpated. The radial artery at the wrist is often used; however, the heart rate is not always identical to the pulse rate determined at the wrist. In some forms of arrhythmia (irregular heartbeats), some beats of the heart may be very weak and thus may not register at the wrist. The radial pulse will, therefore, be slower than the actual heart rate. In fact, even left and right radial pulse rates may differ with vascular disease, such as aortic aneurism (weakening and ballooning of the aortic wall).

As the heart works at pumping blood, the ventricles relax and fill with blood, then contract and eject blood, then repeat the cycle of filling and ejecting. Owing to the nature of the cardiac cycle, the ejection of blood by the ventricles into the arteries is not continuous. Therefore, both blood pressure and blood flow in the arteries are pulsatile, increasing during ventricular systole and decreasing during

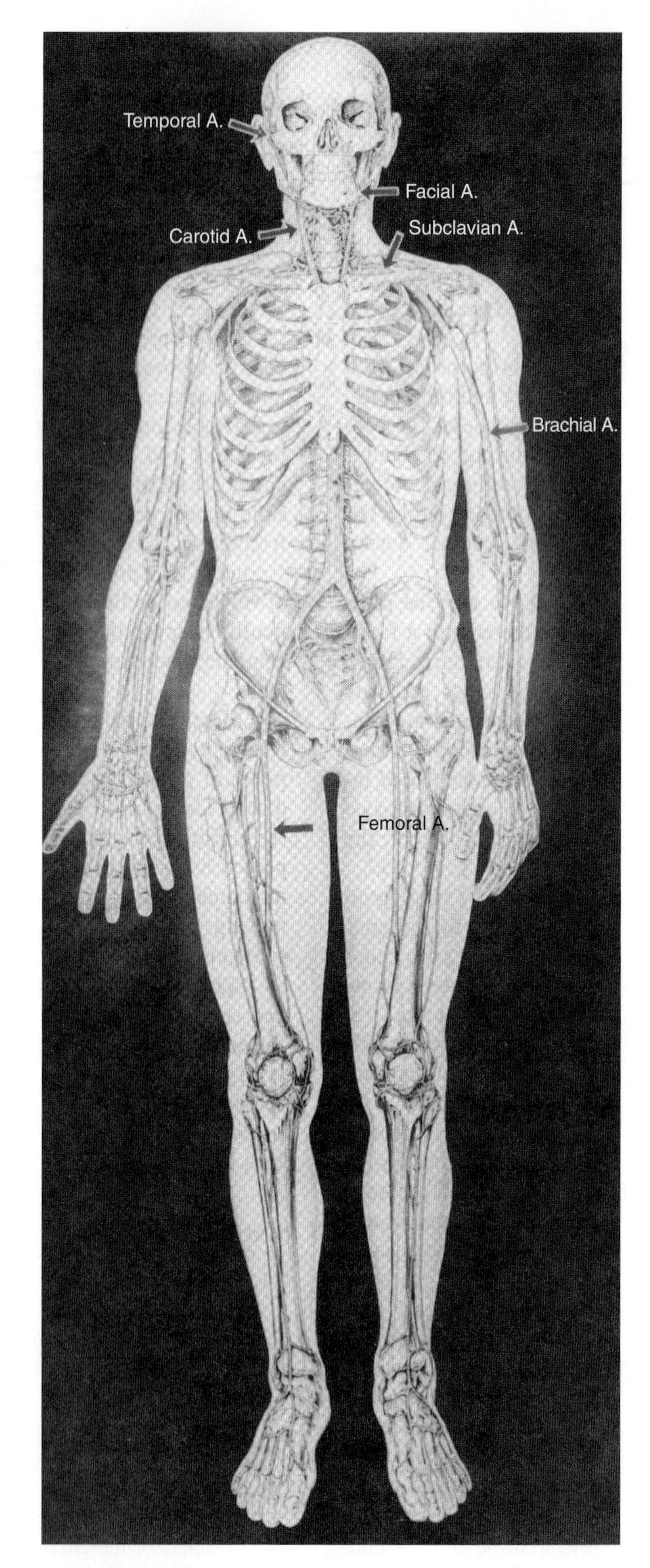

FIGURE 24.4 Pulse pressure points

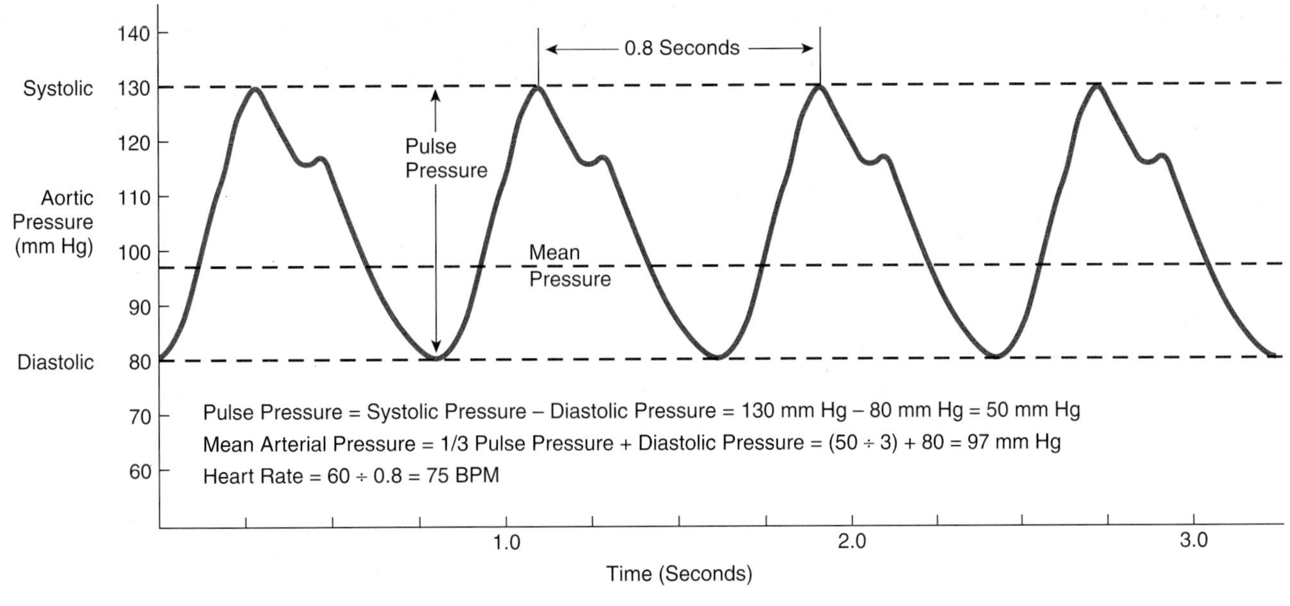

FIGURE 24.5 Direct measurement of systemic blood pressure

ventricular diastole. Figure 24.5 represents a graphic recording of changes in systemic arterial blood pressure measured directly by inserting a small catheter into an artery and attaching the catheter to a pressure measuring and recording device.

Systolic pressure is the force of blood against the walls of arteries during contraction of the ventricles. It is usually recorded as the highest arterial pressure reached during ventricular systole. Normal systolic pressure for a resting adult is less than 120 mm Hg.

Diastolic pressure is the lowest arterial pressure reached during relaxation or diastole of the ventricles. It is recorded at the end of diastole, just before the ventricle begins to contract and raise pressure. Normal diastolic pressure for a resting adult is less than 80 mm Hg.

Chronically elevated blood pressure, that is, resting blood pressure that remains high day after day, increases the risk of cardiovascular disease. A systolic pressure ranging from 120 to 139, *or* a diastolic pressure ranging from 80 to 89 signify **prehypertension.** A prehypertensive person can take steps to prevent high blood pressure by adopting a healthy lifestyle. Prehypertension can lead to outright **hypertension,** or high blood pressure, which begins at a systolic pressure of 140 mm Hg *or* a diastolic pressure of 90 mm Hg. The severity of hypertension falls into two categories: stage 1, and stage 2.

- *Stage 1.* This includes a systolic pressure ranging from 140 to 159, or a diastolic pressure ranging from 90 to 99.
- *Stage 2.* The more severe hypertension, this includes a systolic pressure of 160 or higher, or a diastolic pressure of 100 or higher.

The preceding values for normal and abnormal blood pressure were derived from a revised national classification system announced on May 14, 2003, in the 7th Report by the Joint National Committee on Prevention, Detection, Evaluation, and Treatment of High Blood Pressure and published in the *Journal of the American Medical Association* in May 2003.

Hypertension contributes to atherosclerosis (hardening of the arteries) and increases the work of the heart. It can also lead to congestive heart failure, blindness, kidney disease, and stroke.

Methods of direct measurement of systemic arterial blood pressure are invasive and neither practical nor convenient for routine use; instead, indirect methods are used. The most common indirect method involves the use of a stethoscope or microphone (figure 24.6). Sound is created by the turbulent flow of blood through the compressed vessel. When cuff pressure exceeds systolic arterial pressure, the artery is collapsed, blood flow through it ceases, and no sound is produced (figure 24.6a). As cuff pressure is slowly reduced, blood flow through the artery begins when cuff pressure falls just below systolic arterial pressure (figure 24.6b). At this point, a sharp tapping sound (the *first sound of Korotkoff*) may be heard with the stethoscope or microphone over the artery. The cuff pressure when this sound is first heard is taken as an approximation of systolic pressure. As cuff pressure is further reduced, the sounds increase in intensity (figure 24.6c), then suddenly become muffled (the *second sound of Korotkoff*) at the level of diastolic pressure (figure 24.6d), then disappear. Sounds disappear when the vessel is no longer compressed by the pressure cuff and normal nonturbulent blood flow resumes (figure 24.6e). Because it is easier to determine when the sound disappears than when it becomes muffled, and because only a few millimeters of mercury pressure differential exists between the two, the disappearance of sound is commonly used as an indicator of diastolic pressure.

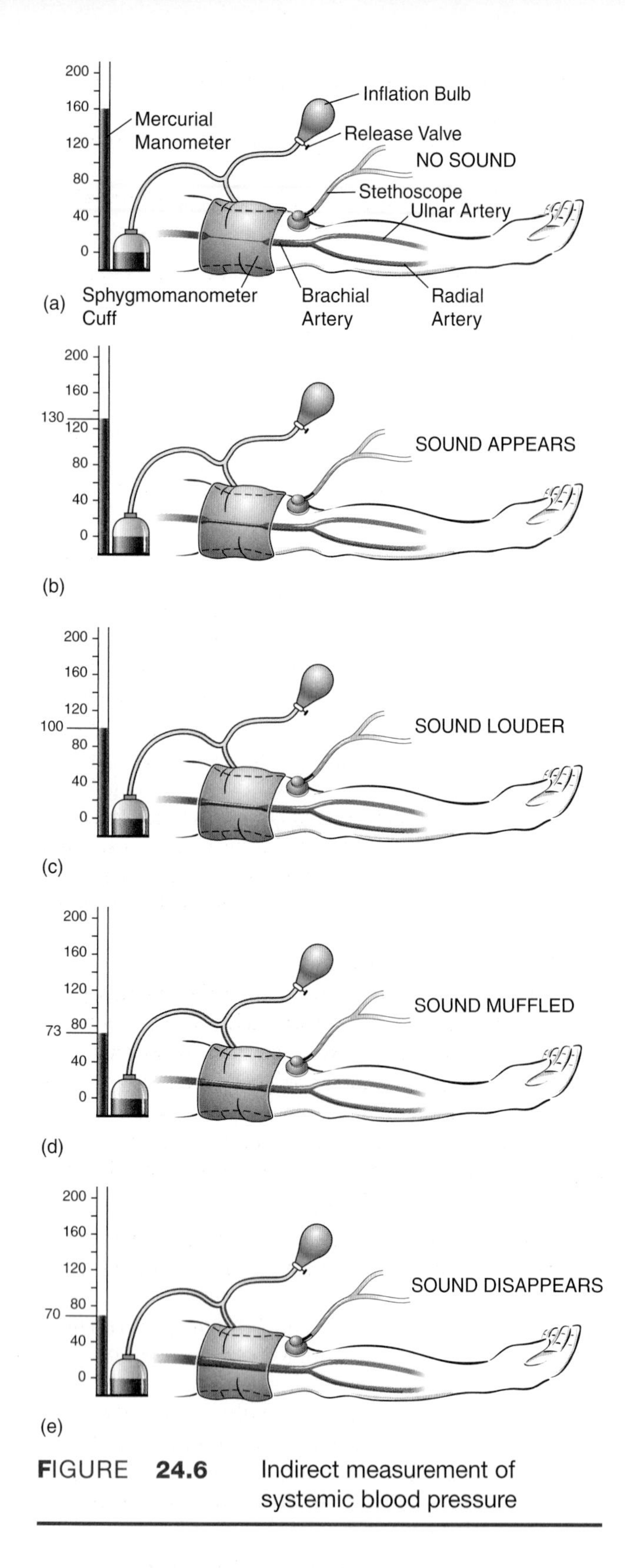

FIGURE 24.6 Indirect measurement of systemic blood pressure

By convention, blood pressures determined by indirect methods are expressed in the form of a ratio: systolic pressure/diastolic pressure. For example, if systolic pressure were measured as 130 mm Hg and diastolic pressure were measured as 70 mm Hg, systemic arterial blood pressure would be expressed as 130/70 and pulse pressure would be 60 mm Hg. If the sound became muffled at 73 mm Hg and disappeared at 70 mm Hg, the systemic arterial blood pressure would be expressed as 130/73–70.

Several types of sphygmomanometer are available for indirect measurement of blood pressure. The two most common types are the aneroid and the mercurial sphygmomanometers. Each type has advantages and disadvantages regarding its use. The *aneroid sphygmomanometer* measures cuff pressure by means of a spring-operated gauge attached to the cuff. Factors that contribute to possible inaccuracy when using the aneroid device include friction and wear associated with internal moving parts, stretch of the spring with age and repeated use, and a relatively small indicator dial that makes reading difficult. However, the aneroid sphygmomanometer is light, compact, and more portable than other sphygmomanometers.

Electronic sphygmomanometers are modified aneroid sphygmomanometers in which the direct-reading pressure gauge has been replaced by a pressure transducer. The transducer converts the pressure signal from the cuff to an electronic signal, which is processed and displayed as a digital readout. Although electronic sphygmomanometers are easier to read and contain fewer moving parts, they are not as accurate as mercurial sphygmomanometers.

The *mercurial sphygmomanometer* measures cuff pressure by means of a vertical column of mercury. Mercurial sphygmomanometers are more accurate, easier to read, and, except for the column of mercury, contain no moving parts; however, they are larger, heavier, and not as portable as the aneroid type, and they are more easily broken, which can result in spillage of mercury, a poisonous metal.

In the systemic circuit (figure 24.1) blood flows out of the left ventricle into systemic arteries and then serially through arterioles, capillaries, venules, and veins before returning to the heart to be pumped through the pulmonary circuit. Flow through a closed circuit such as the systemic circuit is determined by the pressure energy causing the flow and the resistance to flow offered by the blood vessel walls (friction) and the internal viscosity of the blood. The relationship between flow (F), pressure (P) causing the flow, and resistance (R) to the flow is expressed as follows:

$$F = P/R$$

Flow is expressed as liters/minute, pressure is expressed as mm Hg (torr), and resistance is expressed as peripheral resistance units. The pressure (P) is neither systolic nor diastolic but rather a pressure in between the two called *mean arterial pressure* (MAP). Mean arterial pressure converts a pulsatile pressure (systolic/diastolic) into a continuous pressure that determines the average rate of blood flow from the beginning of the circuit (left ventricle) to the end of the circuit (right atrium).

During the cardiac cycle, or one heartbeat, the ventricle spends more time in diastole than it spends in systole. As a result, mean arterial pressure is not the mathematical average of systolic and diastolic pressure but rather an approximation of the geometric mean. Mean arterial pressure can be calculated using either of the following equations:

$$MAP = 1/3 \text{ Pulse pressure} + \text{Diastolic pressure}$$

or

$$MAP = (\text{Systolic pressure} + 2 \text{ Diastolic pressure}) \div 3$$

If systolic pressure is 120 mm Hg and diastolic pressure is 60 mm Hg, mean arterial pressure is 1/3 (60) + 60 = 80 mm Hg.

■ EXPERIMENTAL OBJECTIVES

1. To calculate and compare carotid and radial pulse rates measured on the right and left sides of the same subject under different conditions.
2. To listen to human heart sounds and qualitatively describe them as to intensity or loudness, pitch, and duration.
3. To correlate the human heart sounds with the opening and closing of cardiac valves during the cardiac cycle and with systole and diastole of the ventricles.
4. To use an auscultatory method for an indirect determination of systemic arterial systolic and diastolic blood pressures and to correlate the appearance and disappearance of vascular sound with systolic and diastolic pressures, respectively.
5. To measure, record, and compare systemic arterial blood pressure in the right arm and the left arm of the same subject under identical conditions.
6. To measure, record, and compare systemic arterial blood pressures in the same subject under different experimental conditions of rest and exercise.

Materials

stethoscope

watch (with second hand)

sphygmomanometer

Cardiac patients on cassette. Jules Constant. Little, Brown, and Co., 1982 (optional)

■ EXPERIMENTAL METHODS

Pulse Rate

Using your third and fourth fingers, palpate the radial artery of your laboratory partner, and count the pulse for 15 seconds. Calculate the pulse rate. Repeat the procedure using the radial artery of the other wrist and the right and left carotid arteries. Are the pulse rates identical?

Determine the radial pulse rate of your laboratory partner under the following conditions and record all of the data:

1. After your laboratory partner rests in a supine position for 5 minutes.
2. Immediately after your laboratory partner assumes a sitting position.
3. Immediately after your laboratory partner assumes a standing position.
4. At the end of a prolonged, deep inspiration, before expiration.
5. Immediately after expiration, following a prolonged deep inspiration.

Subjects with a history of cardiovascular disease (high blood pressure, heart attack, etc.) are excused from performing sequences 4 and 5 above.

Heart Sounds

Audiocassette tapes of human heart sounds are available commercially and often used in teaching auscultation. Your laboratory instructor will provide directions regarding their use if available in the laboratory.

Pair with a laboratory partner of the same sex. Locate the apex beat of the heart by palpation and place the bell of the stethoscope over the ventricle, about an inch above the apex. Listen to the sounds that occur during the cardiac cycle, distinguishing two loud sounds ("lub" and "dub"). Describe their loudness, pitch, and duration. Listen to the heart sounds while simultaneously palpating the radial artery. Is the heart rate identical to the pulse rate?

Move the stethoscope bell to each of the four locations on the anterior thorax noted to be best for hearing the sounds produced by valve action (see figure 24.3). Describe the sounds. Are they sharp or muffled? Abrupt or prolonged? Loud or faint?

Record all your observations.

Obtain a sphygmomanometer from the laboratory instructor. A mercurial sphygmomanometer is preferred to the aneroid type. Place the mercurial sphygmomanometer on a flat, level surface and check to see that the column of mercury is exactly zero. With the subject comfortably seated or recumbent, and with the whole forearm at the level of the heart, perform the following sequence to determine and record systemic arterial blood pressure:

1. Wrap the completely deflated cuff snugly around the arm about 3 cm above the elbow, keeping the cuff as flat as possible.
2. Close the valve on the inflating bulb and pump up the pressure in the cuff until the mercury manometer registers 110 mm Hg.
3. Place the bell of the stethoscope directly over the brachial artery (slightly above the elbow crease). When applying the stethoscope, never touch the cuff. Listen for sounds and position the stethoscope bell where sounds are best heard.
4. Increase the cuff pressure until all sounds disappear. Open the valve on the inflating bulb slightly, allowing the cuff to deflate at a rate of 2–3 mm/s while simultaneously listening for sounds with the stethoscope. Record the reading of the manometer when the first sound is heard (systolic pressure).

5. Continue to deflate the cuff at a rate of 2–3 mm/s while listening to the sounds with the stethoscope. Record the reading of the manometer when the sounds become dull and muffled (diastolic pressure) and when the sounds disappear. Make a record of your findings using this form: 120/75–70. If the point at which the sounds become muffled and the point at which the sounds disappear are the same, then record your findings using this form: 120/70–70.
6. After the diastolic pressure has been determined, rapidly deflate the cuff and remove it from the subject. The cuff need not be removed if another determination will be made within a few minutes.

Perform the above procedure twice with the subject's right arm and twice with the left arm. Compute the average systolic and diastolic pressures for each arm. Are they identical? On the chalkboard, record the subject's age, sex, and average blood pressure in each arm. Record all data in the report and compute the class average for both males and females.

Measure the systolic and diastolic blood pressure of the laboratory partner under the following conditions: (1) while the subject is seated at rest, (2) while the subject is reclined at rest, (3) while the subject is seated immediately after he or she has run up and down three flights of stairs or performed the equivalent in exercise.

Record the data and calculate and record mean arterial blood pressures and pulse pressures.

Heart Sounds, Pulse Rate, and Systemic Blood Pressure

REPORT 24

Name: ______________________ Date: ______________________

Lab Section: ______________________

1. Subject's gender __________ Subject's age __________ Subject's height __________ Subject's weight __________
2. Pulse and heart sounds:

a. Pulse rate (BPM):

Artery	Right	Left
Radial		
Carotid		

b. Right radial pulse:

Resting position	Beats/15 sec	BPM
Supine		
Sitting		
Standing		
Inspiration		
Expiration		

c. Heart sounds:

Valve	Sharp	Muffled	Abrupt	Prolonged	Loud	Faint
Mitral						
Tricuspid						
Aortic						
Pulmonary						

3. Blood pressure:

a. Brachial artery resting blood pressure (subject seated or supine):

Right	***Systolic***	***Diastolic***	***Left***	***Systolic***	***Diastolic***
Reading 1			Reading 1		
Reading 2			Reading 2		
Average			Average		
Class average			Class average		

b. Brachial artery blood pressure:

Condition	***Systolic***	***Diastolic***	***Pulse pressure***	***Mean arterial pressure***
Resting, seated				
Resting, supine				
Postexercise				

4. Briefly describe the cause of the turbulence associated with each of the four heart sounds.

a. 1st sound ____________________

____

b. 2nd sound ____________________

____

c. 3rd sound ____________________

____

d. 4th sound ____________________

____

5. Define *systolic murmur* and give one example of a cause. ____________________

____

____

6. Define *diastolic murmur* and give one example of a cause. ____________________

____

____

7. Which of the four heart sounds is loudest? Give a reason. ____________________

____

____

8. Define the following:

a. Systolic pressure ______________________________

____

b. Diastolic pressure ______________________________

____

c. Pulse pressure______________________________

____

d. Mean arterial pressure ______________________________

____

9. Why is the brachial artery used for determining arterial blood pressure? Name one other artery that could be used.

____

____

____

____

10. List four factors that may affect arterial blood pressure.

a. ______________________________

b. ______________________________

c. ______________________________

d. ______________________________

11. Which heart sounds occur during ventricular systole and which occur during ventricular diastole?

a. Ventricular systole ______________________________

b. Ventricular diastole ______________________________

12. How may heart sounds be altered by disease? ______________________________

____

____

13. Is the method of determining heart rate by radial pulse always accurate? Why or why not? What is a "pulse deficit"?

____

____

____

14. List four factors that may affect pulse rate.

a. ______________________________

b. ______________________________

c. ______________________________

d. ______________________________

15. Name four other arteries besides the radial artery that could be palpated for determining pulse rate.

a. ______________________________

b. ______________________________

c. ______________________________

d. ______________________________

16. List four sources of error in the method used to determine arterial blood pressure.

a. ______________________________

b. ______________________________

c. ______________________________

d. ______________________________

CHAPTER 25

Plethysmography and the Peripheral Pressure Pulse

INTRODUCTION

Contraction of the ventricles (ventricular systole) pushes a volume of blood (stroke volume) into arteries. From the left ventricle, the blood goes into the aorta and throughout the rest of the body. Each stroke volume ejected by the ventricles "bumps" the downstream, neighboring section of blood to facilitate blood flow. The aorta and other arteries have elastic walls, which allow them to expand slightly to receive blood during ventricular systole and then elastically recoil during relaxation of the ventricles (ventricular diastole). The elastic recoil of arteries helps to continue the "pushing" of blood through the blood vessels while the ventricles are filling with blood for the next beat.

The pumping action of the ventricles also initiates a pressure wave that is transmitted via the arterial walls. The pressure increases with systole and decreases with diastole. The stiffness of the vessel wall helps transmit the pressure wave. The stiffer the walls, the faster the transmission of the pressure wave but the more work is required of the heart to move the same blood volume. As the human body ages, arterial walls gradually become stiffer, forcing the heart to work harder and blood pressure to rise.

When the pressure wave is transmitted to the periphery (e.g., the fingertip), there is a pulse of increased blood volume. The tissues and organs change in volume as blood vessels dilate or constrict and as pulses of blood pass through the blood vessels during each cardiac cycle. Changes in blood volume of organs may be brought about by the autonomic nervous system acting on the cardiovascular system, by environmental factors such as temperature, by metabolic activity of an organ, and by a variety of other variables.

Note that the actual velocity of blood flow is slower than the speed of transmission of the pressure wave. The aorta has the fastest blood flow in the body (40–50 cm/s, or about 1 mile per hour), whereas the speed of the pressure wave can be much faster. The traveling speed of the pressure wave from the heart to the periphery can be affected by many interrelated factors, among them the heart's ability to contract strongly, blood pressure, the relative elasticity of the arterial walls, and the diameters of systemic arteries and arterioles. These factors change in response to body positions, sympathetic nervous system input, emotional state (anxiety, fear, etc.), and other conditions.

The study of blood volume changes within an organ by using volume displacement techniques is known as *plethysmography.* Changes in tissue or organ volume associated with each cardiac cycle may be mechanically or photoelectrically detected, and the signal may then be transformed or transduced into an electric current, amplified, and recorded as a time record of volume change.

An example of a transducer used in plethysmography is the *piezoelectric crystal* familiar to amateur radio enthusiasts as a component of the "crystal radio." The piezoelectric crystal is pressure-sensitive, emitting a feeble electric signal when the structure of the crystal is deformed. If the crystal is taped to the palmar surface of the index finger, changes in the blood volume of the finger associated with each cardiac cycle mechanically distort the pressure-sensitive surface of the crystal, resulting in the generation of an electric signal, which can then be amplified and mechanically recorded.

Another type of transducer used in plethysmography operates by converting light energy to electric energy and thus is called a photoelectric transducer. The photoelectric transducer works by shining a beam of light through the skin and measuring the amount of light that is reflected. Blood absorbs light in a manner proportional to blood volume. The greater the blood volume, the greater the light absorption and the less light reflection (and the reverse is true also). The photoelectric transducer converts the reflected light into electric signals, which can then be processed and displayed by the recorder.

■ EXPERIMENTAL OBJECTIVES

1. To become familiar with the principle of plethysmography and its usefulness in qualitatively assessing peripheral changes in blood volume.
2. To observe and record changes in peripheral blood volume and pressure pulse under a variety of both experimental and physiologic conditions.

■ EXPERIMENTAL METHODS

Lafayette Minigraph

Materials

Minigraph model 76107 or 76107VS

model 76322 time/marker channel + remote push button

model 76406MG basic amplifier channel

model 76605 piezoelectric pulse sensor or model 76604 photoelectric plethysmograph

sphygmomanometer

ice water and very warm tap water

100-mL beaker

Preparation of the Recorder

Prepare the Minigraph for two-channel recording. Check to make sure that there is sufficient recording paper and the inking system is working properly. Refer to chapter 2, if necessary, and to figure 25.1.

1. Connect model 76604 photoelectric plethysmograph or model 76605 piezoelectric pulse sensor to the basic amplifier channel via the 9-pin amphenol connector. The pulse sensor contains a piezoelectric crystal that is pressure-sensitive. When the structure of the crystal is deformed, it emits a feeble electric signal. If the crystal is taped to the palmar surface of the index finger, changes in blood volume of the finger associated with each cardiac cycle mechanically distort the pressure-sensitive surface of the crystal, resulting in the generation of an electric signal, which can then be amplified and mechanically recorded.
2. Connect the remote handheld push button to the time/marker channel. Set the timer to mark 1-second intervals.
3. Set the amplifier polarity switch to (+), the gain switch to 10, and the sensitivity control on zero (fully counterclockwise).
4. Turn on mainframe power, center the amplifier pen using the pen-position control, then turn off mainframe power.
5. With the subject seated in a chair with the subject's hand resting on the laboratory table, attach the crystal pulse transducer to the subject's palmar surface of the index finger. The transducer should be attached tightly enough so that the pulse may be felt beneath it but not so tight as to occlude circulation. Tape the lead of the

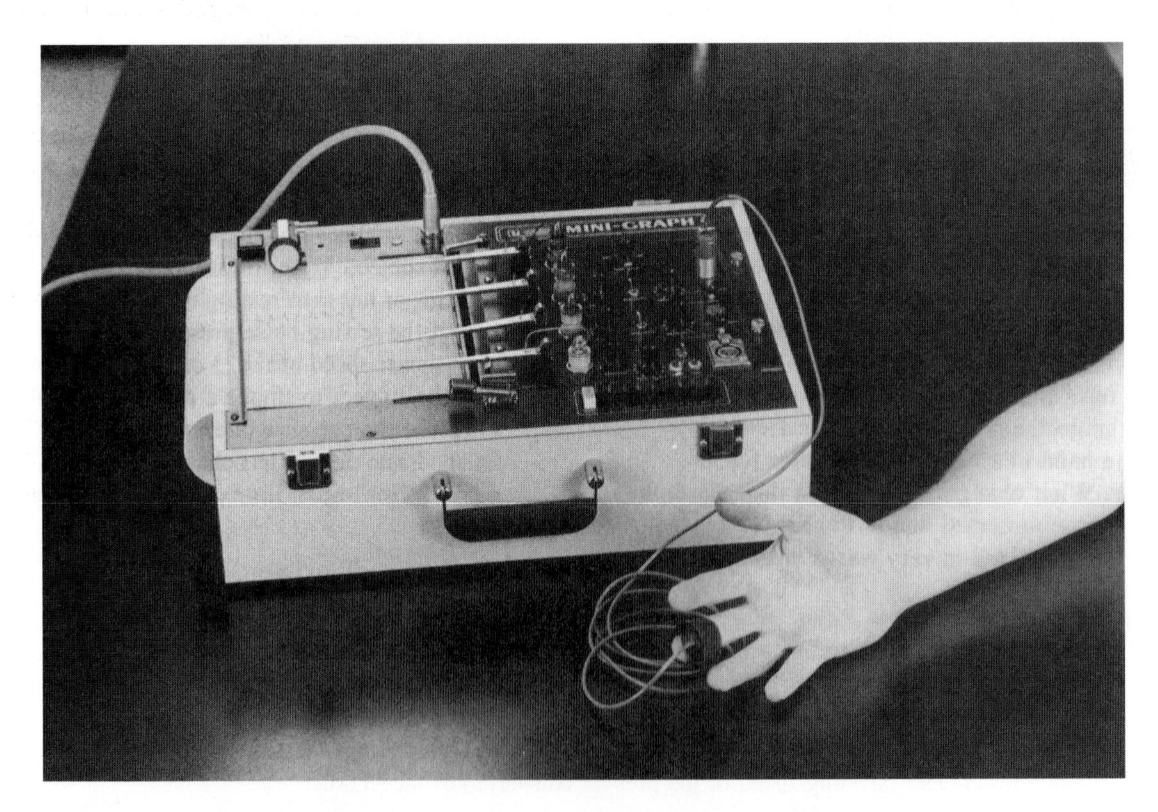

FIGURE 25.1 Minigraph setup for recording peripheral pressure pulse

transducer to the wrist to minimize transducer movement.

6. As the subject's arm rests comfortably on the laboratory table, turn on mainframe and chart drive power and advance the sensitivity and gain controls until a 1–2 cm pen deflection is produced with each pulse. At this time, also check for proper operation of the timer and the event marker. Turn off chart drive power.
7. When recording pressure pulse during different experimental conditions, use the remote handheld push button to record the time of an event simultaneously with the plethysmograph. Also, note experimental conditions directly on the recording paper.
8. Record at a paper speed of 5 mm/s. If a continuously variable speed recorder is being used, a paper speed of 5 mm/s may be approximated by adjusting the speed control until two 1-second time marks fall within a 1 cm^2 square on the recording paper. In between experimental procedures, turn off the chart drive control to conserve recording paper.
9. The piezoelectric crystal is extremely sensitive. Any undue motion in the lead, transducer, or limb will result in erratic pen movement. This is to be avoided. When attaching or detaching the transducer, or in between experimental procedures, the channel amplifier should be off. At the end of the experiment, carefully remove the transducer and gently place it on the laboratory table where it will not be disturbed. Be careful not to bump the crystal against any hard object.

Experimental Procedure

Record the subject's pressure pulse under the following conditions:

1. With the subject seated on a chair with his or her hand resting on the laboratory table, record the normal resting pressure pulse for 10 seconds. Which components of the cardiac cycle are discernible in the record? Calculate resting pulse rate and record data in the report.
2. With the subject standing, record the normal resting pressure pulse for 10 seconds. Compare the record of the standing subject with that of the seated subject. Is the amplitude of the waveform the same? Calculate the pulse rate. Is it the same?
3. Record the standing pressure pulse for 10 seconds, and then, while continuing to record, ask the subject to inhale as deeply as possible and hold the breath for as long as possible (Valsalva maneuver). **Subjects with a history of cardiovascular disease should not perform the Valsalva maneuver.** What changes are noted in the pressure pulse recording during breath holding and shortly after? In terms of venous return and cardiac output, can you account for the changes? Calculate the pulse rate for the 10-second interval immediately prior to the subject's exhalation and compare it with the standing normal rate. Account for the difference.
4. With the subject seated with his or her arm resting on the laboratory table, wrap the sphygmomanometer cuff around the resting arm and pump up the pressure in the cuff until blood flow in the brachial artery is occluded, while continuing to record the pressure pulse. After occlusion for 15 seconds, allow the cuff to deflate at a rate of 5 mm Hg per second while continuing to record the pressure pulse. Are there changes in both the rate and amplitude of the pressure pulse? Describe the nature of the changes.
5. While recording the resting pressure pulse of the seated subject, ask the subject to immerse the opposite hand in a beaker of ice water for 30 seconds. What changes are noted in the recording from the nonimmersed hand? Repeat the procedure, immersing the hand in very warm tap water. Observe the changes in the pressure pulse recording. Describe the nature of the changes and explain them.
6. With the subject seated, have the subject raise the transducer above his or her head (arm extended) for 30 seconds while you are recording the pressure pulse. Continue to record, and have the subject lower the transducer (arm hanging at side) for 30 seconds. Explain the changes in the pressure pulse record.
7. Request a fellow student who smokes cigarettes to be the subject. With the subject seated comfortably with his or her arm resting on the laboratory table, record the resting pulse pressure for 30 seconds at a chartmover speed of 0.025 cm/s. Ask the subject to smoke a cigarette, and continue to record the pressure pulse until the subject has finished smoking. What changes are noted in the record? How much time elapses before the changes occur?

Plethysmography and the Peripheral Pressure Pulse

Name: ______________________ Date: ______________________

Lab Section: ______________________

1. Subject's gender __________ Subject's age __________
2. Data for changes in pulse rate and pulse amplitude:

Condition	*Pulse rate (BPM)*	*Pulse amplitude (mm)*
Resting, sitting		
Resting, standing		
Holding breath		
Brachial artery occlusion		
Warm water immersion		
Cold water immersion		
Arm raised		
Arm lowered		
Smoking a cigarette		

3. Which components of the cardiac cycle are discernible in the pressure pulse record? ______________________

____

____

____

4. What effect did breath holding have on the peripheral pressure pulse? Why? ______________________

____

____

____

5. When the brachial artery was occluded, the peripheral pressure pulse record changed dramatically. Describe and account for the nature of the changes seen during occlusion and as cuff pressure was reduced.

____

____

____

____

____

6. Describe the changes in the rate and/or amplitude of the pressure pulse tracing associated with immersion of the opposite hand in ice water and warm water. How do you account for the observed changes?

____

____

____

____

____

____

7. What changes were noted in the pressure pulse tracing when the arm was raised or lowered? Why?

____

____

____

____

____

____

8. Describe the effect of cigarette smoking on the pressure pulse. From the record, what would you deduce are the effects of nicotine on the cardiovascular system? ______________________________________

____

____

____

____

Electrical Activity of the Heart: The Electrocardiogram

CHAPTER 26

■ INTRODUCTION

The normal heart contracts regularly and continuously throughout life at a rate of 60–100 times each minute. Although the cardiac muscle is supplied with motor nerves that can influence either the rate of contraction or the strength of contraction, the extrinsic nerves play no role in the genesis of the heartbeat. If the extrinsic nerves (sympathetic and parasympathetic) were cut or even if the heart were to be removed from the body, it would continue to beat rhythmically as long as it was supplied with oxygen and vital nutrients. The heart possesses the unique ability to contract by itself without any stimulation from the rest of the body. This property of cardiac muscle is called *inherent rhythmicity* or *automaticity.*

The control and coordination of cardiac muscle's inherent rhythmicity are dependent on a specialized system of conductive tissue within the heart. Before each contraction of the heart can occur, an electric current must first pass through the myocardial fibers. The conduction system of the heart is responsible for generating these electric currents and conveying them in an orderly sequence to all parts of the heart. The *conduction system,* or pacemaker system, consists of the following areas of specialized conducting tissue: the sinoatrial node (SA node), internodal and interatrial pathways, the atrioventricular node (AV node), the bundle of His, right and left bundle branches, and the Purkinje fiber network. Figure 26.1 illustrates each of the components of the pacemaker system.

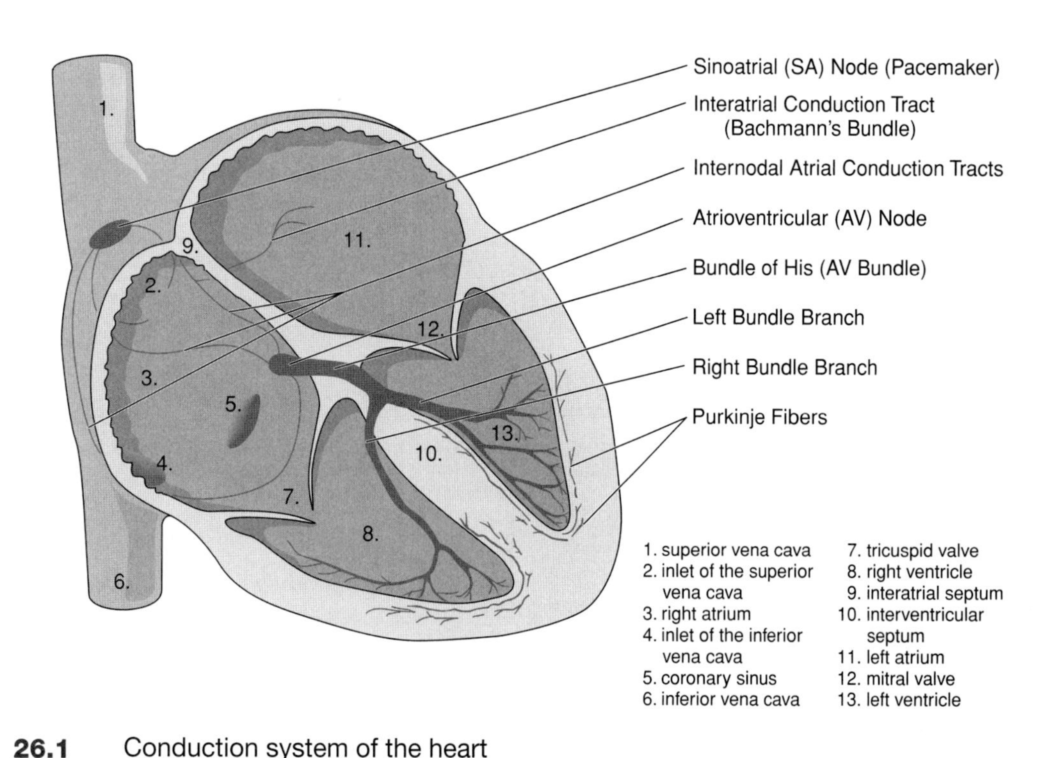

FIGURE **26.1** Conduction system of the heart

The **sinoatrial node (SA node)** is located near the junction of the right atrium and superior vena cava. The electric impulse that initiates each contraction of the heart normally originates from this node. The SA node, without neural or endocrine stimulation, spontaneously depolarizes at a rate of more than 80 times per minute. Normally, its rate of depolarization is 60–80 times per minute because of vagal inhibition. Because the SA node discharges electric impulses at a faster rate than any other part of the conduction system, it paces the electrical and mechanical activities of the entire heart. Therefore, the SA node is commonly called the *pacemaker.*

Once an impulse has been initiated by the SA node, it is transmitted through both atria along the **internodal** and **interatrial pathways,** stimulating atrial muscle to contract. The impulse also spreads to another specialized area of the conduction system—the **atrioventricular node (AV node)**—which is part of the junctional tissue between the right atrium and ventricle. The AV node, driven by the rate of the SA nodal firing, relays the electric impulse toward the ventricles after a slight delay. The delay allows the atria time to contract before excitation of the ventricles occurs. The delay also helps protect the ventricles from rapid atrial impulses.

After passing through the AV node, the impulse is carried to the ventricles through the **bundle of His,** a common bundle of specialized conductive fibers lying along the upper part of the interventricular septum. The bundle of His runs down within the upper interventricular septum and branches into a right and left bundle. The *right bundle branch* carries the impulse to the right ventricle; the *left bundle branch* carries the impulse to the left ventricle. Each bundle branch further subdivides into numerous small conducting fibers called **Purkinje fibers,** which relay the electric impulse directly to ventricular muscle, stimulating the ventricles to contract.

Note that any of the cells of the conduction system may act as pacemaker cells, but atrial and ventricular muscle cells do not normally do so. In damage to the SA node, for example, the AV node may take over as the primary pacemaker for the ventricles, although the intrinsic rate of firing of the AV node (40–60 cycles per minute) is less than the normal firing rate of the SA node (80–100 cycles per minute).

In summary, the contraction of cardiac muscle is associated with an electric impulse initiated at the sinoatrial node, which sweeps over the conduction path of the heart, preceding the mechanical change in the muscle. In each normal cardiac cycle, the electrical events follow a sequence: (1) depolarization and repolarization of the SA node; (2) depolarization and repolarization of atrial muscle; (3) depolarization and repolarization of the AV node and bundle; (4) depolarization and repolarization of the Purkinje network; (5) depolarization and repolarization of ventricular muscle.

The electric current associated with the cardiac cycle may be detected at the surface of the body, amplified, and recorded as a time record of the electrical events occurring during each cardiac cycle. Thus, heart rate can be accurately determined and abnormalities of rhythm and conduction can be identified. The electrical and mechanical device that records the electrical activity of each cardiac cycle is called an **electrocardiograph.** The study of electrocardiograph applications and the interpretation of electrocardiograms is called electrocardiography.

The electric current associated with and generated during the cardiac cycle is detected by placing a positive electrode and a negative electrode on selected areas of the skin surface and recording the electric current changes occurring between the electrodes as the heart beats. The particular arrangement of two electrodes, one positive and the other negative with respect to a third electrode, the ground electrode, is called a **lead.** The *standard bipolar limb leads* (figure 26.2) used in electrocardiography by convention, with the right leg electrode serving as ground, are as follows:

lead I = right arm (–) to left arm (+)

lead II = right arm (–) to left leg (+)

lead III = left arm (–) to left leg (+)

For diagnostic work in cardiology, other leads have been developed to broaden the scope and utility of the electrocardiograph. The electrocardiogram (EKG) of the *chest leads* V_1, V_2, V_3, V_4, V_5, and V_6 is obtained by uniting the standard limb electrodes to a single negative pole called a common terminal (CT). A chest electrode is attached to the positive pole and moved through six standard positions on the chest surface (figure 26.3). The chest leads provide additional information relative to the detection and location of an abnormality in the conduction system of the heart or in cardiac muscle.

Recording from the *unipolar limb leads* (aV_F, aV_R, aV_L) involves uniting two of the standard limb electrodes to a negative pole (CT) and connecting the remaining standard limb electrode to a positive pole (figure 26.4). The unipolar limb leads are as follows:

aV_F = left arm and right arm (–) to left leg (+)

aV_R = left arm and left leg (–) to right arm (+)

aV_L = right arm and left leg (–) to left arm (+)

Additional leads are the *bipolar chest leads* (CR, CL, CF), obtained by making the chest electrode positive and one of the limb electrodes negative. The bipolar chest leads are as follows:

CR = chest lead (+) to right arm (–)

CL = chest lead (+) to left arm (–)

CF = chest lead (+) to left leg (–)

The electrocardiograph records electrical activity of the heart on special graph paper of the following standard dimensions: The horizontal lines represent amplitude in fractions of a millivolt, and the vertical lines represent time

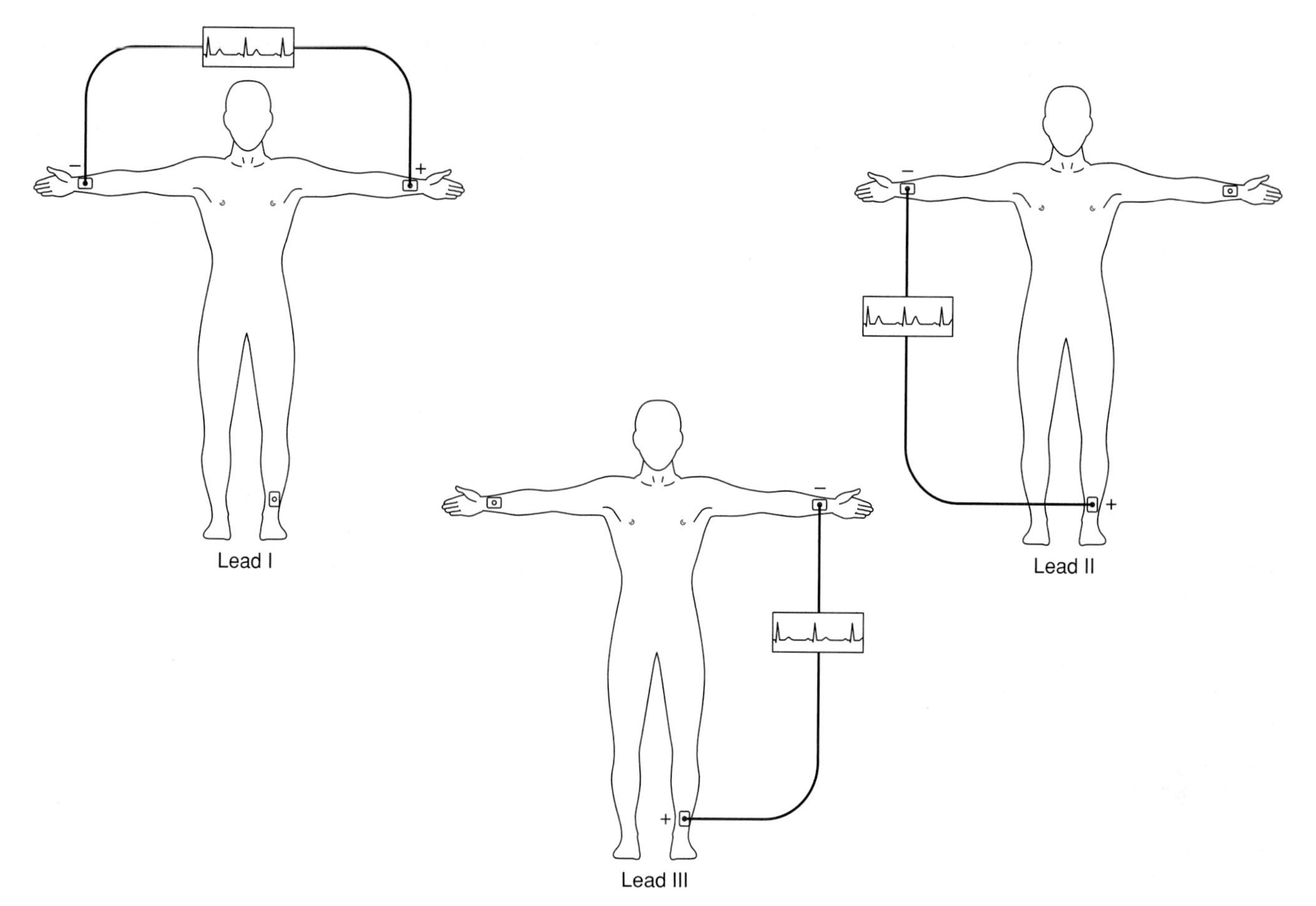

FIGURE **26.2** The standard (bipolar) limb leads I, II, and III

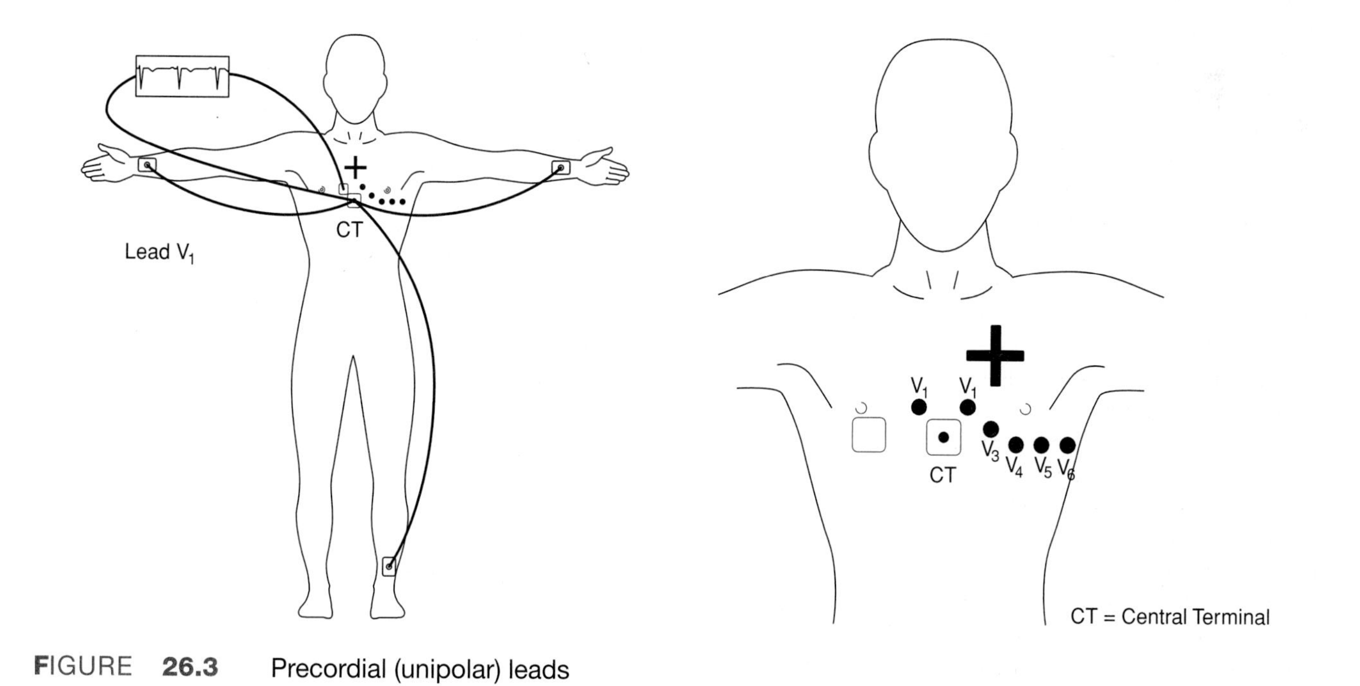

FIGURE **26.3** Precordial (unipolar) leads

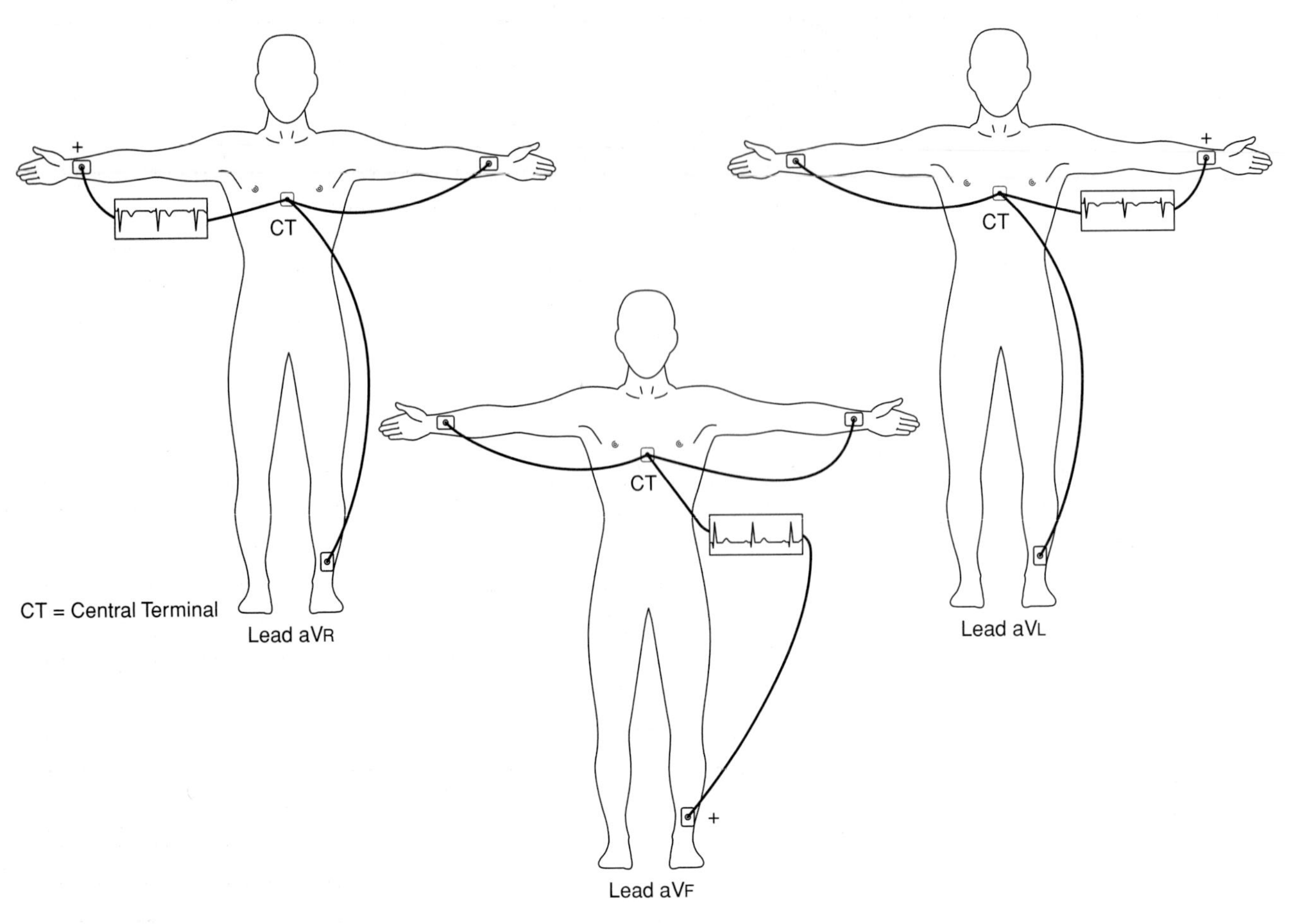

FIGURE 26.4 The augmented (unipolar) leads aVR, aVL, and aVF

in fractions of a second. The recording paper travels at a standard speed of 25 mm/s. The sensitivity of the recorder is set so that a 1-mV input results in a pen deflection of 10 mm on the paper. The interval between two vertical lines is, therefore, 0.04 second; the interval between two horizontal lines is 0.1 mV.

Figure 26.5 is a normal electrocardiogram (EKG), associated with a single cardiac cycle, as recorded from lead II.

The following phases of the EKG complex may be recognized:

1. The *isoelectric line* (baseline) is the point of departure for the P, Q, R, S, and T waves.
2. The *P wave* represents the depolarization of atrial muscle as a wave of negativity spreads from the SA node toward the ventricles. It is normally upright in all three standard limb leads.
3. The *P–R interval* is measured from the beginning of the P wave to the beginning of the R wave or the QRS complex; it represents the interval between the activation of the SA node and the AV node. An abnormal lengthening of the P–R interval suggests interference with conduction to the ventricles.
4. The *P–R segment* is measured from the end of the P wave to the beginning of the R wave or the QRS complex; it represents the interval between atrial depolarization and ventricular depolarization. Although AV nodal delay and depolarization of the AV node, AV bundle, and Purkinje network occur during this segment, no external potentials are recorded.
5. The *QRS complex* represents the spread of excitation through the ventricular myocardium, resulting in depolarization of ventricular muscle. Repolarization of atrial muscle also occurs during this phase of the EKG. The record is complex; its shape, amplitude, and direction depend on the position of the heart in the chest, its size relative to body mass, and the time relationship between right and left ventricular activity. An abnormal lengthening of the duration of the QRS complex suggests interference with the spread of excitation through ventricular muscle, as may occur in Purkinje failure or myocardial infarction.
6. The *S–T segment* represents the interval between the end of the S wave and the beginning of the T wave, the period during which the ventricles are more or less uniformly excited. Normally, it indicates an isoelectric state. Its position and shape are important in the diagnosis of abnormalities of ventricular repolarization.

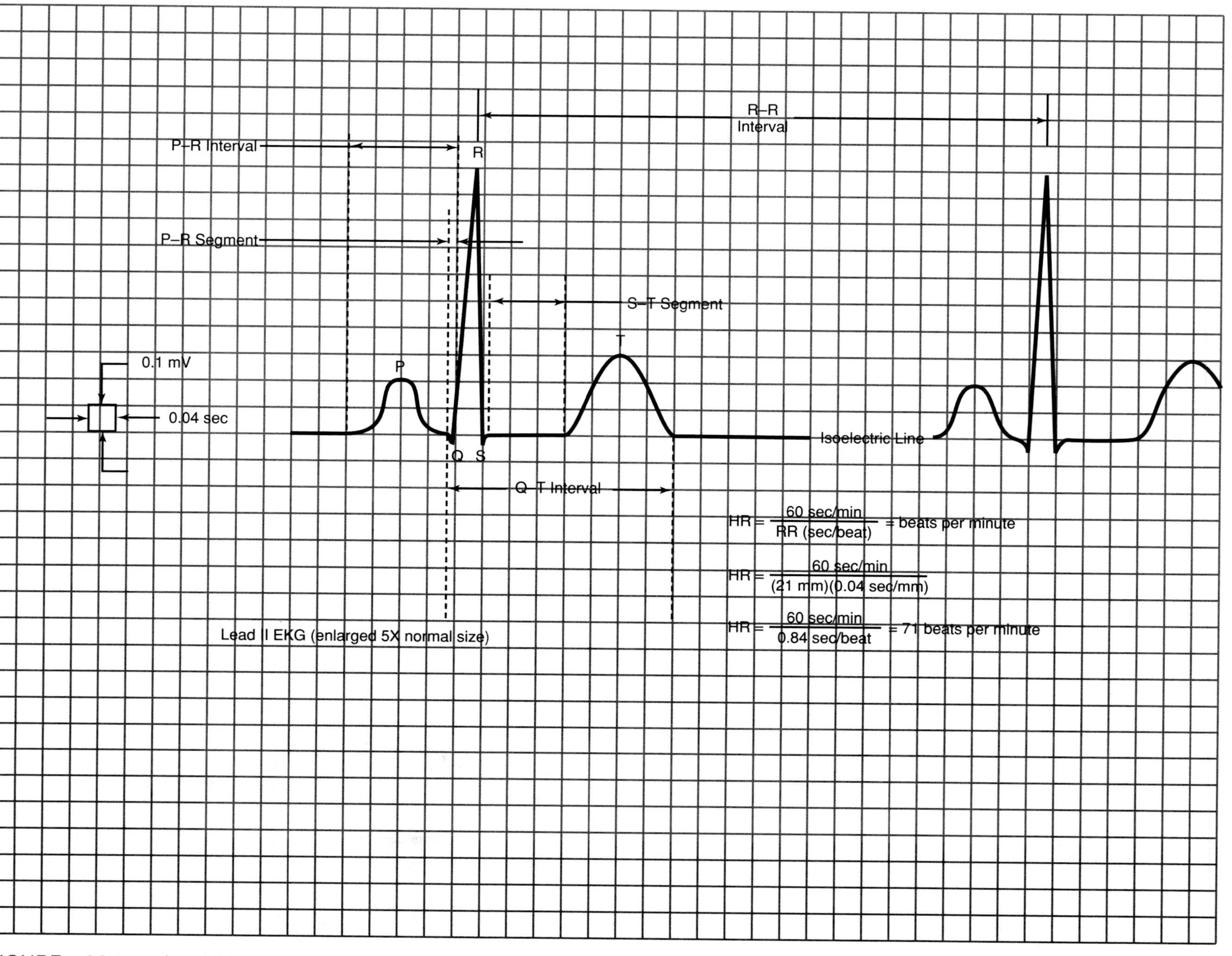

FIGURE 26.5 Lead II EKG

TABLE 26.1 Normal lead II EKG values, heart rate 60–100 BPM

Phase	Amplitude (millivolt)	Duration (second)
P wave	0.02–0.20	0.05–0.10
P–R interval	—	0.12–0.20
P–R segment	—	0.08
QRS complex (R)	0.8–1.2	0.04–0.09
S–T segment	—	0.12
Q–T interval	—	0.28–0.43
T wave	0.2–0.5	0.10–0.25

7. The *T wave* represents the restoration of ventricular myocardium to the resting or excitable state.
8. The *Q–T interval* is measured from the beginning of the QRS complex to the end of the T wave; it represents the time of electrical systole when the ventricular beat is generated. It varies with the heart rate.

Values within normal ranges for the duration and voltage of the different phases of the EKG complex as seen in lead II are indicated in table 26.1.

EXPERIMENTAL OBJECTIVES

1. To become familiar with the electrocardiograph as a primary tool for evaluating electrical events within the heart.
2. To correlate electrical events as displayed on the electrocardiogram with the mechanical events that occur during the cardiac cycle.
3. To observe changes in the electrocardiogram associated with body position, exercise, body size, and age.
4. To anticipate the nature of changes in the electrocardiogram associated with pathology of the heart.

EXPERIMENTAL METHODS

Lafayette Minigraph

Materials

Minigraph model 76107 or 76107VS
model 76322 time/marker channel + remote push button
model 76402MG biopotential amplifier channel
model 76412 EKG electrode box/lead selector
model 76629 deluxe EKG electrode set
model 76621 biogel

Recording the Electrocardiogram

Prepare the Minigraph for single-channel recording. Check to make sure that there is sufficient recording paper and the inking system is working properly. Refer to chapter 2, if necessary, and to figure 26.6.

1. Connect the five-lead patient cable to the EKG electrode box lead selector and plug the EKG electrode box cable into the biopotential amplifier. Set the lead selector switch to standby (STBY).
2. Adjust the amplifier sensitivity to zero. Turn on mainframe power and center the pen using the pen position control.
3. Flip the amplifier run/cal switch to cal. Advance the sensitivity controls and depress the 1 mV-cal button until a pen deflection of 1 cm occurs each time the button is depressed. After this adjustment has been made, do not disturb the sensitivity control. Before recording each lead of the EKG, depress and release the 1 mV-cal button so that the standardization wave (1 cm/mV) is part of the record. When recording the EKG, flip the run/cal switch to the "run" position.
4. Instruct the subject to remove clothing, jewelry, and other accessories from the areas of electrode placement (wrist, ankle). With the subject resting comfortably in a supine position on the laboratory table or a cot, attach the limb electrodes just above the wrists and ankles. Because the skin has a high electric resistance, an electrolyte jelly (EKG sol) is first rubbed on the skin surface to remove oil and dead cells and to form a conducting medium between skin and metal. The electrodes must be snug but must not be secured so tightly as to inhibit arterial circulation to the extremity. Attach lead wires to the proper electrodes (LA = left arm, LL = left leg, etc.). They are color-coded to assist in identification. In this experiment, the chest electrode will not be used.
5. With the subject resting comfortably in a supine position with arms at the side and feet slightly spread, turn the lead selector switch to lead II (position 3), the amplifier run/cal switch to run, and the chart drive control to 25 mm/s (2.5 cm/s). Record lead II for 10 seconds, then turn off the paper-speed control and flip the amplifier run/cal switch to cal. Compare the tracing with figure 26.5. If an adequate record is not obtained, ask the laboratory instructor for assistance.

   If a continuously variable speed recorder is used, paper speed must be adjusted precisely to 25 mm/s before recording the EKG. This adjustment may be accomplished using 1-second time marks and the speed control knob. Once the recorder has been adjusted, do not disturb the speed control setting. Record 1-second time marks along with the EKG so as to be able to periodically verify the standard speed of 25 mm/s.
6. While an electrocardiogram is being recorded the subject must remain quiet. The electrocardiograph is very sensitive to small changes in voltage. Contraction

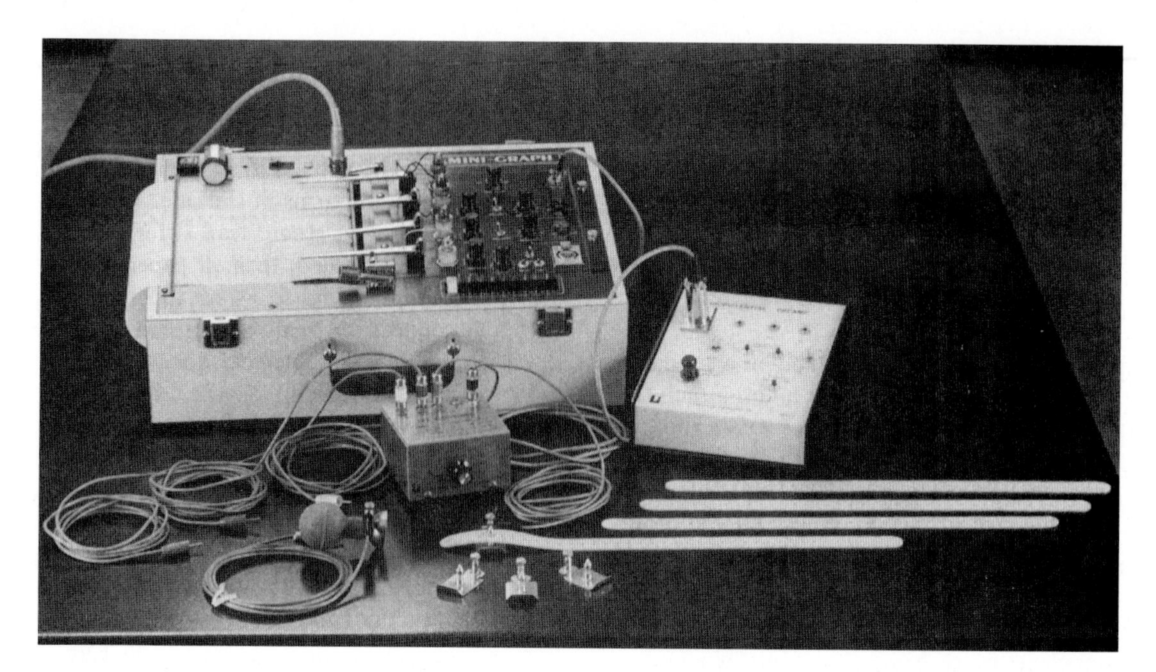

FIGURE 26.6 Minigraph setup for EKG

of skeletal muscle may be detected by the electrocardiograph, resulting in erratic pen movements and unusable electrocardiogram tracings.

7. Whenever the recording leads are changed (via the lead selector switch), always note the change by marking it directly on the electrocardiogram. Similarly, note other changes (e.g., change in position of the subject) at the time of occurrence to facilitate analysis of the electrocardiogram.
8. Turn the amplifier run/cal switch to run and the chart drive control to 25 mm/s and record lead II from the resting subject for 30 seconds. Repeat for lead I (position 2) and lead III (position 4). After recording, turn off the chart drive control and turn the amplifier run/cal switch to cal.
9. Ask the subject to sit upright with legs extended and arms at the side. While the subject remains quiet, record lead II for 15 seconds. When finished recording, turn off chart drive control and flip the amplifier run/cal switch to cal.
10. Disconnect the lead wires from the electrodes but do not remove the electrodes. Ask the subject to exercise by stepping off the table or cot and running in place for 5 minutes. Return the subject to the table or cot and ask the subject to assume a resting supine position. Quickly reattach the lead wires to the proper electrode, flip the amplifier run/cal switch to run and the chart drive control to 25 mm/s, and record lead II for 1 minute. Depress the 1 mV-cal button to record a calibration wave for comparison with initial calibration wave to ensure against a change.
11. Turn off the chart drive and mainframe power. Leave the amplifier run/cal switch on cal and turn the lead selector switch to STBY. Disconnect the lead wires from the electrodes and the electrodes from the subject. Clean the electrodes and coil the patient cable and then place them near the Minigraph.

Analysis of the Electrocardiogram

1. Recall that electrocardiograms are recorded at a standard paper speed of 25 mm/s and a calibrated vertical pen deflection of 1.0 cm/mV. Thus, it is a simple matter to measure the waveform, amplitude (mV), and duration (seconds). One millimeter of horizontal distance represents 1/25 of a second (0.04 second), and 1 mm of vertical distance represents 0.1 mV.
2. Record the following data from the subject: name, age, gender, height, weight, and body build. Enter the data in the report.
3. Begin analysis with a visual survey of leads I, II, and III as recorded from the resting supine subject. Inspect for the presence of regularly occurring P QRS T complexes throughout the recording. Using the peaks of the R waves, calculate the resting heart rate. (Count the number of peaks in 15 seconds and multiply by 4.) The range for the normal resting heart rate is 60–100 beats per minute. Heart rates above 100 beats per minute are known as *tachycardias,* and those below 60 beats per minute are known as *bradycardias.* Under resting conditions, the heart rate is usually 70–90 beats

per minute; however, a conditioned athlete may have a resting heart rate of 46–60 beats per minute due to a more efficient heart pumping a larger stroke volume with a corresponding decrease in rate.

4. Examine the P waves. Are they present or absent—visible in some cases but absent in others? Normal P waves are small, smoothly contoured upright waves in all three limb leads; they indicate an SA node pacemaker. P waves that are peaked, toothed, upside down, absent, or in other ways different from the normal waveform may indicate that some other area (instead of the SA node) is in command as the pacemaker. Variations of P-wave shape also occur in some types of arrhythmias (irregular rhythms). Measure and record the duration and amplitude of the P wave in lead II.
5. Examine each lead relative to the QRS complex. Is each P wave followed by or related to a QRS complex? In some arrhythmias, P waves are not followed by QRS complexes for each cycle (e.g., heart block).
6. Using lead II, measure and record the P–R interval. Normally the P–R interval does not exceed 0.20 second. A prolonged P–R interval indicates an abnormal delay in the spread of the impulse from the SA node to and through the AV node. If the P–R interval exceeds 0.21 second, AV block exists and beats will be dropped.
7. Using lead II, examine the QRS complexes to determine whether the conduction of the impulse through the ventricles is normal. Measure and record the QRS interval. Prolongation of the QRS interval beyond 0.09 second generally indicates a defect or delay in the conduction of the impulse through the ventricles (e.g., Purkinje failure). Examine the shapes of all QRS complexes for a given lead. Are they uniform? Does the QRS waveform change its shape when recording leads are changed? In leads I and III, the amplitude of the QRS complex is less than in lead II.
8. Examine lead II for regularity of rhythm. Mark the distance between two R waves and compare this R–R interval with other R–R intervals in lead II. Is there a slowing down–speeding up rhythm that appears to be correlated with the respiratory cycle? This is commonly and normally seen in young people and is known as sinus arrhythmia.

   If the R–R intervals are equal, the rhythm of the heart is regular. If the R–R intervals are not equal, the rhythm is irregular. Irregular rhythms may be totally irregular and, hence, indicate abnormality, or they may follow a pattern (e.g., sinus arrhythmia) and not necessarily indicate an abnormality.
9. Measure and record the duration of the S–T segment in lead II. Its position should be horizontal along the isoelectric line or slightly ascending. Its duration is normally 0.12 second. The position and shape of the S–T segment are important in the diagnosis of abnormalities of ventricular repolarization, but it is beyond the scope of this book to present an analysis.
10. Measure and record the amplitude and duration of the T wave. It is normally upright in leads I, II, and III, although it sometimes is inverted in lead III. The duration varies normally between 0.10 and 0.25 second, and the amplitude is normally between 0.3 and 0.4 mV.
11. Analyze lead II as recorded from the sitting resting subject and enter the following data in the report:
    (a) P-wave duration and amplitude
    (b) P–R interval
    (c) QRS duration
    (d) R-wave amplitude
    (e) S–T segment
    (f) T-wave duration and amplitude

    Compare the data with that obtained by other laboratory groups.
12. Compare lead II as recorded from the supine resting subject with lead II as recorded from the sitting subject. Are there changes in waveform amplitude and/or duration of waveforms, intervals, and segments?
13. Compare lead II as recorded from the supine resting subject with lead II as recorded immediately after the subject exercised. What changes are noted in the shape, amplitude, and duration in waveforms, intervals, and segments? Note these changes in the report. Calculate the postexercise heart rate and compare it with the resting rate.

Electrical Activity of the Heart: The Electrocardiogram

REPORT 26

Name: ______________________________ Date: ______________

Lab Section: ______________________________

1. Data:

a. Subject's initials _____ Age _____ Gender _____

Height _____ Weight _____ Body build _____

b. Lead II electrocardiogram

Measurement	***Supine***	***Sitting***	***Postexercise***
Heart rate (BPM)			
P-wave amplitude (mV)			
P-wave duration (s)			
P–R interval (s)			
QRS interval (s)			
R-wave amplitude (mV)			
S–T segment (s)			
T-wave amplitude (mV)			
T-wave duration (s)			

2. Compare the rhythm and rate observed above with similar data obtained by other groups in the laboratory. Does there appear to be any correlation between heart rate and age, gender, or body build?

____

____

____

____

3. Compare the EKG data with those obtained from other groups in the laboratory. Do there appear to be discernible differences in the EKG complex based on age, gender, weight, height, or body build?

____

____

____

____

____

____

____

4. List three differences in the EKG complex as displayed in the recording of leads I, II, and III.

   a. ______________________________

   b. ______________________________

   c. ______________________________

5. Compare the EKG complex recorded immediately after exercise with that recorded under resting conditions. Are there any changes in the intervals of the complex? If so, describe their nature and account for them.

   ____

   ____

   ____

   ____

   ____

   ____

   ____

6. What changes in the lead II electrocardiogram would you expect to see associated with the following conditions?

   a. 3:1 AV block ______________________________

   ____

   ____

   b. Premature atrial contractions ______________________________

   ____

   ____

   c. Premature ventricular contractions ______________________________

   ____

   ____

   d. Increased AV nodal delay ______________________________

   ____

   ____

   e. Tachycardia ______________________________

   ____

   ____

Electrical Activity of the Heart: Vectorcardiography

CHAPTER 27

■ INTRODUCTION

The *anatomic axis* of the heart is the angle of the heart in the body from base to apex. The normal anatomic axis is around 55°. That is, if an imaginary cross were drawn through the center of the heart with the horizontal arms parallel to the ground representing 0° left and +180° right and +90° at the bottom of the cross, the apex, or tip, of the heart would be located along the 55° radial. In other words, the apex of the heart normally points toward the lower left rib cage.

The electrical axis of the heart *(mean electrical axis)* is the preponderant direction of current flow during the cardiac cycle. Typically, the mean electrical axis is around +60° for a 70-kg adult. The normal range is 0° to +90°. The mean electrical axis is influenced by the size of the heart, the anatomic position of the heart, and electrical activity of the conduction system and cardiac muscle.

Willem Einthoven developed a "string galvanometer" in 1901 that could record the electrical activity of the heart. Although it was not the first such recorder, it was a breakthrough in that it was accurate enough to allow anybody to duplicate the results on the same patient. His work established a standard configuration for recording the EKG and won him the Nobel Prize in 1924. Since that time, the EKG has become a very powerful tool in diagnosing disorders of the heart. It should be noted that the clinical interpretation of the EKG is quite empirical in practice and has evolved from a long history of reference to and correlation with known cardiac disorders.

Recall from chapter 26 that the cardiac cycle, or heartbeat, begins with an electric impulse (a wave of depolarization followed by a wave of repolarization) generated by and conducted away from the sinoatrial (SA) node. This impulse spreads throughout the atria, stimulating atrial muscles to contract. Depolarization of the atria is recorded as the P wave of the electrocardiogram. The impulse spreading through the right atrium reaches the atrioventricular (AV) node, where impulse conduction is much slower, thereby delaying conduction of the impulse to the ventricles so that the atria have time to complete their contraction. After AV node delay, the impulse is rapidly conducted down the AV bundle (bundle of His), bundle branches, and Purkinje fibers, which stimulate ventricular muscles to contract. Depolarization of the ventricles is recorded as the QRS complex of the electrocardiogram. Repolarization of the ventricles is recorded as the T wave.

Because the current spreads along specialized pathways and depolarizes parts of the pathway in sequence, the electrical activity has directionality or a spatial orientation represented by an electrical axis. The magnitude of the recorded voltage in the EKG is proportional to the amount of tissue being depolarized. Because the ventricles make up the majority of the heart mass, the largest recorded depolarization waveform (QRS) reflects depolarization of the ventricles. The left ventricle has significantly more mass than the right ventricle; therefore, more of the QRS complex reflects the depolarization of the left ventricle, resulting in the axis's being shifted slightly more toward the left ventricle.

The body contains fluids with ions that allow for electrical conduction, which makes it possible to measure electrical activity in and around the heart from the surface of the skin (assuming good electrical contact with the body via electrodes) and also allows the legs and arms to act as simple extensions of points in the torso. Measurements from the leg approximate those occurring in the groin, and measurements from the arms approximate those from the corresponding shoulder (figure 27.1).

It is desirable to place the electrodes on the ankles and wrists for convenience to the subject undergoing the EKG evaluation. In order for the EKG recorder to work properly, a ground reference point on the body is required. This ground is obtained from an electrode placed on the right leg above the ankle.

The electrocardiogram is a record of the overall spread of electric current through the heart as a function of time in the cardiac cycle. The direction of polarity (+ or –) of the waveforms obtained depends upon the location of the recording electrodes on the surface of the body and whether the electrical activity is directed toward or away from the surface electrode. In general, if a wave of depolarization is approaching a positive electrode, a positive voltage will be seen by that electrode relative to a grounded reference electrode. If the wave of depolarization is traveling toward a negative electrode, a negative voltage will be seen. The opposite is true for repolarization—that is, repolarization waves approaching a positive electrode produce negative voltages.

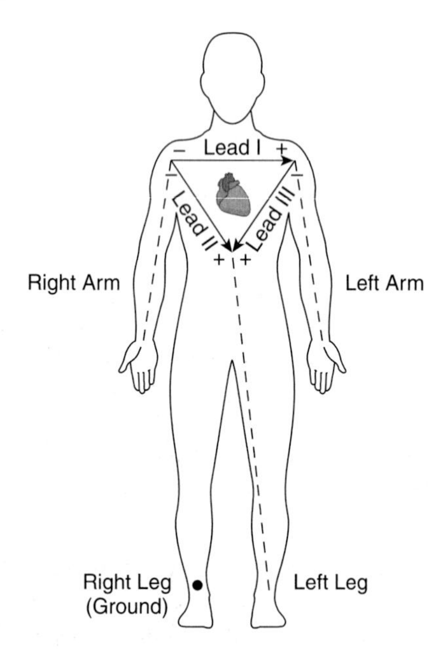

FIGURE 27.1 Einthoven's triangle and the bipolar limb leads

Recall from chapter 26 that the term *lead* is defined as a spatial arrangement of two electrodes on the body. One lead is labeled + and the other –. The electrode placement defines the recording direction of the lead, which is called the *lead axis* or angle. The axis is determined by the direction when going *from the negative to positive electrode.* The EKG recorder computes the voltage difference (magnitude) between the positive and negative electrodes and displays the changes in voltage difference with time.

The standard bipolar limb leads are:

lead I = right arm (–) to left arm (+)

lead II = right arm (–) to left leg (+)

lead III = left arm (–) to left leg (+)

The three standard bipolar limb leads may be used to construct an equilateral triangle, called **Einthoven's triangle,** at the center of which lies the heart (figure 27.1). Each side of the triangle represents one of the bipolar limb leads, and each lead forms a 60° angle with the two opposite leads.

At any given moment during the cardiac cycle (during the QRS complex, for example), the net electrical activity seen by a lead may be represented by a vector. A **vector** is an entity represented by an arrow that has both *magnitude* and *direction.* The vector for a particular lead is plotted on the axis for that lead, with the arrowhead pointing in the positive direction. The length of the arrow is proportional to the magnitude of the lead.

Figure 27.2 shows another way to look at Einthoven's triangle. You can move each axis horizontally or vertically and still have the same representation. This illustration makes it a little easier to visualize the mean electrical axis of the heart.

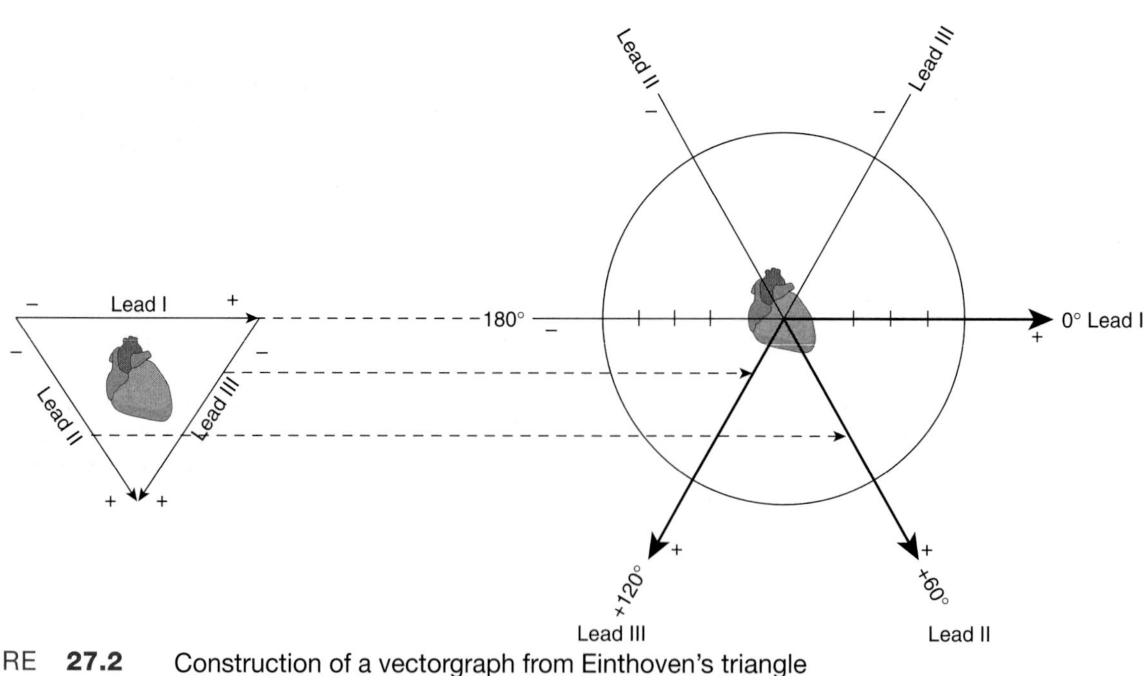

FIGURE 27.2 Construction of a vectorgraph from Einthoven's triangle

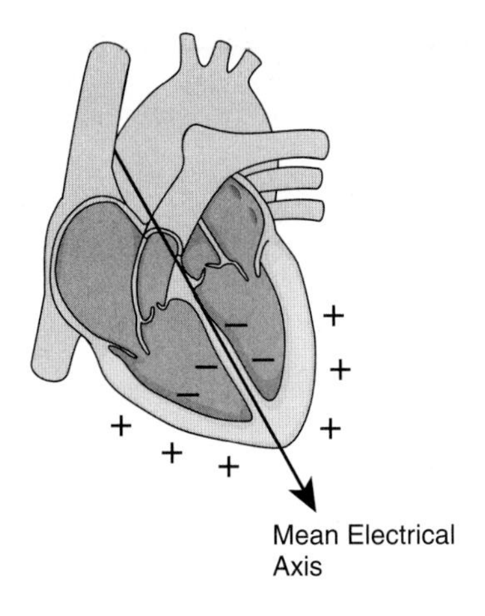

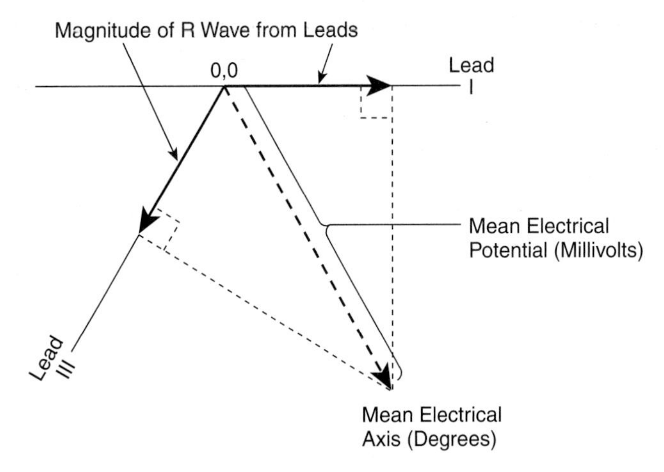

FIGURE **27.3** Approximation of the mean electrical axis of the ventricles

The mean electrical axis of the heart is the summation of all the vectors occurring in a cardiac cycle. Because the QRS interval caused by ventricular depolarization represents the majority of the electrical activity of the heart, we can approximate the mean electrical axis by looking only in this interval.

A further approximation can be made by looking only at the peak of the R wave, which makes up the largest magnitude in the cardiac cycle. To define the mean electrical axis precisely, you would need to define it in three dimensions (*x*, *y*, and *z*). This definition is done in practice by using a standardized set of 12 leads. Three of these leads—the ones previously defined—allow the mean electrical axis to be calculated in the frontal plane. For this chapter we will be interested in only the frontal plane axis.

One way to approximate the mean electrical axis in the frontal plane is to plot the magnitude of the R wave from lead I and lead III (figure 27.3), as follows:

1. Measure the amplitude of the R wave in lead I and in lead III of the subject's electrocardiogram.
2. Plot the data on the respective leads of the vectorgram.
3. Draw an arrow from the center of the vectorgram (0,0) to the marked data point along lead I. Repeat for lead III.
4. Draw a perpendicular line through the tip of the arrow in lead I. Repeat for lead III.
5. Mark the point where the two perpendicular lines intersect.
6. Draw an arrow from the center of the vectorgram (0,0) to the intersection point of step 5.
7. Extend the line of the arrow in step 6 to intersect the circumference of the vectorgram and record the degree at the intersection. This degree is the approximate direction of the *mean electrical axis* of the heart.
8. Divide the length of the arrow in step 6 into subdivisions equal to the subdivisions marked for lead I. Each subdivision represents 0.10 mV.
9. Count the number of subdivisions from the center of the vectorgram (0,0) to the tip of the arrow and multiply the number by 0.10 mV. The result is an approximation of the *mean electrical potential* of the heart.

A more accurate method of approximating the mean electrical axis is to algebraically add the Q, R, and S potentials for each lead instead of using just the magnitude of the R wave to plot the lead vectors. This method will be discussed later.

Clinically, using sophisticated equipment, vectorcardiography involves the continuous recording of electrical activity of the heart and the plotting of electrical vectors in two and even three dimensions. It allows a continuous display of the depolarization–repolarization process as it sweeps over the heart and is used to reveal abnormalities of the conduction process. In this experiment, we will focus on the determination of the mean electrical axis and mean potential of the ventricles.

■ EXPERIMENTAL OBJECTIVES

1. To review procedures for electrocardiographic recording of the standard bipolar limb leads I, II, and III.
2. To determine the mean electrical axis of the ventricles using vectors derived from the amplitude and polarity of the QRS complex in two of the three bipolar limb leads.
3. To determine the mean electrical potential of the ventricles.

■ EXPERIMENTAL METHODS

Lafayette Minigraph

Materials

Minigraph model 76107 or 76107VS
model 76322 time/marker channel + remote push button
model 76402MG biopotential amplifier channel
model 76412 EKG electrode box/lead selector
model 76629 deluxe EKG electrode set
model 76621 biogel

Recording the Electrocardiogram

1. Review the procedures in chapter 26 for setting up the recorder and recording the electrocardiogram.
2. Remember to record at the standard speed of 25 mm/s and to record the calibration wave (1 mV/cm) before recording each lead.
3. Record standard bipolar limb leads I, II, and III from a resting supine subject.
4. After obtaining the electrocardiogram, disconnect the subject, turn off mainframe power, and proceed with determination of the mean electrical axis and mean electrical potential.

Data Analysis

Determining Mean Electrical Axis and Mean Electrical Potential

1. Measure the amplitude (millivolt) and polarity (+ or –) of the Q, R, and S waves in lead I and again in lead II. Measure from the isoelectric line. Waveforms above the isoelectric line are positive (+), and waveforms below the isoelectric line are negative (–). Record the data in the report. If the record for either lead I or lead II is poor, use data from lead III and plot the data on the lead III axis of the vectorgram.
2. Algebraically, add the Q, R, and S potentials for one lead and then the other to obtain the net potential for the QRS complex. For example:

Lead I	*Lead II*
Q – 0.05 mV	Q – 0.10 mV
R + 0.35 mV	R + 0.45 mV
S – 0.10 mV	S – 0.05 mV
QRS net + 0.20 mV	QRS net + 0.30 mV

3. Each of the net QRS potentials represents a vector, having both magnitude (e.g., 0.2 mV) and direction (e.g., +) and can be used to determine the mean electrical axis and mean electrical potential of the ventricles. For an example, examine figure 27.4, in which the sample data from step 2 are plotted. The net vector of lead I QRS is +0.20 mV; therefore, it is plotted in the (+) direction (to the right of center) along the axis of lead I. Had the vector been negative, it would have been plotted in the negative direction (to the left of center). The length of the lead I vector is 0.20 mV.

   The net vector of lead II is +0.30 mV. It is plotted in the positive direction (below center) along the axis of lead II. The length of the lead II vector is 0.30 mV.
4. At the tip of each vector arrowhead, draw a line perpendicular to the lead axis. Mark the point where

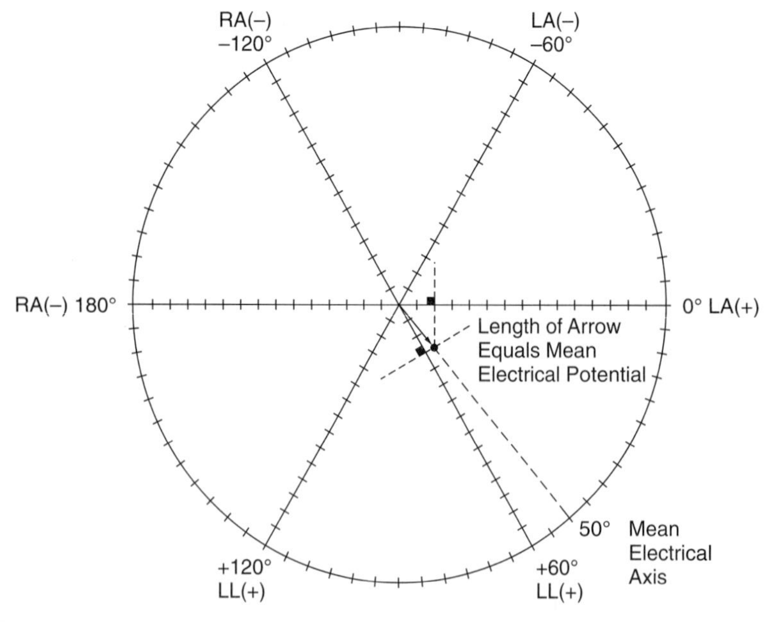

FIGURE 27.4 Plotting the mean electrical axis vector

the two perpendicular lines intersect. Draw an arrow from the center point to the point of the intersection. The direction of the resultant vector (52°) represents the mean electrical axis of the ventricles. The length of the resultant vector (using the same 0.1-mV graduations as the lead axes) represents the mean potential (0.32 mV) of the ventricles during depolarization.

5. Plot your data from steps 1 and 2 on the blank vectorgram in the report. Follow the procedure as outlined in steps 3 and 4 to determine the mean electrical axis and the mean potential of the ventricles.
6. Use only the data for R-wave amplitudes from step 1 to plot the vector (figure 27.5) for determining the mean electrical axis and the mean electrical potential. Plot the data using a different color to distinguish this vector from the vector plotted using the QRS data.

Interpreting the Vectorgram

The normal range of the mean electrical axis of the ventricles is approximately −30° to +90°. This broad range of normal values is due to variations in body size, heart mass, and orientation of the heart within the thorax as well as individual variations in the anatomic distribution of the atrioventricular conduction system. In the normal individual, the mean electrical axis may shift slightly with a change in body position (e.g., standing versus supine) or an increase in physical exercise.

Deviations from the normal range may occur to the right (+90° to +180°) or to the left (−30° to −90°) and hence are referred to as right axis deviations or left axis deviations, respectively.

Right axis deviations may occur if the right ventricle takes longer than normal to depolarize. Right ventricular hypertrophy (enlargement), damage to the right AV bundle branch, and damage to the myocardium or Purkinje system of the right ventricle may cause right axis deviation.

Conversely, *left axis deviations* may occur if the left ventricle takes longer than normal to depolarize. For example, systemic hypertension (high blood pressure) or stenosis (narrowing) of the aortic semilunar valve causes the left ventricle to hypertrophy and shifts the mean electrical axis to the left. Damage (due to drugs, coronary blockage, etc.) to the left ventricular myocardium, left AV bundle branch, or left Purkinje network also results in slower depolarization of the left ventricle and hence left axis deviation.

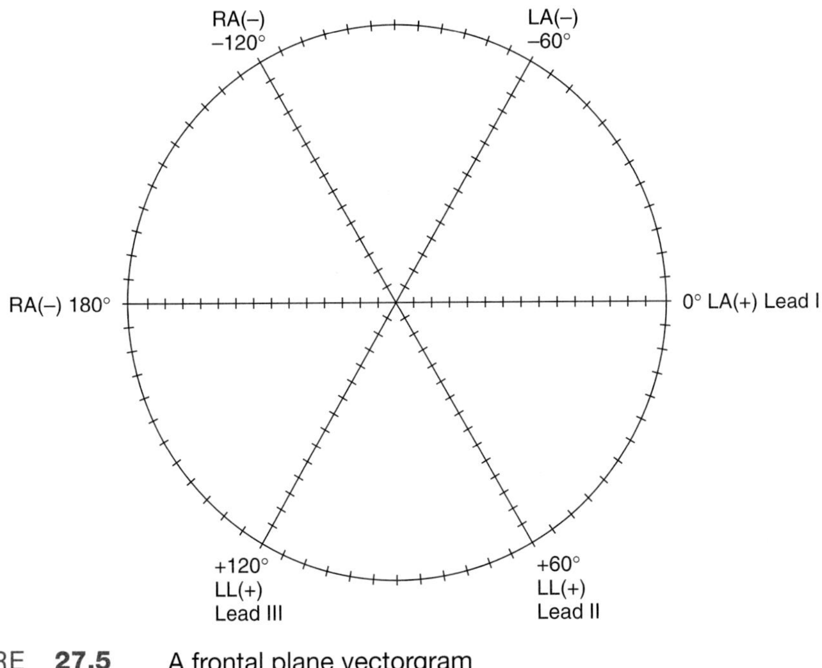

Lead I = RA(–) LA(+)
Lead II = RA(–) LL(+)
Lead III = LA(–) LL(+)
Radial Graduations = 0.1 mV

FIGURE 27.5 A frontal plane vectorgram

Electrical Activity of the Heart: Vectorcardiography

REPORT 27

Name: ______________________ Date: ______________________

Lab Section: ______________________

1. Data: Append a copy of the experimental record to this report.

a. Bipolar limb leads of the electrocardiogram

Lead I			Lead II			Lead III		
Wave	***+ or –***	***Amplitude***	***Wave***	***+ or –***	***Amplitude***	***Wave***	***+ or –***	***Amplitude***
Q		mV	Q		mV	Q		mV
R		mV	R		mV	R		mV
S		mV	R		mV	S		mV
Net		mV	Net		mV	Net		mV

b. Vectorgram

Plotting Data	***Mean Electrical Axis***	***Mean Electrical Potential***
Net QRS	degrees	mV
R wave	degrees	mV

2. Define the following:

a. Mean electrical axis ______

b. Left axis deviation ______

c. Right axis deviation ______

3. List three possible causes of left axis deviation.

a. ______

b. ______

c. ______

The Respiratory Cycle

CHAPTER 28

■ INTRODUCTION

External respiration refers to processes that move air into and back out of the lungs for the purposes of supplying oxygen (O_2) to the blood and removing carbon dioxide (CO_2) from the blood. External respiration includes the alternating processes of *inspiration* and *expiration,* which together, make up the **respiratory cycle.** Elementary gas laws help us understand the mechanics and modifications of the respiratory cycle.

Boyle's law states that the pressure and the volume of a gas vary inversely with one another if the temperature of the gas is constant. In other words, the product of the pressure and the volume is a constant for a given temperature:

$$(P)\ (V) = K\ (T \text{ constant})$$

As gas temperature is held constant and gas volume increases, the pressure of the gas decreases, and vice versa. Because air (a gas mixture) moves from an area of higher pressure to an area of lower pressure, an application of Boyle's law to the respiratory cycle clarifies the mechanical processes of moving air into and out of the lungs.

During **inspiration,** skeletal muscles such as the diaphragm and external intercostals contract, thereby increasing volume within the thorax and lungs. As volume within the air spaces of each lung *(intrapulmonic volume)* increases, air pressure within the lung *(intrapulmonic pressure)* falls below atmospheric pressure and air rushes into the lung. During **expiration,** the inspiratory muscles relax, causing the volume of the thorax and lungs to be reduced. The reduction in intrapulmonic volume is accompanied by an increase in intrapulmonic pressure above atmospheric pressure, forcing pulmonary gas back into the atmosphere. Normally, unlabored expiration at rest is a passive event determined by relaxation of inspiratory muscles. When an increase in pulmonary ventilation is required, such as during exercise, expiration becomes an active event dependent upon contraction of expiratory muscles that pull down the rib cage and compress the lungs.

During one respiratory cycle, a specific volume of air is drawn into the respiratory system and then pushed back out. This volume, first inspired, then expired, is known as **tidal volume (TV).** The actual value of tidal volume varies in direct proportion to the depth of inspiration. During normal quiet breathing *(eupnea)* at rest, adult tidal volume is about 500 mL.

A normal rate of breathing at rest is about 15 respiratory cycles per minute (RR = 15 cpm). The respiratory rate varies with changes in body activity. During exercise, respiratory rate increases, but it decreases when body activity is reduced, as in rest.

The product of tidal volume and respiratory rate (TV × RR) equals the rate of **pulmonary ventilation,** also known as *minute respiratory volume.* During conditions of body rest, an adult rate of pulmonary ventilation is approximately 7.5 L/min (500 mL × 15 cpm/1000).

The rate and strength of contraction of respiratory muscles, and hence the rate and depth of respiration, are controlled by **primary respiratory centers** *(inspiratory and expiratory)* located in the medulla oblongata at the base of the brain stem. The primary centers are inherently rhythmic, alternating their activity to produce inspiration and then expiration. During normal quiet breathing at rest the expiratory center acts to limit and then inhibit the inspiratory center, thereby producing a passive expiration. In contrast, the inspiratory center always acts to produce an active inspiration. When respiratory depth increases, as in exercise, both inspiration and expiration are active processes controlled by their respective medullary centers.

To adjust respiratory rate and depth according to the body's needs, the *medullary centers* receive inputs from higher neural centers (e.g., *pons, cerebellum, cerebral cortex*) and from peripheral receptors such as *chemoreceptors* in aortic and carotid bodies; *stretch receptors* in joints, muscles, and tendons; and somatic *sensory receptors* for pain and thermal stimuli (figure 28.1). For example, cerebral control of the medullary respiratory centers may be

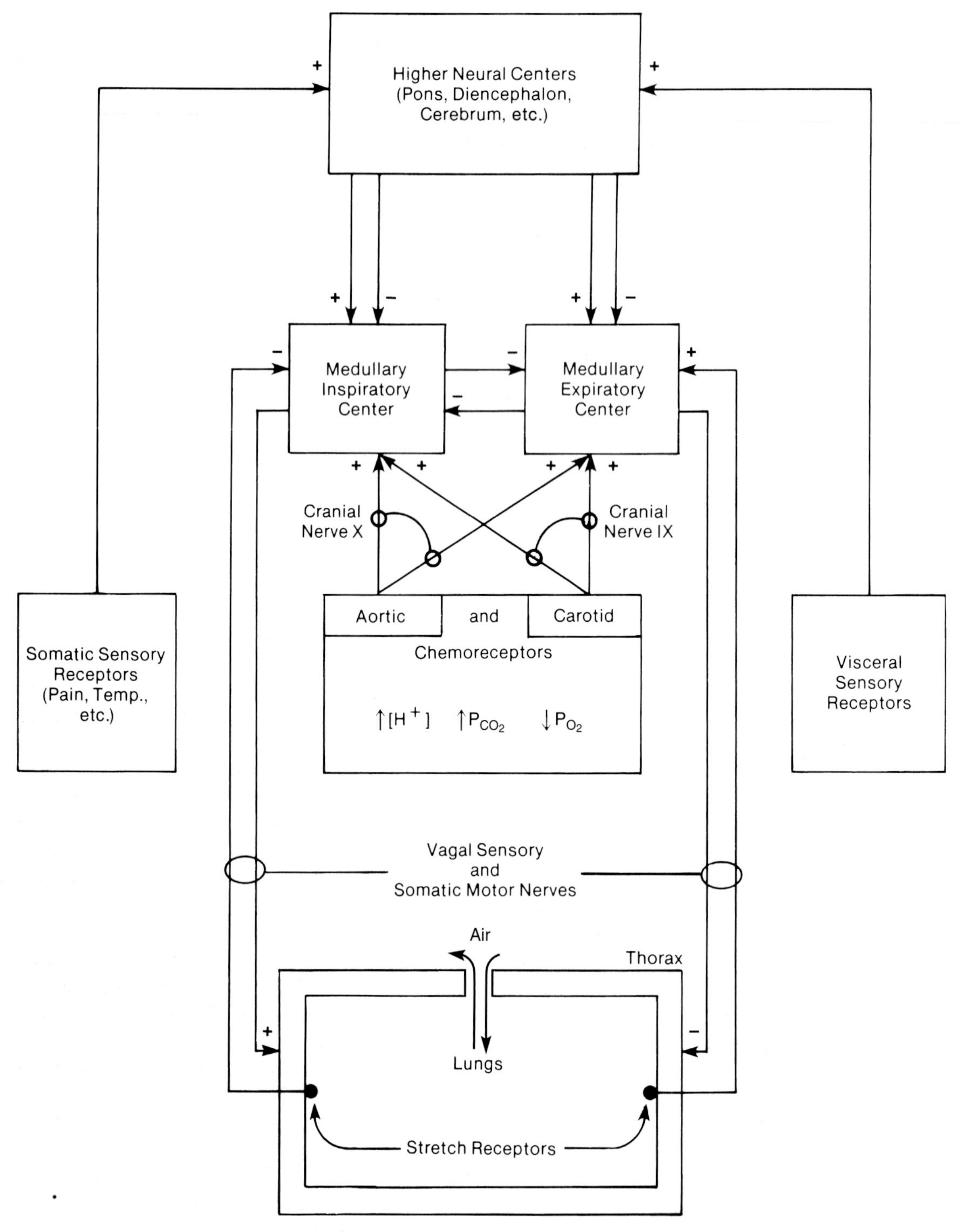

FIGURE **28.1** Regulation of pulmonary ventilation

evidenced by observing the modification of the respiratory cycle as a subject attempts to thread a needle. The cycle temporarily ceases in order to minimize body movement so that the needle may be threaded more easily. Other modifications of the respiratory cycle occur as a result of changes in oxygen, carbon dioxide, or hydrogen ion levels in plasma and cerebrospinal fluid. Respiratory rate and depth are functions of chemical regulation. Blood bathes chemoreceptors located in the aortic and carotid bodies. Chemoreceptors sense changes in arterial P_{CO_2}, H^+, and P_{O_2} and send impulses to the respiratory centers in the medulla oblongata.

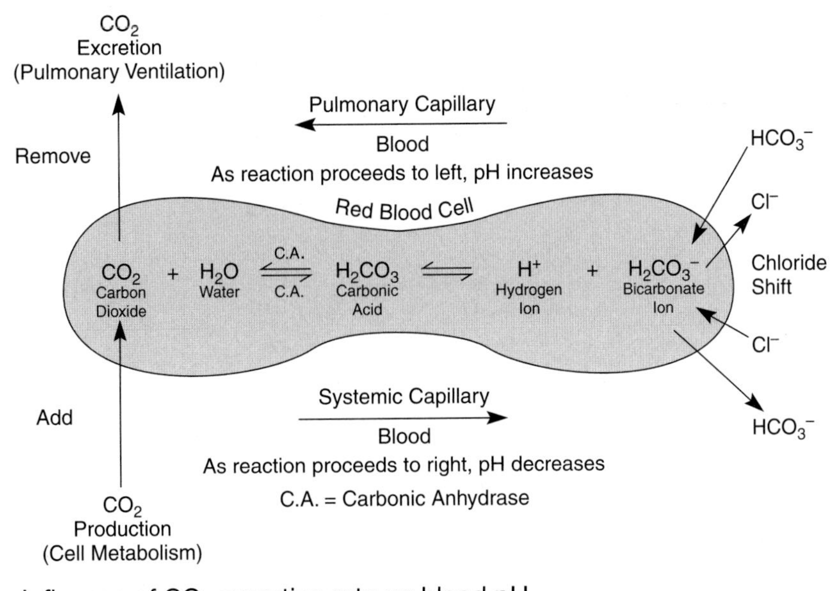

FIGURE 28.2 Influence of CO_2 excretion rate on blood pH

$\uparrow P_{CO_2}$, $\uparrow[H^+]$, and/or $\downarrow P_{O_2}$ stimulate rate and depth of respiration.

$\downarrow P_{CO_2}$, $\downarrow[H^+]$, and/or $\uparrow P_{O_2}$ tend to inhibit respiratory activity.

Changes in arterial carbon dioxide content exert by far the strongest influence on respiratory drive. By virtue of its solubility in body fluids, CO_2 is able to influence respiratory activity by directly affecting chemoreceptors on the ventral surface of the medulla oblongata that are bathed by cerebrospinal fluid. Also, being lipid-soluble and able to cross the *blood-brain barrier* (BBB), CO_2 can exert its effect by changing the hydrogen ion concentration of the brain cerebrospinal fluid (figure 28.2).

P_{O_2} changes are important in regulation of respiration mainly because they change the relative sensitivity of chemoreceptors to P_{CO_2}. If the arterial P_{O_2} is very low, chemoreceptors become more sensitive to changes in arterial P_{CO_2}.

One of the functions of the respiratory system is to eliminate carbon dioxide from body fluids. Carbon dioxide is being continually produced during cellular metabolism. Carbon dioxide diffuses into systemic capillary blood, where some of it reacts with water to form carbonic acid, which dissociates into hydrogen ion and bicarbonate ion (figure 28.3). These reactions are reversible and are accelerated by an enzyme within the red blood cell called *carbonic anhydrase.*

In the systemic capillary, CO_2 is added to the reaction, driving the reaction to the right, forming more H^+ and lowering the pH of the blood.

In the pulmonary capillary, CO_2 is removed from the reaction, pulling the reaction to the left, reducing the amount of H^+ and elevating the pH of the blood.

Normally, the lungs eliminate CO_2 at the same rate as it is being produced by cells. Under such conditions, the reaction sequence (figure 28.3) moves to the right and then back to the left an equal amount (in a state of equilibrium) and no net change in hydrogen ion concentration or carbon dioxide content occurs.

Elimination of carbon dioxide from body fluids at a rate faster than it is being produced drives the reaction sequence more to the left, thereby reducing the amount of H^+ in the body fluids and raising the pH. This occurs in **alveolar hyperventilation.** The process is called respiratory alkalosis, and the resultant condition of elevated blood pH is called **alkalemia.**

If the hyperventilation is voluntary, excess CO_2 will be removed from blood *(hypocapnia),* tending to depress breathing until normal CO_2 and H^+ levels are restored. The temporary cessation of breathing after voluntary hyperventilation is known as *apnea vera.*

Elimination of carbon dioxide from the body fluids at a rate slower than it is being produced drives the reaction sequence more to the right, thereby increasing the amount of H^+ in body fluids and lowering the pH. This occurs in **alveolar hypoventilation.** The process is called **respiratory acidosis,** and the resultant condition of decreased blood pH is called **acidemia.**

If the hypoventilation is voluntary, excess CO_2 will accumulate in the blood *(hypercapnia),* tending to stimulate breathing until normal CO_2 and H^+ levels are restored.

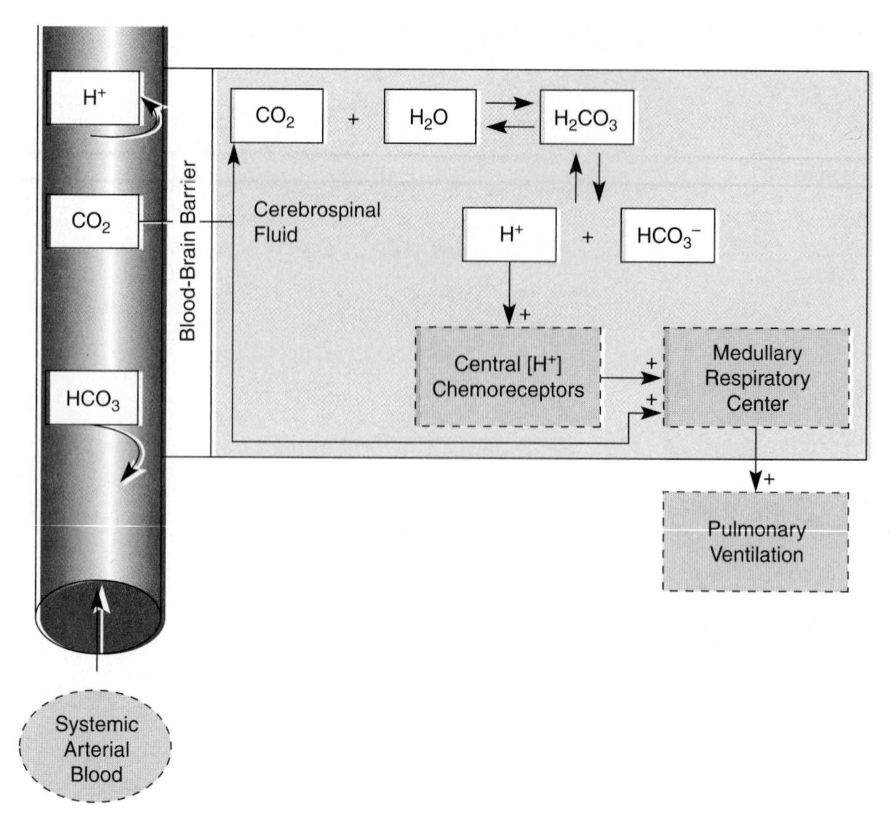

FIGURE **28.3** Direct and indirect effects of carbon dioxide on pulmonary ventilation

In the experiments that follow, physiologic modification of the respiratory cycle as it occurs under a variety of conditions will be observed.

■ EXPERIMENTAL OBJECTIVES

1. To observe and record normal respiratory rate and depth utilizing a bellows pneumograph or an impedance pneumograph.
2. To observe and record modifications of the normal respiratory cycle due to cerebral influence and chemoreceptor influence on the medullary control centers.

■ EXPERIMENTAL METHODS

Lafayette Minigraph

Materials

paper cup

drinking water

fine sewing needles

thread

plastic or paper bag with mouthpiece

nose clip

Minigraph model 76107 or 76107VS

model 76322 time/marker channel + remote push button

model 76406MG basic amplifier channel

model 76607 pneumo bellows and transducer

Preparation of the Recorder

Prepare the Minigraph for two-channel recording. Check to make sure there is sufficient recording paper and the inking system is working properly. Refer to chapter 2, if necessary, and to figure 28.4.

1. Connect the pneumo chest assembly to the pneumograph control box. Turn the vent control on the box fully clockwise to the open position.
2. Connect the pneumograph control box to the channel amplifier via the 9-pin amphenol connector.
3. Connect the remote control push button to the time/marker channel and set the timer to mark 1-second intervals.
4. Set the amplifier polarity switch to (+), the gain switch to 10, and the sensitivity control on zero.
5. Turn on mainframe power, center the amplifier pen using the pen-position control, and then turn off mainframe power.

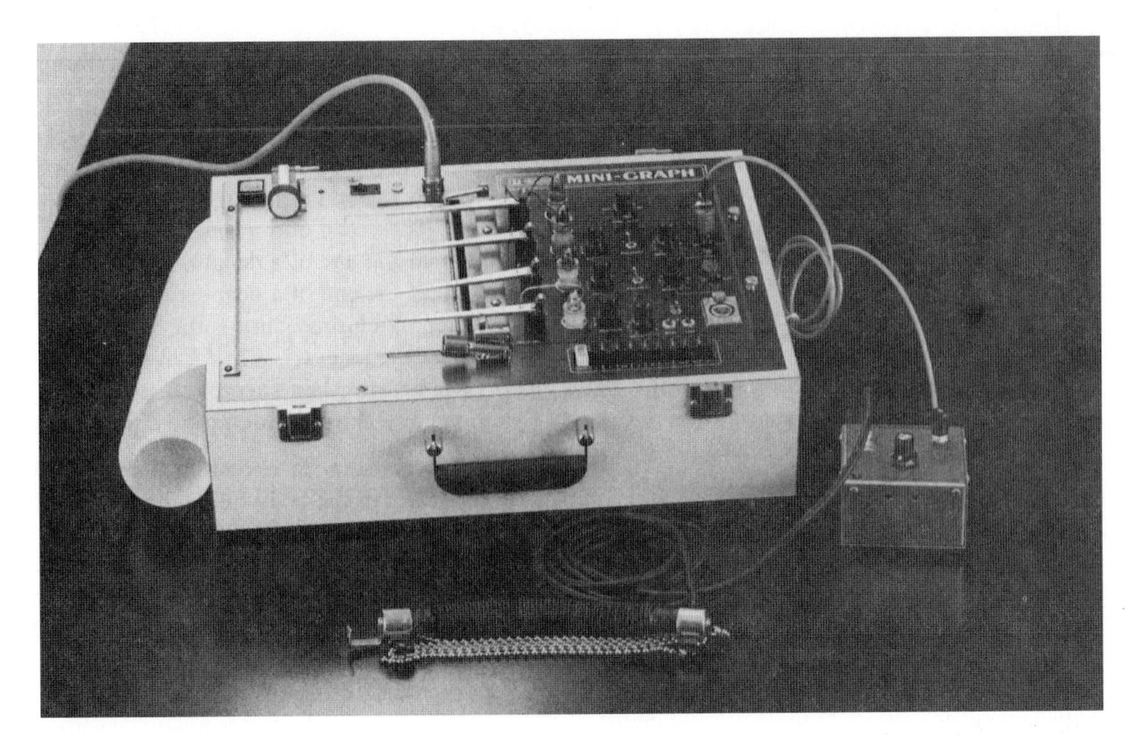

FIGURE **28.4** Minigraph setup for respiration experiments

6. Attach the pneumo chest assembly around the seated subject's chest, just above the breasts. The bellows should be just above the breasts with the connecting chain running under the arms and around the back. Secure the bellows so that they are slightly expanded at the end of the subject's normal expiration.
7. With the subject seated facing away from the recorder (the subject should not observe the record as it is being made), ask the subject to breathe normally. Turn on mainframe power and adjust the chart speed control to 5 mm/s.
8. At the beginning of inspiration, turn the vent control fully counterclockwise to the closed position. Gradually increase the amplifier sensitivity until a pen excursion of approximately 1 cm occurs with each respiratory cycle. Inspiration should be recorded by upward pen movement and expiration by downward pen movement. If necessary, readjust the pen position using the centering control on the model 76607 coupler. At this time, also check for proper operation of the timer and event marker.
9. When recording respiration during different experimental conditions, use the handheld remote push button to record the time of an event. Also note experimental conditions directly on the recording paper.
10. Record at a paper speed of 5 mm/s. If a continuously variable speed recorder is being used, a paper speed of 5 mm/s may be approximated by adjusting the speed control until two 1-second time marks fall within a 1-cm square on the recording paper. In between experimental procedures, turn off the chart drive power to conserve recording paper.
11. Whenever the pneumo chest assembly is attached or detached, the vent control must be open (turned clockwise) to prevent erratic pen movement. When it is desirable to begin or continue recording the respiratory cycle, close the vent as per the directions in item 8.

■ EXPERIMENTAL PROCEDURE

Record respiratory activity under the following conditions:

1. *Eupnea:* While the subject is seated quietly, record respiration for 10 seconds. Label the inspiratory and expiratory phases and calculate the duration of each. Calculate the respiratory rate. Notice the depth of resting respiration as reflected in the amplitude of the recording. Record your observations and data in the report.
2. *Coughing:* While recording eupneic breathing, instruct the subject to cough once. What modifi cations occur in the respiratory cycle? Describe the modifications as they are related to respiratory rate, depth, and the duration of inspiration and expiration.
3. *Reading aloud:* While recording eupneic breathing, instruct the subject to read aloud from the textbook. Record respiration during the reading and shortly

thereafter. Describe the modifications of the respiratory cycle.

4. *Concentration:* While recording normal resting respiration, instruct the subject to concentrate on saying the alphabet backward (not aloud) for 1 minute. Observe and describe the modifications of respiration.
5. *Swallowing:* Instruct the subject to swallow half a glass of water during the inspiratory phase of eupneic breathing. After swallowing, does the subject complete inspiration or does expiration immediately follow? Repeat the experiment with the subject swallowing half a glass of water during expiration. Describe the modification of the respiratory cycle.
6. *Neuromuscular coordination:* While recording normal respiratory cycles, instruct the subject to thread a fine needle as quickly as possible. What happens to the normal respiratory cycle as the subject attempts to thread the needle? Why does respiratory modification occur?
7. *Apnea vera:* Record 10 normal respirations and then instruct the subject to breathe as deeply as possible with an open mouth at a rate of 15 times per minute for 2 minutes, after which the subject permits his or her breathing to return to normal frequency and depth. Continue recording until respiration returns to a normal rate and depth. Was apnea vera observed? Why?
8. *Hypoventilation:* Hypoventilation of the lungs results in a net gain of carbon dioxide in the body fluids because the lungs fail to remove carbon dioxide as rapidly as it is being formed. A net gain of hydrogen ions results from the formation of carbonic acid, and respiration is stimulated. Elevation of respiratory rate and depth occurs until carbon dioxide and hydrogen ion levels are restored to normal. Hypoventilation may be produced by voluntarily holding the breath (voluntary apnea). While recording normal respiratory cycles, instruct the subject to hold his or her breath at the following times:
   (a) at the end of a normal inspiration
   (b) at the end of a normal expiration
   (c) at the end of a deep inspiration
   (d) at the end of a deep expiration

   Determine the length of time the breath can be held (duration of voluntary apnea) in each case. Allow for adequate recovery after each procedure. Record your observations in the report.

   After which procedure is the duration of voluntary apnea the longest? Why? Children frequently hold their breath when angered to spite their parents. Is there any danger in doing this? Why or why not?
9. *Rebreathing:* An increase in the carbon dioxide tension (partial pressure) in the body fluids results in stimulation of the medullary respiratory centers. Stimulation of respiratory rate and depth due to an elevated partial pressure of carbon dioxide is caused by the resultant increase of hydrogen ions in cerebrospinal fluid. Pulmonary ventilation will increase until carbon dioxide levels and extracellular-fluid pH return to normal. Increases in the carbon dioxide content (hypercapnia) and hydrogen ion content of arterial plasma also stimulate respiration by way of the chemoreceptor mechanisms of the aortic and carotid bodies.

   Experimental increases in arterial carbon dioxide can be affected by rebreathing into a closed space. Adjust the paper speed to 0.1 cm/s. Close the subject's nose with a nose clip and, while recording, instruct the subject to breathe into and out of the specially prepared plastic bag. The subject must not breathe into the atmosphere and should continue rebreathing from the bag until rebreathing can no longer be tolerated. After the cessation of rebreathing, continue recording until the subject's recovery is complete.

   As the alveolar carbon dioxide content became elevated above normal, what changes in respiratory rate and depth occurred? Why? Rebreathing from a closed bag also results in arterial hypoxia (low oxygen content), which stimulates respiration. Which stimulus (hypoxia or hypercapnia) is more powerful in its effect on respiratory center activity?

The Respiratory Cycle

REPORT 28

Name: ______________________________ Date: ______________________

Lab Section: ______________________________

1. Data:

a. Eupnea

Measurement	Inspiration	Expiration
Duration	sec	sec
Depth	mm	mm

Respiratory rate = number of inspirations per 10 s × 6 = breaths per minute (bpm)

Respiratory rate = __________bpm

b. Cycle modification in:

(1) Coughing ______________________________

(2) Reading aloud ______________________________

(3) Concentration ______________________________

(4) Swallowing ______________________________

(5) Threading a needle ______________________________

c. Duration of apnea vera ______________________________

d. Duration of voluntary apnea:

(1) At the end of a normal inspiration __________ seconds

(2) At the end of a normal expiration __________ seconds

(3) At the end of a deep inspiration __________ seconds

(4) At the end of a deep expiration __________ seconds

2. Describe and account for the effects of hyperventilation on the respiratory cycle. ______

3. Define apnea vera. Why does this occur after voluntary hyperventilation? ______

4. People in anxiety states may subconsciously and chronically hyperventilate. What effect does this have on the central nervous system and the pH of body fluids? ______

5. Describe the effects of rebreathing on the respiratory cycle. ______

6. Explain how the following factors influence respiratory rate and depth:
   a. Increased alveolar P_{CO_2}: ______
   b. Decreased alveolar P_{O_2}: ______
   c. Decreased blood pH: ______
   d. Increased blood pH: ______

7. Define the following terms:
   a. Eupnea ______
   b. Dyspnea ______
   c. Hyperventilation ______

d. Hypoventilation ___

e. Respiration acidosis ___

f. Respiratory alkalosis ___

g. Hypocapnia ___

h. Hypercapnia ___

i. Hypoxia ___

j. Hyperoxia ___

8. Explain why, at the end of voluntary apnea, the subject is no longer able to hold his or her breath even though the attempt is made. ___

9. In an abnormal type of respiratory cycle called Cheyne-Stokes breathing, the patient hyperventilates, then hypo ventilates, then hyperventilates, and so on. On the basis of your knowledge of the effects of hyperventilation and hypoventilation on respiratory centers, how do you account for Cheyne-Stokes breathing? ___

Pulmonary Function Tests: Volumes and Capacities

CHAPTER 29

INTRODUCTION

The volume of air a person inhales (inspires) and exhales (expires) can be measured with a **spirometer** (*spiro* = breath, *meter* = to measure). A bell spirometer (figure 29.1) consists of a double-walled cylinder in which an inverted bell filled with oxygen-enriched air is immersed in water to form a seal. The bell is attached by a pulley to a recording pen that writes on a drum rotating at a constant speed. During inspiration, air is removed from the bell and the pen rises, recording an inspired volume. As expired air enters the bell, the pen falls and an expired volume is recorded. The resultant record of volume change versus time is called a **spirogram.**

The volume of air inspired or expired during a single breath is called **tidal volume (TV).** When a resting person breathes normally, tidal volume is approximately 500 mL. During exercise, tidal volume can be more than 3 L.

Tidal volume is one of four nonoverlapping primary compartments of total lung capacity (see figure 29.1). The other three primary lung volumes are inspiratory reserve volume, expiratory reserve volume, and residual volume.

Inspiratory reserve volume (IRV) is the volume of air that can be maximally inhaled at the end of a tidal inspiration. Resting IRV is approximately 3300 mL and 1900 mL in young adult males and females, respectively.

Expiratory reserve volume (ERV) is the volume of air that can be maximally exhaled at the end of a tidal expiration. Resting ERV is approximately 1000 mL and 700 mL in young adult males and females, respectively.

Residual volume (RV) is the volume of gas remaining in the lungs at the end of a maximal expiration. In contrast to IRV, TV, and ERV, residual volume does not change with exercise. Average adult values for RV are 1200 mL (male) and 1100 mL (female). Residual volume reflects the fact that after we take our first breath at birth and inflate the lungs, we never completely empty the lungs during any subsequent respiratory cycle.

A *pulmonary capacity* is the sum of two or more primary lung volumes. There are five pulmonary capacities: inspiratory capacity, expiratory capacity, functional residual capacity, vital capacity, and total lung capacity.

Inspiratory capacity (IC) = TV + IRV

Expiratory capacity (EC) = TV + ERV

Functional residual capacity (FRC) = ERV + RV

Vital capacity (VC) = IRV + TV + ERV

Total lung capacity (TLC) = IRV + TV + ERV + RV

Each of these capacities is represented graphically in figure 29.1.

Table 29.1 summarizes the terms, symbols, and definitions for the standard divisions of lung volume, and table 29.2 lists normal adult values.

Pulmonary volumes and capacities are generally measured in the clinical assessment of a variety of pulmonary disorders. In general, chronic pulmonary diseases may be classified into two physiologic categories: (1) *obstructive pulmonary disorders,* such as emphysema and bronchial asthma, and (2) *restrictive pulmonary disorders,* such as pulmonary fibrosis and other chronic diseases of the lung interstitium.

In a chronic obstructive pulmonary disease (COPD) such as bronchial asthma, excessive mucus secretion partially blocks airways, increasing airway resistance and thus making breathing more difficult. The asthmatic may take longer to inspire and expire, but pulmonary volumes may be normal or near normal.

In a restrictive pulmonary disease, the ability to change lung volume is decreased. For example, in silicosis (grinder's disease), a disorder caused by chronic inhalation of stone dust, sand, or flint, the lungs lose distensibility and become stiffer.

In restrictive pulmonary diseases, lung capacities and volumes are generally reduced (e.g., decreased vital capacity),

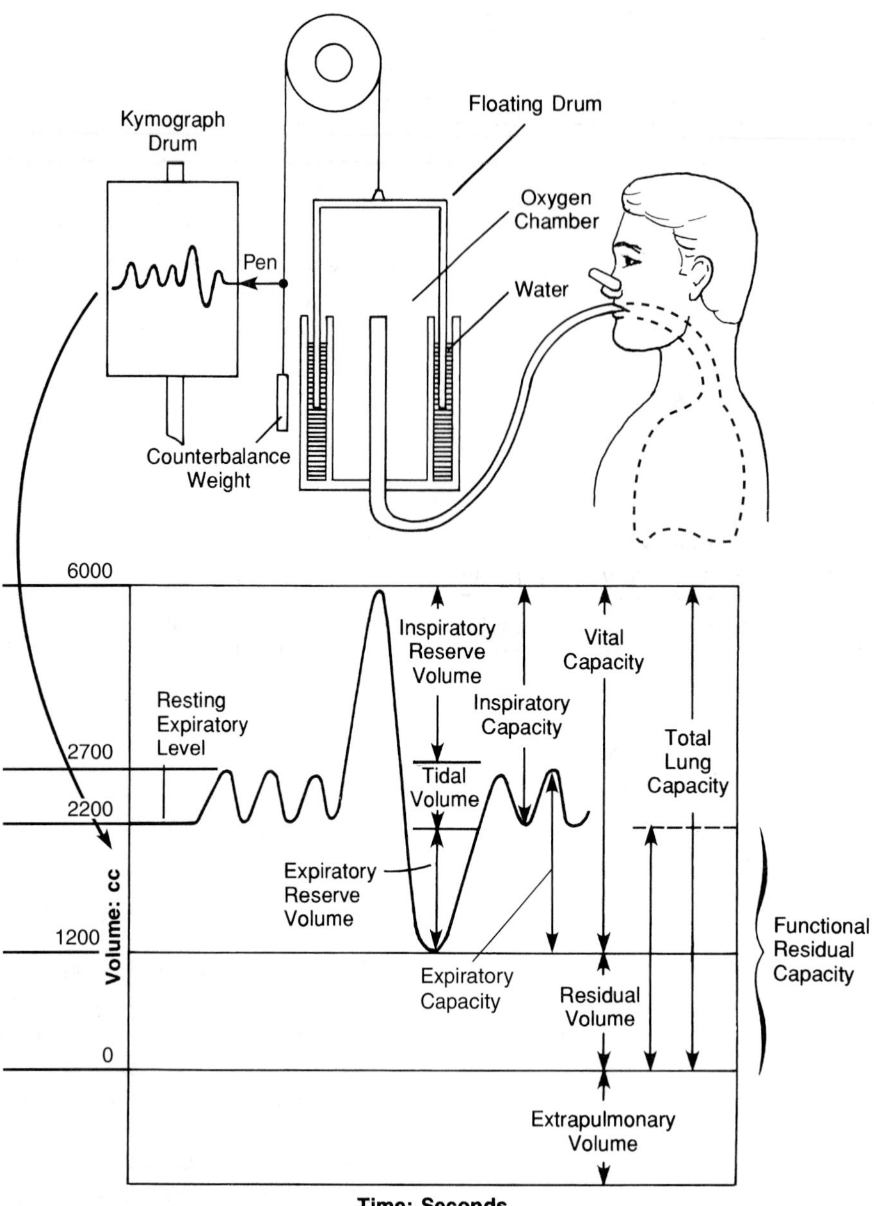

FIGURE **29.1** Bell spirometer and spirogram

and in obstructive pulmonary diseases, pulmonary airflow is generally reduced. Obstructive and restrictive pulmonary diseases often coexist (e.g., combined pulmonary emphysema and fibrosis).

Restrictive pulmonary diseases may be diagnosed, in part, by determining the lung capacities and volumes that will be measured in the experiments in this chapter. Obstructive pulmonary diseases usually require measurements of pulmonary flow rates, which will be measured in the experiments in chapter 30.

EXPERIMENTAL OBJECTIVES

1. To observe experimentally, record, and/or calculate selected pulmonary volumes and capacities.
2. To compare the observed values of volume and capacity with predicted normals.
3. To assess by comparison different methods of determining vital capacity.
4. To compare the normal values of pulmonary volumes and capacities of subjects differing in gender, age, weight, and height.

TABLE 29.1 Pulmonary volumes and capacities

Standardized term	Symbol	Definition
Inspiratory reserve volume	IRV	Maximal volume of gas that can be inspired from end tidal inspiration
Tidal volume	TV	Volume of gas inspired or expired during each respiratory cycle
Expiratory reserve volume	ERV	Maximal volume of gas that can be expired from resting expiratory level
Residual volume	RV	Volume of gas remaining in the lungs at the end of maximal expiration
Total lung capacity	TLC	Amount of gas contained in the lungs at the end of a maximal inspiration
Vital capacity	VC	Maximal amount of gas that can be expelled from the lungs after a maximal inspiration
Inspiratory capacity	IC	Maximal amount of gas that can be inspired from the resting expiratory level
Functional residual capacity	FRC	Amount of gas remaining in the lungs at the resting end-expiratory level

TABLE 29.2 Components of lung volume (adult)

	Males (mL)	Females (mL)
Tidal volume (TV)	500	500
Inspiratory reserve volume (IRV)	3300	1900
Expiratory reserve volume (ERV)	1000	700
Residual volume (RV)	1200	1100
Inspiratory capacity (IC)	3800	2400
Functional residual capacity (FRC)	2200	1800
Total lung capacity (TLC)	6000	4200

EXPERIMENTAL METHODS

Propper Spirometry

Materials

Propper compact spirometer

disposable spirometer mouthpieces

nose clip

recording paper (if required for spirometer)

clinical scale (weight and height)

Measuring Tidal Volume

Obtain from the laboratory instructor a Propper compact spirometer (figure 29.2) and determine the subject's tidal volume as follows:

1. Set the index to 1000 by rotating the knurled bezel ring until the pointer is coincident with the 1000 mark.
2. Attach a clean disposable mouthpiece to the spirometer and a nose clamp to the subject.
3. Ask the subject to breathe normally for about a minute. After inhaling a normal breath, the subject places the mouthpiece of the spirometer between the lips and exhales in a normal, unforced way into the spirometer.
4. Determine the volume change by reading the value on the dial that the pointer now indicates and subtracting 1000 cc from the number. For example, if the pointer moves to 1400 cc, tidal volume is 1400 cc – 1000 cc = 400 cc.
5. Record the average of three trials as the subject's tidal volume.

Measuring Expiratory Reserve Volume and Expiratory Capacity

1. Set the index to 1000 as in the tidal volume experiment.
2. Ask the subject to stand and breathe normally for about a minute. After a normal exhalation the subject places the mouthpiece between the lips and forcibly exhales all the additional air possible.
3. Determine the volume change by reading the value on the dial that the pointer now indicates and subtracting 1000 cc from the number.
4. Record the average of three trials as the subject's expiratory reserve volume.
5. Calculate the subject's expiratory capacity by adding the values for tidal volume and expiratory reserve volume. Record the calculated value.

Measuring Inspiratory Reserve Volume and Inspiratory Capacity

1. Set the index to zero by rotating the knurled bezel ring until the pointer is coincident with the red zero mark.
2. Ask the subject to stand and breathe normally for about a minute.
3. Ask the subject to inhale as deeply as possible, then insert the mouthpiece and exhale normally without forcing the air out.
4. Determine the volume change by reading the value on the dial that the pointer now indicates. This value is the subject's inspiratory capacity.

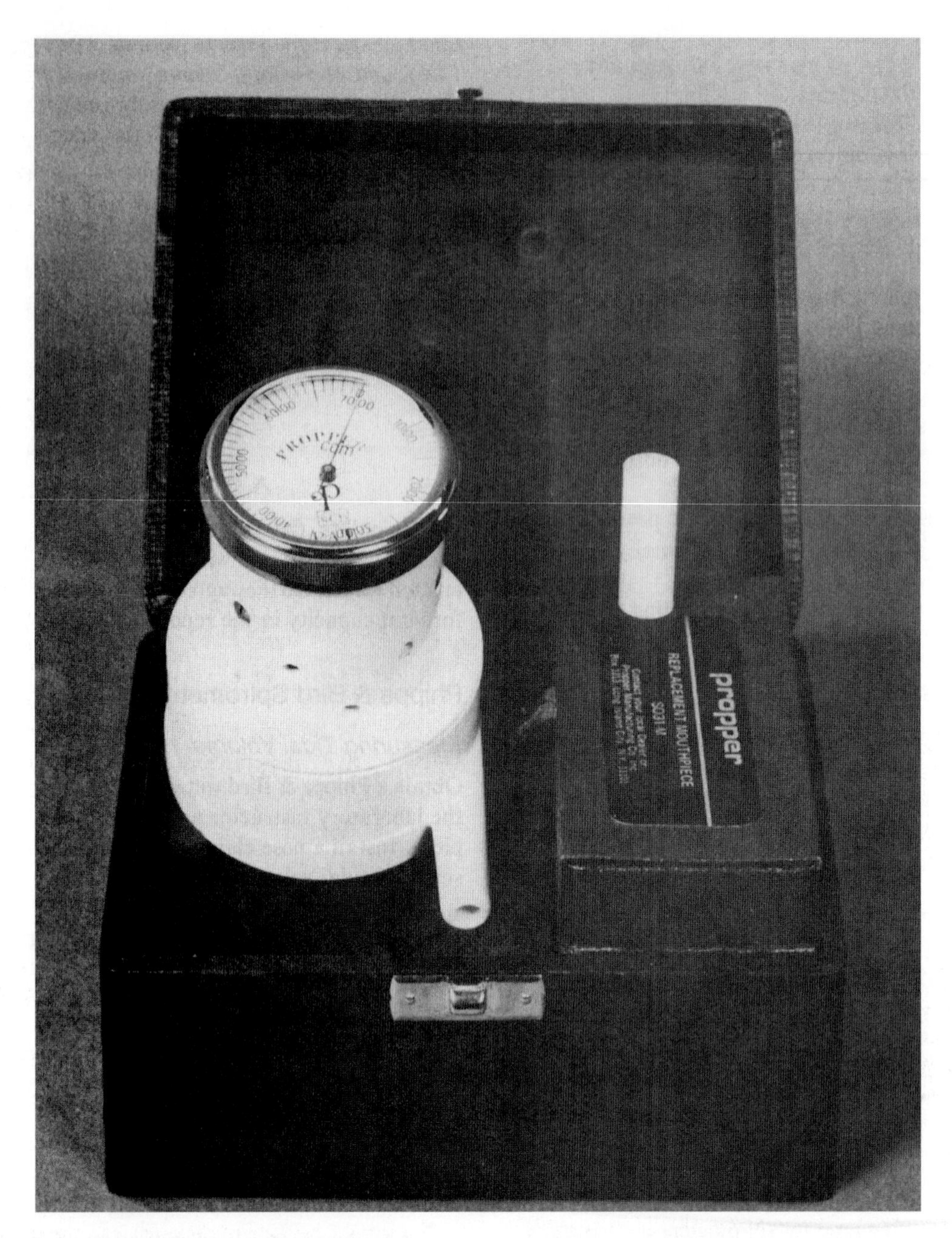

FIGURE **29.2** Propper compact spirometer

5. Record the average of three trials as the subject's inspiratory capacity.
6. Calculate the subject's inspiratory reserve volume by subtracting the value for tidal volume (as previously determined) from the value for inspiratory capacity. Record the calculated value.

Measuring Vital Capacity

1. Set the index to zero by rotating the knurled bezel ring until the pointer is coincident with the red zero mark.
2. Attach a clean disposable mouthpiece to the spirometer and a nose clamp to the subject.
3. Ask the subject to stand and take two or three forced inspirations and expirations without using the spirometer and then to inspire as deeply as possible from the atmosphere and exhale completely into the spirometer. The expiration into the spirometer should be regular, even, and complete. It is not necessary to blow hard.
4. After the subject has completely expired into the spirometer, note the position of the pointer and read the vital capacity in cubic centimeters from the dial. Record the average of three trials. Using the appropriate Propper vital capacity table (tables 29.3 and 29.4), determine the percentage of normal vital

capacity. Record both the measured vital capacity and the percentage of the normal as obtained from the tables. Proceed to the section titled Calculating Vital Capacity.

■ EXPERIMENTAL METHODS

Phipps & Bird Spirometry

Materials

Phipps & Bird wet spirometer

disposable spirometer mouthpieces

nose clip

recording paper

clinical scale (weight and height)

Measuring Tidal Volume

Obtain a Phipps & Bird wet spirometer (figure 29.3) from the laboratory instructor and install a disposable mouthpiece. Install a nose clamp on the subject. Determine the subject's tidal volume as follows:

1. Set the pointer on the horizontal scale to zero. Ask the subject to sit near the spirometer and breathe normally for about a minute.

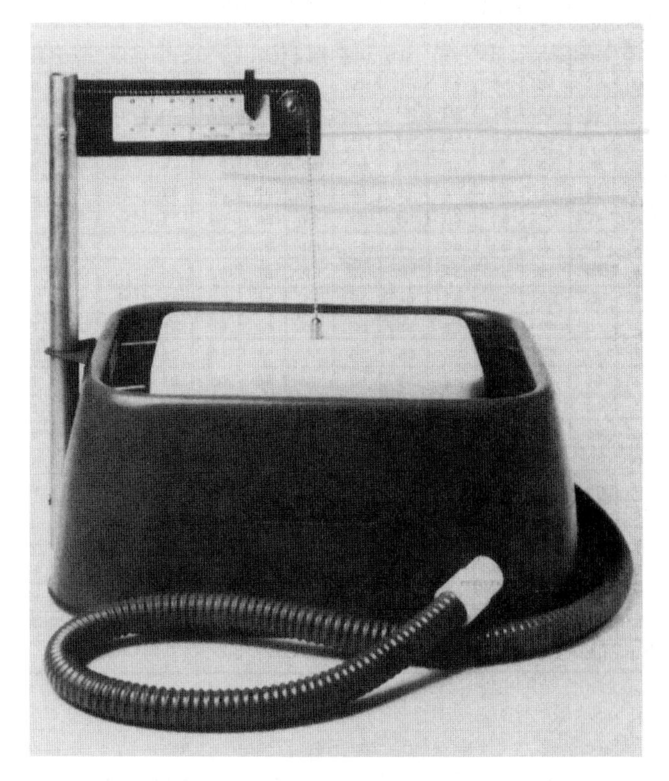

FIGURE **29.3** Phipps & Bird wet spirometer

2. After inhaling a normal breath, the subject places the mouthpiece of the spirometer between the lips and exhales in a normal, unforced way into the spirometer.
3. Determine the volume change by reading the value on the horizontal scale that the pointer now indicates.
4. Record the average of three trials as the subject's tidal volume, making sure that the pointer is reset to zero before each reading.

Measuring Expiratory Reserve Volume and Expiratory Capacity

1. Set the pointer on the horizontal scale to zero. Ask the subject to stand near the spirometer and breathe normally for about a minute.
2. After a normal exhalation, the subject places the mouthpiece between the lips and forcibly exhales all the additional air possible into the spirometer.
3. Determine the volume change by reading the value on the horizontal scale that the pointer now indicates.
4. Record the average of three trials as the subject's expiratory reserve volume, making sure that the pointer is reset to zero before each reading.
5. Calculate the subject's expiratory capacity by adding the values for tidal volume and expiratory reserve volume. Record the calculated value.

Measuring Inspiratory Reserve Volume and Inspiratory Capacity

1. Set the pointer on the horizontal scale to zero. Ask the subject to stand near the spirometer and breathe normally for about a minute.
2. Ask the subject to inhale as deeply as possible, then insert the mouthpiece and exhale normally without forcing the air out.
3. Determine the volume change by reading the value on the horizontal scale that the pointer now indicates. This value is the subject's inspiratory capacity.
4. Record the average of three trials as the subject's inspiratory capacity, making sure the pointer is reset to zero before each reading.
5. Calculate the subject's inspiratory reserved volume by subtracting the value for tidal volume (as previously determined) from the value for inspiratory capacity. Record the calculated value.

Measuring Vital Capacity

1. Set the pointer on the horizontal scale to zero. Ask the subject to stand near the spirometer and breathe normally for about a minute.
2. Ask the subject to inhale as deeply as possible, then insert the mouthpiece and exhale as deeply as possible into the spirometer.

TABLE 29.3 Percentage of vital capacity—men (calculated from standing height)

Standing height	Vital capacity in cubic centimeters																								
in cm	700	800	900	1000	1100	1200	1300	1400	1500	1600	1700	1800	1900	2000	2100	2200	2300	2400	2500	2600	2700	2800	2900	3000	3100	3200
144.8	19	22	25	28	30	33	36	39	41	44	47	50	52	55	58	61	64	66	69	72	75	77	80	83	86	88
147.3	19	22	24	27	30	33	35	38	41	43	46	49	52	54	57	60	63	65	68	71	73	76	79	82	84	87
149.8		21	24	27	30	32	35	37	40	43	45	48	51	53	56	59	61	64	67	69	72	75	77	80	83	85
152.4		21	24	26	29	32	34	37	39	42	45	47	50	53	55	58	60	63	66	68	71	74	76	79	81	84
154.9		21	23	26	28	31	34	36	39	41	44	47	49	52	54	57	59	62	65	67	70	72	75	78	80	83
157.4		20	23	25	28	30	33	36	38	41	43	46	48	51	53	56	58	61	64	66	69	71	74	76	79	81
160.0		20	23	25	28	30	33	35	38	40	43	45	48	50	53	56	58	60	63	65	68	70	73	75	78	80
162.5		20	22	25	27	30	32	35	37	39	42	44	47	49	52	54	57	59	62	64	67	69	71	74	76	79
165.1			22	24	27	29	32	34	36	39	41	44	46	48	51	53	56	58	61	63	66	68	70	73	75	78
167.6			21	24	26	29	31	33	36	38	41	43	45	48	50	52	55	57	60	62	64	67	69	71	74	76
170.2			21	24	26	28	31	33	35	38	40	42	45	47	49	52	54	56	59	61	63	66	68	71	73	75
172.7			21	23	26	28	30	32	35	37	39	42	44	46	49	51	53	56	58	60	63	65	67	70	72	74
175.5			21	23	25	27	30	32	34	37	39	41	43	46	48	50	52	55	57	59	62	64	66	68	71	73
177.8			20	22	25	27	29	31	34	36	38	40	43	45	47	49	52	54	56	58	61	63	65	67	70	72
180.4			20	22	24	27	29	31	33	35	38	40	42	44	47	49	51	53	55	58	60	62	64	67	69	71
182.9			20	22	24	26	28	31	33	35	37	39	42	44	46	48	50	53	56	57	59	61	63	66	68	70
185.4			19	22	24	26	28	30	32	35	37	39	41	43	45	47	50	52	54	56	58	60	63	65	67	69
188.0			19	21	23	26	28	30	32	34	36	38	40	43	45	47	49	51	53	55	57	60	62	64	66	68
190.5				21	23	25	27	29	32	34	36	38	40	42	44	46	48	50	53	55	57	59	61	63	65	67
193.0				21	23	25	27	29	31	33	35	37	39	42	44	46	48	50	52	54	56	58	60	62	64	66

TABLE 29.4 Percentage of vital capacity—women (calculated from standing height)

Standing height	Vital capacity in cubic centimeters																			
in cm	600	700	800	900	1000	1100	1200	1300	1400	1500	1600	1700	1800	1900	2000	2100	2200	2300	2400	2500
139.7	21	23	29	32	36	39	43	47	50	54	57	61	64	68	71	75	79	82	86	89
142.2	21	23	28	32	35	39	42	46	49	53	56	60	63	67	70	74	77	81	84	88
144.8	21	24	28	31	35	38	41	45	48	52	55	59	62	66	69	72	76	79	83	86
147.3	20	24	27	31	34	38	41	44	47	51	54	58	61	64	68	71	75	78	81	85
149.8	20	23	27	30	33	37	40	43	47	50	53	57	60	63	67	70	73	77	80	83
152.4	20	23	26	30	33	36	39	43	46	49	52	56	59	62	66	69	72	75	79	82
154.9		23	26	29	32	36	39	42	45	48	52	55	58	61	65	68	71	74	77	81
157.4		22	25	29	32	35	38	41	44	48	51	54	57	60	63	67	70	73	76	79
160.0		22	25	28	31	35	38	41	44	47	50	53	56	59	62	66	69	72	75	78
162.5		22	25	28	31	34	37	40	43	46	49	52	55	58	62	65	68	71	74	77
165.1		21	24	27	30	33	36	39	42	46	49	52	55	58	61	64	67	70	73	76
167.6		21	24	27	30	33	36	39	42	45	48	51	54	57	60	63	66	69	72	75
170.2		21	24	27	29	32	35	38	41	44	47	50	53	56	59	62	65	68	71	74
172.7		20	23	26	29	32	35	38	41	44	46	49	52	55	58	61	64	67	69	72
175.3		20	23	26	29	31	34	37	40	43	46	49	51	54	57	60	63	66	68	71
177.8		20	22	25	28	31	34	37	39	42	45	48	51	53	56	59	62	65	67	70
180.4			22	25	28	31	33	36	39	42	44	47	50	53	55	58	61	64	66	69
182.9			22	25	27	30	33	36	38	41	44	47	49	52	55	57	60	63	66	68
185.4			22	24	27	30	32	35	38	40	43	46	49	51	54	57	59	62	65	67

TABLE 29.3 cont.

Vital capacity in cubic centimeters																									
3300	3400	3500	3600	3700	3800	3900	4000	4100	4200	4300	4400	4500	4600	4700	4800	4900	5000	5100	5200	5300	5400	5500	5600	5700	5800
91	94	97	99	102	106	108	110	113	116	119	121														
90	92	95	98	100	103	106	109	111	114	117	120	122													
88	91	93	96	99	101	104	107	109	112	115	117	120	123												
87	89	92	95	97	100	102	105	108	110	113	115	118	121	123											
85	88	90	93	96	98	101	103	106	108	111	114	116	119	121	124										
84	86	89	91	94	97	99	102	104	107	109	112	114	117	119	122										
83	85	88	90	93	96	99	101	103	105	108	111	113	115	118	120	123									
81	84	86	89	91	95	98	100	103	106	108	110	113	116	118	120	123									
80	82	85	87	90	92	95	97	99	102	104	107	109	111	114	116	119	121	123							
79	81	83	86	88	91	93	95	98	100	102	105	107	110	112	114	117	119	121	124						
78	80	82	85	87	89	92	94	97	99	101	104	106	108	111	113	115	118	120	122						
76	79	81	83	86	88	90	93	95	97	100	102	104	107	109	111	114	116	118	120	123					
75	78	80	82	84	87	89	91	94	96	98	100	103	105	107	109	112	114	116	119	121	123				
74	76	79	81	83	86	88	90	92	94	97	99	101	103	106	108	110	112	115	117	119	121	124			
73	75	78	80	82	84	86	89	91	93	95	98	100	102	104	106	109	111	113	115	117	120	122	124		
72	74	77	79	81	83	85	88	90	92	94	96	98	101	103	105	107	109	111	114	116	118	120	122	125	
71	73	76	78	80	82	84	86	89	91	93	95	97	99	101	103	106	108	110	112	114	117	119	121	123	
70	72	75	77	79	81	83	85	87	89	92	94	96	98	100	102	104	106	108	111	113	115	117	119	121	123
69	71	73	76	78	80	82	84	86	88	90	92	95	97	99	101	103	106	107	109	111	113	116	118	120	122
68	70	72	75	77	79	81	83	85	87	89	91	93	95	97	99	101	104	106	108	110	112	114	116	118	120

Used with permission of Propper Manufacturing Co., Inc., 10-34 44th Drive, Long Island City, New York 11101.

TABLE 29.4 cont.

Vital capacity in cubic centimeters																			
2600	2700	2800	2900	3000	3100	3200	3300	3400	3500	3600	3700	3800	3900	4000	4100	4200	4300	4400	4500
93	96	100	103	107	111	114	118	121											
91	95	98	102	105	109	112	116	119	123										
90	93	97	100	103	107	110	114	117	121										
88	91	95	98	102	105	108	112	115	118	122									
87	90	93	97	100	103	107	110	113	117	120									
85	88	92	95	98	102	105	108	112	115	118	121								
84	87	90	94	97	100	103	106	110	113	116	119	122							
83	86	89	92	95	98	102	105	108	111	114	118	121							
81	84	88	91	94	97	100	103	106	109	112	116	119	122						
80	83	86	89	92	95	98	101	105	108	111	114	117	120						
79	82	85	88	91	94	97	100	103	106	109	112	115	118	121					
78	81	84	87	90	93	96	98	101	104	107	110	113	116	119	122				
77	79	82	85	88	91	94	97	100	103	106	109	112	115	118	120	123			
75	78	81	84	87	90	93	96	98	101	104	107	110	113	116	119	121			
74	77	80	83	85	88	91	94	97	100	102	105	108	111	114	117	120	123		
73	76	79	82	84	87	90	93	96	98	101	104	107	110	112	115	118	121		
72	75	78	80	83	86	89	91	94	97	100	102	105	108	111	114	116	119	122	
71	74	77	79	82	85	88	90	93	96	98	101	104	107	109	112	115	118	120	
70	73	76	78	81	84	86	89	92	94	97	100	102	105	108	111	113	116	119	121
69	72	75	77	80	83	85	88	90	93	96	99	101	104	106	109	112	114	117	120

Used with permission of Propper Manufacturing Co., Inc., 10-34 44th Drive, Long Island City, New York 11101.

3. Determine the volume change by reading the values on the horizontal scale that the pointer now indicates.
4. Record the average of three trials of the subject's vital capacity, making sure the pointer is reset to zero before each reading. Proceed to the section titled Calculating Vital Capacity.

EXPERIMENTAL METHODS

Jones Pulmonor Spirometry

Materials

Jones Pulmonor waterless spirometer

disposable spirometer mouthpieces

nose clip

recording paper

clinical scale (weight and height)

Preparing the Spirometer

Obtain a Jones Pulmonor waterless spirometer (figure 29.4) from the laboratory instructor. The laboratory instructor will explain the correct procedures for the use and care of the spirometer. Do not use the spirometer until instructions have been given and they are thoroughly understood. Always use a new disposable mouthpiece for each subject and disinfect the spirometer before allowing another subject to use it. Disinfectant procedures are outlined in appendix E.

Always perform the following steps:

1. Turn off the timer at the end of recording.
2. Remove the stylus from its writing position before removing the timer plate and before inserting the timer plate.
3. Check to ensure the proper placement of spirograph paper on the timer plate before recording.
4. Check to make sure the stylus is in its writing position before recording data from the subject.

To check that the spirometer has been properly calibrated for recording, push the T-bar toward the spirometer cabinet until it stops, and then release it. The stylus should return exactly to the zero baseline on the spirogram. If it does not, ask the laboratory instructor to adjust the spirometer. To disengage the T-bar from the zero stylus position, ask the subject to sharply exhale into the spirometer hose.

Recording Tidal Volume and Reserve Volumes

1. Adjust the timer plate so that the stylus is at the beginning of the spirograph paper. Adjust the stylus point to the recording position.
2. Disengage the T-bar from the zero stylus position by having the subject sharply exhale into the spirometer hose.
3. Pull the T-bar out until the stylus is centered on the spirograph paper and hold it there manually.
4. Instruct the nose-clamped subject to stand next to the spirometer and inspire normally (tidal inspiration) from the atmosphere, then to expire normally (tidal expiration) into the spirometer, and then to continue thereafter to inspire from and expire into the spirometer. As the subject expires into the spirometer, release the T-bar. Note the direction of pen movement as the subject inspires and expires. Inspiration is recorded in a downward position on the spirogram, and expiration in an upward position.
5. Turn on the timer and record tidal respiration for about 7 seconds.
6. With the timer plate moving, ask the subject to inspire normally and immediately expire maximally and then resume normal breathing. Turn off the timer.
7. As the subject breathes normally into and out of the spirometer, turn on the timer and ask the subject to expire normally, immediately inspire maximally, and then resume normal breathing.
8. Turn off the timer, remove the timer plate after disengaging the stylus from its writing position, and compare the record with figure 29.5.
9. From the spirogram, measure tidal volume, expiratory capacity, inspiratory capacity, expiratory reserve volume, and inspiratory reserve volume. Use figure 29.5 as a guide in determining volumes and capacities.
10. Repeat the test twice more and record the data and averages of three trials in the report.

Recording Single-Stage Vital Capacity

1. Use one of the previously recorded spirograms to record single-stage vital capacity.
2. Insert the timer plate with attached spirogram into the timer box.
3. Set the recording stylus to the zero baseline by pushing the T-bar toward the spirometer until it stops and then rebase it.
4. Adjust the timer plate so that the stylus is at the 3-second line. (Each vertical line on the spirograph represents 1 second.)
5. Instruct the subject to maximally inhale laboratory air and immediately exhale as completely as possible into the spirometer. Because the timer plate is not moving, vital capacity will be recorded as a straight line on the graduated spirogram paper.
6. Repeat the determination of vital capacity twice more and record the data and average in the report. With each determination, set the stylus to the zero baseline and manually move the timer plate a distance equal to 1 second. Proceed to the section titled Calculating Vital Capacity.

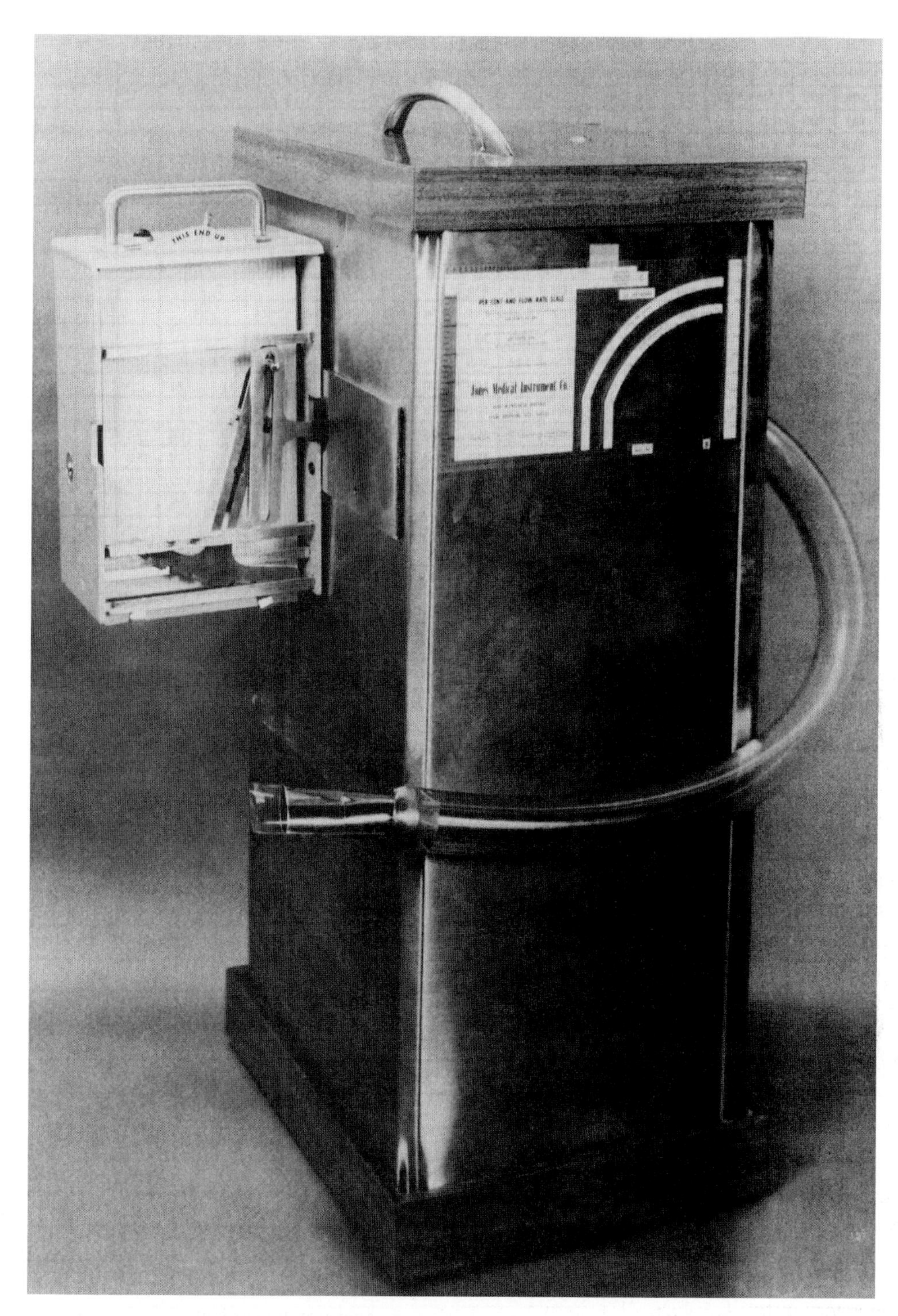

FIGURE **29.4** Jones Pulmonor waterless spirometer

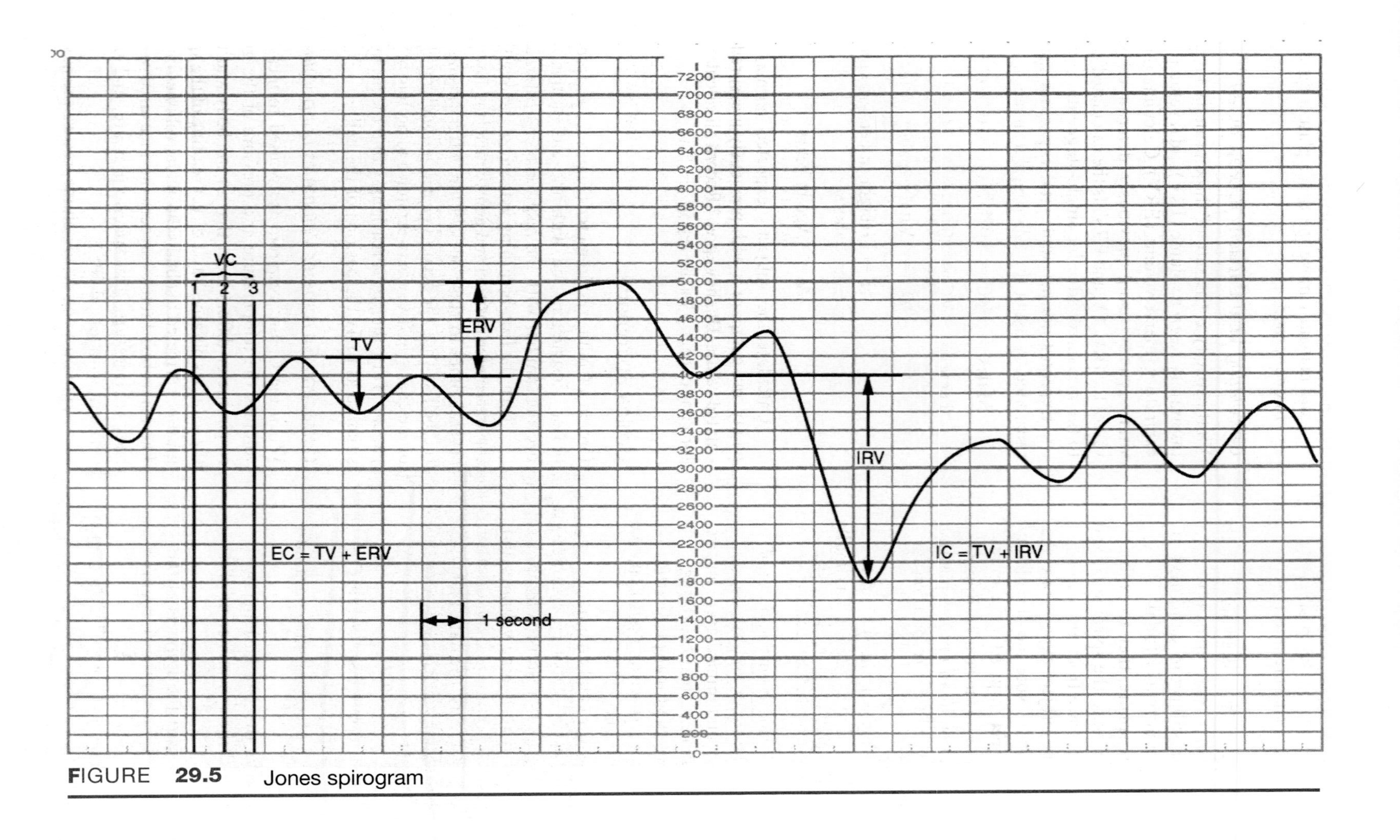

FIGURE 29.5 Jones spirogram

EXPERIMENTAL METHODS

Collins Spirometry

Materials

Collins respirometer (9 L or 13.5 L)
disposable spirometer mouthpieces
nose clip
recording paper (if required for spirometer)
clinical scale (weight and height)

Preparing the Respirometer

Obtain a Collins respirometer (figure 29.6) from the laboratory instructor, and with the aid of the instructor become familiar with the respirometer's external parts.

Turn the free-breathing valve near the mouthpiece to the "spirometer" position and gently raise and lower the bell several times to rinse out old air. After rinsing, position the bell so that the recording pen is touching the center of the chart (halfway between the top and the bottom of the chart paper). Turn the free-breathing valve to the "bypass" position so as to fix the position of the bell.

Correcting Spirometer Volume to Lung Volume

When the Collins respirometer is used to measure lung volumes, the temperature and pressure of the gas in the respirometer differ from the temperature and pressure of the air in the body, and therefore the volumes obtained by a direct reading of the spirogram must be corrected to obtain the actual volumes in the lung. The correction factor corrects the measured volume to body temperature, atmospheric pressure, and saturation with water vapor (Body Temp Pressure Saturation).

To obtain a corrected volume, record the gas temperature of the respirometer, obtain the appropriate correction factor from table 29.5, and multiply the factor times the recorded volume from the spirogram. For example, if the spirometer temperature is 25° C and measured tidal volume on the spirogram is 500 mL, the actual pulmonary tidal volume is 500 mL × 1.075 = 537.5 mL.

Recording Tidal Volume and Reserve Volumes

1. Ask the subject to install a new disposable mouthpiece on the spirometer hose and to wear a nose clamp so as to prevent nasal breathing. The subject should stand comfortably next to the spirometer, and the position of the spirometer hose should be adjusted for the subject.
2. Set the free-breathing valve to the open, or bypass, position and instruct the subject to begin breathing through the mouthpiece.
3. Turn on the respirometer drive to the medium speed position (160 mm/min), turn the free-breathing valve to the closed position, and allow the subject to breathe

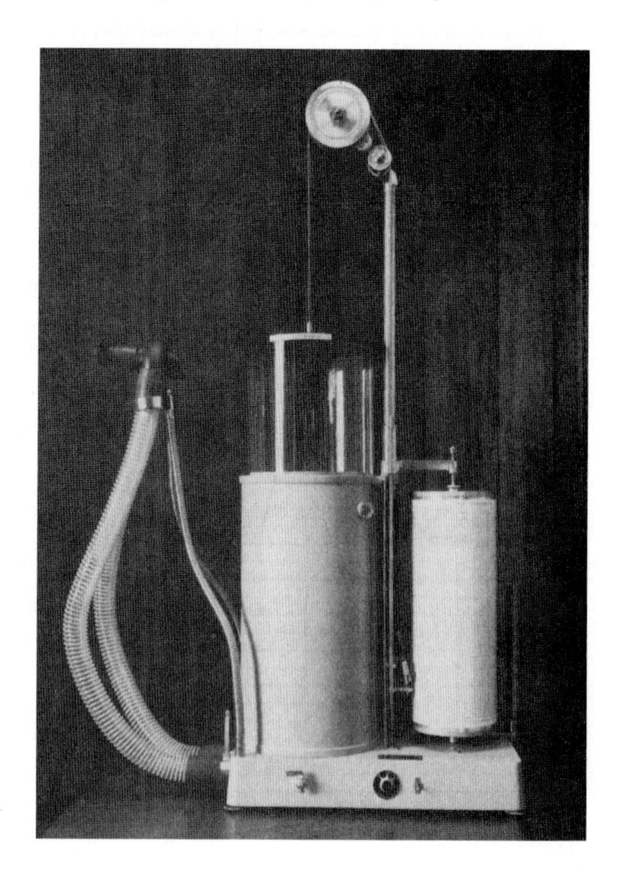

FIGURE 29.6 Collins respirometers

TABLE 29.5 Factors to convert gas volumes from room temperature, saturated, to 37° C, saturated

Factor to convert volume to 37° C, saturated	***When gas temperature °C is***	***With water vapor pressure (mm Hg)* of***
1.102	20	17.5
1.096	21	18.7
1.091	22	19.8
1.085	23	21.1
1.080	24	22.4
1.075	25	23.8
1.068	26	25.2
1.063	27	26.7
1.057	28	28.3
1.051	29	30.0
1.045	30	31.8
1.039	31	33.7
1.032	32	35.7
1.026	33	37.7
1.020	34	39.9
1.014	35	42.2
1.007	36	44.6
1.000	37	47.0

*H_2O vapor pressures from *Handbook of Chemistry and Physics* (28th ed., Cleveland: Chemical Rubber Publishing Co., 1944), p. 1802.

Note: These factors have been calculated for barometric pressure of 760 mm Hg. Since factors at 22° C, for example, are 1.0904, 1.0910, and 1.0915, respectively, at barometric pressures 770, 760, and 750 mm Hg, it is unnecessary to correct for small deviations from standard barometric pressure.

normally and in a relaxed manner for about a minute. Note the direction of recording pen movement as the subject inspires from and expires into the spirometer bell. The volume change that occurs during one respiratory cycle is the tidal volume.

4. Ask the subject to inspire normally and immediately expire maximally and then resume normal resting respiration. The volume change recorded from the end of a normal inspiration to the end of a maximal expiration is known as expiratory capacity. Subtracting tidal volume from expiratory capacity yields expiratory reserve volume (figure 29.7).
5. As the subject continues to breathe normally, ask the subject to expire normally, immediately inspire maximally, and then resume normal resting respiration. The volume change recorded from the end of a normal expiration to the end of a maximal inspiration is known as inspiratory capacity. Subtracting tidal volume from inspiratory capacity yields inspiratory reserve volume (figure 29.7).
6. Turn off the respirometer drive. Turn the free-breathing valve to the bypass (open) position and allow the subject to breathe laboratory air.
7. Repeat steps 3, 4, 5, and 6 twice more and record the data and averages of the three trials in the report.
8. Instruct the subject to remove the mouthpiece from the mouth and breathe normally.

Recording Single-Stage Vital Capacity

1. Turn the free-breathing valve to the closed position.
2. Ask the subject to inspire as deeply as possible from the laboratory air and then to maximally and forcefully exhale into the spirometer. The recorded volume change represents the subject's single-stage vital capacity.
3. Repeat the determination of vital capacity twice more and record the data and average of the three trials in the report. With each determination, manually rotate the recording drum about 10 mm so as to properly record each volume change.

■ CALCULATING VITAL CAPACITY

Vital capacity (VC) is the sum of three primary lung volumes: inspiratory reserve volume (IRV), tidal volume (TV), and expiratory reserve volume (ERV). Calculate the subject's vital capacity using previously determined spirogram values for IRV, TV, and ERV. Is the calculated value equal to the measured value for single-stage vital capacity?

Expected normal adult vital capacities may also be calculated by using the following formulas:

Males $\mathrm{VC} = 0.052H - 0.022A - 3.60$

Females $\mathrm{VC} = 0.041H - 0.018A - 2.69$

where VC = vital capacity in liters

H = height in centimeters (without shoes)

A = age in years

Calculate the subject's expected normal vital capacity and compare it with previously measured and calculated values. Do they agree? Record all calculated values for vital capacity in the report.

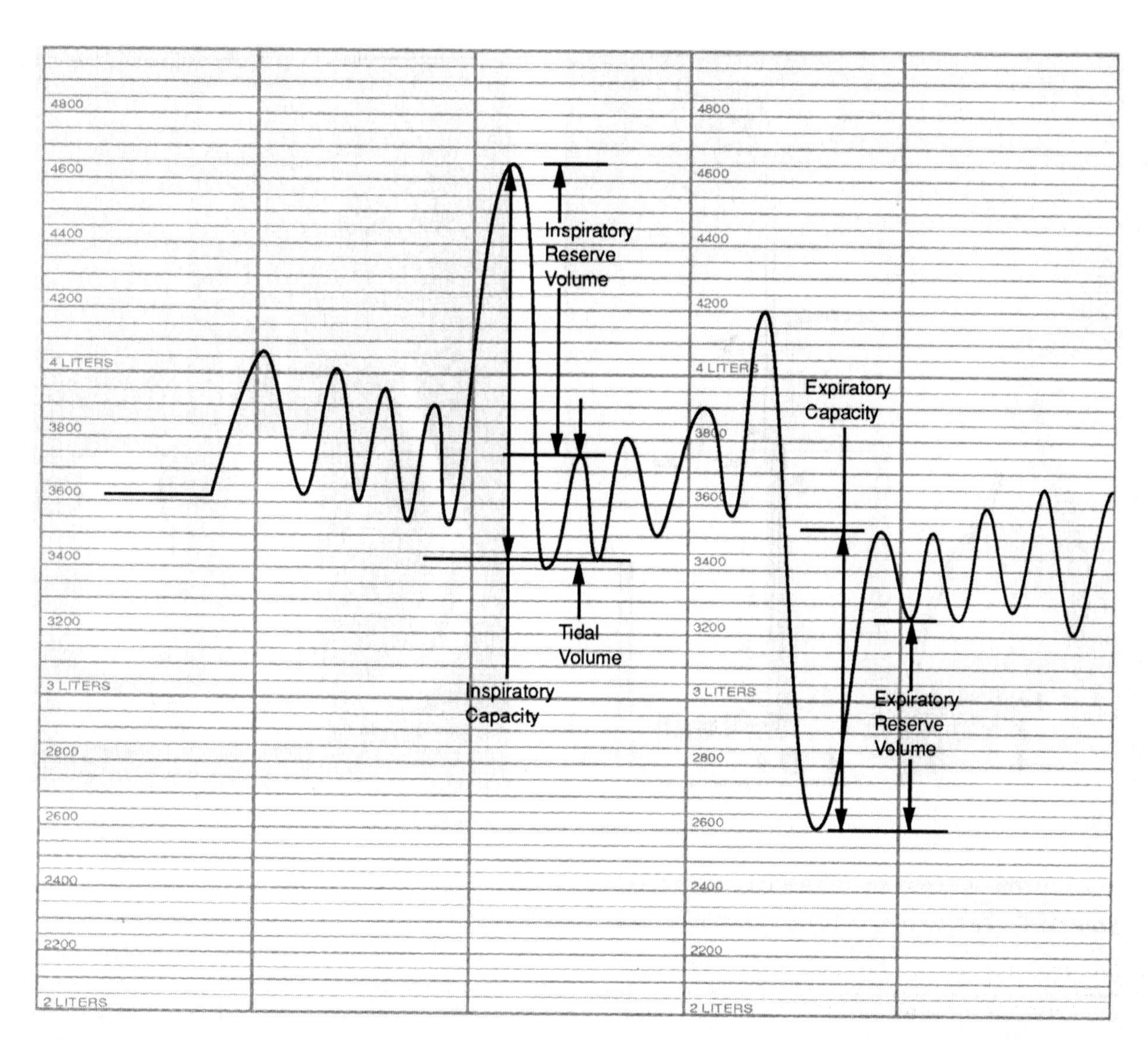

FIGURE 29.7 Collins spirogram

REPORT 29

Pulmonary Function Tests: Volumes and Capacities

Name: ______________________ Date: ______________________

Lab Section: ______________________

1. Data:

a. Subject's initials ______ Age _____ Gender ______

Height _____ Weight _____

b. Pulmonary volumes and capacities:

Trial	*TV (mL)*	*IRV (mL)*	*ERV (mL)*	*IC (mL)*	*EC (mL)*	*VC (mL)*
1						
2						
3						
Average						

c. Calculation of vital capacity:

Male: VC = $0.052H - 0.022A - 3.60$ = _____ – _____ – 3.60 = _____ liters

Female: VC = $0.041H - 0.018A - 2.69$ = _____ – _____ – 2.69 = _____ liters

2. Explain the difference between a pulmonary volume and a pulmonary capacity. ______________________

3. Vital capacity and expiratory reserve volume tend to decrease with age, but functional residual capacity normally remains constant (for ages 20–65). Why? ______________________

4. Explain the difference between obstructive pulmonary disease and restrictive pulmonary disease.

5. Bronchial asthma is an obstructive pulmonary disease in which vital capacity (single-stage) may be normal. Explain.

____

____

____

6. Adult vital capacity gradually decreases with age. Give two reasons why.

a. ______________________________

____

____

b. ______________________________

____

____

Pulmonary Function Tests: Forced Expiratory Volume and Maximal Voluntary Ventilation

CHAPTER 30

INTRODUCTION

Measures of pulmonary volumes, pulmonary capacities, and pulmonary airflow rates are often used in diagnosing and assessing chronic disease of the respiratory system. In general, chronic pulmonary diseases may be classified into two physiologic categories: chronic obstructive pulmonary disease (COPD) and chronic restrictive pulmonary disease.

In restrictive pulmonary disease, the subject's ability to inflate and deflate the lungs is reduced, and, as a result, some lung volumes and capacities are below normal. For example, in pulmonary fibrosis, such as occurs in coal miner's disease, silicosis, or other chronic diseases of the lung interstitium in which the lungs become less distensible, vital capacity is reduced because of reductions in inspiratory and expiratory reserve volumes.

In obstructive pulmonary disease, such as asthma or emphysema, airflow into and out of the lungs is reduced. In asthma, inflammation of the linings of the airways and heavy mucous secretion reduce airway diameters and increase airway resistance. This results in a wheezing sound characteristic of asthmatic breathing and a reduction in the volume of air flowing per minute into and out of the lungs.

It is not uncommon for a person to have both restrictive and obstructive pulmonary disease at the same time. For example, a person may suffer from emphysema and fibrosis of the lung at the same time, even though each disease may have a different origin and begin at a different time.

Restrictive pulmonary diseases are diagnosed, in part, by determining lung volumes and capacities as outlined in chapter 29. However, obstructive pulmonary diseases usually require measurements of pulmonary flow rates. This chapter presents two tests to measure pulmonary flow rates: (1) forced expiratory volume and (2) maximal voluntary ventilation.

Vital capacity (VC) is the maximum volume of gas that can be expired from the end of a maximal inspiration. In the determination of *single-stage vital capacity,* only the volume of expired air is measured, and the subject can take as long as is necessary to complete the maximal exhalation.

Forced vital capacity (FVC), also called *timed vital capacity,* is determined by requiring the subject to forcibly and rapidly exhale the vital capacity volume into a recording spirometer. Both the volume of the expired air and the time required to exhale 100% of the vital capacity are measured. FVC measurement provides more useful information about pulmonary function than does measurement of single-stage VC because a volume change over time measurement provides information about air flow and factors that affect it. FVC is determined primarily by four factors: (1) strength of the chest and abdominal muscles, (2) airway resistance, (3) lung size, and (4) elastic properties of the lung.

The percentage of forced vital capacity expired at the end of the first, second, or third second of maximal expiratory effort is termed **forced expiratory volume (FEV).** Normal adults are able to expire about 83% of their vital capacity in 1 second (**$FEV_{1.0}$**), 94% of their vital capacity in the second second (**$FEV_{2.0}$**), and 97% of their vital capacity by the end of the third second (**$FEV_{3.0}$**). In obstructive pulmonary diseases, FEV is reduced. A person with asthma may have a normal or near-normal vital capacity as measured in a simple one-stage test so long as the subject can take as long as necessary to maximally inhale and exhale. However, when an asthmatic exhales vital capacity with maximal effort, FEV measurements are all reduced because heavy mucous secretion reduces airway diameter and it takes longer to completely exhale vital capacity against increased airway resistance. Normal FEV values are tabulated in table 30.1.

TABLE 30.1 Forced expiratory volumes (adult)

$FEV_{1.0}$ =	83% of vital capacity in one second
$FEV_{2.0}$ =	94% of vital capacity in second second
$FEV_{3.0}$ =	97% of vital capacity in third second

TABLE **30.2** Predicted maximal voluntary ventilation—males, 1/min (Cournand and Richards)

Age (years)	Body surface area (m^2) 1.40	1.45	1.50	1.55	1.60	1.65	1.70	1.75	1.80	1.85	1.90	1.95	2.00	2.05	2.10
16–19	103	112	116	120	124	128	132	135	139	143	147	151	155	159	163
20–24	105	109	113	116	120	124	128	131	135	139	143	146	150	154	158
25–29	101	105	109	112	116	120	123	126	130	134	138	141	145	148	152
30–34	98	101	105	108	112	115	119	122	126	129	133	136	140	143	147
35–39	94	97	101	104	108	111	114	117	121	124	128	131	134	137	141
40–44	91	94	97	100	103	106	110	113	116	119	123	126	129	132	136
45–49	87	90	93	96	99	102	108	112	115	118	121	124	127	130	130
50–54	83	86	89	92	95	98	101	104	107	110	113	115	118	121	124
55–59	79	82	85	88	91	93	96	99	102	105	108	110	113	116	119
60–64	76	78	81	83	86	89	92	97	100	103	105	108	111	113	113
65–69	72	74	77	79	82	84	87	90	92	95	98	100	103	106	108
70–74	68	70	73	75	78	80	83	85	88	90	93	95	98	100	102
75–79	64	67	69	71	74	76	78	80	83	85	88	90	92	94	97

TABLE **30.3** Predicted maximal voluntary ventilation—females, 1/min (Cournand and Richards)

Age (years)	Body surface area (m^2) 1.40	1.45	1.50	1.55	1.60	1.65	1.70	1.75	1.80	1.85	1.90	1.95	2.00	2.05	2.10
16–19	88	91	94	98	101	104	107	110	113	116	120	123	126	129	132
20–24	85	88	92	95	98	101	104	106	109	112	116	119	122	125	128
25–29	82	85	88	91	94	97	100	102	105	108	111	114	117	120	123
30–34	79	81	84	87	90	93	96	98	101	104	107	109	112	115	118
35–39	75	77	80	83	86	89	91	94	97	99	102	105	108	111	113
40–44	72	74	77	79	82	85	87	90	93	95	98	101	103	106	108
45–49	68	70	73	75	78	81	83	85	88	91	93	96	98	101	103
50–54	65	67	70	72	74	77	79	81	84	86	88	91	93	96	98
55–59	62	64	66	68	71	73	75	77	79	81	84	86	88	91	93
60–64	59	61	63	65	67	69	71	73	75	77	79	81	84	86	88
65–69	55	57	59	61	63	65	67	69	71	73	75	77	79	81	83
70–74	52	54	56	58	59	61	63	65	67	69	71	72	74	76	78
75–79	48	50	52	54	55	57	59	61	62	64	66	68	69	71	73

Maximal voluntary ventilation (MVV), also known as maximal breathing capacity (MBC), measures both volume and flow rates and is used to assess overall function with respect to pulmonary ventilation. In performing this test, the subject inspires and expires as deeply and as rapidly as possible while the tidal volume and the respiratory rate are measured. The product of the average volume per respiratory cycle (liters) times the number of cycles per minute equals MVV (L/min). Normal values vary with gender, age, and body size, as shown in tables 30.2 and 30.3. MVV tends to be reduced in both restrictive and obstructive pulmonary diseases.

EXPERIMENTAL OBJECTIVES

1. To observe experimentally, record, and/or calculate forced expiratory volume (FEV) and maximal voluntary ventilation (MVV).
2. To compare observed values of FEV and MVV with predicted normals.
3. To compare normal values of pulmonary flow rates of subjects differing in gender, age, and body surface area.

EXPERIMENTAL METHODS

Jones Pulmonor Spirometry

Materials

Jones Pulmonor waterless spirometer
disposable spirometer mouthpieces
nose clip
recording paper
clinical laboratory scale

Determining Body Surface Area

Using the clinical laboratory scale, record the subject's body weight and height (without shoes). Subtract 1 kg from measured body weight to account for clothing, jewelry, and other accessories (glasses, barrettes, etc.).

Using the body surface area nomogram based on height and weight (figure 30.1), determine the subject's body surface area in square meters. Record the subject's body weight, height, and surface area in the report.

Experimental Procedure

Forced Expiratory Volume

Obtain a Jones Pulmonor waterless spirometer from the laboratory instructor, who will explain the correct procedures for its use and care. Do not use the spirometer until instructions have been given and they are thoroughly understood. Always use a new disposable mouthpiece for each subject and disinfect the spirometer before allowing another subject to use it. Disinfectant procedures are outlined in appendix E. The following procedure is used to measure forced expiratory volume:

1. Set the stylus to the zero baseline by pushing the T-bar toward the spirometer until it stops and then releasing it.
2. Change the spirograph paper on the timer plate and adjust the timer plate so that the stylus is at the beginning of the spirograph paper.
3. Secure the nose clip on the subject and instruct the subject to inspire maximally from the atmosphere and then, as maximally and as forcibly as possible, to expire into the spirometer.
4. As the subject begins to place his or her mouth over the mouthpiece, turn on the timer.
5. As the subject exhales into the spirometer, the stylus will move. When the stylus shows no further increase in volume, turn the timer off.
6. Disengage the stylus from its writing position, and remove the timer plate for measurement of FEV.

With the aid of figure 30.2, determine forced expiratory volumes using Method A or Method B.

Method A

1. Examine the left margin of the Jones percentage and flow rate scale and find a percentage scale (0–100), at the bottom of which is an arrow and the letter A.
2. Position the scale so that the letter A is located at the beginning of the expiratory curve on the zero baseline of the spirogram and the 100% mark intersects the horizontal line, indicating the maximal volume on the expiratory curve.
3. Each vertical line on the spirogram denotes a 1-second time interval. To determine the percentage of vital capacity expired in 1 second ($FEV_{1.0}$), draw a vertical line 1 second after point A until the line intersects with the expiratory curve. At the point where it intersects, draw a horizontal line to the scale and read the percentage directly from the scale at that point. Compare the values obtained with the normal values tabulated in table 30.1.
4. Similarly, determine the percentage of vital capacity expired in 2 seconds ($FEV_{2.0}$) and in 3 seconds ($FEV_{3.0}$).
5. Record these data in the report.

Method B

1. Determine vital capacity (mL) by subtracting the starting point volume from the maximal volume expired. Use the volume scale in the center of the spirogram.
2. Determine the volume expired at the end of 1 second by subtracting the starting point volume from the volume expired at the end of 1 second.
3. Divide the volume obtained in step 2 by the volume obtained in step 1 and multiply the answer by 100 to obtain $FEV_{1.0}$.
4. Repeat step 2 and step 3 for volumes expired at the 2-second and 3-second intervals to obtain $FEV_{2.0}$ and $FEV_{3.0}$, respectively.
5. Compare the values obtained with normal values tabulated in table 30.1.
6. Record your data in the report.

Maximal Voluntary Ventilation (MVV)

The following procedure is used to record maximal voluntary ventilation:

1. Adjust the timer plate so that the stylus is at the beginning of the spirograph paper.
2. Secure the nose clip to the subject and gently hold the T-bar at about the 2400 cc position.
3. Instruct the subject to inhale maximally from the atmosphere, then to exhale maximally into the spirometer, and to continue to inhale from and exhale into the spirometer as deeply and as rapidly as possible.

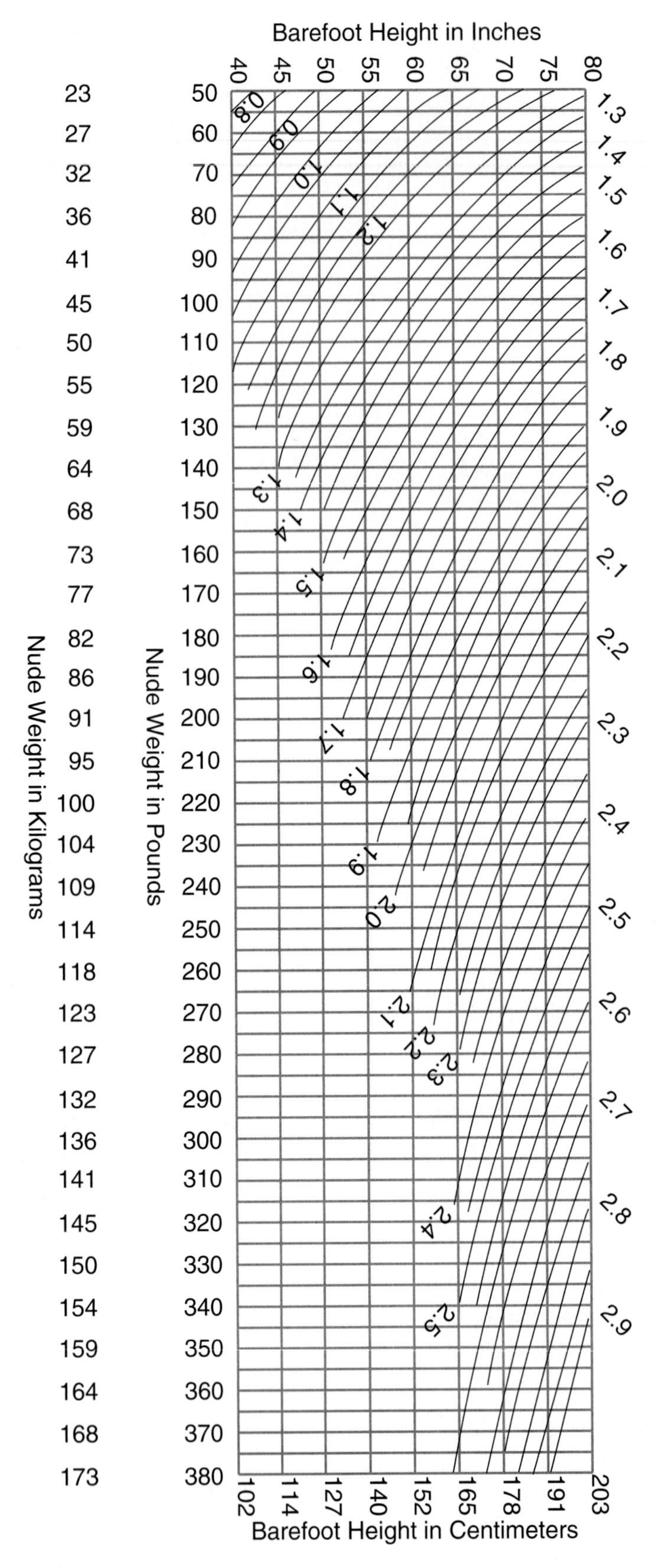

FIGURE **30.1** Body surface area nomogram

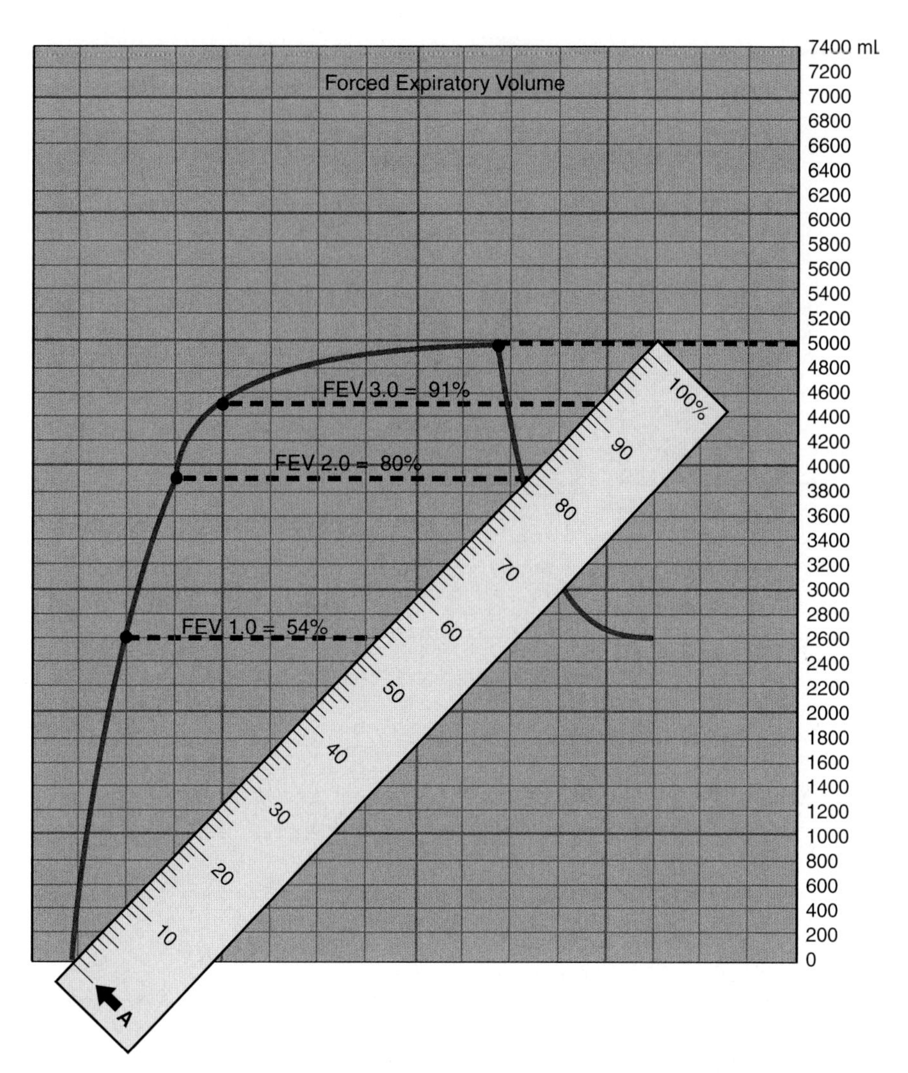

FIGURE 30.2 Measurement of FEV: Jones spirogram

4. Turn on the timer and record the ventilation for 15 seconds (from the end to the middle of the spirograph paper).

With the aid of figure 30.3, compute the maximal voluntary ventilation (MVV) in liters per minute as follows:

1. Draw a horizontal line across the peaks of the inspiratory and expiratory waveforms. From the volume scale in the center of the spirogram, determine the average volume inspired and expired during each respiratory cycle.
2. Count the number of respiratory cycles in a 12-second interval (each vertical line represents a 1-second interval) by counting the peaks of each cycle.
3. Compute the maximal voluntary ventilation by the following formula:

$$\text{MVV} = \frac{(\text{RR})\,(\text{AVPC})}{1000} = \text{liters per minute, where}$$

MVV = maximal voluntary ventilation (liters per minute)

RR = respiratory rate (cycles per minute)

AVPC = average volume per cycle (milliliters)

4. Compare the calculated value with the predicted normal values indicated in table 30.2 or 30.3.
5. Record these data in the report.

EXPERIMENTAL METHODS

Collins Spirometry

Materials

Collins respirometer (9 L or 13.5 L)
disposable spirometer mouthpieces
nose clip
recording paper
clinical laboratory scale

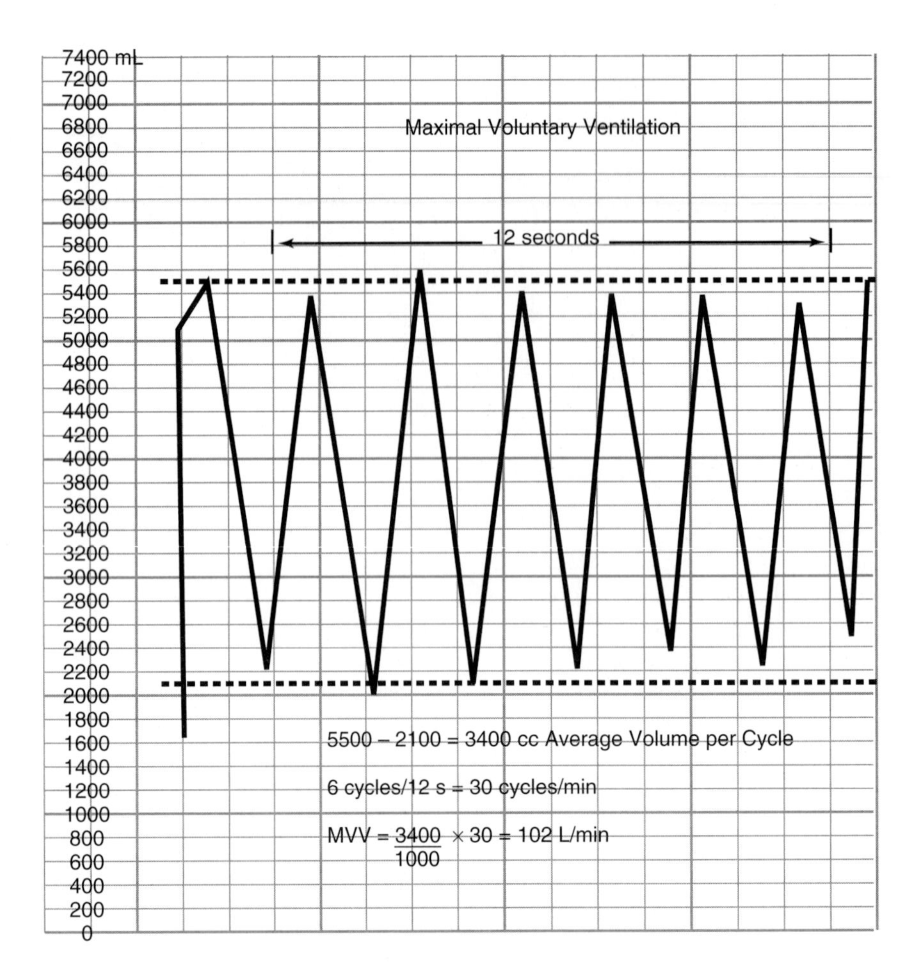

FIGURE 30.3 Measurement of MVV: Jones spirogram

Determining Body Surface Area

Using the clinical laboratory scale, record the subject's body weight and height (without shoes). Subtract 1 kg from measured body weight to account for clothing, jewelry, and other accessories (glasses, barrettes, etc.).

Using the body surface area nomogram based on height and weight (figure 30.1), determine the subject's body surface area in square meters. Record the subject's body weight, height, and surface area in the report.

Experimental Procedure

Forced Expiratory Volume (FEV)

Obtain a Collins respirometer from the laboratory instructor and, with the aid of the instructor, become familiar with the respirometer's external parts. Always use a new disposable mouthpiece for each subject and completely empty spirometer air before allowing another subject to use it. The forced expiratory volume test (timed vital capacity) measures the percentage of vital capacity exhaled in 1 second ($FEV_{1.0}$), 2 seconds ($FEV_{2.0}$), and 3 seconds ($FEV_{3.0}$).

Here are the directions for performing the test:

1. Turn the bypass valve on the mouthpiece to the closed position and adjust the height of the spirometer bell so that the recording pen rests on the 4-L line of the recording paper.
2. Open the bypass valve so that the bell maintains position.
3. Adjust the drum speed to maximum (1920 mm/min). At this speed, each vertical line on the recording paper represents 1 second. Do not turn on the drum.
4. Manually rotate the drum until the recording pen rests on one of the vertical lines.
5. Instruct the subject to stand comfortably near the spirometer and adjust the mouthpiece height for comfortable insertion between the subject's lips.
6. Attach a nose clip to the subject and instruct the subject to place one hand on the mouthpiece and the other hand on the drum on/off switch.
7. Turn the bypass valve to the closed (spirometer) position.
8. Instruct the subject to take two or three deep breaths from laboratory air. The subject then maximally

inspires, places the mouthpiece between the lips, and simultaneously turns on the drum and exhales as rapidly, forcibly, and completely as possible into the spirometer.

9. Turn off the drum, readjust the bell position, open the bypass valve, and remove the spirogram.
10. The FEV curve should resemble that shown in figure 30.4.
11. Note that the curve is recorded downward and to the left. Determine vital capacity by subtracting the volume at the lowest part of the curve from the volume at the beginning of the curve. If a 9-L respirometer is used, the volumes can be read directly from the spirogram. If a 13.5-L spirometer is used, the volume values must be doubled. In figure 30.4, vital capacity = (4000 mL – 1900 mL) × 2 = 4200 mL.
12. Determine the expired volume at the 1-second, 2-second, and 3-second intervals. Divide each of these volumes by the vital capacity and multiply each by 100 to obtain the percentage of vital capacity expired at the end of 1, 2, or 3 seconds ($FEV_{1.0}$, $FEV_{2.0}$, $FEV_{3.0}$). For example, in figure 30.4, $FEV_{2.0}$ = [(4000 mL – 2150 mL) × 2] ÷ 4200 × 100 = 88%. The normal value as indicated in table 30.1 is 94%. (The subject was a smoker.)
13. Calculate $FEV_{1.0}$, $FEV_{2.0}$, and $FEV_{3.0}$ from the recorded data and enter the results in the report. Remember to double volume values if using a 13.5-L respirometer.

TABLE 30.4 Factors to convert gas volumes from room temperature, saturated, to 37° C, saturated

Factor to convert volume to 37° C, saturated	*When gas temperature °C is*	*With water vapor pressure (mm Hg)* of*
1.102	20	17.5
1.096	21	18.7
1.091	22	19.8
1.085	23	21.1
1.080	24	22.4
1.075	25	23.8
1.068	26	25.2
1.063	27	26.7
1.057	28	28.3
1.051	29	30.0
1.045	30	31.8
1.039	31	33.7
1.032	32	35.7
1.026	33	37.7
1.020	34	39.9
1.014	35	42.2
1.007	36	44.6
1.000	37	47.0

*H_2O vapor pressures from *Handbook of Chemistry and Physics* (28th ed., Cleveland: Chemical Rubber Publishing Co., 1944), p. 1802.

Note: These factors have been calculated for barometric pressure of 760 mm Hg. Since factors at 22° C, for example, are 1.0904, 1.0910, and 1.0915, respectively, at barometric pressures 770, 760, and 750 mm Hg, it is unnecessary to correct for small deviations from standard barometric pressure.

Maximal Voluntary Ventilation

Because the temperature and pressure of the gas in the respirometer differ from that in the body, the volumes obtained by a direct reading of the spirogram must be corrected to obtain the actual volumes in the lungs. The correction factor corrects the measured volume to body temperature, atmospheric pressure, and saturation with water vapor. To obtain a corrected volume, record the temperature of the respirometer, obtain the appropriate correction factor from table 30.4, and multiply the factor times the recorded volume from the spirogram.

To record maximal voluntary ventilation, proceed as follows:

1. Turn the bypass valve on the mouthpiece to the closed position and adjust the height of the spirometer bell so that the pen rests approximately in the middle of the drum.
2. Open the bypass valve so that the bell maintains position.
3. Adjust the drum speed to intermediate (160 mm/min). At this speed, each vertical line on the recording paper represents 12 seconds. Do not turn on the drum.
4. Instruct the subject to stand comfortably near the spirometer and adjust the mouthpiece height for comfortable insertion between the subject's lips.
5. Attach a nose clip to the subject.
6. Turn the bypass valve to the closed position and turn on the drum.
7. Instruct the subject to inhale and exhale out of and into the spirometer as rapidly and as deeply as possible for about 15–20 seconds.
8. Turn off the drum, open the bypass valve, and remove the spirogram. The spirogram should resemble figure 30.5.
9. Record the temperature of the spirometer and use table 30.4 to determine the gas volume correction factor.
10. From the MVV spirogram, determine the average volume moved per respiratory cycle by drawing a line across the peaks of the inspiratory and expiratory waveforms. Correct the average volume to BTPS. Remember to double volumes if using a 13.5-L spirometer.
11. Count the number of respiratory cycles that occurred in the 12-second interval and multiply this by 5 to obtain the number of respirations per minute.

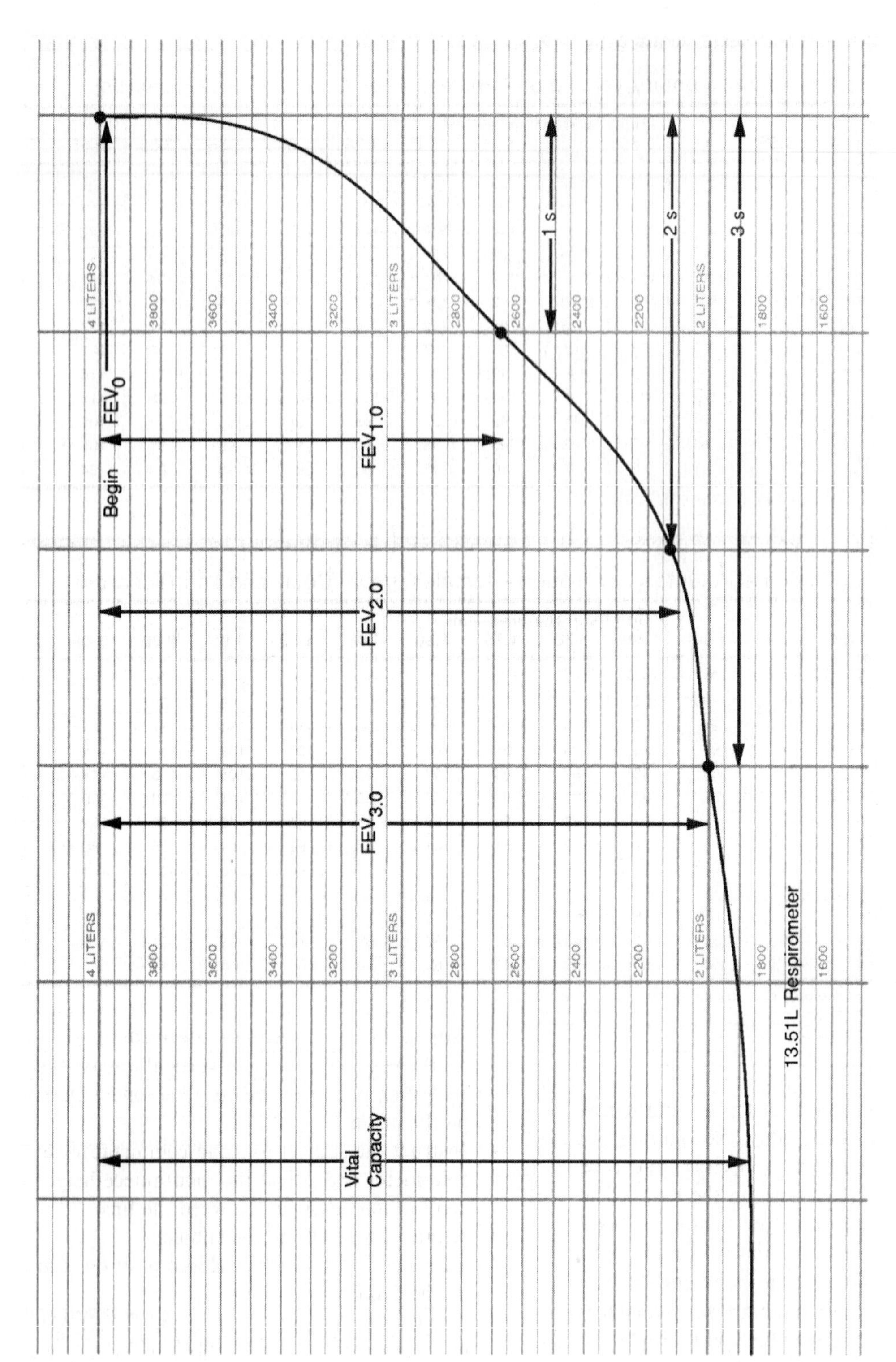

FIGURE 30.4 Measurement of FEV: Collins spirogram

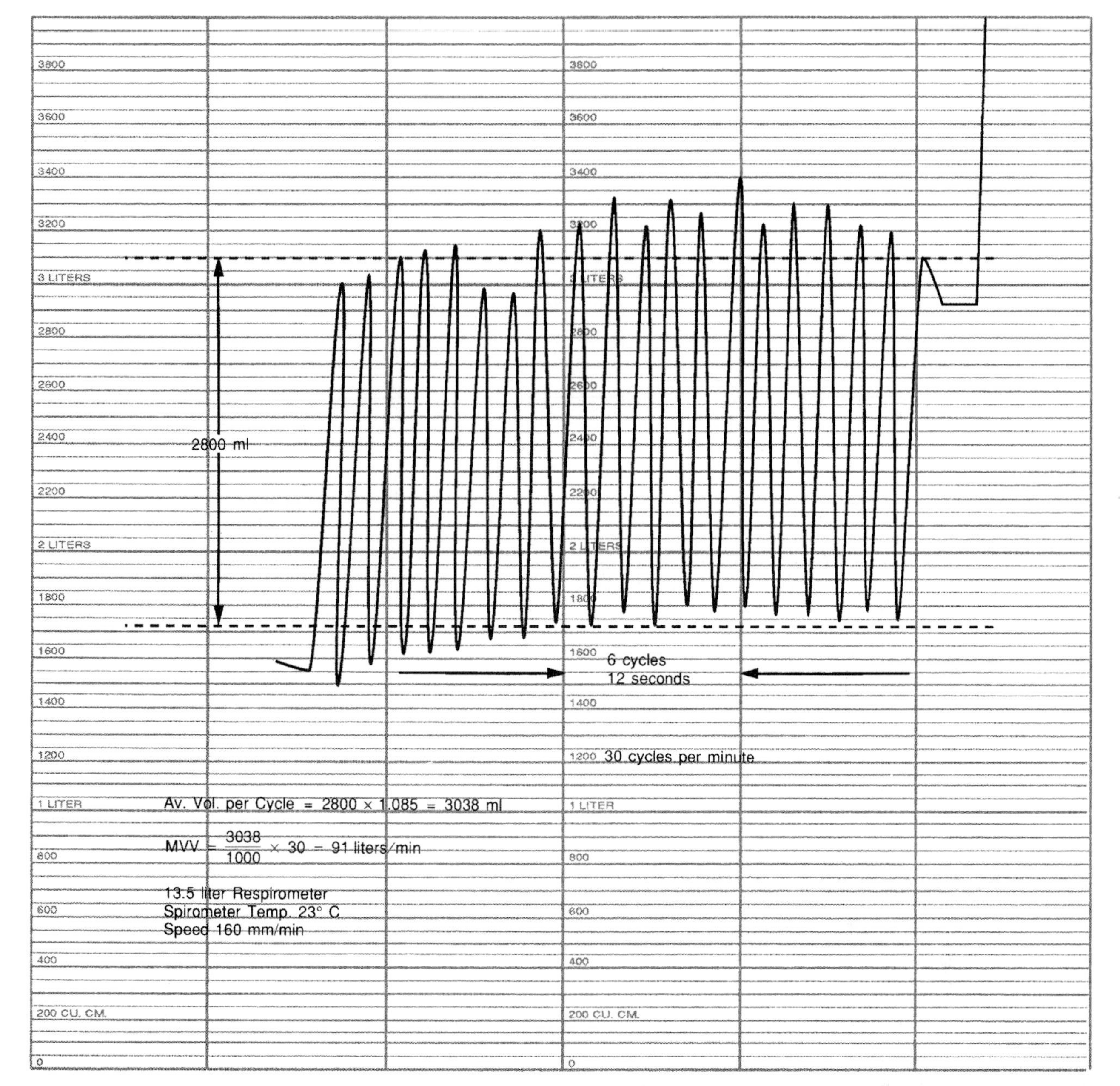

FIGURE 30.5 Measurement of MMV: Collins spirogram

12. Multiply the respiratory rate (cycles per minute) times the average volume moved per cycle (milliliters/cycle) and divide by 1000 to obtain MVV in liters per minute.

$$\text{MVV} = \frac{(\text{RR})\ (\text{AVPC})}{1000} = \text{liters per minute, where}$$

MVV = maximal voluntary ventilation (liters per minute)

RR = respiratory rate (cycles per minute)

AVPC = average volume per cycle (milliliters)

13. Record the calculated MVV in the report and compare it with normal values in table 30.2 or 30.3.

Pulmonary Function Tests: Forced Expiratory Volume and Maximal Voluntary Ventilation

REPORT 30

Name: ______________________ Date: ______________________

Lab Section: ______________________

1. Data:

a. Subject's Initials _____ Age_____ Gender _____

Height _____ Weight _____ Body Surface Area (BSA) ______ m^2

b. Forced expiratory volumes:

Measurement	***Observed***	***Expected normal***
$FEV_{1.0}$	%	%
$FEV_{2.0}$	%	%
$FEV_{3.0}$	%	%
FVC	mL	mL

c. Maximal voluntary ventilation:

Respiratory rate = _____ cycles per 12 seconds × 5 = _____ cycles per minute

Average volume of air moved into or out of lungs per cycle (AVPC) _____ mL

MVV = (RR) _____ cpm (AVPC) _____ mL ÷ 1000 = _____ liters per minute

Predicted MVV based on gender, age, and BSA = _____ liters per minute

2. Is it possible for a subject to have a vital capacity (single-stage) within normal range but a value for $FEV_{1.0}$ below normal range? Why or why not? ______________________

____

____

____

3. Maximal voluntary ventilation decreases with age. Why? ______________________

____

____

____

4. According to tables 30.2 and 30.3, for any given age and body surface area, men have a greater MVV than women. Give one anatomic or physiologic reason why. ______________________________

____

____

____

5. What effect would you expect smoking to have on a subject's FEV and MVV? Explain. ______________________________

____

____

____

CHAPTER 31

Analysis of Urine

■ INTRODUCTION

The kidneys play an essential role in regulating the volume and chemical composition of the plasma, interstitial fluid, and lymph. They are primarily concerned with the regulation of extracellular concentrations of water, electrolytes, and certain end products of metabolism; therefore, they regulate not only the concentrations of chemicals in the extracellular fluids but also the pH and osmotic pressure.

The primary functional unit of the kidneys is the **nephron** (figure 31.1). Each kidney contains approximately 1 million nephrons. Urine is formed in the nephron by the combined processes of glomerular filtration, tubular reabsorption, and tubular secretion. Excretion by the kidney

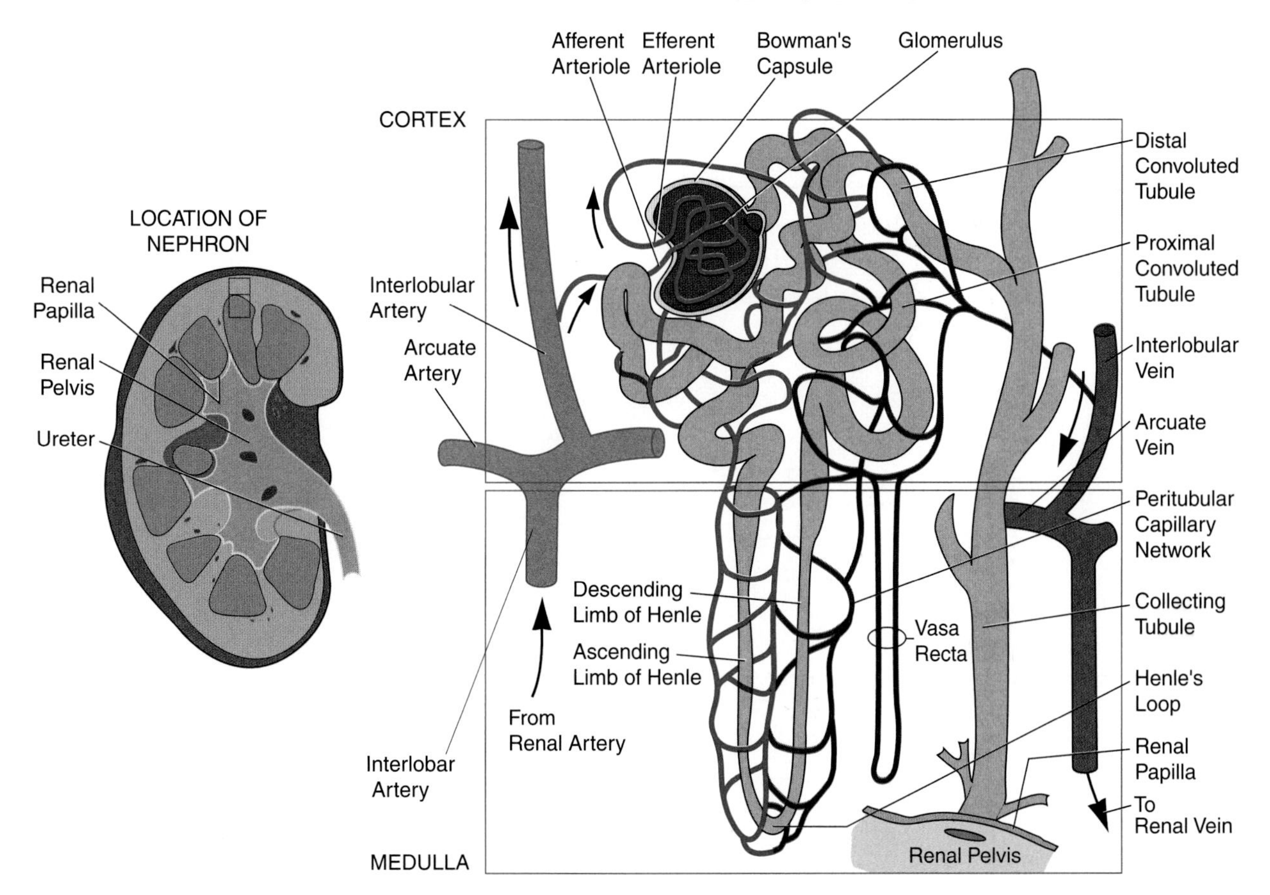

FIGURE 31.1 A corticomedullary nephron

refers to the process of eliminating urine into the *ureter,* a thin-walled tube that conveys urine to the urinary bladder, where urine is stored until it can be voided from the body (micturition). The glomerular filtrate is an *ultrafiltrate* of the plasma. This means that the filtrate contains substances of small molecular weight (glucose, amino acids, urea, water, etc.) in essentially the same concentrations as the plasma, but colloidal material and substances of high molecular weight (plasma proteins, erythrocytes, leukocytes, etc.) are normally absent from the filtrate. As the filtrate passes through the nephron, most of the constituents of the glomerular filtrate are reabsorbed to varying degrees, depending on the kidneys' need to adjust and maintain the chemical concentrations of the extracellular fluids. In addition, some substances (e.g., potassium, hydrogen ions, and drugs) are secreted by tubular cells into the urine. The quantity of urine formed and its chemical composition vary, depending on such factors as exercise, dietary intake, and pregnancy.

A typical 70-kg adult normally forms about 1500 mL of urine per 24-hour period. This 1500 mL of urine normally contains about 60 g of solids. Approximately 30 g of this is *urea* (a nitrogenous waste product), and about 15 g is sodium chloride. The remaining 15 g consists of various organic and inorganic constituents, such as creatinine, amino acids, uric acid, ammonia, sulfates, phosphates, potassium, and magnesium. Under unusual or pathologic conditions, substances normally absent or present in only trace amounts in the urine may be detected. These include proteins, glucose, ketone bodies, bile, hemoglobin, and blood cells. Table 31.1 indicates the chemical composition of normal urine.

TABLE 31.1 Chemical and physical composition of normal urine

Color	Light to dark amber
Quantity	800–2300 mL/24 hours
Odor	Variable, usually ammoniacal
pH	4.8–7.5
Sugar	0.015%
Nitrogen	Total 7–20 g/24 hours
Amino acids	0.15–0.30 g/24 hours
Urea	12–35 g/24 hours
Ammonia	0.6–1 to 2 g/24 hours
Creatinine	0.8–2.0 g/24 hours
Uric acid	0.3–0.8 g/24 hours
Hippuric acid	0.7 g/24 hours
Chlorides	10–15 g/24 hours
Phosphorus	1.2 g/24 hours
Sulfur	1.2 g/24 hours
Sodium	2.5–4.0 g/24 hours
Calcium	0.1–0.3 g/24 hours
Potassium	1.5–2.0 g/24 hours
Magnesium	0.1–0.2 g/24 hours
Hormones, enzymes, vitamins	Variable

What the kidneys contribute to homeostasis cannot be understood by examination of the urine alone. However, examination of the urine *(urinalysis)* is useful when performed in conjunction with other diagnostic methods for assessing renal function and diagnosing various types of pathophysiologies involving not only the kidneys but other organs and tissues as well.

EXPERIMENTAL OBJECTIVES

1. To become familiar with routine laboratory tests for screening urine.
2. To examine urine for the presence of normal and abnormal constituents.
3. To compare urine samples of normal persons as to specific gravity, pH, and chemical constituents.
4. To acquire an appreciation for the value of urinalysis as a diagnostic aid.

Materials

Equipment

compound light microscope
water bath
watch with second hand
test-tube rack
centrifuge

Supplies

clean microscope slides
coverslips
urinometer
forceps
hydrion paper
six disposable test tubes (16 × 25 mm)
wire test-tube grippers
Pasteur pipettes and bulbs
filter paper
small plastic funnel
droppers
Ictotest tablet and mat
Acetest tablets and chart
Hemastix
Albustix and chart
Bili-labstix and chart
Clinistix and chart
Multistix 10 SG
urine specimen jars and lids

Solutions

50 mL of fresh urine (subject's own)

concentrated nitric acid

methylene blue solution

Sulkowitch reagent (2.5 g oxalic acid + 2.5 g ammonium oxalate + 5 mL glacial acetic acid + distilled water to make a total of 150 mL)

Fouchet's reagent (10 g ferric chloride + 100 mL distilled water) + (25 g trichloroacetic acid + 100 mL distilled water)

3% silver nitrate solution

concentrated ammonium hydroxide solution

concentrated hydrochloric acid

10% barium chloride solution

Benedict's reagent

sodium nitroprusside powder

saturated ammonium sulfate solution

■ EXPERIMENTAL METHODS

CAUTION

In this experiment you will obtain and analyze a sample of your own urine. Handle only your own urine. Dispose of all supplies (disposable test tubes, microscope slides, coverslips, etc.) that come into contact with your urine by placing them in properly marked containers. Always treat all body fluids and supplies as infectious. Nondisposable items that come into contact with your urine are to be washed with a detergent and tap water, then rinsed successively with tap water, bleach solution, distilled water, and acetone. Disinfect your laboratory table or laboratory bench top by wiping it clean with a bleach solution before obtaining your urine sample and again at the end of the experiment.

Quantity of Urine

Increase your fluid intake the day before and on the day of urinalysis so as to increase urine output. If possible, avoid emptying the urinary bladder 2–4 hours before obtaining a urine sample.

Obtain a urine specimen jar and lid from the laboratory instructor. Go to the restroom and collect a midstream urine sample in the container. (Void the first part of the urine sample before collecting the urine specimen.) Return to the laboratory bench and mark the specimen jar for identification.

The average adult voids approximately 1500 mL of urine per day. The exact amount varies between 800 and 2300 mL per day, depending on the intake of liquid and food. The elimination of an increased quantity of urine (beyond the normal range) is called *polyuria* and occurs in a number of diseases, such as diabetes mellitus, diabetes insipidus, adrenal diabetes, and chronic nephritis. The elimination of a decreased quantity of urine *(oliguria)* occurs in fever, diarrhea, thermal stress, and other states.

Color, Transparency, and Odor of Urine

Normally, the color of urine is amber (yellow) with the intensity of the color (light to dark) varying with the amount of urine voided. The color is due to the presence of pigments, such as urochrome, urobilin, and hematoporphyrin, which are normally present in urine. The presence of abnormal constituents may change the color drastically. For example, the presence of hemoglobin will give the urine a brown to red color. Freshly voided urine is transparent, but after a few hours it becomes slightly opaque because of the separation and settling of epithelial cells, leukocytes, and mucus. After standing in the air, urine becomes alkaline, owing to the conversion of urea to ammonia, and cloudy, owing to the precipitation of substances such as phosphates. Cloudiness in freshly voided urine may indicate the presence of pus, blood, or bacteria from urinary tract infections.

The odor of urine varies with dietary intake. The excretion of waste products associated with the metabolism of certain foodstuffs (e.g., asparagus, cabbage) or the excretion of drugs by the kidneys often causes the urine to have an offensive odor. The decomposition and evaporation of urine usually results in an ammoniacal odor.

Examine the urine specimen for color, transparency, and odor. Record your observations in the report.

pH of Urine

The chemical reaction of normal urine varies from pH 4.8 to 7.5, depending primarily on dietary intake. Usually the pH is slightly acid (about 6) because the metabolism of the foodstuffs contained in the average diet results in the production of excess acids, which must be eliminated from the body fluids. The acidity of urine also increases in acidosis (metabolic and respiratory) and during fever. An alkaline urine may be produced by letting it stand or by storing it in the urinary bladder. In both cases, the alkaline pH is due to the conversion of urea to ammonia. Other causes of alkaline urine include excessive dietary intake of certain foods (e.g., fruits), the ingestion of alkaline substances (e.g., sodium bicarbonate), and various states of alkalosis (metabolic and respiratory).

Test the chemical reaction (acidity or alkalinity) of the specimen by using hydrion papers. Hydrion paper contains chemical indicators that change color when wetted with solutions of various pH: red to orange colors indicate pH 1–6; green indicates neutral, pH 7; and dark green to blue colors indicate pH 8–11. Compare the color reaction with the color chart on the side of the hydrion paper dispenser. Record the data in the report and on the chalkboard. Compare the reaction of your urine sample with the results reported by others in the class and note the variation.

Specific Gravity of Urine

The specific gravity of a liquid is determined by comparing the weight of a given volume of the liquid with the weight of an identical volume of pure water. Specific gravity is expressed as a simple number: the ratio of two numbers. By definition, the specific gravity of water is 1.000. The normal range for the specific gravity of urine is from 1.010 to 1.030. Higher values for the specific gravity indicate a more concentrated urine, and lower values indicate a more dilute urine.

The specific gravity of urine tends to be low in diabetes insipidus and after excessive quantities of liquid have been taken. It may be high during fever and in various diseases, such as diabetes mellitus and adrenal diabetes.

Adhering to the precautions outlined by the laboratory instructor, determine the specific gravity of the specimen by using the urinometer (figure 31.2). Fill the urinometer cylinder to three-fourths capacity with the urine specimen and insert the float. If the float does not float, add more urine until it does. Read the specific gravity from the stem of the float where the urine sample intersects. Record the specific gravity in the report and on the chalkboard. **Empty the urine back into the specimen jar** and then carefully wash the urinometer with soap and rinse it with distilled water, bleach solution, and acetone. Compare the specific gravity of your urine sample with the results of others in the class and note the variation.

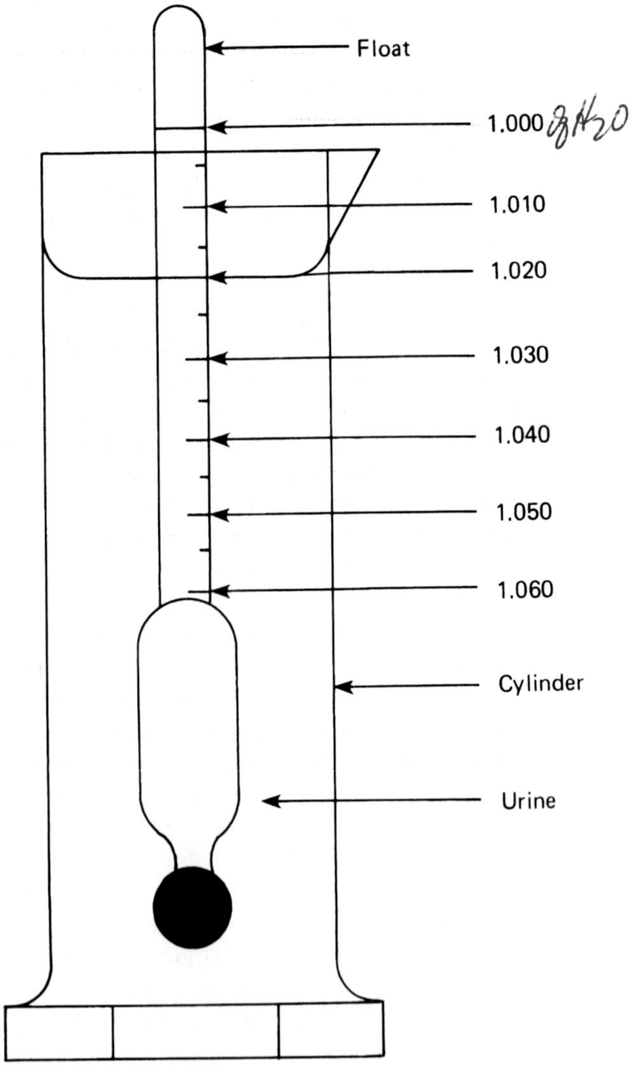

FIGURE 31.2 The urinometer

Estimation of Amount of Urinary Solids

The concentration of urinary solids in grams per liter can be approximated by multiplying the last two digits of the specific gravity by 2.66 (Long's coefficient). For example, if the specific gravity of the sample were 1.022, the approximate concentration of urinary solids would be $22 \times 2.66 = 58.5$ g/L. Estimate the amount of urinary solids in the sample and record the data in the report.

Normal Inorganic Constituents of Urine

1. *Chlorides:* Next to urea, chlorides are the most abundant substances found in normal urine. They are derived primarily from the metabolism of food, and their quantity in the urine fluctuates, depending on dietary intake. The principal chloride in urine is sodium chloride. Fevers and various forms of nephritis tend to decrease the renal output of chloride.

   Test for the presence of chlorides in the sample by pouring approximately 10 mL of urine into a clean test tube. Add 3 drops of nitric acid to prevent the precipitation of phosphates, and then add 3 drops of 3% silver nitrate solution. If chlorides are present, a white, curdy precipitate will form. If the chloride content of the urine is low, the urine will appear only milky. Record your observations in the report.
2. *Phosphates:* Phosphates are derived chiefly from food, although small amounts are produced during cellular metabolism. Large amounts of phosphates are present in the mineral complex of bone; hence, in bone diseases, such as rickets or osteomalacia, there is an increase in the renal excretion of phosphates.

   Test for the presence of phosphates by pouring approximately 10 mL of fresh urine into a clean test tube. Add ammonium hydroxide until the urine is alkaline. If phosphates are present in the urine, precipitates of calcium and magnesium phosphate will form. Record your observations in the report.
3. *Sulfates:* Most of the sulfur present in urine originates from the dietary intake of protein and from the cellular metabolism of protein compounds. Ninety percent of it is present in an inorganic form combined with sodium, potassium, calcium, and magnesium.

   Test for the presence of sulfates by pouring approximately 10 mL of fresh urine into a clean test tube. Add 1 drop of hydrochloric acid and shake well.

Add 3 drops of barium chloride. If sulfates are present in the urine, a white precipitate will form. Record your observations in the report.

4. *Calcium:* Calcium is involved in a number of important physiologic processes in the body, such as blood coagulation, nerve conduction, and muscle contraction. Thus, the extracellular concentration of calcium must be maintained within fairly narrow limits. Approximately 99% of the total body calcium is located in bone. There, in the form of mineral complexes, it helps give bone its rigidity. Total plasma calcium is normally about 10 mg/dL.

   Two primary forms of calcium in plasma are filterable calcium and protein-bound calcium. Filterable calcium consists of a free calcium ion (Ca^{++}) and calcium complexed with an anion such as phosphate. Free calcium is the physiologically active form of calcium.

   Approximately 99% of calcium filtered by the kidneys is reabsorbed, mostly in the proximal tubule and loop of Henle. Normal urinary excretion of calcium is 0.1–0.3 g every 24 hours. An increase in urinary calcium is observed in conditions associated with hypercalcemia (above-normal plasma calcium levels), such as hyperparathyroidism and hypervitaminosis D. A decrease in urinary calcium is observed in conditions associated with hypocalcemia (below-normal plasma calcium levels), such as hypoparathyroidism and hypocalcemic tetany of skeletal muscle.

   Urinary excretion of calcium can be easily and rapidly assessed by Sulkowitch's test, often used to quickly diagnose hypocalcemic tetany.

   Mix equal volumes of urine and Sulkowitch's reagent. A white precipitate of calcium oxalate is formed in a quantity proportionate to the urine calcium present. A fine white precipitate is indicative of a normal blood level; a heavy precipitate suggests hypercalcemia; and the absence of a precipitate indicates hypocalcemia.

Abnormal Constituents of Urine

1. *Glucose:* Normally, all of the glucose filtered out of the glomerulus is reabsorbed by the proximal convoluted tubule. The reabsorption of glucose is an active process requiring energy and involving a carrier-mediated transport system in the tubular cell. The transport system has a maximum capacity (T_m), which is normally not exceeded. However, when the glucose level in the plasma exceeds 180 mg/dL (glucose threshold), the transport capacity is exceeded and glucose begins to appear in the urine *(glucosuria)*. The normal blood level of glucose is 60–120 mg/dL, well below the renal threshold for glucose. When the glucose level exceeds the normal range *(hyperglycemia)*, as occurs in diabetes mellitus, adrenal diabetes, or the excessive intake of sugar, glucosuria will result if the renal threshold is exceeded.

   Glucose may also be excreted even at a normal plasma glucose concentration. Renal glucosuria is a benign, inherited abnormality in which the proximal tubules' transport capacity for glucose is reduced.

   Benedict's Test

   Test for the presence of glucose by adding 5 mL of Benedict's reagent to a clean, dry test tube. Heat the test tube contents to boiling by placing the test tube in a boiling water bath. Now add 8 drops of fresh urine and allow the contents to boil for 5 minutes. Remove the test tube from the water bath and set it aside to cool. If glucose is present in concentrations of 0.05% or higher, the reagent is reduced and a greenish to red-brown precipitate forms, depending on the concentration of glucose. Record your observations in the report.

   Clinistix

   Remove a test strip from the bottle and replace the cap. Dip the test area of the strip in urine and remove it immediately. Draw the edge of the strip against the rim of the urine container to remove excess urine. Compare the test area with the Clinistix color chart exactly 10 seconds after it is moistened. Ignore color changes that occur after 10 seconds. Record your observations in the report.

2. *Albumin* and *globulin:* Under normal circumstances, only a small amount of protein is present in the glomerular filtrate, and most of any filtered protein is reabsorbed. The "albumin" of the urine is actually a mixture of serum albumin and serum globulin. Excess protein in the urine *(proteinuria)* reflects an abnormal leakiness or severe damage of the glomerular membrane or both. Various types of nephrosis and nephritis due to infection, vascular degeneration, and other causes may result in proteinuria.

   Nitric Acid Test

   Test for the presence of protein in the urine by adding 10 mL of fresh urine to a clean test tube. Heat to boiling by placing the tube in a boiling water bath. Now add 5 drops of concentrated nitric acid. If a white precipitate forms on boiling and does not disappear when nitric acid is added, protein is present in the urine. Record your observations in the report.

   Albustix

   Remove a test strip from the bottle and replace the cap. Dip the test area of the strip in urine and remove immediately. Draw the edge of the strip against the rim of the urine container to remove excess urine. Immediately compare the test area with the Albustix color chart. Record your observations in the report.

3. *Ketone bodies:* Ketone bodies, such as acetic acid, acetoacetic acid, and β-hydroxybutyric acid, are normally present only in trace amounts in the urine. However, the excessive metabolism of fats due to a high dietary intake of fat or a dependence of the cells on lipid metabolism to produce energy because of an inadequate glucose uptake will result in the presence of larger amounts of ketone bodies in the urine.

   Rothera's Test

   Test for the presence of ketone bodies in the urine by placing about 1 g of powdered sodium nitroprusside crystals in a test tube. Add 5 mL of urine, then 5 mL of saturated ammonium sulfate. Mix. Carefully "layer" concentrated ammonium hydroxide on the mixture. If a purple ring appears at the junction of the two fluids, acetoacetic acid is present in the urine.

   Acetest

   Test for the presence of acetone bodies by placing an Acetest tablet on a clean laboratory napkin. Put a drop of fresh urine on the tablet. Compare the color change at exactly 30 seconds with the color chart. The test result is negative if the tablet does not change color or turns cream-colored from wetting. The test is positive if the tablet color changes from lavender to purple. Record your observations in the report.

4. *Bilirubin:* Approximately 250 billion erythrocytes are destroyed and replaced each day in a healthy adult human. Aged erythrocytes are destroyed by the reticuloendothelial tissues of the spleen, liver, and bone marrow. In the process, hemoglobin is broken down and bilirubin is formed as an end product of heme metabolism. Bilirubin is released into the plasma, bound to albumin, and transported to the liver. The hepatic cells conjugate bilirubin with glucuronic acid and excrete it in the bile as bilirubin glucuronide. In the large intestine, it is converted to stercobilinogen and urobilinogen by bacteria. Some of the urobilinogen is then reabsorbed and eventually eliminated into the urine by the kidney. The remainder is excreted in the feces. Unconjugated bilirubin is not excreted by the kidney. However, conjugated bilirubin is normally excreted in very small amounts, because there is usually only a small amount in the serum. Bilirubin appears in the urine *(bilirubinuria)* when there is partial or complete obstruction of the extrahepatic biliary ducts, hepatitis, cirrhosis, or other types of destructive liver disease.

   Fouchet's Test

   Test for the presence of bilirubin by mixing 5 mL of urine with 5 mL of 10% barium chloride solution. Filter the mixture. The precipitated bile will remain on the filter paper. Allow the filter paper to dry. Add 2 drops of Fouchet's reagent to the area of the filter paper containing the residue. A green color appears in the presence of bilirubin. Record your observations in the report.

   Ictotest

   Place an Ictotest mat on a paper towel and place 10 drops of urine on the mat. Use forceps to remove an Ictotest tablet (do not touch reagent tablet with your fingers) from the bottle and place the tablet in the center of the moistened area on the mat. Immediately recap the bottle. Add one drop of water to the tablet, wait 5 seconds, then add another drop so that the water runs off onto the mat. Wait 1 minute, then observe the color of the mat around the tablet. A blue or purple color indicates the presence of bilirubin. A pink or light red color indicates negative for bilirubin. Record your observations in the report.

5. *Hemoglobin* and *myoglobin:* Aged erythrocytes are destroyed by the cells of the reticuloendothelial system, but some of the hemoglobin molecules escape metabolism and are released into the plasma (free hemoglobin). Normally, the amount of free hemoglobin in the plasma is very small; however, when the plasma level of free hemoglobin exceeds 100 mg/dL, hemoglobin appears in the urine *(hemoglobinuria).* Hemoglobinuria occurs when there is an extensive or rapid destruction of erythrocytes at a rate that is too fast to allow for the adequate storage or metabolism of free hemoglobin. The causative factors include several types of hemolysis, burns, crushing injuries, transfusion reactions, and poisons (e.g., snake venoms, mushrooms). *Myoglobin* is a red respiratory pigment found in skeletal muscle, where the pigment performs in a manner similar to hemoglobin of the blood. Normally, myoglobin is absent from the plasma and, therefore, from the urine. However, free myoglobin may appear in the plasma and, hence, in the urine, after an extensive crushing injury to muscle, necrotic diseases of muscle, certain types of infections (e.g., *Clostridium*), and occasionally after severe exercise.

   Hemastix

   Test for the urinary presence of hemoglobin and myoglobin by dipping the test end of a Hemastix into a fresh specimen of urine and immediately removing the dipstick. While removing the dipstick, run the edge of it against the rim of the urine container to remove excess urine. Wait 60 seconds and compare the test area with the color chart provided. The absence of a blue color in the test area after 1 minute indicates a negative test. A positive test is indicated by the appearance of a blue color on the test area in 1 minute or less. The intensity of the color change is proportional to small, moderate, or large amounts of excreted hemoglobin or myoglobin. Record your observations in the report. The Hemastix test does not differentiate hemoglobinuria from myoglobinuria; however, such a differentiation can be done by spectrophotometry.

6. *Nitrite:* Normally, no nitrite is detectable in the urine. A positive test for nitrite indicates the presence of infectious microorganisms, such as bacteria, that can reduce dietary (urinary) nitrate to nitrite. A negative test for nitrite does not rule out the presence of bacteria because some bacteria lack the enzyme reductase needed to convert nitrate to nitrite.
7. *Urobilinogen:* Bilirubin, a bile pigment formed during the breakdown of hemoglobin following red blood cell destruction, is excreted by the liver into the small intestine, where bacteria convert it to urobilinogen. Some of the urobilinogen is absorbed from the intestine into the blood, converted to urobilin, and excreted by the kidneys into the urine. Urobilin is a pigment imparting a yellow-orange color to the urine. A normal range of urobilinogen value as detected by the Multistix 10 SG strip discussed below is 0.2 to 1.0 mg/dL.
8. *Leukocytes:* Leukocytes generally are absent from urine, although trace amounts of white blood cells may not be associated with disease. A small or greater positive test for leukocytes in the urine may be indicative of renal disease or disease of the ureters, urinary bladder, or urethra.

Multistix 10 SG

Bayer Multistix 10 SG reagent strips for urinalysis are dip-and-read tests for glucose, bilirubin, ketone, specific gravity, blood, pH, protein, urobilinogen, nitrite, and leukocytes in the urine. All 10 individual test reagent pads are on a single plastic strip, but are not all read at the same time. Note the time intervals to be followed in reading the test strip.

1. Remove a test strip from the bottle and replace the cap. Completely immerse reagent areas of the strip in fresh urine and remove immediately to avoid dissolving out reagents.
2. While removing, run the edge of the entire length of the strip against the rim of the urine container to remove excess urine. Hold the strip in a horizontal position to prevent mixing of chemicals from adjacent reagent areas.
3. Compare the reagent areas with the corresponding color chart **at the time specified.** Color changes that occur after 2 mintues are of no diagnostic value. Hold the strip close to color blocks and match carefully. *Avoid laying the strip directly on the color chart because doing so will cause the urine to soil the chart.*

Time	*Test*
Immediately up to 2 minutes	pH and protein
30 seconds	glucose and bilirubin
40 seconds	ketone
45 seconds	specific gravity
60 seconds	urobilinogen, blood (hemoglobin, myoglobin), nitrite
120 seconds	leukocytes

Record your observations in the report. Discard the reagent strip, following directions from your laboratory instructor.

Microscopic Examination of Urine Sediment

A complete microscopic examination of urine involves both a visual assessment of the sediment and a bacteriologic analysis. In health, the urine contains small numbers of cells and other formed elements from the whole length of the genitourinary tract: casts and epithelial cells from the nephron; epithelial cells from the pelves, ureters, bladder, and urethra; mucus threads and spermatozoa from the prostate. A few erythrocytes and leukocytes may also be present from any part of the urinary tract. We will perform only a simple qualitative analysis of urine using the compound light microscope.

Place 5 mL of fresh urine in a clean centrifuge tube and centrifuge for 5 minutes. Pour off the clear urine and mix the remaining sediment by shaking the test tube. Place a drop of the sediment on a clean microscope slide, cover it with a clean coverslip, and examine it under the microscope while it is still wet. Always use subdued light; use the condenser and diaphragm to reduce the lighting. Examine the slide using first the low-power objective and then the high-power objective. Scan the sample for various cells, casts (protein formations), crystals, and mucus threads (figure 31.3).

Casts may be distinguished from other contaminants, such as mucus fibers or crystals, by their regular parallel walls and "squared" ends. They consist of densely packed cells or cellular elements, or both, and include hyaline casts, red cell casts, leukocyte casts, epithelial cell casts, granular casts, fatty casts, and others.

A variety of *crystals* may be found in normal urine, and generally the type of crystal depends on the urinary pH. Clinically, the presence of crystals in urine is of little significance, except in certain metabolic diseases and drug intoxications.

Erythrocytes (red blood cells) come from anywhere in the genitourinary tract. They are hemolyzed in dilute urine and crenated in concentrated urine. Crenated red cells and red cell casts come from the kidney and indicate serious renal disease. Persistent findings of even small numbers of erythrocytes should be thoroughly investigated.

Leukocytes (white blood cells) may come from anywhere in the genitourinary tract. Leukocyte casts always come from the kidney and suggest an infective or noninfective inflammatory disease of the kidney. Urine cultures are used to differentiate infective from noninfective disease.

After scanning the sample with low- and high-power objectives, add a drop of methylene blue to the sample and scan again. Methylene blue helps to delineate the cells. Record your observations in the report.

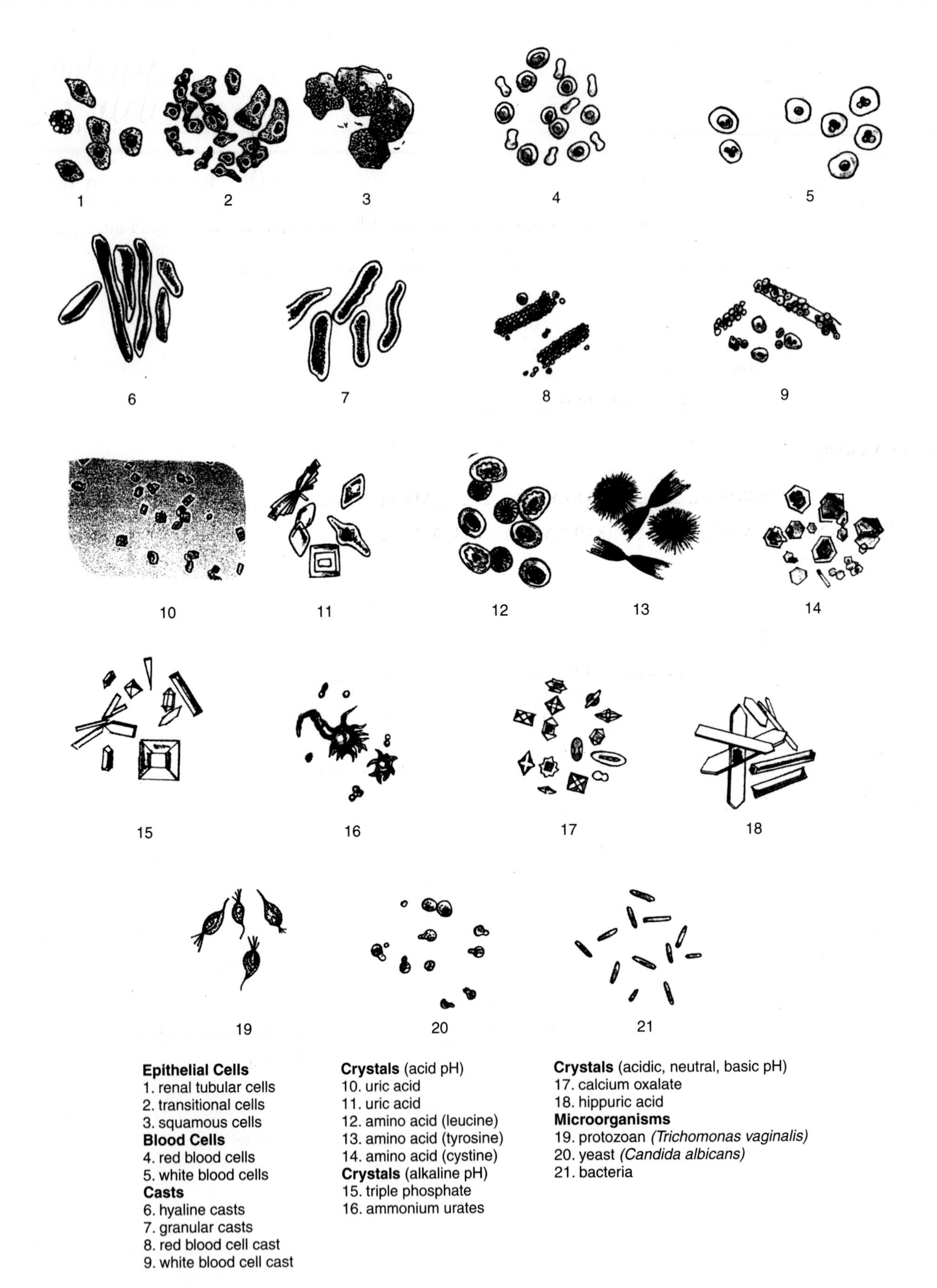

FIGURE **31.3** From *Atlas of Urine Sediment* (Bayer Corporation, 1999).

Digestion

■ INTRODUCTION

The digestive system consists of the *alimentary canal* and the *accessory organs* of digestion. Subdivisions of the alimentary canal and accessory organs are listed in table 32.1 and shown in figure 32.1.

The general functions of the digestive system may be divided into two categories of activity: (1) mixing and propulsion of the contents of the alimentary canal and (2) digestion and assimilation of gastrointestinal contents.

The motility of the alimentary canal is controlled from the esophagus to the anus by means of activity in the *myenteric plexus,* a nerve network between the layers of smooth muscle in the wall of the alimentary canal. Myenteric activity is, in turn, controlled by the autonomic nervous system. The principal propulsive movement of the alimentary canal is *peristalsis* (figure 32.2), a wave of relaxation followed by a wave of contraction of the intrinsic muscles. The normal stimulus for peristalsis is distension of the alimentary canal by its contents.

TABLE 32.1 Gastrointestinal system

Alimentary canal	Accessory organs
Oral cavity	Tongue
Pharynx	Teeth
Esophagus	Muscles of mastication and swallowing
Stomach	Salivary glands:
Small intestine:	Parotid glands
Duodenum	Submaxillary glands
Jejunum	Sublingual glands
Ileum	Pancreas
Large intestine:	Liver and gallbladder
Cecum and appendix	
Colon: ascending, transverse, descending, sigmoid	
Rectum	
Anal canal	

In the small and large intestines, several types of nonpropulsive smooth muscle contraction also occur. *Segmentation,* characteristic of the small intestine, is produced by the alternate contraction and relaxation of smooth muscle in several parts of the small intestine, which breaks up the intestinal contents and mixes them with digestive enzymes. A similar type of movement, *haustration,* is characteristic of the large intestine, where the contents are compacted and water is absorbed. In addition, periodic strong peristaltic waves pass through the large intestine, moving waste material toward the rectum. Such movements are referred to as mass peristalsis.

The process of digestion refers to the degradation of large ingested molecules of foodstuffs into smaller absorbable molecules by the action of enzymes. The types of enzymatic secretions of the alimentary canal and accessory organs and the activity of the enzymes are listed in table 32.2.

The absorption of nutrients from the alimentary canal occurs in the stomach, the small intestine, and the large intestine. Although the stomach actually plays a minimal role in the absorption of nutrients, liquids are rapidly absorbed there. Most of the absorption of digestive end products (amino acids, monosaccharides, fatty acids, and glycerol) occurs in the small intestine, where the mucosa of the small intestine presents a large surface area for absorption. The large intestine is primarily concerned with absorbing the remaining water and electrolytes.

The digestive end products of protein and carbohydrate catabolism are absorbed into the blood of the submucosal vasculature, but those of lipid catabolism are absorbed into intestinal lymphatic vessels (lacteals) and transported to the systemic venous circulation.

In humans, the digestion of fat by intestinal and pancreatic lipase is assisted by the emulsification action of *bile salts.* Bile is excreted by the liver into the duodenum via

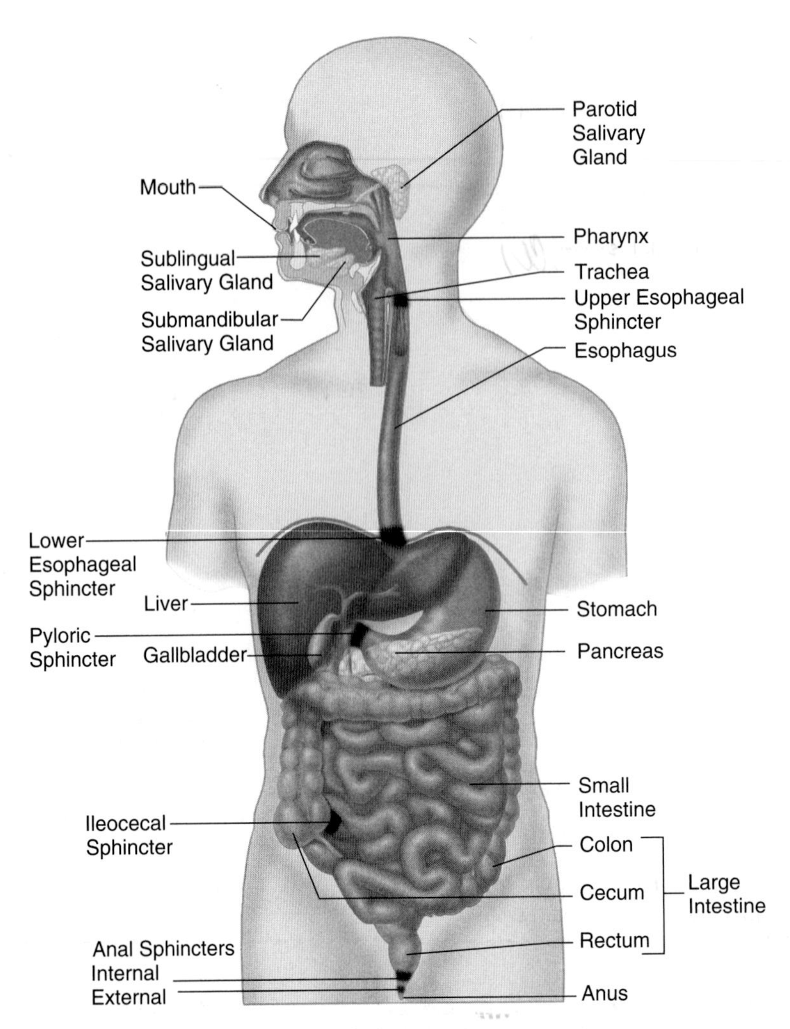

FIGURE 32.1 The alimentary canal and accessory organs. The liver overlies the gallbladder and a portion of the stomach, and the stomach overlies part of the pancreas.

the biliary duct system. The *gallbladder,* which is attached to the biliary duct system, acts as a reservoir for bile until it is needed in the duodenum. Some animals do not have a gallbladder, nor do they need one if their dietary intake of lipids is very low.

The biliary duct system empties into the duodenum by way of the common bile duct. Joining the common bile duct immediately before it enters the duodenum is the main pancreatic duct. The *exocrine pancreas* secretes a number of digestive enzymes, as well as large quantities of sodium bicarbonate, a nonenzymatic secretion that serves to buffer hydrochloric acid from the stomach.

Before ingested foodstuffs can be absorbed by the gastrointestinal mucosa, they must be degraded into chemically less complex molecules. Carbohydrates must be degraded into component monosaccharides, proteins into component amino acids, and lipids into glycerol and fatty acids. The process of breaking down complex carbohydrates, proteins, and lipids into simpler molecules is known as *digestion.*

Digestion occurs because of the chemical activities of the enzymes secreted into the lumen of the alimentary canal. All digestive enzymes function by chemically breaking the bonds that hold parts of complex molecules together. The process is analogous to destroying the mortar joints that hold bricks together in a wall. When the mortar is destroyed, the wall crumbles into its component parts.

The major component parts of large carbohydrate molecules are simple, single sugars called *monosaccharides.* There are three principal monosaccharides: glucose, fructose, and galactose. Monosaccharides may be linked in pairs to form *disaccharides,* or double sugars. Three examples of disaccharides are maltose (glucose + glucose); sucrose, or table sugar (glucose + fructose); and lactose, or milk sugar (glucose + galactose). Monosaccharides and disaccharides may be linked in varying combinations to form larger and more complex carbohydrate polymers. In the process of enzymatic digestion, complex carbohydrates must be reduced to component monosaccharides and disac-

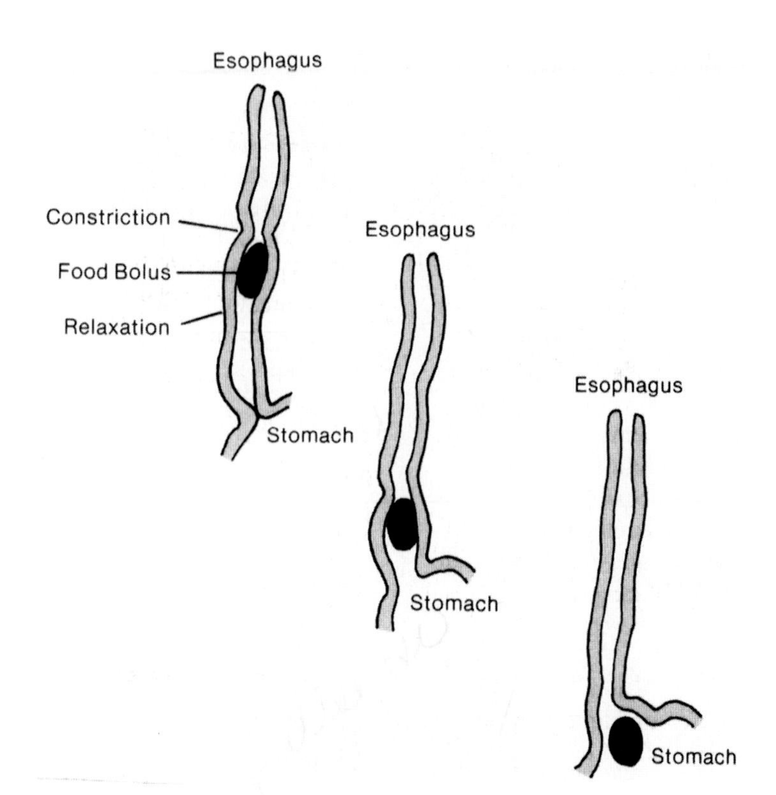

FIGURE 32.2 Peristalsis

TABLE 32.2 Gastrointestinal enzymes

Enzyme	Source	Optimum pH	Activity
Amylase	Salivary glands Pancreas	6.8–7.0	Hydrolyzes starch and glycogen to maltose and glucose
Pepsin*	Gastric glands	1.5–2.5	Hydrolyzes proteins to proteoses, peptones, polypeptides
Trypsin*	Pancreas	7.8	Hydrolyzes proteoses and peptones to polypeptides
Chymotrypsin*	Pancreas	7.8	Hydrolyzes polypeptides to dipeptides
Peptidases*	Pancreas	7.8–8.0	Hydrolyzes polypeptides and dipeptides to amino acids
Lipase	Pancreas	7.0–9.0	Degrades lipids to glycerol and free fatty acids

*Secreted in the form of an inactive precursor, which becomes activated in the lumen.

charides before mucosal absorption can occur. For example, *glycogen,* or animal starch, is a complex carbohydrate polymer made up of many glucose molecules linked together in the form of a long, branching chain. *Amylase* (salivary and pancreatic) digests starch into maltose, maltotriose (three glucose molecules linked together), and other oligosaccharides. *Maltase* splits maltose into two molecules of glucose. *Sucrase* splits sucrose into glucose and fructose, and *lactase* splits lactose into glucose and galactose.

Proteins are very large, complex molecules consisting essentially of chains of subunits known as *L-amino acids.* There are 20 types of amino acid. The structural and chemical properties of a protein are determined, in part, by the types of amino acids making up the protein and by the sequence of those amino acids. The component amino acids of a protein are joined together by a chemical bond known as a *peptide bond.* Two amino acids joined together form a dipeptide. Several dipeptides joined together form a polypeptide, a small chain of amino acids. Polypeptides may, in turn, be joined together to form peptones (intermediate-size chains of amino acids) or proteoses (large chains of amino acids). In the processes of protein digestion, all proteins, large or small, must be broken down into their component amino acids before mucosal absorption can occur.

Lipids consist essentially of two basic kinds of building blocks: glycerol and fatty acids. The combination of one glycerol molecule and three fatty acids forms a neutral fat called a *triglyceride.* Compound lipids, such as phospholipid, glycolipid, and lipoprotein, are formed from various combinations of neutral fat with other components. In the process of lipid digestion, fats must be broken down into component glycerol and free fatty acids before mucosal absorption can occur.

The rate of enzymatic digestion of ingested foodstuffs depends on the amount of enzyme present in relation to the amount of substrate (material to be digested), the temperature, and the activity of hydrogen ions (pH) in the medium in which the enzyme is acting. Normally, digestion occurs within the lumen of the alimentary canal, where the temperature is kept at a relatively constant 37° C. The concentration of hydrogen ions (pH), however, varies considerably from one part of the alimentary canal to another. All enzymes are affected by changes in pH, and each enzyme functions best within a certain range of pH values. Table 32.2 lists the optimum pH for selected digestive enzymes.

In the experiments that follow, the action of digestive enzymes on proteins, carbohydrates, and lipids will be observed, in addition to the influence of pH on enzyme activity.

■ EXPERIMENTAL OBJECTIVES

1. To examine human saliva with respect to its physical characteristics and chemical composition.
2. To deduce, from experimental observation, the primary function of salivary amylase.
3. To observe the effects of pepsin on protein digestion.
4. To observe the effects of pancreatic lipase on lipid digestion.
5. To observe the influence of pH on digestive enzymatic activity.

Materials

compound light microscope
paraffin stick
microscope slide
coverslip
watch glass
15-mL disposable test tubes
test-tube rack
porcelain plate
water bath (37° C)
water bath (100° C)
polyethylene funnel
hydrion paper
droppers
litmus cream (dairy cream + powdered litmus)
distilled water
starch paste
Benedict's solution
iodine
5% glucose solution
10% acetic acid
silver nitrate solution
concentrated hydrochloric acid
10 N sodium hydroxide
5% pepsin
2% pancreatic lipase
boiled egg white
bile salts
vegetable oil

■ EXPERIMENTAL METHODS

CAUTION

In this experiment you will obtain and analyze a sample of your own saliva. Handle only your own saliva. Dispose of all supplies (disposable microscope slides, coverslips, test tubes, etc.) that come into contact with your saliva by placing them in properly marked containers. Always treat all body fluids as infectious. Nondisposable items that come into contact with your saliva are to be washed with a detergent and tap water, then rinsed successively with tap water, bleach solution, distilled water, and acetone. Disinfect your laboratory table or laboratory bench top by wiping it clean with a bleach solution before obtaining your saliva sample and again at the end of the experiment.

Protein Digestion by Pepsin

Pepsin is a proteolytic enzyme secreted in an inactive form (pepsinogen) by the gastric glands. In the gastric lumen, pepsinogen is activated by hydrochloric acid, and active pepsin is formed. The peptic digestion of proteins results mostly in the formation of proteoses, peptones, and polypeptides. These are eventually digested into component amino acids in the small intestine under the influence of trypsin, chymotrypsin, carboxypeptidase, and other proteolytic enzymes. Pepsin activity is most effective at a pH of 2.0. Because the gastric glands also secrete large quantities of hydrochloric acid, the pH of gastric juice is very low (as low as 0.9), providing a favorable environment for optimal pepsin activity. To observe the activity of pepsin and the influence of pH on peptic digestion, follow these directions:

1. Place three clean test tubes in a test-tube rack, and add 5 mL of 5% pepsin solution to each. Number each tube consecutively 1–3.
2. Add a drop of 10 N sodium hydroxide to test tube 1.

3. Add a drop of concentrated hydrochloric acid to test tube 2.
4. Add a drop of distilled water to test tube 3.
5. Using hydrion paper, determine the pH of the contents of each tube. Record the data in the report.
6. Cut equal-sized pieces (about 0.5 square centimeter) of boiled egg white, and add 1 piece to each tube.
7. Incubate all 3 tubes in a 37° C water bath for 45 minutes.
8. At the end of the incubation period, determine the pH in each tube and record your observations in the report. Remember that as protein is hydrolyzed, amino acids and amino acid chains are produced, and they are acidic. Observe the effect of the peptic digestion of egg white (albumin) and compare the effect of pH on peptic activity.

Emulsification of Fat and Lipid Digestion by Pancreatic Lipase

Pancreatic lipase degrades triglycerides to monoglycerides and free fatty acids. The optimum pH for activity on pancreatic lipase depends on the types of fatty acids contained in the triglyceride molecule and ranges from pH 7.0 to 9.0. Because the pancreas also secretes large amounts of sodium bicarbonate into the small intestine, the luminal fluid is usually alkaline, thereby allowing for optimal lipase activity.

Lipids are insoluble in water and therefore tend to aggregate in large clumps in the aqueous fluid of the gastrointestinal lumen, much as large droplets of oil float on the surface of water. For lipase to be most effective, large aggregates of fat must be dispersed into smaller droplets, which can then be more readily attacked by lipase. The dispersion of large aggregates of fat into smaller particles is called *emulsification,* which is the primary function of bile salts. In the absence of bile salts, the lipase digestion of fat can still occur, but the process takes much longer.

To observe the emulsification of fat by bile salts, follow these directions:

1. Place two clean test tubes in a test-tube rack and label one A and the other B.
2. Add 3 mL of vegetable oil to each tube.
3. Add 3 mL of distilled water to each tube. Note how the oil floats in large droplets on the surface of the water.
4. Add a tiny amount of bile salts to tube A.
5. Shake each tube for about 1 minute and note the results. Are the fat droplets in tube A smaller and more dispersed than those in tube B?

To observe the activity of pancreatic lipase, follow these directions:

1. Place two clean test tubes in a test-tube rack and number them 4 and 5.
2. Add 3 mL of pancreatic lipase to each tube.
3. Add 3 mL of distilled water to each tube.
4. Add 3 mL of litmus cream (dairy cream to which powdered litmus has been added for blue coloration) to each tube.
5. Add a small quantity of bile salts to test tube 5, agitate each tube for approximately 1 minute, record the pH of each tube's contents, and incubate each tube in a 37° C water bath for 15 minutes.
6. After incubation, note the color change (blue litmus will turn pink as acids are released during the digestion of fat) and test the contents of each tube for pH. Record the data in the report.

Properties of Saliva and the Activity of Salivary Amylase (Ptyalin)

Analyze the physical and chemical characteristics of saliva by performing the following:

1. Obtain a paraffin stick from the laboratory instructor and chew it. Chewing paraffin (which is chemically inert) will increase salivary secretion without altering the chemical composition of saliva. Obtain 15 ml of saliva in a test tube.
2. Examine the physical characteristics of saliva as to color, viscosity, pH, and turbidity. Record your observations in the report.
3. Place a drop of saliva on a clean microscope slide, cover with a coverslip, and examine under low and high power. Are cells present? If so, what type?
4. Dilute 7 ml of saliva with 7 ml of distilled water. Add 4 drops of 10% acetic acid. Notice the formation of a precipitate. The precipitate is *mucin,* a glycoprotein. What is the function of mucin? Filter the diluted salivary solution.
5. Put 3 ml of the filtrate in a watch glass and test for the presence of chlorides by adding 4 drops of silver nitrate. The formation of a white precipitate indicates the presence of chlorides. Are chlorides present? What other inorganic constituents are present in saliva?

Normal saliva contains a digestive enzyme called *ptyalin* (salivary amylase). Ptyalin initiates the digestion of starch and glycogen into glucose. The digestive action of ptyalin on starch (a carbohydrate) may be observed by checking the rate of starch disappearance from a starch solution mixed with saliva.

During digestion of starch by ptyalin, erythrodextrin is formed, as indicated by a red color when iodine is added. Achromodextrin is also formed (it is colorless), and finally maltose and glucose (reducing sugars) are formed. The presence of a reducing sugar is indicated by a positive Benedict's test (formation of a red precipitate).

Perform the following control tests to observe positive test results before analyzing saliva:

6. *Control test for starch:* Add 1 ml of starch paste to a drop of iodine in a porcelain plate. Observe the color.

Blue indicates the presence of starch. Add 2 ml of starch paste to 2 ml of Benedict's solution in a test tube. Place the test tube in a 100° C water bath for 10 minutes. If a reducing sugar is present, a precipitate will form. Does a precipitate form?

7. *Control test for reducing sugars (maltose, glucose, fructose, etc.):* Add 2 ml of 5% glucose solution to 2 ml of Benedict's solution in a test tube. Place the test tube in a 100° C water bath for 10 minutes. Remove the test tube from the water bath and allow it to cool in a test-tube rack. The presence of a red-to-brown precipitate indicates the presence of a reducing sugar.

Perform the following tests for analysis of ptyalin activity:

8. Add 3 ml of saliva to 5 ml of starch solution and mix them well in a test tube. The mixture is called a digest. Place the test tube in a 37° C water bath and record the time. After 3 minutes, remove 2 drops of digest and test it with 2 drops of iodine on the porcelain plate. Compare it with the control test for starch. Is the test positive or negative? Why? Record the data in the report.
9. Remove the digest tube from the 37° C water bath after 30 minutes. Add 2 ml of digest to 2 ml of Benedict's solution in a test tube and place it in a 100° C water bath for 10 minutes. Remove the test tube, allow it to cool, and compare it with the control test for reducing sugar. Is the test positive or negative? Why? What happened to the original starch? Record your observations in the report.

Digestion

REPORT 32

Name: Nancy Thomas Date: 11/18/06

Lab Section: Tim Strief

1. Data:

a. Peptic Digestion

	Temp	Tube	Initial pH	Final pH	Observations
NaOH	37° C	1	8	9	
HCL	37° C	2	1	1	
H_2O	37° C	3	6	6	

b. Lipase Digestion

	Temp	Tube	Initial pH	Final pH	Observations
w/o bile	37° C	4	7	7	
w/ bile	37° C	5	6	6	looks different - clear

pH should go down w/digestion - more acidic

c. Salivary Analysis

Color	Viscosity	Turbidity	pH	Mucin	Cl⁻	Cells present
milky	yes	yes	7.5	+	+	yes

d. Analysis of Salivary Amylase Activity

Temp	Elapsed time	Iodine test for starch (+ or –)
37° C	3 minutes	

Temp	Elapsed time	Benedict's test for sugar (+ or –)
37° C	30 minutes	

Starch is not sugar - tests negative for starch

2. After you performed the control test for starch, did a precipitate form when Benedict's solution was added? Why or why not? yes. Starch became sugar. Saliva has enzyme (amalyase) breaks it down.

3. Does pepsin contribute significantly to the digestion of protein in the chyme after it leaves the stomach and enters the small intestine? Why or why not?

No. Ph goes up. pepsin no longer works.

4. What influence does bile have on the intestinal digestion of fats?

emulsifies the fats - larger globs become less hydrophilic

5. Is the process of emulsification an enzymatic activity? If not, how does emulsification occur?

No.

6. List the major digestive end products absorbed by each of the following: Carbs, proteins, + fats.

a. Stomach

b. Small intestine amino acids, simple sugars, fatty acids, glycerol

c. Large intestine

7. Which organ—the stomach or the small intestine—presents the greater surface area for absorption? Why?

Small intestine - needed for absorption

8. Gastric contents, usually acidic, are propelled into the duodenum by gastric peristalsis, yet duodenal contents are usually alkaline. Why?

Body releases sodium bicarbonate.

9. In the space provided, design an experiment to test the influence of pH on the digestive activity of salivary amylase.

Metabolic Rates: Indirect Calorimetry

CHAPTER 33

■ INTRODUCTION

Energy from the oxidation of foodstuffs is used by the body's cells to perform useful work. Nerve conduction, muscle contraction, cellular secretion, and other body processes depend on an adequate supply and release of energy; in fact, all activities of the living cell are energy dependent. Ultimately, all energy that is released and used by the cell derives from the combustion of foodstuffs, such as carbohydrate, fat, and protein.

When foodstuffs are metabolized by the body, oxygen is consumed and energy is released, along with water and carbon dioxide:

$$\text{Food} + O_2 \rightarrow CO_2 + H_2O + \text{Energy}$$

Some of the released energy is used by the cell to generate *adenosine triphosphate (ATP),* a compound containing high-energy bonds, which the cell may later break to release energy and perform work. Much of the energy escapes as heat, in amounts proportional to the energy content of the foodstuff. Measurement of the total amount of heat (via **calorimetry**) released by the body per unit of time provides an approximation of the body's overall metabolic rate.

The release of body heat and, hence, the metabolic rate may be measured directly *(direct calorimetry)* by placing the subject in a whole-body calorimeter and measuring the amount of heat given off by the body per unit of time. This method, although accurate, is practical and convenient only for a research laboratory. A more universally practical and convenient method of approximating metabolic rate is to measure the release of the body heat indirectly *(indirect calorimetry),* using the concept of the dependency of heat production on oxygen consumption.

The amount of heat released during metabolism is expressed in terms of a thermal unit of energy called the calorie. By definition, a *calorie* is the amount of heat necessary to raise the temperature of 1 g of pure water from 14.5° C to 15.5° C. Commonly, the unit is called a **small calorie (c).** Because the number of calories produced during the oxidation of foodstuffs is very large, it is often more convenient to use a larger unit of thermal energy, the *kilocalorie* (kcal), or **large calorie (C),** which is simply the equivalent of 1000 small calories.

When glucose is completely oxidized (in a calorimeter), according to the equation below, 673 kilocalories of energy are released. In terms of the amount of oxygen consumed during the complete combustion (6 moles × 22.4 L/mole = 134.4 L O_2), the caloric production is 5.01 kcal per liter of O_2 consumed. The caloric equivalents of other foodstuffs are 5.06 kcal/L O_2 for starches, 4.70 kcal/L O_2 for fats, and 4.60 kcal/L O_2 for proteins.

$$C_6H_{12}O_6 + 6O_2 \rightarrow 6CO_2 + 6H_2O + 673 \text{ kcal}$$

At any given moment, the typical human body is oxidizing a mixture of carbohydrate, fat, and protein, rather than using any single foodstuff as the sole source of energy. Based on an average utilization of all three foodstuffs, the average release of energy (in terms of oxygen consumed) is 4.825 kcal per liter of O_2. Metabolic rates under varying conditions may, therefore, be measured by determining oxygen consumption in liters per hour and multiplying by the caloric equivalent of 4.825 kcal per liter of oxygen consumed.

Oxygen consumption, relative to an indirect determination of metabolic rate, can be measured by a respirometer, such as the one diagrammed in figure 33.1. The apparatus consists of a recording spirometer filled with oxygen. Within the spirometer is a canister of soda lime. As the subject inspires, pure oxygen is inhaled. As the subject expires, carbon dioxide is removed from expired air by the soda lime, and the remainder of the expired air is allowed to reenter the spirometer bell. During inspiration, the spirometer bell falls; during expiration, the bell rises. The movements of the spirometer bell are recorded on a chart that is graduated in milliliters of oxygen per minute. As the subject inspires oxygen from the bell, some of the inspired oxygen will be consumed; therefore, the subsequent elevation of the bell during expiration will be slightly less than it

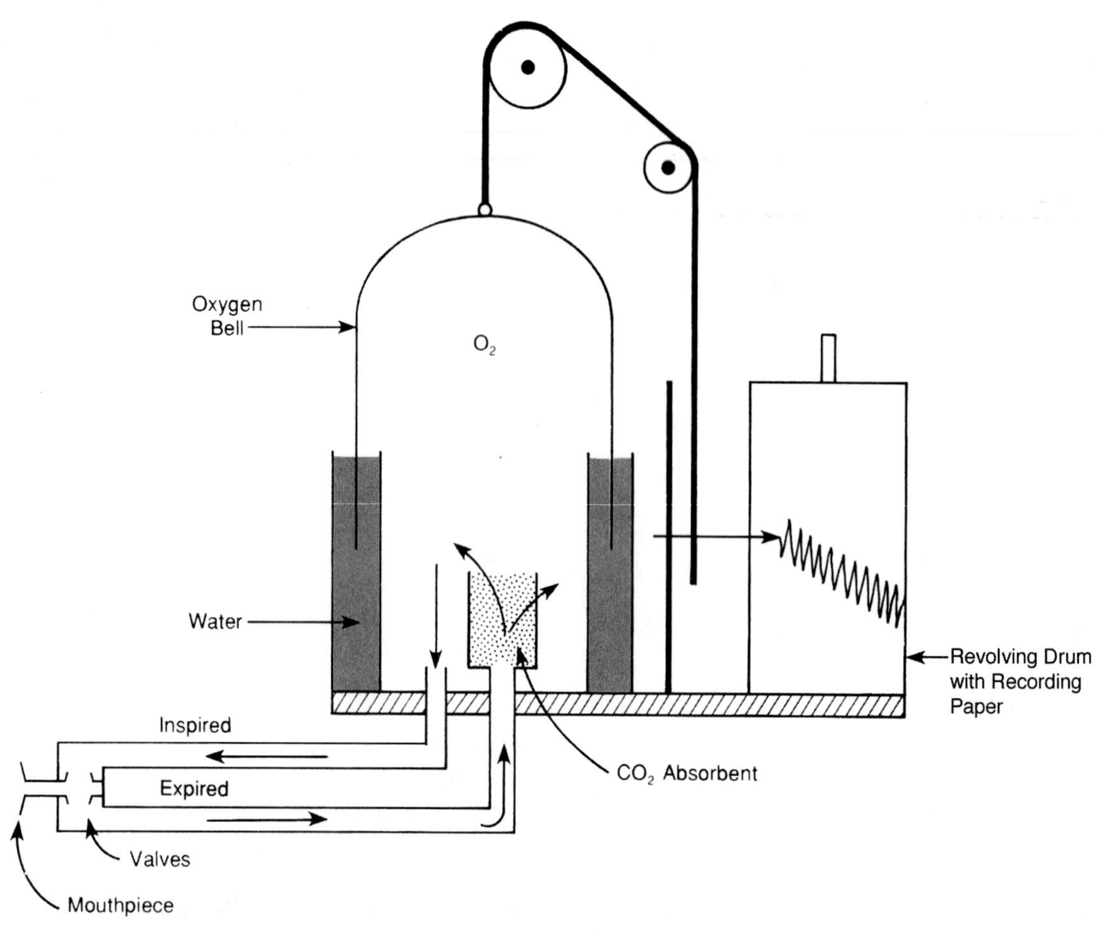

FIGURE **33.1** Respirometer used for indirect calorimetry

was with the preceding expiration. As a result, the spirogram tracing will be sloped rather than horizontal. The slope of the tracing can be used to calculate the amount of oxygen consumed from the bell per unit of time.

Several factors influence the metabolic rate of a healthy person. Exercise increases metabolic rate, the energy expenditure generally being directly proportional to the severity of the exercise. Table 33.1 indicates energy expenditure during different types of activity for a 70-kg man. In part, the ingestion of food elevates metabolic rate (the effect is called the specific dynamic action of food) because some of the end products of digestion (e.g., amino acids) stimulate cellular metabolism. The release of *calorigenic hormones,* such as thyroxine and growth hormone, increases metabolic rate, as does increased activity of the sympathetic nervous system via release of norepinephrine and epinephrine. Climate, nutrition, age, gender, body surface area, and many other factors influence metabolic rate. Table 33.2 indicates the influence of age and gender on metabolic rates measured under basal conditions.

Basal metabolic rate (BMR) refers to body metabolism as measured under a set of standard basal conditions designed to minimize the effects of as many influencing factors as possible. Basal conditions are as follows:

1. The subject must ingest no food for at least 12 hours prior to the test.
2. The subject must be mentally and physically relaxed. Most metabolic tests are performed early in the morning after the subject has reclined in a quiet, semidark room for about 30 minutes.
3. The subject must not be febrile. Fever elevates BMR.
4. The temperature of the room air must be comfortable (65°–80° F).

For comparative purposes, basal metabolic rates are ordinarily expressed in kilocalories per hour per square meter of body surface area. The test result is compared with a predicted normal (average) value for the subject's age and gender, and a percent deviation from normal is calculated as follows:

$$\text{Percent deviation of BMR} = \frac{\text{Measured BMR} - \text{Predicted BMR}}{\text{Predicted BMR}} \times 100$$

A negative percent deviation will occur if the test result is below the predicted normal value. A positive percent deviation will occur if the test result is higher than the predicted normal value.

Elevated basal metabolic rates occur in hyperthyroidism, acromegaly, gigantism, various forms of leukemia,

TABLE **33.1** Energy expenditure per hour during different types of activity for a 70-kg man

Form of activity	*Kilocalories per hour*
Sleeping	65
Awake lying still	77
Sitting at rest	100
Standing relaxed	105
Dressing and undressing	118
Tailoring	135
Typewriting rapidly	140
"Light" exercise	170
Walking slowly (2.6 miles per hour)	200
Carpentry, metalworking, industrial painting	240
"Active" exercise	290
"Severe" exercise	450
Sawing wood	480
Swimming	500
Running (5.3 miles per hour)	570
"Very severe" exercise	600
Walking very fast (5.3 miles per hour)	650
Walking up stairs	1100

Extracted from data compiled by Professor M. S. Rose. From Guyton, A. C. *Textbook of Medical Physiology,* 5th ed. (Philadelphia, PA.: W. B. Saunders Co., 1976).

TABLE **33.2** Metabolic rates by age and gender

Age, years	*Males, Kilocalories/ m²/hr*	*Females, Kilocalories/ m²/hr*
10–12	49.5	45.8
12–14	47.8	43.4
14–16	46.0	41.0
16–18	43.0	38.5
18–20	41.0	37.6
20–30	40.5	36.8
30–40	39.5	36.5
40–50	38.0	35.3
50–60	36.9	34.4
60–70	35.8	33.6
70–80	34.5	32.6

From Langley, L. L. *Physiology of Man,* 4th ed. (New York: Van Nostrand Reinhold Co., 1971).

Hodgkin's disease, polycythemia vera, pregnancy, and several other pathologic or physiologic states.

Depressed basal metabolic rates occur in hypothyroidism, anorexia nervosa, and numerous other disorders.

Because abnormal BMRs occur in a wide variety of pathologic states, the usefulness of BMR determination as a diagnostic test is very limited. Again, as with the majority of diagnostic tests, the value of testing basal metabolic rate is that it provides an additional piece of information, which allows for a more complete picture of the pathophysiologic state when considered with other test results.

EXPERIMENTAL OBJECTIVES

1. To become familiar with the use of the respirometer in the indirect measurement of metabolic rate.
2. To measure metabolic rates under basal and other conditions.
3. To compare measured metabolic rates with predicted normals.
4. To observe the influence of exercise on metabolic rate.

Materials

Collins respirometer (9 L or 13.5 L)

disposable mouthpiece

pressurized oxygen tank with regulator and hose

nose clip

clinical laboratory scale

EXPERIMENTAL METHODS

Basal Metabolic Rate

Subjects for this experiment must refrain from eating for at least 12 hours prior to the test. If the test will be performed in the morning, the subject should report to the laboratory without having eaten breakfast. If the test is to be performed in the afternoon, a light breakfast is permissible, although the subject's BMR will be slightly higher than if no food had been eaten.

On reporting to the laboratory, subjects will be given instructions about the required 30-minute rest period prior to the test. The subject must recline in a comfortable, quiet environment and be allowed to relax physically and mentally.

Before the administration of the test, the observer makes certain that the spirometer bell has been filled with pure oxygen, that a new disposable mouthpiece has been installed on the breathing valve, and that the respiration pen and recorder have been properly serviced. It must be remembered that the observer, not the subject, administers the test.

The procedure to be followed in administering the test is as follows:

1. Turn the free-breathing valve to the bypass position so that the subject will breathe room air through the mouthpiece.
2. Instruct the subject to insert the mouthpiece into the mouth between the lips and the teeth. Apply a nose clip

so that the only route of inspired and expired air is through the mouthpiece. Allow the subject to become accustomed to the mouthpiece.

3. Observe the subject's respiratory pattern. When the subject completes an expiration and just before the subject starts to inspire, turn the free-breathing valve to the position that connects the subject with the spirometer bell. The spirometer bell should begin moving down and then up with each respiratory cycle. Allow the subject to respire into and out of the spirometer for approximately 1 minute.
4. Turn on the recorder switch to the slow speed (32 mm/min) and record the rate of oxygen consumption for 6 minutes.
5. At the end of the 6-minute period, turn off the recorder switch, turn the free-breathing valve to the bypass position, and allow the subject to respire room air for several minutes. The subject remains at rest, with the spirometer mouthpiece in place. Record the spirometer temperature, room temperature, and barometric pressure.
6. Record oxygen consumption for a second 6-minute period by repeating steps 3 and 4. At the end of the second recording period, turn off the recorder, turn the free-breathing valve to the bypass position, and disconnect the spirometer from the subject. Record the spirometer temperature.

Postexercise Metabolic Rate

After the basal metabolism test has been completed, determine the same subject's metabolic rate after performing moderate exercise. The procedure is as follows:

1. Instruct the subject to exercise moderately by running in place for approximately 5 minutes. After the exercise period, allow the subject to assume a comfortable, reclined position, as during the BMR test, and to rest approximately 8–10 minutes.
2. During the subject's postexercise rest period, replenish the oxygen in the spirometer bell and adjust the recorder paper for proper recording.
3. Repeat steps 2, 3, and 4 of the basal metabolism test outlined previously.
4. At the end of the recording period, turn off the recorder, turn the free-breathing valve to the bypass position, disconnect the spirometer from the subject, and remove the spirogram for analysis. Record the spirometer temperature.

Analysis of the Spirogram and Calculation of Metabolic Rate

1. Oxygen consumption is measured from the slope of the spirogram tracing. Draw a straight line connecting the bottom of all respiratory cycles within the 6-minute recording period (figure 33.2). Draw a horizontal line from the end of the slope at the 6-minute mark to the vertical line denoting volumes. Draw another horizontal line from the beginning of the slope to the vertical volume line. The oxygen consumption is represented by the difference in volume between the two horizontal lines. If using a 9-L Collins respirometer, read the volume change directly. If using a 13.5-L Collins respirometer, double the volume reading. Record the measured oxygen consumption for both basal and postexercise metabolic rate tests in the laboratory report.
2. Gas volumes recorded by the spirometer during metabolic rate tests must be corrected to standard temperature (0° C), pressure (760 mm Hg), and dryness. Table 33.3 gives the necessary correction factors for reducing the volume of moist gas at ambient barometric pressure and temperature to the volume occupied by dry gas at 0° C and 760 mm Hg pressure. Using the values for ambient barometric pressure (ask the laboratory instructor) and the spirometer temperature, find the appropriate correction factor in the table and multiply it by the recorded volumes. Enter the volumes corrected to STPD in the report.
3. Using clinical laboratory scales, determine the subject's height (without shoes) and body weight (subtract 1 kg for clothing). Using the body surface area nomogram (figure 33.3), determine and record the subject's body surface area.
4. Calculate the basal metabolic rate (kcal/m^2/hr) from the following formula (use gas volumes corrected to STPD):

$$\text{Metabolic rate} = \frac{(\text{Liters of } O_2 \text{ consumed/6 min})(10)(4.825 \text{ kcal/L})}{\text{Body surface area in square meters}}$$

5. Enter the calculated basal metabolic rate in the laboratory report, and compare the value with expected normal (table 33.2).
6. Use the following formula to compute the percent deviation from the normal:

$$\text{Percent deviation of BMR} = \frac{\text{Measured BMR} - \text{Predicted BMR}}{\text{Predicted BMR}} \times 100$$

Enter the percent deviation from normal in the report (±10% deviation is considered within the normal range).

7. Calculate the postexercise metabolic rate using the formula in step 4. Record the value in the report.
8. Calculate the percent increase of postexercise metabolic rate over basal metabolic rate using the following formula:

$$\text{Percent increase} = \frac{\text{Postexercise MR} - \text{BMR}}{\text{BMR}} \times 100$$

Record the percent increase in the report.

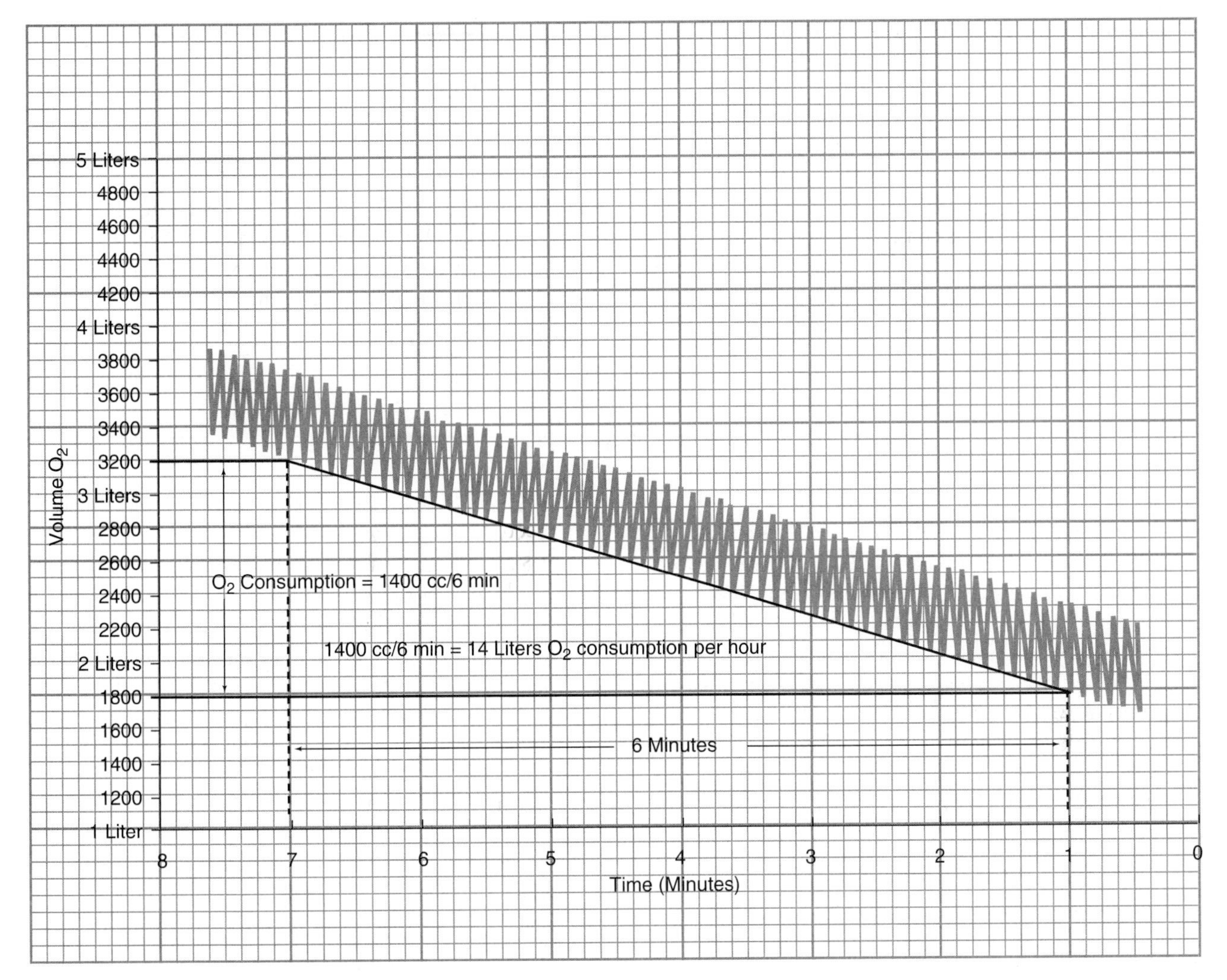

FIGURE 33.2 Measurement of O_2 consumption from spirogram

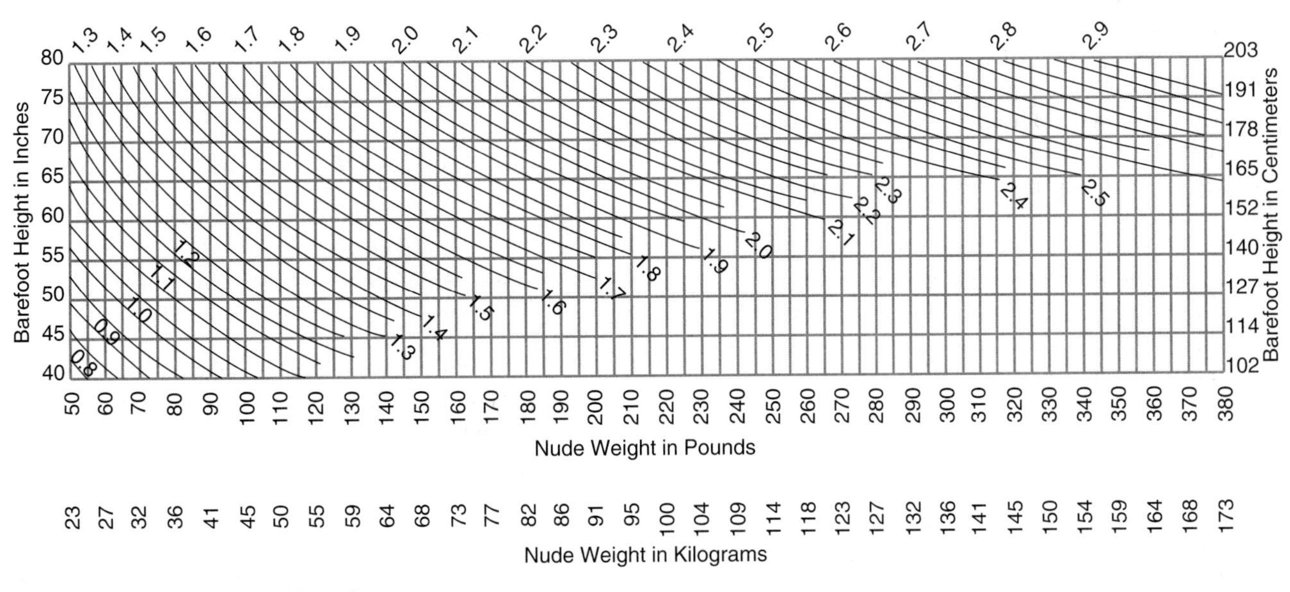

FIGURE 33.3 Body surface area nomogram

TABLE **33.3** Factors for reducing volume of moist gas to volume occupied by dry gas at 0° C, 760 mm Hg

Observed barometric reading, uncorrected for temperature	***15°***	***16°***	***17°***	***18°***	***19°***	***20°***	***21°***	***22°***	***23°***	***24°***	***25°***	***26°***	***27°***	***28°***	***29°***	***30°***	***31°***	***32°***
700	0.855	851	847	842	838	834	829	825	821	816	812	807	802	797	793	788	783	778
702	857	853	849	845	840	836	832	827	823	818	814	809	805	800	795	790	785	780
704	860	856	852	847	843	839	834	830	825	821	816	812	807	802	797	792	787	783
706	862	858	854	850	845	841	837	832	828	823	819	814	810	804	800	795	790	785
708	865	861	856	852	848	843	839	834	830	825	821	816	812	807	802	797	792	787
710	867	863	859	855	850	846	842	837	833	828	824	819	814	809	804	799	795	790
712	870	866	861	857	853	848	844	839	836	830	826	821	817	812	807	802	797	793
714	872	868	864	859	855	851	846	842	837	833	828	824	819	814	809	804	799	794
716	875	871	866	862	858	853	849	844	840	835	831	826	822	816	812	807	802	797
718	877	873	869	864	860	856	851	847	842	838	833	828	824	819	814	809	804	799
720	880	876	871	867	863	858	854	849	845	840	836	831	826	821	816	812	807	802
722	882	878	874	869	865	861	856	852	847	843	838	833	829	824	819	814	809	804
724	885	880	876	872	867	863	858	854	849	845	840	835	831	826	821	816	811	806
726	887	883	879	874	870	866	861	856	852	847	843	838	833	829	824	818	813	808
728	890	886	881	877	872	868	863	859	854	850	845	840	836	831	826	821	816	811
730	892	888	884	879	875	871	866	861	857	852	847	843	838	833	828	823	818	813
732	895	890	886	882	877	873	868	864	859	854	850	845	840	836	831	825	820	815
734	897	893	889	884	880	875	871	866	862	857	852	847	843	838	833	828	823	818
736	900	895	891	887	882	878	873	869	864	859	855	850	845	840	835	830	825	820
738	902	898	894	889	885	880	876	871	866	862	857	852	848	843	838	833	828	822
740	905	900	896	892	887	883	878	874	869	864	860	855	850	845	840	835	830	825
742	907	903	898	894	890	885	881	876	871	867	862	857	852	847	842	837	832	827
744	910	906	901	897	892	888	883	878	874	869	864	859	855	850	845	840	834	829
746	912	908	903	899	895	890	886	881	876	872	867	862	857	852	847	842	837	832
748	915	910	906	901	897	892	888	883	879	874	869	864	860	854	850	845	839	834
750	917	913	908	904	900	895	890	886	881	876	872	867	862	857	852	847	842	837
752	920	915	911	906	902	897	893	888	883	879	874	869	864	859	854	849	844	839
754	922	918	913	909	904	900	895	891	886	881	876	872	867	862	857	852	846	841
756	925	920	916	911	907	902	898	893	888	883	879	874	869	864	859	854	849	844
758	927	923	918	914	909	905	900	896	891	886	881	876	872	866	861	856	851	846
760	930	925	921	916	912	907	902	898	893	888	883	879	874	869	864	859	854	848
762	932	928	923	919	914	910	905	900	896	891	886	881	876	871	866	861	856	851
764	936	930	926	921	916	912	907	903	898	893	888	884	879	874	869	864	858	853
766	937	933	928	924	919	915	910	905	900	896	891	886	881	876	871	866	861	855
768	940	935	931	926	922	917	912	908	903	898	893	888	883	878	873	868	863	858
770	942	938	933	928	924	919	915	910	905	901	896	891	886	881	876	871	865	860
772	945	940	936	931	926	922	917	912	908	903	898	893	888	883	878	873	868	862
774	947	943	938	933	929	924	920	915	910	905	901	896	891	886	880	875	870	865
776	950	945	940	936	931	927	922	917	912	908	903	898	893	888	883	878	872	867
778	952	948	943	938	934	929	924	920	915	910	905	900	895	890	885	880	875	869
780	955	950	945	941	936	932	927	922	917	912	908	903	898	892	887	882	877	872

From Peters and Van Slyke, *Quantitative Clinical Chemistry,* vol. 11. (Methods) (Baltimore: Williams and Wilkins, 1932, reprinted 1956).

REPORT 33

Metabolic Rates: Indirect Calorimetry

Name: Nancy Thomas Date: 11/18/06

Lab Section: Jim Strief

1. Data:

Subject's initials __________ Age __________ Gender __________

Height __________ Weight __________ Body surface area (BSA) __________

Room temperature __________ Ambient barometric pressure __________

Spirometer temperature __________

Basal O_2 consumption for 6 minutes __________

Volume correction factor (table 33.3) __________

Basal corrected O_2 consumption for 6 minutes __________

Postexercise corrected O_2 consumption for 6 minutes __________

Calculation of basal metabolic rate (show calculations here):

Basal metabolic rate = __________ $Kcal/m^2/hr$

Percent deviation of BMR from normal = __________ %

Calculation of postexercise metabolic rate (show calculations here):

Postexercise metabolic rate = __________ $Kcal/m^2/hr$

Percent deviation from BMR = __________ %

2. The indirect method of determining metabolic rate allows for only an approximation of actual metabolic rate. Why is it not more accurate? Don't know their diet.

all increase

3. Do the following hormones tend to increase or decrease metabolic rate? How?

 a. Triiodothyronine

 b. Thyroxine

 c. Norepinephrine

 d. Adrenaline (epinephrine)

 e. Growth hormone (somatotropin)

 f. Why is the BMR test limited in its use as a diagnostic tool? In flux always moving.

4. List two examples of pathophysiology in which BMR is elevated above normal and two examples in which it is below normal.

 a. Elevated BMR hyperthyroidism

 Pg330

 b. Depressed BMR MC hypothy., anorexia

Maintenance and Regulation of Body Temperature

CHAPTER 34

■ INTRODUCTION

Ectotherms (*ecto* = outside, *therm* = heat) are animals that produce metabolic heat at comparatively low rates and have bodies that conduct heat at comparatively high rates. As a result, ectotherms rely on heat exchange with the environment to regulate their body temperature, which may vary over a wide range of acceptable values.

Animals in which body temperature tends to change in the direction of changing ambient air or water temperatures are called poikilotherms. *Poikilotherms* regulate their body temperature principally by behavioral means, such as sunning on a rock, moving into the shade, or swimming to a preferred water temperature. Fish, amphibians, and reptiles are poikilotherms.

Endotherms (*endo* = within, *therm* = heat) are animals that generate their own body heat through heat production as a by-product of metabolism. Sufficient heat can be generated metabolically to elevate body temperature several degrees centigrade above ambient temperatures.

Animals whose body temperature tends to remain stable regardless of the difference and changes in environmental temperatuare are called *homeotherms.* Homeotherms regulate their core body temperature to be within a very narrow range of acceptable values, usually within one degree of a desired temperature. Birds and mammals are homeotherms.

In general, the more biologically complex an organism is, the greater the need for careful regulation of body temperature. This is because cellular biochemistry is temperature dependent; that is, the rate of biochemical reactions increases as the temperature rises and decreases as the temperature falls. Changes in body temperature, therefore, may greatly influence nerve and muscle physiology, hormone- and enzyme-dependent reactions, and most biochemical and biophysical phenomena.

Humans regulate their body temperature close to a mean value of 37° C. The mean value represents the mean temperature in deep central areas (brain, heart, lungs, abdominal viscera) and is commonly referred to as core temperature. The maintenance of body temperature is accomplished through a balance of heat loss and heat gain. Table 34.1 lists several factors involving heat gain and heat loss.

TABLE **34.1** Mechanisms of heat gain and loss

Heat gain	***Heat loss***
Increased BMR	Evaporation
Shivering	Radiation
Increased hormone secretion	Conduction
Piloerection (air trapping)	Convection

Cellular metabolism results in the continual production of heat, because most chemical reactions are *exothermic* (give off heat energy). Body temperature may be raised by increasing the rate of metabolism. Many hormones are known to be calorigenic. Examples include thyroxine, thyroid-stimulating hormone, adrenocorticotropic hormone, and glucocorticoids. Increasing the synthesis and release of calorigenic hormones elevates the basal metabolic rate and, subsequently, the body temperature. Shivering, a type of skeletal muscle contraction in which most of the muscle's chemical energy is converted into heat rather than work, also results in increased heat production. Piloerection (erection of body hair) results in the trapping of air near the skin surface. Air that is not moving is an excellent insulator; therefore, less body heat is lost through radiation, conduction, and so on when body hair is erect. Because of the relative scarcity of body hair on the human, however, piloerection is of minimal value in conserving body heat. For other primates and vertebrates (such as apes, monkeys, dogs, cats), piloerection is a very important mechanism for conserving body heat, evidenced in part by the increase in the thickness of an animal's fur in the winter.

The principal route for heat loss is the evaporation of water on the skin surface. A large amount of thermal energy is required to convert liquid water to its gas phase. Because this heat is derived from body heat, it results in heat loss. The amount of heat that can be lost by evaporation is, however, inversely related to the humidity of the surrounding air.

If skin temperature is higher than environmental temperature, heat will be lost by radiation from areas of the body exposed to the atmosphere. If skin temperature is lower than environmental temperature, heat will be gained by radiation. Conduction is heat exchange between two objects of different temperatures (e.g., the body and a chair in which the body is seated), including heat transfer between air and liquid. The amount of heat gained or lost through conduction is directly proportional to the temperature difference between objects. Normally, little heat is lost from the body through conduction unless the body is in cold water. Convection refers to the transfer of heat from one area of a gas or liquid to another by air or fluid current. The movement of air over the surface of the body assists in convective heat loss (e.g., fanning on a warm day), as does the movement of fluid over the body surface (e.g., taking a cool shower). At normal room temperature, conduction, convection, and radiation play a minimal role in the regulation of body temperature.

Body temperature is controlled by the *hypothalamus,* which may be thought of as containing a "heat-gain" center and a "heat-loss" center. Physiologic adjustments made by the heat-gain center tend to elevate body temperature. Conversely, physiologic adjustments made by the heat-loss center tend to decrease body temperature. Both centers operate to maintain body temperature at a mean value of 37° C ± 0.5° C by effecting heat gain and conservation when body temperature begins to fall and by effecting heat loss when body temperature begins to rise. Some of the physiologic adjustments made in regulating body temperature are listed in table 34.2.

The widest range of core temperatures observed in normal humans has a lower limit of 35° C and an upper limit of 41° C. Body temperatures outside this range suggest an impaired ability to effectively regulate body temperature.

TABLE 34.2 Physiologic adjustments controlled by the hypothalamus

Heat gain center	***Heat loss center***
Increased metabolism	Decreased metabolism
Secretion of calorigenic hormones	Sweating
Shivering	Peripheral vasodilation
Peripheral vasoconstriction	Panting
Piloerection (air trapping)	Lethargy

Although core body temperature is carefully regulated by the hypothalamus, body temperature is not the same everywhere in or on the body (figure 34.1). Skin temperatures average about 4° C lower than rectal temperatures, even though both may be used as indicators of core temperature. Body temperature also exhibits a circadian (*circa* = about, *dies* = day) rhythm—a 24-hour cycle during which core temperature fluctuates ±0.5° C around the normal mean core temperature. Body temperature is lowest in early morning and highest in midafternoon. Age influences body temperature, as evidenced by the fact that children have slightly higher oral (37.5° –38.0° C) and rectal temperatures than adults, and gender influences body temperature. Basal body temperature falls at the time of ovulation in the female and then elevates slightly during the postovulatory phase of the menstrual cycle.

In the experiments that follow, several aspects of body temperature, its regulation, and its measurement will be examined.

■ EXPERIMENTAL OBJECTIVES

1. To learn the procedures for obtaining oral and axillary temperatures.
2. To compare oral and axillary temperatures.
3. To observe the effects on oral temperature of time and environmental change.
4. To observe the effects of evaporation, conduction, and convection on body temperature.

Materials

500-mL beakers

100-mL beakers

0° –100° C thermometer

ice

paraffin oil

ether

ring stand

acetone

digital oral thermometer (Marshall model 4B)

mercury-free oral thermometer, dual scale (Gerathorm)

disposable probe covers for digital thermometers, model 9Prob (Omron Healthcare, Inc.)

liquid crystal strip thermometer (20°–40° C) with strap (Carolina Biological Supply Co.)

paper cup

rubber bands

cotton

brass and glass rods

double clamps

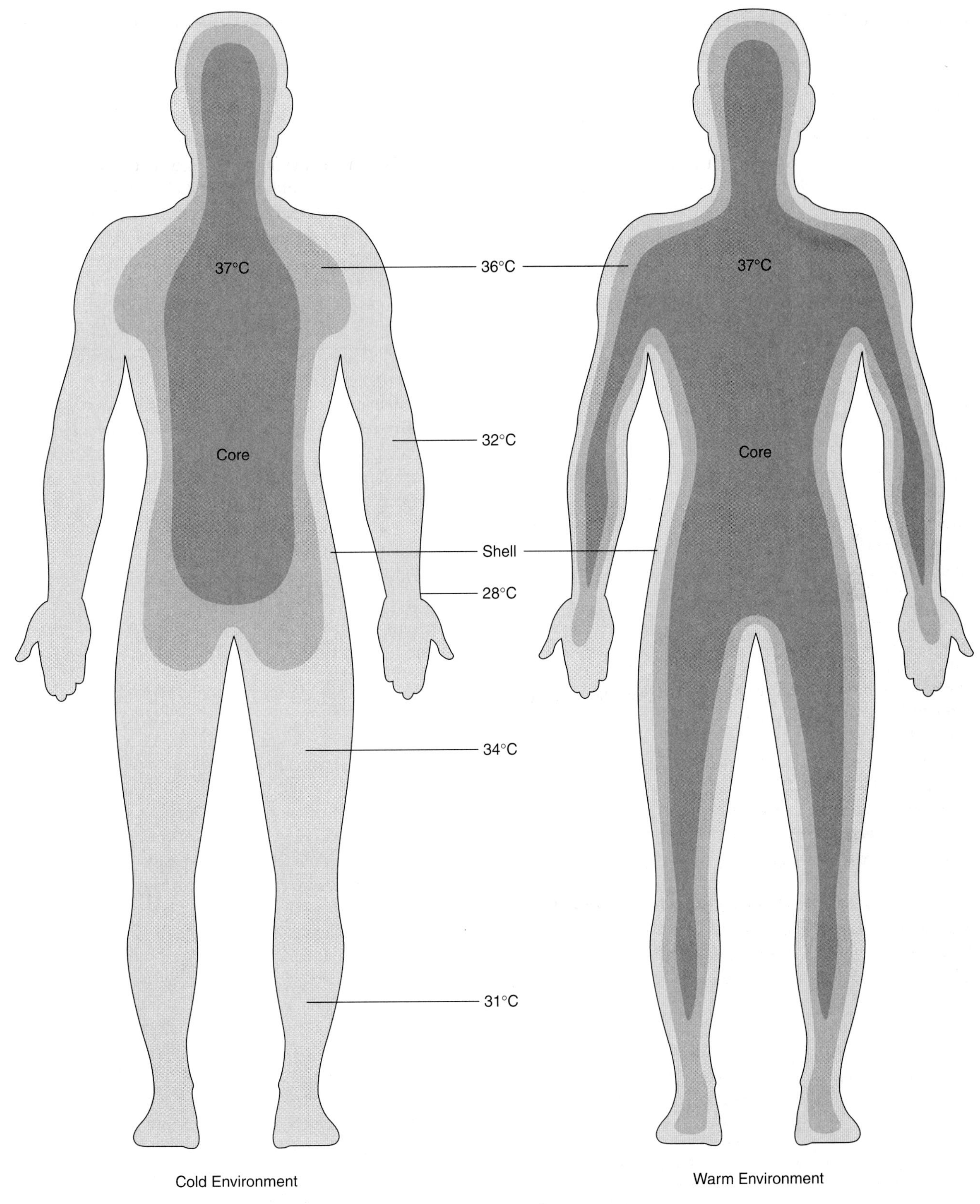

FIGURE 34.1 Body temperature versus environmental temperature

■ EXPERIMENTAL METHODS

1. Obtain a digital fever thermometer from the laboratory instructor.

   Mercury-Free Oral Thermometer

   Firmly grasp the end opposite the probe and gently shake the alcohol down into the bulb. Install a disposable probe cover. Insert the bulb of the thermometer under your tongue so that the tip of the thermometer is at the base of the tongue (figure 34.2) and close your mouth. Remove the thermometer after exactly 30 seconds and record the temperature. Carefully shake the thermometer so that the alcohol collects in the bulb. Repeat the observation, allowing the thermometer to remain in the oral cavity exactly 1 minute. Obtain a series of readings for ½, 1, 1-½, 2, 2-½, and 3 minutes. Record the data.

   Are all of the oral temperature values the same? Why or why not? What is the minimum time the oral thermometer should remain in a closed mouth? Remove and discard the probe cover.

   Digital Thermometer

   Install a disposable probe cover. Press the on/off button to turn the unit on. The display window will display "98.6 + 0.2° F" for about 1 second, then it will display an L or an H with a flashing "° F." The unit is ready for temperature measurement. Place the thermometer probe tip under your tongue at the base of the tongue (figure 34.2) and close your mouth. Remove the thermometer after exactly 30 seconds and record the temperature. Press the on/off button to turn the unit off, and again to turn it back on (this maneuver clears the previous measurement), and repeat the observation, allowing the thermometer to remain in the oral cavity exactly 1 minute. Obtain a series of readings for ½, 1, 1-½, 2, 2-½, and 3 minutes. Record the data.

   Are all of the oral temperature values the same? Why or why not? What is the minimum time the oral thermometer should remain in a closed mouth?

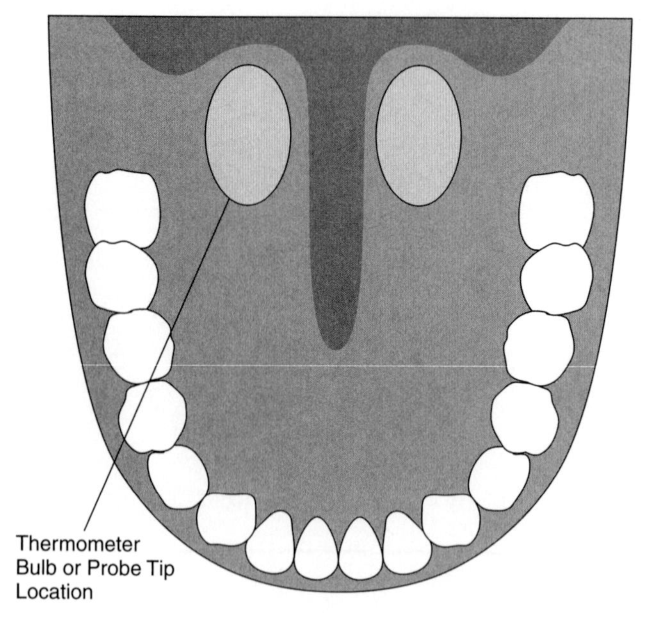

FIGURE 34.2 Placement areas for oral thermometer/temperature probe

2. Take a drink of cold water and immediately record your oral temperature as outlined in method 1. Obtain a series of readings for ½, 1, 1-½, 2, 2-½, and 3 minutes, taking a drink at the beginning of each recording interval. Record the data. Does it take longer for a stable oral temperature to be recorded?
3. Place the bulb or probe tip of the thermometer in your closed axilla and determine (as above) the minimum time to obtain a maximal reading. Record all data and compare the axillary temperatures with the oral temperatures obtained before. Typically axillary temperatures are about a degree lower than oral temperatures. Remove and discard the probe cover.
4. Obtain a liquid crystal strip thermometer from the laboratory instructor. The temperature band is a flexible mylar strip with linearly variable temperature sensitivity. Liquid crystal numerals display temperatures from 20° C to 40° C. Use the temperature band to measure skin temperature on the following parts of your body: forehead, right and left arms, forearms, palms, and right and left legs (between knee and ankle). Record temperature data in the report and compare the data with figure 34.1. Is skin temperature the same everywhere on the body? Why or why not?
5. Working in groups, wrap a thin layer of cotton around the bulb of a chemical thermometer (0°–100° C), and fasten it with a rubber band. Place two chemical thermometers (one wrapped, one plain) into a beaker containing water at approximately 40° C. Permit the thermometers to remain in the water until both read the same temperature, then remove both thermometers and suspend them from double clamps attached to a ring stand. Record the temperature in each thermometer at 1-minute intervals for 15 minutes. Enter the data in the report. Do the temperatures of the two thermometers fall at the same rate? Why not? What principles of heat loss are illustrated by this experiment?
6. Wet one hand completely by holding it under a laboratory faucet for a few minutes. Now expose the wet hand and the dry hand and blow air alternately over each one. Is a sensation of equal warmth or coolness felt from both hands? Why or why not? What principles of heat loss are illustrated by this experiment?
7. Place 2 drops of each of the following solutions at room temperature on the inner surface of the wrist in

the following order: paraffin oil, water, acetone, ether. Which drop disappears first? Which gives rise to the warmest temperature sensation? Why? Explain the phenomena in view of the fact that the temperatures of all of the applied solutions were identical. Record all observations.

8. Obtain a piece of glass rod (2.5 cm in diameter) and a piece of brass rod (2.5 cm in diameter) from the laboratory instructor. Holding the glass rod at one end, place the length of the rod on the inner aspect of your forearm for about 5 sec and note any sensation of temperature. Repeat the procedure using the brass rod. Are the temperature sensations in each case identical? Why or why not? Explain the phenomenon in view of the fact that the two rods had identical temperatures (room temperature), diameters, and lengths. What principle of heat loss does this experiment illustrate?
9. Using your personal clinical thermometer at home or a thermometer provided by the instructor, record oral temperature every 2 hours from the time of arising until retiring during one full day and every 3 hours from the time of retiring to the time of arising the next day. Record the data and plot temperature versus time in the report (figure 34.3). At which time during the 24-hour period is temperature highest? Lowest? Why?

Maintenance and Regulation of Body Temperature

REPORT 34

Name: ______________________ Date: ______________________

Lab Section: ______________________

1. Oral temperature data for initial experiment: In older mercury thermo -

½ minute ________ 1½ minutes ________ 2½ minutes ________

1 minute ________ 2 minutes ________ 3 minutes after 3 mins - normal reached

2. What practical point about taking someone's oral temperature is indicated by the above data?

____

3. Oral temperature data after a drink of cold water:

½ minute ________ 1½ minutes ________ 2½ minutes ________

1 minute ________ 2 minutes ________ 3 minutes ________

4. What practical point do the above data indicate about taking someone's oral temperature?

wait for ≈ 5 mins.

5. Axillary temperature data: - Not as high as oral usu. 1° lower than oral

½ minute ________ 1½ minutes ________ 2½ minutes ________

1 minute ________ 2 minutes ________ 3 minutes ________

6. Temperature data from the skin temperature experiment:

Ambient air ________ Forehead ________

Right arm ________ Left arm ________

Right forearm ________ Left forearm ________

Right palm ________ Left palm ________

Right leg ________ Left leg ________

7. Give two reasons why skin temperatures vary from one part of the body to another. ________

extremities would be lower than core

25° C

8. Temperature data from the thermometer temperature experiment:

	Thermometer Temperature	
Minute	***Wrapped***	***Unwrapped***
1		
2		
3		
4		
5		
6		
7		
8		
9		
10		
11		
12		
13		
14		
15		

9. Analysis of the above data indicates that the temperature of each thermometer falls at a different rate. Why? What principles of heat loss are illustrated? ______________________

10. After you wet one hand and then blew air over both the wet and dry hand, why were the sensations of warmth or coolness not the same for each hand? What principles of heat loss does this experiment illustrate?

convection & evaporation

11. According to the results obtained in step 7 of the experiment list, rank the following solutions relative to the sensation of coolness perceived (1 = cold, 4 = warm):

a. Water ______________ **c.** Ether ______________

b. Acetone ______________ **d.** Paraffin oil ______________

12. Explain the results of question 11. ______________________

Rate of evaporation feels colder.

13. Which rod (glass or brass) resulted in the coldest sensation when placed on the forearm? Why? What principle of heat loss does this experiment illustrate? ______________________________

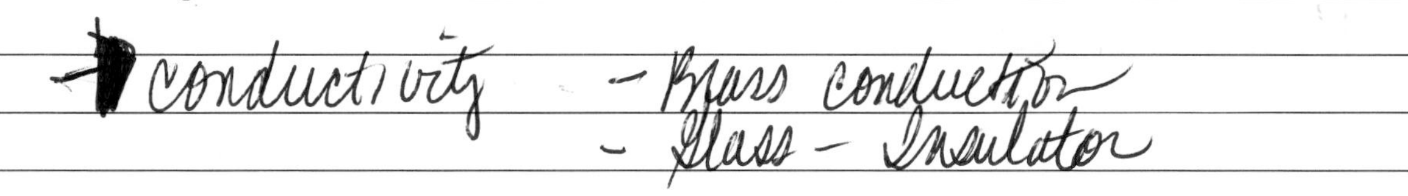

14. Oral temperature data from the home-based experiment:

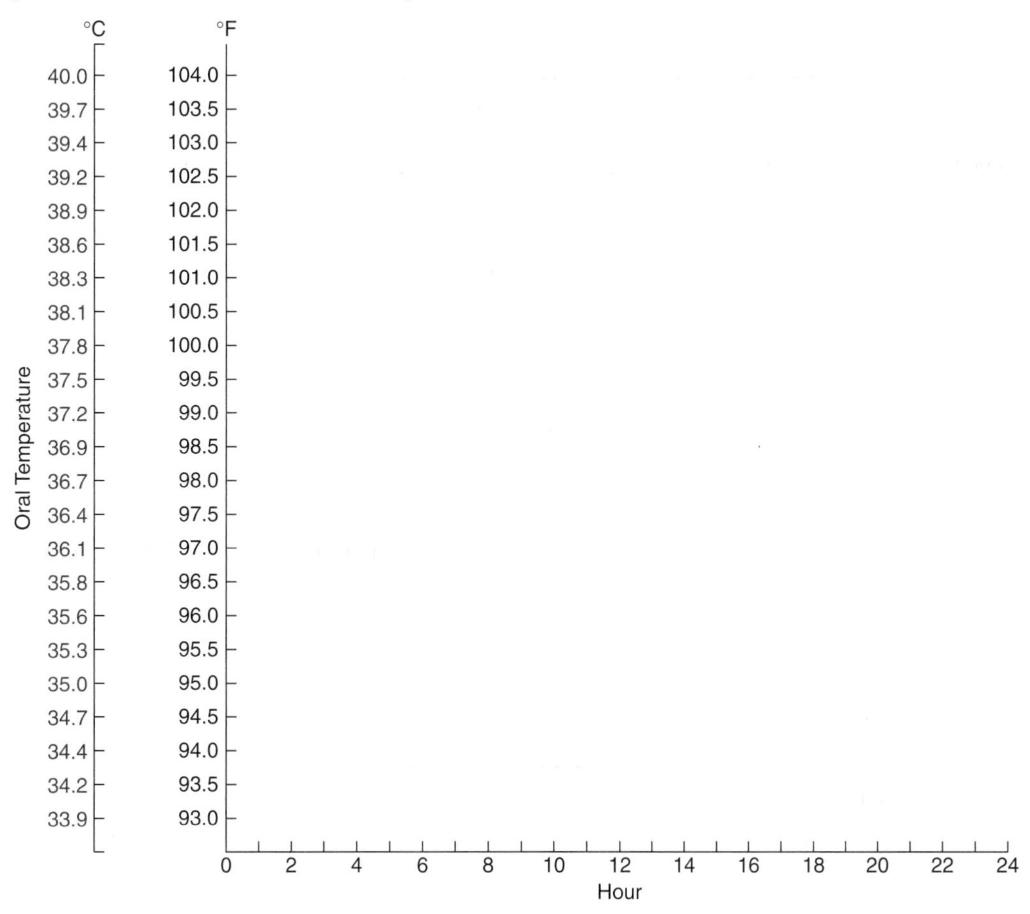

FIGURE **34.3** Body temperature versus time

15. What does the term *relative humidity* mean? Why is it more difficult to regulate body temperature via sweating on a warm, humid day than on a warm, dry day? — ↓ rate of evaporation

16. It is a common practice for people to "warm up" on cold winter days by drinking alcohol (e.g., whiskey) because of the sensation of warmth that is perceived. Actually, drinking alcohol tends to accelerate heat loss rather than facilitate heat gain. How? ______

Vasodilation

CHAPTER 35

Glucose Tolerance

INTRODUCTION

Monosaccharides formed from the breakdown of carbohydrates, amino acids formed from the breakdown of proteins, and fatty acids and glycerol formed from the breakdown of lipids are the principal sources of energy for cellular metabolism. However, the preferred source of energy for cellular metabolism by many of the body's cells is a monosaccharide—*glucose*. Indeed, some cells (e.g., neurons) cannot derive sufficient energy from glycerol, fatty acids, or amino acids and, therefore, require glucose to maintain the normal production of cellular energy.

The concentration of glucose in the blood is regulated within a normal range of 60–120 mg/dL by the endocrine portion of the pancreas, known as the *islets of Langerhans*. A decrease in the plasma concentration of glucose stimulates the *alpha cells* (α) of the islets to secrete glucagon, a polypeptide hormone that elevates the plasma glucose level by stimulating *glycogenolysis* (breakdown of glycogen to form glucose) and *gluconeogenesis* (formation of glucose from lipid or protein). An increase in the plasma concentration of glucose stimulates the *beta cells* (β) of the islets to secrete insulin, a polypeptide hormone that lowers the plasma glucose level by stimulating *glycogenesis* (conversion of glucose to glycogen) and increasing cellular uptake of glucose.

The hypersecretion of glucagon and the hyposecretion of insulin elevate the plasma glucose level above the normal range *(hyperglycemia)*. The hypersecretion of insulin and the hyposecretion of glucagon depress the plasma glucose level below the normal range *(hypoglycemia)*. Normal pancreatic function serves to maintain plasma glucose at a normal level through a balance of insulin and glucagon secretion, as illustrated in figure 35.1.

Failure of the pancreas to secrete an adequate amount of insulin in response to elevated plasma glucose results in diabetes mellitus (sugar diabetes), a potentially fatal disease characterized by hyperglycemia, glucosuria, accelerated lipid and protein metabolism, ketoacidosis, polyuria (increased urine flow), polydipsia (excessive thirst), and polyphagia (excessive eating).

In the diagnosis of diabetes mellitus, a *glucose tolerance test* is often performed. The test may be administered orally or intravenously. The oral glucose tolerance test (GTT) is sometimes used as a screening test because it is simple and noninvasive. The test consists of prescribing a regular mixed diet containing at least 250 g of carbohydrate daily for 3 days prior to the test. On the fourth day, following an overnight fast, blood and urine specimens are obtained and analyzed for glucose. An oral dose of 80–100 g of glucose dissolved in cold water or a commercially prepared carbonated and flavored beverage containing the dose is then administered. After the ingestion of glucose, blood and urine samples are taken at intervals of 30, 60, 90, 120, 150, and 180 minutes and analyzed for glucose. A normal glucose tolerance test would give approximately the following levels of blood glucose; after fasting, the level would be 80 mg/dL; at 30 minutes, 150 mg/dL; at 60 minutes, 140 mg/dL; at 90 minutes, 75 mg/dL; at 120–180 minutes, 80 mg/dL. A normal glucose tolerance curve and two abnormal glucose tolerance curves are illustrated in figure 35.2. All urine samples are normally negative for glucose.

A diminished tolerance to glucose, indicated by an elevated curve, most commonly occurs in diabetes mellitus but may also be seen in patients with myasthenia gravis, hyperthyroidism, severe liver damage, emotional stress, and Cushing's disease. cortisol

An increased tolerance to glucose, indicated by a depressed curve, may occur in conjunction with pancreatic islet tumors, hypothyroidism, Addison's disease, hypofunction of the adenohypophysis, and diseases characterized by poor gastrointestinal absorption of carbohydrates (e.g., sprue, celiac disease).

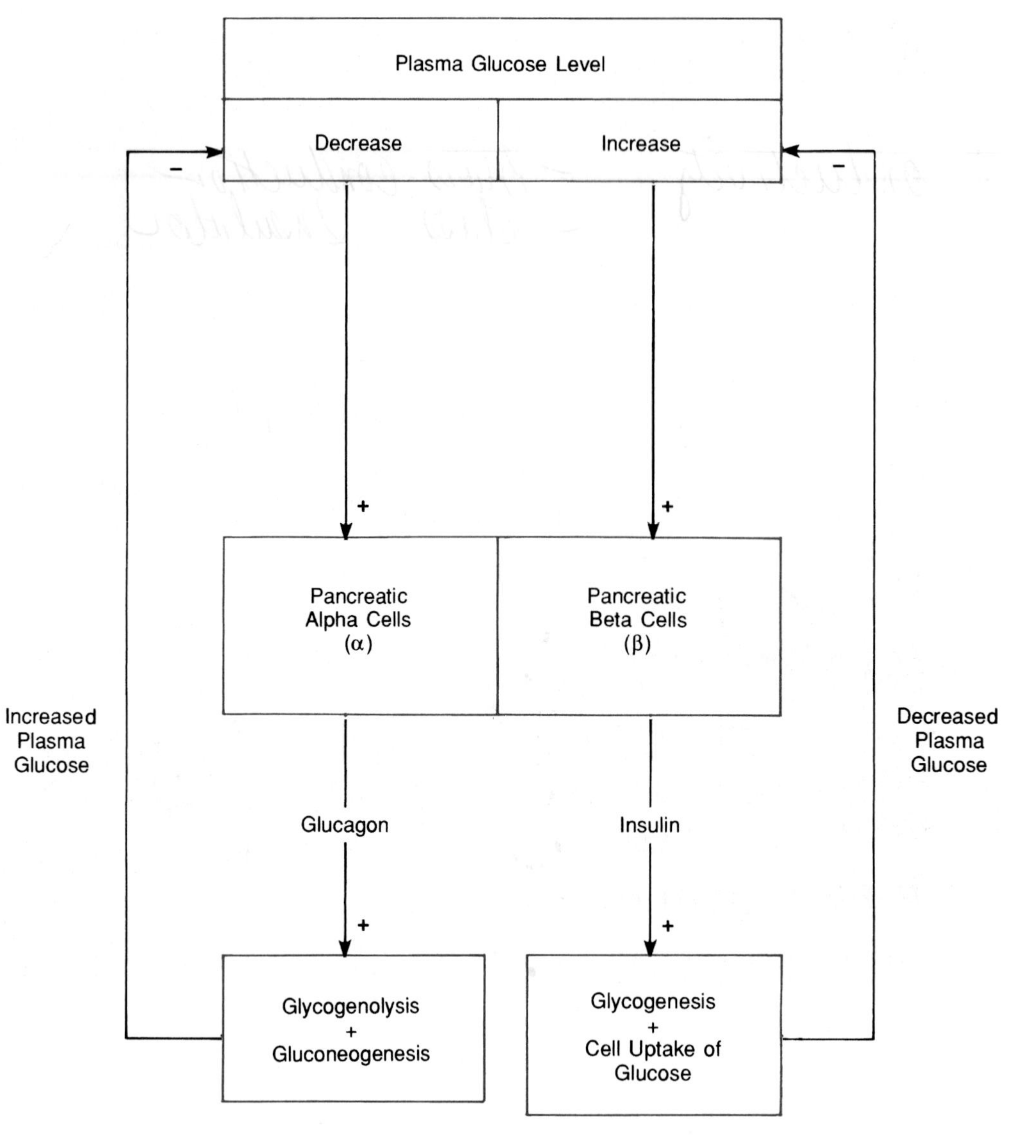

FIGURE 35.1 Regulation of plasma glucose by the endocrine portion of the pancreas

EXPERIMENTAL OBJECTIVES

1. To become familiar with an in vitro semiquantitative test for glucose in whole blood.
2. To perform and record the results of a glucose tolerance test using an in vitro semiquantitative method for glucose analysis.
3. To become familiar with the normal values of plasma glucose concentration.
4. To compare the glucose tolerance curves of adults of each gender and varying age.
5. To appreciate the role of the pancreas in maintaining the plasma glucose level.

EXPERIMENTAL METHODS

CAUTION

If you are diabetic or suffer from hypoglycemia (low blood sugar), you are not to be the subject of this experiment. Any history of other endocrine disorders or diseases of blood coagulation also preclude your volunteering as the subject. If you have any questions about your suitability as a subject, ask the instructor or your family physician.

In this experiment you may be asked to obtain a sample of blood from your fingertip as well as to obtain urine samples. If you have any history of a blood-clotting disorder or if you are being treated with anticoagulant medicine, inform your laboratory instructor and you will be excused from obtaining samples of your blood. Do not use samples from another student. Avoid contact with the bloods of other persons. Handle only your own blood.

Properly dispose of all sharps (lancets, needles, blades, etc.) by discarding them in appropriately marked hazardous waste containers as soon as possible after use. Nondisposable items that come into contact

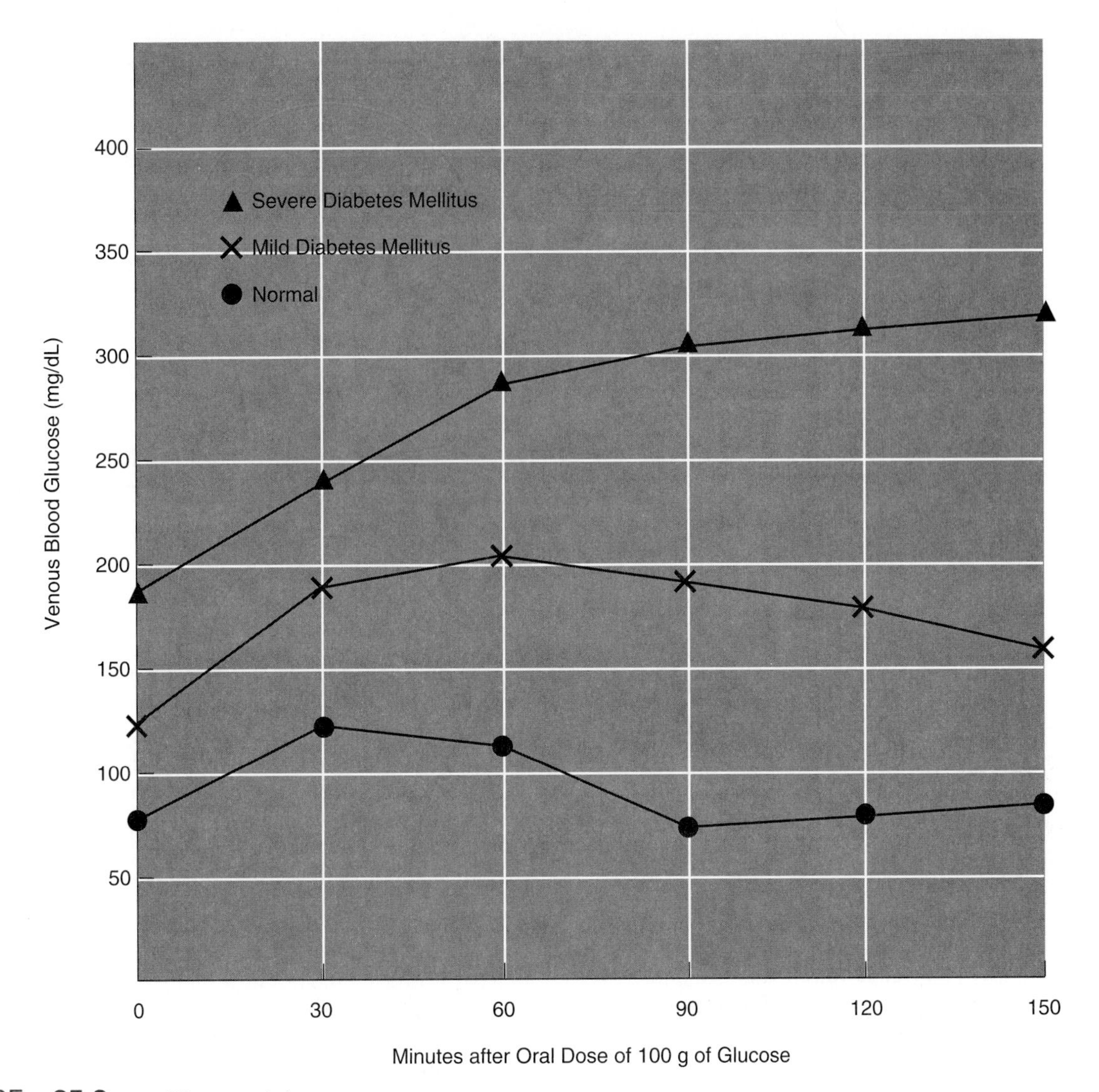

FIGURE 35.2 Glucose tolerance curves

with your blood or urine are to be washed with a detergent and tap water, then rinsed successively with tap water, bleach solution, distilled water, and acetone.

Disinfect your laboratory table or laboratory bench top by wiping it clean with a bleach solution before the beginning of the experiment and again at the end of the experiment.

Technique for Obtaining Fingertip Blood

Materials

lancets (sterile, prepackaged, single-use, manual or automatic) such as Single-Let (Bayer), or EZ-Lets II (Palco Labs, Inc.)

alcohol preps

gauze pads

finger bandages

disposable examination gloves

NOTE: Read all steps before proceeding. Ask questions if you do not understand the directions.

1. Clean the palmar surface of the third or fourth finger with a sterile gauze pad soaked with 70% alcohol or a sterile disposable alcohol prep. Save the pad.
2. Allow the skin to dry. Do not blow on it to make it dry faster.
3. Obtain an EZ-Lets II sterile, single-use, capillary blood sampling device from your laboratory instructor. Your laboratory instructor may require use of another capillary blood sampling device; if so, follow the instructor's directions carefully regarding procedures for its use.
   (a) Push the yellow stem into the body of the device (figure 35.3a). **Do not use the device if the yellow stem is missing.**
   (b) Hold the device securely with one hand while twisting the yellow stem with the other hand (figure 35.3b). Twist the stem completely off.

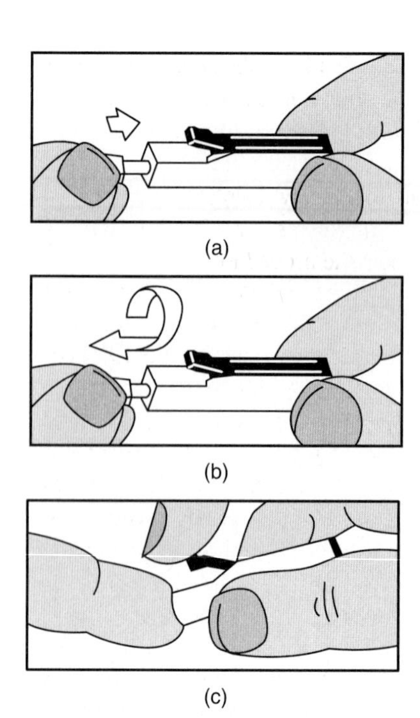

FIGURE **35.3** Procedure for obtaining a sample of fingertip capillary blood

(c) Place the opening against the cleansed, dry surface of the fingertip and firmly depress the end of the green lever (figure 35.3c). The needle automatically retracts after use.

4. Allow a good-sized drop to form before using the blood sample. Do not "milk" the finger or squeeze it to increase flow, as doing so will alter the composition of the sample. Do not allow blood to drip from your finger. Place the sample where it belongs, as directed in the experiments that follow.
5. Discard the lancet in the appropriately marked hazardous waste container for sharps. If another attempt at obtaining a fingertip sample is necessary, always use a new sterile lancet.
6. After obtaining the sample, compress the gauze pad over the cut until the bleeding ceases. Discard the gauze pad or alcohol prep in the appropriately marked hazardous waste container.

NOTE: Several methods for determining whole-blood glucose are described. The method used will be determined by your laboratory instructor. Regardless of the method used, follow directions precisely in order to obtain accurate data. Pay particular attention to the time requirements in the testing procedure.

Dextrostix Analysis

Materials

Dextrostix (Ames Company)
chilled carbonated glucose solution (Trutol or Glucola)
color charts
distilled water in plastic wash bottle
timer (stopwatch or watch with second hand)
urine specimen jar and lid

Experimental Procedure

The Dextrostix analysis of whole-blood glucose is based on the action of glucose oxidase, an enzyme that catalyzes the oxidation of blood glucose by oxygen in the atmosphere, producing gluconic acid and hydrogen peroxide. Peroxidase then catalyzes the reaction of hydrogen peroxide with a chromogen system, producing oxidized chromogens. The oxidized chromogens impart color to the reagent strip. The hue, intensity, and shade of the color are proportional to the amount of glucose present in the blood.

1. Select a subject from your laboratory group and have the subject perform the following test sequence before the ingestion of the glucose solution to determine the fasting level of blood glucose:
   (a) Remove a reagent strip from the Dextrostix container. Replace the cap immediately and tightly after removing the reagent strip. Do not touch the test area of the reagent strip, and keep the reagent strip away from contaminating substances on the laboratory table. Do not remove the desiccant from the Dextrostix container.
   (b) Compare the test area on the reagent strip closely with the "O" block on the color chart. If the reagent area on the strip does not closely match the "O" block, discard the strip.
   (c) Follow the technique for obtaining fingertip blood outlined previously in this chapter.
   (d) Apply a large drop of blood so that it covers the entire reagent area on the printed side of the strip.
   (e) Wait exactly 60 seconds.
   (f) Quickly (in 1 or 2 seconds) wash the blood off the strip with a sharp stream of water using a wash bottle. Do not overwash; otherwise, low readings will result.
   (g) Within 1 or 2 seconds after washing, hold the strip close to the color chart and read the result. Interpolate the result if the color produced falls between two color blocks. Record the data in the report.
2. Obtain a urine specimen jar with lid and mark it with your name. Collect a fasting midstream urine sample and analyze it for the presence of glucose using the Clinistix procedure (see chapter 31). Record the data in the report.
   Empty the urine specimen jar, rinse the jar and lid with tap water, and save for reuse.
3. Ingest the chilled glucose solution (in the form of a carbonated beverage) and note the time. The solution

may be gradually ingested within a period of 5–7 minutes.

4. Repeat the test procedure for blood glucose at 30, 60, 90, and 120 minutes after ingestion and the test procedure for midstream urine glucose at 60 and 90 minutes. Other fingers may be used to obtain the blood sample. Record the data and plot blood glucose versus time in the report. Compare with figure 35.2.

When interpreting the results, keep in mind that whole-blood glucose is being determined rather than plasma glucose. Also, mixed capillary blood, which normally has approximately 25 mg/dL more glucose than venous blood, is being assayed for glucose, so the glucose tolerance curve that you obtain will normally be higher than the normal curve illustrated in figure 35.2. Expected blood glucose values for capillary whole blood in normal, nonpregnant adults are as follows:

Fasting values between 60 mg/dL and 120 mg/dL

2-hour oral GTT values less than 140 mg/dL

Chemstrip bG Analysis

Materials

Chemstrip bG (Boehringer Mannheim)

Clinistix

chilled carbonated glucose solution (Trutol or Glucola)

color charts

cotton balls

timer (stopwatch or watch with second hand)

urine specimen jar and lid

Experimental Procedure

The determination of whole-blood glucose with Chemstrip bG test strips is based on actions of the enzymes glucose oxidase and peroxidase. Both test pads on the reagent strip contain chromogens (color indicators), glucose oxidase, and peroxidase. When blood is placed on the test pad, the glucose is oxidized, forming hydrogen peroxide, which oxidizes the indicators in the presence of peroxidase and causes a change of color of the pad. The hue, intensity, and shade of the color are proportional to the amount of glucose present in the blood.

1. Select a subject from your laboratory group and have the subject perform the following test sequence before the ingestion of the glucose solution to determine the fasting level of blood glucose.
   (a) Remove a test strip from the vial. Replace the cap immediately and tightly after removing the reagent strip. Do not touch the test pads on the reagent strip, and keep the reagent strip away from contaminating substances on the laboratory table. Note the lot number on the vial.
   (b) Compare the pads on the unused strip with the "unused" color block on the vial label. If the pads do not closely match the unused color block, discard the test strip, obtain a new test strip, and repeat the procedure.
   (c) Follow the technique for obtaining fingertip blood outlined previously in this chapter.
   (d) Touch the blood drop to the test pads, ensuring that both test pads are covered entirely with blood. Begin timing at once.
   (e) After exactly 60 seconds (do not stop timing), carefully wipe the blood from the test pads with a cotton ball. Wipe gently several times until all blood is removed. Use a clean part of the cotton ball for each wipe. Continue timing.
   (f) After another 60 seconds (2 minutes after adding blood to pads), match the colors on your test strip to the color scale on the vial your test strip came from (check the lot number on the vial to make sure it matches the one you noted earlier).
   (g) If the colors on your test strip match or are darker than those for 240 mg/dL, wait another 60 seconds (3 minutes after adding blood to pads) and then compare the final reacted colors with the color scale.
   (h) If the color match is not perfect, interpolate as follows:
      - Find the closest match to the lower pad on your test strip. *Example:* 80 mg/dL.
      - Next, find the closest match to the upper pad on your test strip. *Example:* 120 mg/dL.
      - Add the two numbers together. *Example:* 80 mg/dL + 120 mg/dL = 200 mg/dL.
      - Divide by 2 for estimated blood glucose level. *Example:* 200 mg/dL ÷ 2 = 100 mg/dL.

      Record the estimated fasting level of blood glucose in the report.
2. Obtain a urine specimen jar with lid and mark it with your name. Collect a fasting midstream urine sample and analyze it for the presence of glucose using the Clinistix procedure (see chapter 31). Record the data in the report.

   Empty the urine specimen jar, rinse the jar and lid with tap water, and save for reuse.
3. Ingest the chilled glucose solution and note the time. The solution may be gradually ingested within a period of 5–7 minutes.
4. Repeat the test procedures for blood glucose at 30, 60, 90, and 120 minutes after ingestion and the test procedures for midstream urine glucose at 60 and 90 minutes. Other fingers may be used to obtain the blood sample. Record the data and plot blood glucose versus time in the report. Compare with figure 35.2.

When interpreting the results, keep in mind that whole-blood glucose is being determined rather than plasma glucose. Also, mixed capillary blood, which normally has approximately 25 mg/dL more glucose than venous blood, is being assayed for glucose, so the glucose tolerance curve that you obtain will normally be higher than the normal curve illustrated in figure 35.2. Expected blood glucose values for capillary whole blood in normal, nonpregnant adults are as follows:

Fasting values between 60 mg/dL and 120 mg/dL

2-hour oral GTT values less than 140 mg/dL

Glucometer 3/Glucofilm System

Materials

Glucometer 3/Glucofilm system (Ames)

model #5485 Glucometer

Glucofilm test strips #238399

Glucometer 3 check paddle #95001018

Glucofilm normal control #5451A

chilled carbonated glucose solution (Trutol or Glucola)

Clinistix

timer (stopwatch or watch with second hand)

urine specimen jar and lid

Experimental Procedure

Glucofilm test strips are used to monitor blood glucose concentrations in whole blood. The test strip is a plastic strip with a reagent test pad on one end. The test pad reaction is based on the actions of the enzymes glucose oxidase and peroxidase. The glucose oxidase catalyzes the oxidation of glucose in blood in the presence of oxygen in the atmosphere, producing gluconic acid and hydrogen peroxide. In the presence of peroxidase, the hydrogen peroxide oxidizes tetramethylbenzidine in the test pad, turning the test pad blue with an intensity proportional to the glucose concentration. The glucometer compares the test pad color with a set of preprogrammed color/glucose-concentration values and displays the blood glucose concentration in mg/dL. The operating range of the Glucometer 3 is 20–500 mg/dL, with an accuracy of approximately 3.5%. Before determining blood glucose concentrations during the oral glucose tolerance test, become familiar with the Glucometer 3/Glucofilm system and proper procedures for use by (1) checking the meter, (2) running a quality control test, and (3) measuring the subject's preingestion blood glucose level.

Checking the Meter

1. Find the check (✓) paddle stored in the plastic box in your kit. Note and record in the lab report the values for the initial range as printed on the check paddle box. Do not touch the small end of the check paddle. Always handle the check paddle by the large end.
2. Press the button once to turn on the meter. A full display appears followed by a program number. This number does not need to be changed for a check paddle test.
3. Open the test slide at the bottom of the meter by pushing it to the left. The number 60 should appear in the register. Push the button again to activate the timer. A beep will signify the beginning of a 60-second countdown. Audible beeps again will occur at 22, 21, and 20 seconds.
4. At 20 seconds insert the narrow end of the check paddle fully into the test slot and be sure the side with the (✓) is facing the display screen.
5. Close the test slide and wait for display of the test result. The test value should be within the initial range recorded in step 1. Record the test value in the report.
6. Open the test slide. When the letter *d* is displayed, push the button. This will delete from memory the test paddle result and turn off the meter.
7. Remove the check paddle from the test slot and close the slide. Store the check paddle in the plastic container to protect the test pad surface.
8. The first test result obtained with the check paddle is an electronic baseline value. Any other check paddle testing with that paddle and the same meter should have results within 6 mg/dL of the initial test value. For example, if the initial test value was 71 mg/dL, then other tests should fall between 65 mg/dL and 77 mg/dL. If a test result is outside this 12 mg/dL range, check the test paddle end for fingerprints or marring, clean if necessary, and repeat the test, making sure the paddle is inserted fully into the test slot with the (✓) facing the display screen.
9. If a new check paddle is ever used with your meter, steps 2–7 will need to be repeated in order to establish a new baseline value.

Running a Quality Control Test

1. Remove one test strip from the Glucofilm test strip container and place it on a clean paper napkin that has been folded in half. Close the bottle cap tightly. Note the program number printed on the bottle.
2. Press the button to turn on the meter and wait for a program number from 1 to 8 to appear. Press the button repeatedly until the displayed program number matches the printed program number on the test strip container.
3. Open the test slide. The number 60 will appear and hold in the display.
4. Remove the cap from the control solution bottle. Hold the test strip by the handle and apply a drop of control solution to the test pad, covering the test pad. Immediately press the button for the 60-second countdown to begin.
5. Pick up the napkin and prepare to use it to wipe the solution from the test pad by folding it over the handle.
6. As the countdown continues, audible beeps will occur at 22, 21, and 20 seconds. At exactly 20 seconds,

quickly wipe the solution from the test pad. Use one firm stroke, applying pressure to the napkin as in cleaning the blade of a table knife.

7. Insert the test strip immediately and fully into the test slot and make sure the test pad is facing the display screen. Close the test slide immediately (before the countdown reaches 1) and wait for the test result to be displayed.
8. Compare the test result with the range listed in the Glucofilm control package insert. Record the range and the test result in your laboratory report.
9. Open the test slide. When the letter *d* is displayed, press the button to delete the control test result from storage and to turn off the meter.
10. Remove the test strip from the slot and close the slide.

Measuring Blood Glucose and Urine Glucose

1. Select a subject from your laboratory group and have the subject perform the following test sequence before the ingestion of the glucose solution to determine the fasting level of blood glucose:
   (a) Lay a clean napkin on the laboratory table and fold it in half. Remove one test strip from the Glucofilm test strip container and place it on the napkin. Close the bottle cap tightly.
   (b) Press the button to turn on the meter. A full display of 888 shows all symbols and segments working properly.
   (c) When the program number appears, match the number on the Glucofilm test strip container by pressing the button to advance to the correct program number.
   (d) Open the test slide. The number 60 will appear and hold in the display.
   (e) Follow the technique for obtaining fingertip blood outlined previously in this chapter.
   (f) Hold the test strip by the handle and apply blood to completely cover the test pad. Immediately press the button to begin the 60-second countdown.
   (g) At 25 seconds, pick up the folded napkin and fold it around the test strip handle getting ready to wipe the blood off the test pad. Beeps at 22 and 21 seconds alert you to get ready to wipe.
   (h) At 20 seconds, quickly wipe the blood from the test pad using one firm stroke, applying pressure to the napkin as in cleaning the blade of a table knife. Do not blot the pad.
   (i) Insert the test strip immediately and fully into the test slot, and make sure that the test pad is facing the display screen. Close the test slide immediately (before the countdown reaches 1) and wait for the test result to be displayed. Record the test value in the lab report.
   (j) Open the test slide. When the letter *d* is displayed, press the button to delete the test result from memory and to turn off the meter. Remove the test strip from the slot and close the slide.
2. Obtain a urine specimen jar with lid and mark it with your name. Collect a fasting midstream urine sample and analyze it for the presence of glucose using the Clinistix procedure (see chapter 31). Record the data in the report.

   Empty the urine specimen jar, rinse the jar and lid with tap water, and save for reuse.
3. Ingest the chilled glucose solution (in the form of a carbonated beverage) and note the time. The solution may be gradually ingested within a period of 5–7 minutes.
4. Repeat the Glucometer 3/Glucofilm test procedures (steps a–j) for blood glucose at 30, 60, 90, and 120 minutes after ingestion and the Clinistix procedure for midstream urine glucose at 60 and 90 minutes. Other fingers may be used to obtain the blood sample. Record the data and plot blood glucose versus time in the report. Compare with figure 35.2.

When interpreting the results, keep in mind that whole-blood glucose is being determined rather than plasma glucose. Also, mixed capillary blood, which normally has approximately 25 mg/dL more glucose than venous blood, is being assayed for glucose, so the glucose tolerance curve that you obtain will normally be a little higher than the normal curve illustrated in figure 35.2. Expected blood glucose values for capillary whole blood in normal, nonpregnant adults are as follows:

Fasting values between 60 mg/dL and 120 mg/dL

2-hour oral GTT values less than 140 mg/dL

Accu-Chek Instant System

Materials

Accu-Chek Instant blood glucose monitor (Boehringer Mannheim)

Accu-Chek Instant glucose test strips and control solution

chilled carbonated glucose solution (Trutol or Glucola)

Clinistix

timer (stopwatch or watch with second hand)

urine specimen jar and lid

Experimental Procedure

The determination of whole-blood glucose with the Accu-Chek Instant system is based on actions of the enzymes glucose oxidase and peroxidase. The yellow test pad on the test strip contains chromogens (color indicators), glucose oxidase, and peroxidase. When blood is placed on the test pad, glucose is oxidized, forming hydrogen peroxide, which oxidizes the indicators in the presence of peroxidase and causes the pad to change color. The hue, intensity, and shade of the color are proportional to the amount of glucose present in the blood.

The monitor photometrically compares the color of the used test pad with known glucose concentrations and displays the matching blood glucose concentration on the monitor screen. Before determining blood glucose concentrations during the oral glucose tolerance test, become familiar with the Accu-Chek Instant system and proper procedures for use by (1) checking the meter, (2) running a quality control test, and (3) measuring the subject's preingestion blood glucose level.

Checking the Meter

1. Turn on the monitor by pressing and releasing the on/off button (center right). A battery symbol should appear briefly in the lower left corner of the display (figure 35.4), then disappear. If the battery symbol does not disappear, contact your laboratory instructor.
2. After the display check, the word "CODE" will flash alternately with a four-digit code (or _ _ _ _ if the monitor has never been coded).
3. Check to make sure the code on the display matches the code printed in the blue box on the side of the vial of test strips you will be using. If the codes do not match or if the monitor has never been coded, press and hold the rocker button (center left) for 2 seconds until you see the word "CODE" and the code symbol. Use the rocker button (up or down) to select a matching code number, then press the on/off button one time to store the code.
4. The word "CODE" and the new four-digit code number should be flashing alternately on the display. If they do not, turn off the monitor and then repeat steps 1, 2, and 3.
5. When the test strip symbol appears in the lower right corner of the display, you may continue with a test.

Running a Quality Control Test

1. Take a test strip out of the vial and immediately replace the vial cap, making sure it is firmly in place. Do not touch the yellow test pad or round window on the back of the strip when handling the strip.
2. Compare the back of the strip with the "unused" color circle on the test strip vial. If the test strip is discolored, discard it and obtain a new one as in step 1.
3. With the test strip symbol flashing in the lower right corner of the display, insert the test strip into the slot of the test strip guide until it locks in place. The arrows of the test strip should be facing up and pointing toward the monitor when the test strip is inserted.
4. On the display, a blood drop symbol will appear and flash next to the test strip symbol, signaling that you may apply a drop of control solution.
5. Apply a hanging drop of control solution to the center of the yellow test pad. Do not touch the pad with the tip of the bottle of the control solution. Do not apply a second drop of control solution to the test pad. Do not smear the solution with the tip of the bottle of control solution.
6. After 12 seconds, the glucose control result is displayed. Compare the result with the acceptable

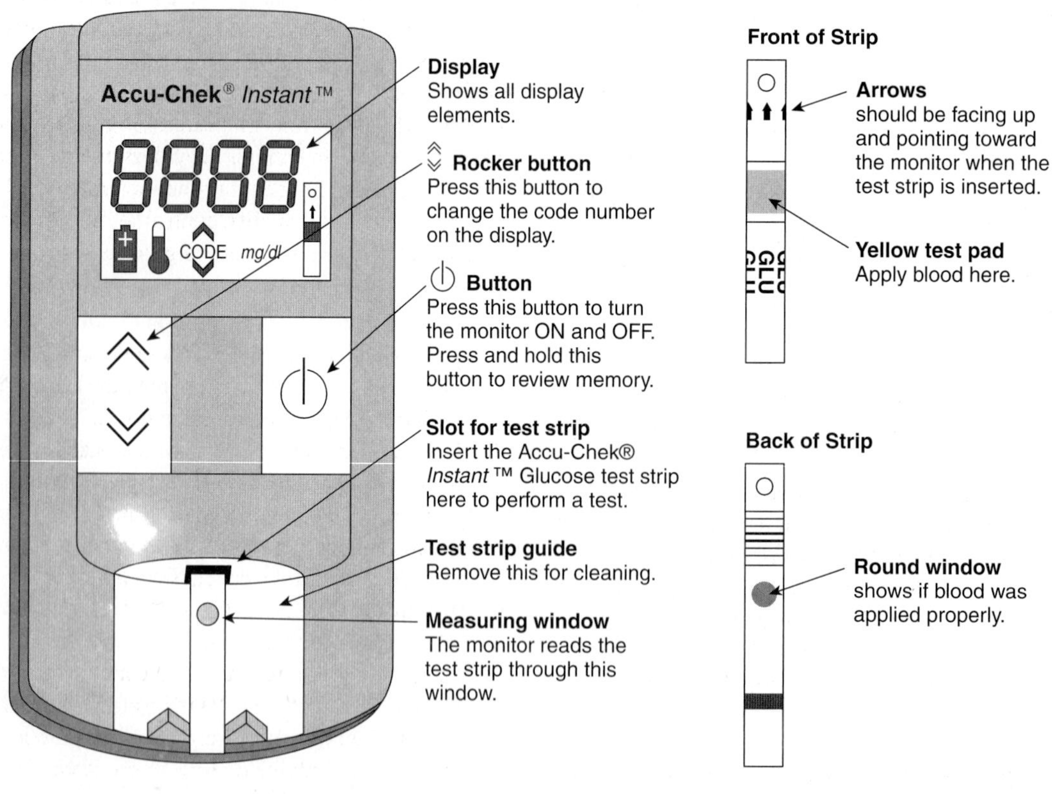

FIGURE 35.4 The Accu-Check® Instant Glucometer

range printed on the side of the test strip vial for the solution you used. If the result is not within the acceptable range, contact your laboratory instructor.

7. Turn off the monitor. Remove and discard the used test strip into a hazardous waste container.

Measuring Blood Glucose and Urine Glucose

1. Select a subject from your laboratory group and have the subject perform the following test sequence before the ingestion of the glucose solution to determine the fasting level of blood glucose:
   (a) Turn the monitor on, check that the display code matches the strip code, and compare the back of the test strip with the "unused" color circle on the vial. Be sure the test strip vial lid is secure after removing the test strip.
   (b) Insert the test strip into the monitor until it locks in place. Wait for the flashing blood drop icon to appear on the display.
   (c) Follow the technique for obtaining fingertip blood outlined previously in this chapter.
   (d) Apply a hanging drop of blood to the center of the yellow test pad. After 12 seconds, the blood glucose value is displayed. Record the measured value in the report.
   (e) Turn off the monitor. Remove the test strip and check the round window on the back of the test strip for even and complete color development. If the round window shows lighter color around the edges, too little blood was applied to the strip. This will result in a reading that is lower than the actual value. If this is the case, repeat the measurement using a new test strip and a larger drop of blood.
   (f) Discard the used test strip into a hazardous waste container.
2. Obtain a urine specimen jar with lid and mark it with your name. Collect a fasting midstream urine sample and analyze it for the presence of glucose using the Clinistix procedure (see chapter 31). Record the data in the report.

   Empty the urine specimen jar, rinse the jar and lid with tap water, and save for reuse.
3. Ingest the chilled glucose solution (in the form of a carbonated beverage) and note the time. The solution may be gradually ingested within a period of 5–7 minutes.
4. Repeat the Accu-Chek test procedures (steps a–f) for blood glucose at 30, 60, 90, and 120 minutes after ingestion and the Clinistix procedure for midstream urine glucose at 60 and 90 minutes. Other fingers may be used to obtain the blood sample. Record the data and plot blood glucose versus time in the report. Compare with figure 35.2.

When interpreting the results, keep in mind that whole-blood glucose is being determined rather than plasma glucose. Also, mixed capillary blood, which normally has approximately 25 mg/dL more glucose than venous blood, is being assayed for glucose, so the glucose tolerance curve that you obtain will normally be a little higher than the normal curve illustrated in figure 35.2. Expected blood glucose values for capillary whole blood in normal, nonpregnant adults are as follows:

Fasting values between 60 mg/dL and 120 mg/dL

2-hour oral GTT values less than 140 mg/dL

Glucometer Elite System

Materials

Glucometer Elite Meter with Check Strip (Bayer)

Glucometer Elite blood glucose test strips and code strip

normal control solution

chilled carbonated glucose solution (Trutol or Glucola)

Clinistix

timer (stopwatch or watch with second hand)

urine specimen jar and lid

Experimental Procedure

The determination of whole-blood glucose with the Glucometer Elite system is based on the actions of the enzyme glucose oxidase and potassium ferricyanide, reagents on the test strip. The blood sample is drawn into the tip of the test strip through capillary action. Glucose in the sample reacts with glucose oxidase and potassium ferricyanide. Electrons are generated, producing a current that is proportional to the glucose in the sample. After the reaction time, the glucose concentration in the sample is displayed on the meter screen. Before determining blood glucose concentrations during the oral glucose tolerance test, become familiar with the Glucometer Elite system and proper procedures for use by (1) checking the meter, (2) calibrating the meter, (3) running a quality control test, and (4) measuring the subject's preingestion blood glucose level.

Checking the Meter

1. Remove the Check Strip package insert from the glucometer storage case and note the *Check Strip Range* printed on the package insert.
2. Remove the Check Strip from the package and insert the Check Strip fully (until it comes to a complete stop) into the glucometer (with the tab pointing right). A beep sounds, and a full display appears. Another beep sounds and the check Strip test result is displayed. A checkmark symbol appears on the display when the Check Strip test is done.
3. Compare the test result with the acceptable range of values listed on the package insert. If the test result is outside the acceptable range, see your laboratory instructor for advice on how to proceed.
4. Remove the Check Strip, place it in its storage package, and return it to the glucometer storage case.

Calibrating the Meter (Matching the Meter to the Reactivity of the Test Strip)

1. Each carton of test strips contains a code strip. Remove the code strip from its protective packet (save the packet) and note the function number (for example, F-4) printed on the code strip.
2. Insert the code strip fully into the test slot (tab pointing to the right). A beep sounds and a full display appears briefly. Another beep sounds and a function number (for example, F-4) appears in the display.
3. The function number appearing in the display *must* match the number appearing on:
   (a) the code strip
   (b) the front of the code strip packet
   (c) the back of the test strip packet
   If the function numbers match as indicated above, the meter may now be used to run a control test or to measure blood glucose. If they do not match, consult your laboratory instructor for advice on how to proceed.
4. Store the code strip in the clear plastic packet and return it to the test strip container.

Running a Quality Control Test

1. Open a foil test strip packet by peeling back the foil until the test strip is completely exposed.
2. Remove the test strip. Holding the round end, insert the strip fully into the meter. Save the empty foil packet for use in step 4.
3. A beep sounds and a full display appears followed by a function number. The function number and the previous test result begin flashing alternately.
4. Gently squeeze a small drop of control solution onto the flattened inside edge of the empty foil packet.
5. Touch and hold the *test end* of the test strip in the drop until after the meter beeps. A small amount of control solution is automatically drawn into the test strip, and the internal timer begins counting down from 29 seconds.
6. After 29 seconds, the control test result appears in the display. Compare the result with the range listed on the end flap of the test strip carton.
7. Using a tissue, remove the test strip and discard it. The meter will turn off automatically.

Measuring Blood Glucose and Urine Glucose

1. Select a subject from your laboratory group and have the subject perform the following test sequence before the ingestion of the glucose solution to determine the fasting level of blood glucose.
   (a) Obtain six foil packets of test strips. Make certain the meter has been properly coded for the test strips selected and that the function numbers match. The function number printed on the code strip, code strip packet, and test strip foil must match the function number displayed on the meter. Open a foil test strip packet by peeling back the foil until the test strip is completely exposed.
   (b) Remove the test strip. Holding the round end, insert the strip fully into the meter. A beep sounds, and a full display appears followed by a function number. The function number and the previous test result begin flashing alternately.
   (c) Follow the technique for obtaining fingertip blood outlined previously in this chapter. Remember to discard all materials coming into contact with your blood into the hazardous waste container and to minimize exposure of your blood to other students in the laboratory.
   (d) Touch and hold the test end (tip) of the test strip to the drop of blood until after the meter beeps. Blood is automatically drawn into the test strip. The internal timer begins counting down from 29 seconds.
   (e) After 29 seconds, the blood glucose result appears in the display. Record the result in the laboratory report.
   (f) Using a tissue, remove the test strip and discard both into the hazardous waste container.
2. If you are the subject, obtain a urine specimen jar with a lid and mark it with your name. Go to the restroom and collect a fasting midstream urine sample. Return to the laboratory and analyze the sample for the presence of glucose, using the Clinistix procedure (see chapter 31). Record the data in the report.
3. Empty the urine specimen jar, rinse the jar and lid with tap water, and save for reuse.
4. Ingest the chilled glucose solution (in the form of a carbonated beverage) and note the time. The solution may be gradually ingested within a period of 5–7 minutes.
5. Repeat the Glucometer Elite test procedures for blood glucose at 30, 60, 90, and 120 minutes after ingestion and the Clinistix procedure for midstream urine glucose at 60 and 90 minutes. Other fingers may be used to obtain the blood sample. Record the data and plot blood glucose versus time in the report. Compare with figure 35.2.

When interpreting the results, keep in mind that whole-blood glucose is being determined rather than plasma glucose. Also, mixed capillary blood, which normally has approximately 25 mg/dL more glucose than venous blood, is being assayed for glucose, so the glucose tolerance curve that you obtain will normally be a little higher than the normal curve illustrated in figure 35.2. Expected blood glucose values for capillary whole blood in normal, nonpregnant adults are as follows:

Fasting values between 60 mg/dL and 120 mg/dL

2-hour oral GTT values less than 140 mg/dL

Glucose Tolerance

REPORT 35

Name: ______________________ Date: ______________________

Lab Section: ______________________

1. Glucose tolerance data: blood and urine glucose:

a. Subject's initials __________ Age __________ Gender __________ Height __________ Weight __________

b. Blood glucose method __________

c. Urine glucose method __________

Blood Glucose		***Urine Glucose***	
Time	***Concentration***	***Present***	***Absent***
Fasting	mg/dL		
30 minutes	mg/dL		
60 minutes	mg/dL		
90 minutes	mg/dL		
120 minutes	mg/dL		

d. Plot the results on the glucose tolerance graph (figure 35.5)

2. Define the following terms:

a. Insulin ______________________

b. Glucagon ______________________

c. Hyperglycemia ______________________

d. Hypoglycemia ______________________

e. Diabetes mellitus ______________________

f.. Glycogenolysis ______________________

g. Gluconeogenesis ______________________

h. Glycogenesis ______________________

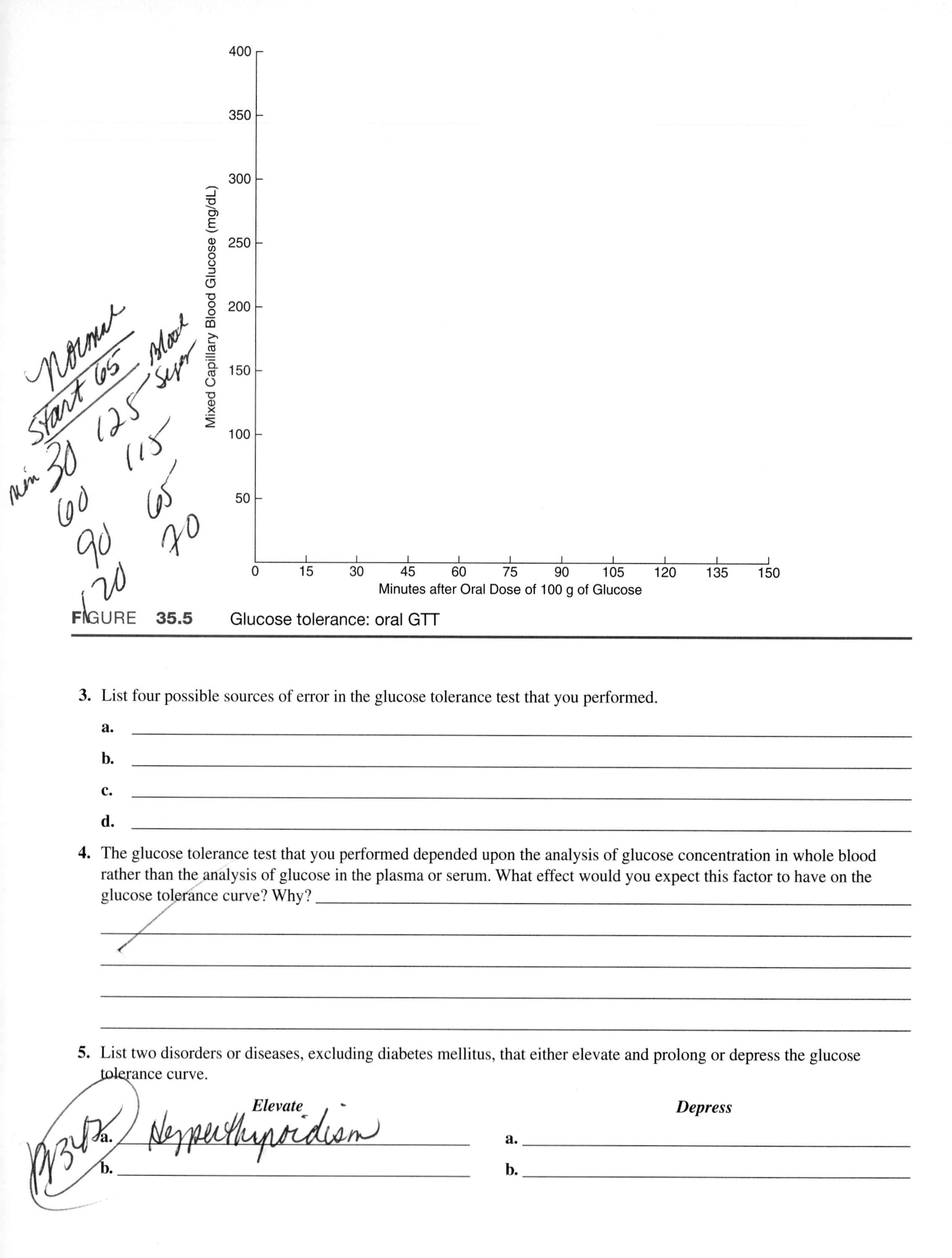

FIGURE 35.5 Glucose tolerance: oral GTT

3. List four possible sources of error in the glucose tolerance test that you performed.

a. ______________________

b. ______________________

c. ______________________

d. ______________________

4. The glucose tolerance test that you performed depended upon the analysis of glucose concentration in whole blood rather than the analysis of glucose in the plasma or serum. What effect would you expect this factor to have on the glucose tolerance curve? Why? ______________________

5. List two disorders or diseases, excluding diabetes mellitus, that either elevate and prolong or depress the glucose tolerance curve.

Elevate	*Depress*
a. ______	**a.** ______
b. ______	**b.** ______

Preparation of Agar Plates and Pipettes

Place 20 g of Noble agar in an 1800-mL Erlenmeyer flask and add distilled water to the 1000-mL mark. Boil the solution until the agar dissolves (about 40 minutes).

Pour level amounts into petri plates. Use a Bunsen burner to flame out any air bubbles and allow the agar to harden. Punch a 1-cm well in the center of the plate.

Use the agar solution to prepare 10 mL and 1 mL agar pipettes. Add a few drops of phenol red indicator until the agar solution turns bright yellow. Use a "Pi-Pump" or other pipetting pump to draw the agar solution up into the pipettes. Cover the tips of the pipettes with Parafilm and store in the refrigerator until used. When the agar pipettes are to be used in chapter 4, file a slot at the first graduation nearest the tip of the pipette and break off the tip.

Preparation of the Egg Osmometer

Obtain the following items: fresh chicken egg (from grocer), a 30-inch piece of glass tubing, Devcon 5-minute epoxy cement, a meterstick, scotch tape, a flat-base laboratory stand, burette clamp, double clamp, dissection kit, distilled water, and two glass beakers (50 mL and 500 mL). Refer to figure 4.4

1. Tape the glass tube to the meterstick so that 5 cm of the tube extends beyond the zero mark.
2. Clamp the meterstick–glass tube to the flat-base stand as shown in figure 4.4
3. Use the large end of a blunt probe to gently crack the egg's shell at the large end of the egg. Use needle-nose forceps to remove bits of shell until a circle about the size of a nickel is formed. Do not pierce the membrane that lies beneath the shell. If the membrane is violated, use another egg.
4. Turn the egg over and make a hole at the top of the smaller end slightly larger than the diameter of the glass tube. Remove shell and membrane.
5. Rest the egg, large end down, on the mouth of a 50-mL beaker resting inside an empty 500-mL beaker. Adjust the position of the meterstick–glass tube and gently insert the free end of the glass tube into the small end of the egg. Be careful not to puncture the yolk sac.
6. Use Devcon 5-minute epoxy to seal the hole where the glass tube enters the egg. This seal must be watertight.
7. Carefully add distilled water to the 500-mL beaker until the water level rises to the middle of the egg. Allow 30 minutes and check to see if fluid appears in the glass tube. If no fluid appears after 30–60 minutes, the osmometer will not function and another egg must be prepared.

APPENDIX C

Pithing the Frog

1. Grasp the frog firmly with your left hand, placing its snout between your index finger and middle finger. Flex its head with your index finger, and run the tip of a straight dissecting needle (probe) over the frog's head in the midline, until a depression is felt at the rear of the skull.
2. Insert the point of the dissecting needle at this location, which will sever the spinal cord from the brain. Aim it forward, after insertion, and push it forward, rotating the tip from side to side to destroy the brain.
3. Always check for the complete destruction of the brain by touching the cornea of the eye with the point of the probe. If the lower eyelid blinks (corneal reflex), the pithing is not complete. The brain-pithing is complete if the corneal reflex is absent. If in doubt about completeness of the brain-pithing and the animal's ability to perceive pain, see the laboratory instructor.

A swift, accurate, more complete alternative to brain-pithing is the complete removal of the upper half of the frog's head. This procedure is recommended in lieu of brain-pithing, especially if the student is uncertain about being able to pith the frog.

1. Grasp the frog firmly with your left hand, placing your index finger between the upper and lower jaws.
2. Insert the lower blade of a medium-size pair of laboratory scissors at the rear angle of the frog's mouth.
3. Firmly and quickly sever the head as far posterior to the eyes as possible.

Because the spinal cord is left intact, the preparation is referred to as a spinal frog. The spinal frog will display skeletal muscle activity of a reflex nature, but it should not be interpreted that the frog is aware and can still perceive pain. Without the brain, awareness or perception is impossible.

Preparing the Freshwater Turtle for Heart Experiments

This procedure requires two persons, a turtle board, strong twine, paper napkins, frog Ringer's solution, a small animal surgery kit, cotton thread, an electric drill, a 2-inch hole saw, and a claw hammer.

1. Place the animal on the laboratory table with the plastron against the table surface. Hold the carapace with one hand and, using the brass hook from the turtle board, extract the turtle's head, holding it at the edge of the laboratory table and against the table surface.
2. Using the claw hammer, strike the top of the turtle's skull, destroying the brain. Test for the absence of corneal reflexes and repeat the procedure if necessary.
3. Place the turtle carapace down on the turtle board and secure the legs and head to appropriate binding posts. Wrap the turtle's head with paper napkins but leave the neck exposed.
4. Place the pilot drill bit of the hole saw in the midline of the plastron between the forelegs and carefully cut a 2-inch hole in the plastron. Do not exert undue pressure on the hole saw. Rock the hole saw back and forth until the plastron (about ¼ inch thick) is penetrated and then complete the drilling procedure.
5. Remove the 2-inch diameter piece of plastron by scraping off the attached connective tissue with a scalpel. Cover the exposed heart and pericardial sac with gauze pads soaked in frog Ringer's solution until ready for experimentation.
6. Using forceps and small surgical scissors, make a lateral longitudinal incision through the skin of the neck from the base of the skull to the base of the neck. Using blunt dissection, separate the muscles and fascia, exposing the carotid artery, jugular vein, and vagus nerve, all of which run together. The artery is most noticeable, and the vagus nerve can be seen as a white cord accompanying it.
7. Use fine forceps and a glass probe to carefully separate the vagus nerve from surrounding fascia. Loop a piece of white thread around the vagus and allow the nerve to settle back among the muscles. The thread will be used to retrieve the nerve for stimulation. Retract the skin back over the incision and cover the incision with a paper napkin soaked in frog Ringer's solution.

Jones Pulmonor Disinfectant Procedure

The Jones Pulmonor must be disinfected after each student uses it before another student can use it. To clean the Pulmonor, use the iodine solution *Wescadyne* (Catalog #JP1009) available from the Jones Instrument Company.

1. Mix 1.5 oz of Wescadyne with 5 L of tap water. Store the solution in a 5-gallon plastic carboy. The solution should be brownish purple and can be reused until its color changes to orange.
2. Place the Pulmonor on the laboratory table and the carboy on top of the Pulmonor. Bring the Pulmonor hose up to the carboy spout and allow about 3 L of the iodine solution to flow through the hose and into the Pulmonor. Let stand for 2 minutes.
3. Place the carboy below the Pulmonor and allow the Pulmonor hose to drain into the mouth of the carboy.
4. Remove the Pulmonor hose from the machine and thoroughly rinse with hot tap water. Reattach hose.

INDEX

C

D

E

F

G

H

N

O

P

S

T

W

X

Y

Z